D1548669

PROGRESS IN CLINICAL AND BIOLOGICAL RESEARCH

Vol 1: **Erythrocyte Structure and Function,** George J. Brewer, *Editor*

Vol 2: **Preventability of Perinatal Injury,** Karlis Adamsons and Howard A. Fox, *Editors*

Vol 3: **Infections of the Fetus and the Newborn Infant,** Saul Krugman and Anne A. Gershon, *Editors*

Vol 4: **Conflicts in Childhood Cancer,** Lucius F. Sinks and John O. Godden, *Editors*

Vol 5: **Trace Components of Plasma: Isolation and Clinical Significance,** G.A. Jamieson and T.J. Greenwalt, *Editors*

Vol 6: **Prostatic Disease,** H. Marberger, H. Haschek, H.K.A. Schirmer, J.A.C. Colston, and E. Witkin, *Editors*

Vol 7: **Blood Pressure, Edema and Proteinuria in Pregnancy,** Emanuel A. Friedman, *Editor*

Vol 8: **Cell Surface Receptors,** Garth L. Nicolson, Michael A. Raftery, Martin Rodbell, and C. Fred Fox, *Editors*

Vol 9: **Membranes and Neoplasia: New Approaches and Strategies,** Vincent T. Marchesi, *Editor*

Vol 10: **Diabetes and Other Endocrine Disorders During Pregnancy and in the Newborn,** Maria I. New and Robert H. Fiser, *Editors*

Vol 11: **Clinical Uses of Frozen-Thawed Red Blood Cells,** John A. Griep, *Editor*

Vol 12: **Breast Cancer,** Albert C.W. Montague, Geary L. Stonesifer, Jr., and Edward F. Lewison, *Editors*

Vol 13: **The Granulocyte: Function and Clinical Utilization,** Tibor J. Greenwalt and G.A. Jamieson, *Editors*

Vol 14: **Zinc Metabolism: Current Aspects in Health and Disease,** George J. Brewer and Ananda S. Prasad, *Editors*

Vol 15: **Cellular Neurobiology,** Zach Hall, Regis Kelly, and C. Fred Fox, *Editors*

Vol 16: **HLA and Malignancy,** Gerald P. Murphy, *Editor*

Vol 17: **Cell Shape and Surface Architecture,** Jean Paul Revel, Ulf Henning, and C. Fred Fox, *Editors*

Vol 18: **Tay-Sachs Disease: Screening and Prevention,** Michael M. Kaback, *Editor*

Vol 19: **Blood Substitutes and Plasma Expanders,** G.A. Jamieson and T.J. Greenwalt, *Editors*

Vol 20: **Erythrocyte Membranes: Recent Clinical and Experimental Advances,** Walter C. Kruckeberg, John W. Eaton, and George J. Brewer, *Editors*

Vol 21: **The Red Cell,** George J. Brewer, *Editor*

Vol 22: **Molecular Aspects of Membrane Transport,** Dale Oxender and C. Fred Fox, *Editors*

Vol 23: **Cell Surface Carbohydrates and Biological Recognition,** Vincent T. Marchesi, Victor Ginsburg, Phillips W. Robbins, and C. Fred Fox, *Editors*

Vol 24: **Twin Research,** Proceedings of 2nd International Congress on Twin Studies
Walter E. Nance, *Editor*
Published in 3 Volumes:
Part A: Psychology and Methodology
Part B: Biology and Epidemiology
Part C: Clinical Studies

Vol 25: **Recent Advances in Clinical Oncology,** Tapan A. Hazra and Michael C. Beachley, *Editors*

Vol 26: **Origin and Natural History of Cell Lines,** Claudio Barigozzi, *Editor*
Vol 27: **Membrane Mechanisms of Drugs of Abuse,** Charles W. Sharp and Leo G. Abood, *Editors*
Vol 28: **The Blood Platelet in Transfusion Therapy,** G.A. Jamieson and Tibor J. Greenwalt, *Editors*
Vol 29: **Biomedical Applications of the Horseshoe Crab (Limulidae),** Elias Cohen, *Editor-in-Chief*
Vol 30: **Normal and Abnormal Red Cell Membranes,** Samuel E. Lux, Vincent T. Marchesi, and C. Fred Fox, *Editors*
Vol 31: **Transmembrane Signaling,** Mark Bitensky, R. John Collier, Donald F. Steiner, and C. Fred Fox, *Editors*
Vol 32: **Genetic Analysis of Common Diseases: Applications to Predictive Factors in Coronary Disease,** Charles F. Sing and Mark Skolnick, *Editors*
Vol 33: **Prostate Cancer and Hormone Receptors,** Gerald P. Murphy and Avery A. Sandberg, *Editors*
Vol 34: **The Management of Genetic Disorders,** Constantine J. Papadatos and Christos S. Bartsocas, *Editors*
Vol 35: **Antibiotics and Hospitals,** Carlo Grassi and Giuseppe Ostino, *Editors*
Vol 36: **Drug and Chemical Risks to the Fetus and Newborn,** Richard H. Schwarz and Sumner J. Yaffe, *Editors*
Vol 37: **Models for Prostate Cancer,** Gerald P. Murphy, *Editor*
Vol 38: **Ethics, Humanism, and Medicine,** Marc D. Basson, *Editor*
Vol 39: **Neurochemistry and Clinical Neurology,** Leontino Battistin, George A. Hashim, and Abel Lajtha, *Editors*
Vol 40: **Biological Recognition and Assembly,** David S. Eisenberg, James A. Lake, and C. Fred Fox, *Editors*
Vol 41: **Tumor Cell Surfaces and Malignancy,** Richard O. Hynes and C. Fred Fox, *Editors*
Vol 42: **Membranes, Receptors, and the Immune Response: 80 Years After Ehrlich's Side Chain Theory,** Edward P. Cohen and Heinz Köhler, *Editors*
Vol 43: **Immunobiology of the Erythrocyte,** S. Gerald Sandler, Jacob Nusbacher, and Moses S. Schanfield, *Editors*
Vol 44: **Perinatal Medicine Today,** Bruce K. Young, *Editor*

BIOLOGICAL RECOGNITION AND ASSEMBLY

BIOLOGICAL RECOGNITION AND ASSEMBLY

Proceedings of the ICN-UCLA Symposium
held at Keystone, Colorado
March 4—March 9, 1979

Editors

DAVID S. EISENBERG
University of California
Los Angeles, California

JAMES A. LAKE
University of California
Los Angeles, California

C. FRED FOX
University of California
Los Angeles, California

Alan R. Liss, Inc. • New York

Address all inquiries to the publisher:
Alan R. Liss, Inc., 150 Fifth Avenue, New York, NY 10011

Printed in the United States of America

Library of Congress Cataloging in Publication Data

Main entry under title:

Biological recognition and assembly.

(Progress in clinical and biological research; v. 40)
Includes index.
1. Viruses – Morphology – Congresses. 2. Cellular recognition – Congresses. 3. Binding sites (Biochemistry) – Congresses. 4. Molecular biology – Congresses. I. Eisenberg, David S. II. ICN Pharmaceuticals, inc. III. California. University. University at Los Angeles. IV. Series.
QR450.B56 574.87 80-7797
ISBN 0-8451-0040-8

Pages 1–208 of this volume are reprinted from Journal of Supramolecular Structure, Volumes 10, 11, and 12. The Journal is the only appropriate literature citation for these articles. The page numbers in the table of contents, author index, and subject index of this volume correspond to the page numbers at the foot of these pages.

The **table of contents** does not necessarily follow the actual pattern of the plenary sessions but, rather, reflects the thrust of the meeting as it evolved from the combination of plenary and poster sessions culminating in the final collection of invited and submitted papers. The actual order of the articles appearing in this volume does not follow the order in which the articles are cited in the table of contents. Many articles were submitted for consideration by Journal of Supramolecular Structure and are reprinted here. These articles are included in the text in the order in which they were accepted for publication and then published in the Journal, and they are followed by invited papers which were submitted solely for publication in the proceedings and did not receive editorial review.

Contents

Preface

In 1977, when we started planning for this meeting, it was clear from the rapid accumulation of new information being obtained by structural and cell biologists that the time was right to bring these two groups together and to focus on problems of common interest. We believed that both fields had advanced to the point where the gap between disciplines could be bridged.

Prior to this meeting on Biological Recognition and Assembly, no comparable meeting had been held in the ICN-UCLA Symposia series since the 1974 meeting on Assembly Mechanisms. In the intervening years between 1974 and 1979, the field had changed dramatically with a shift in emphasis in both areas. Biological structure had moved from emphasizing techniques for solving structures to emphasizing the biological implications of molecular structure, and cell biology had become more molecular in its approach. Structural and cellular biologists then found themselves in a position where they could interact with each other and share common research interests. By focusing on recognition processes, we hoped that a natural alliance might occur between the two groups.

The conference was held in a spirit of learning about other fields and approaches and in transferring this information to one's own work. In this sense, the meeting was a great success and exceeded even our optimistic expectations. In a continuation of this goal of sharing information between fields, we have assembled many of the lectures and poster session presentations from the meeting and combined them into this volume. We hope that the spirit of excitement and discovery that prevailed at the meeting is conveyed in the present volume.

Annual co-sponsorship of the ICN-UCLA Symposia is provided by the Life Sciences division of ICN Pharmaceuticals. The Biological Recognition and Assembly meeting received additional contributions from Phillips Electronics and JEOL to defray partial program expenses. The National Institutes of Health (National Institute of Allergy and Infectious Diseases and Fogarty International Center) also awarded Contract No. 263-MO-912643 in partial support of the symposium. We wish to express our thanks to these agencies that provided support and to Leslie Rubin and Betty Handy for their assistance in the compilation of this volume.

David S. Eisenberg
James A. Lake
C. Fred Fox

Journal of Supramolecular Structure 10:443–455 (1979)
Biological Recognition and Assembly 1–13

The Accessibility of Antigenic Determinants of Ribosomal Protein S4 In Situ

Donald Winkelmann and Lawrence Kahan

Department of Physiological Chemistry, University of Wisconsin, Madison, Wisconsin 53706

Antibodies to Escherichia coli ribosomal protein S4 react with S4 in subribosomal particles, eg, the complex of 16S RNA with S4, S7, S8, S15, S16, S17, and S19 and the RI* reconstitution intermediate, but they do not react with intact 30S subunits. Antibodies were isolated by three different methods from antisera obtained during the immunization of eight rabbits. Some of these antibody preparations, which contained contaminant antibodies directed against other ribosomal proteins, reacted with subunits, but this reaction was not affected by removal of the anti-S4 antibody population. Other antibody preparations did not react with subunits. It is concluded that the antigenic determinants of S4 are accessible in some protein deficient subribosomal particles but not in intact 30S subunits.

Key words: ribosomes, 30S subunit structure, immunochemistry

Antibodies to individual ribosomal proteins have been utilized to delineate the functional roles and structural locations of these proteins in the intact subunit. These studies have generally assumed that each of the proteins studied has exposed antigenic determinants [1, 2] and that the observed functional inhibition or antibody binding could be ascribed to the protein which had been chosen as the immunogen [3]. Ribosomal protein S4 has been assigned a role in several functional sites on the basis of inhibition of activity by antibody putatively specific for S4 [1] and has been assigned locations on the 30S subunit surface by immunoelectron microscopy [4, 5]. In the case of S4, however, the reactivity of anti-S4 antibody with subunits was originally described as weak, and demonstrable only in certain assays [2]. Subsequently it was found that the replacement of Escherichia coli S4 with its functional homolog from Bacillus stearothermophilus only partly inhibited the reaction of anti-E coli S4 with hybrid 30S subunits, although no cross-reaction could be demonstrated with the purified proteins [4]. Therefore, we felt it important to test the hypothesis that S4 was, in fact, not available to antibody in the intact subunit and that the observed functional inhibition and map locations were the result

Received April 4, 1979; accepted April 18, 1979.

0091-7419/79/1004-0443$02.60

of the reaction of minor amounts of contaminating antibodies directed against other ribosomal proteins. In order to test the greatest possible variety of antibody populations, we have immunized eight rabbits, collected serum throughout the course of long immunization programs (up to 32 weeks of immunization), and purified anti-S4 IgG preparations by three different procedures. We have tested these preparations under a variety of reaction conditions and found that none of these antibody preparations gave detectable reaction with S4 in the intact 30S subunit. Thus the functional and structural roles ascribed to S4 must be considered unproven until an S4-specific reaction of antibody with subunits can be convincingly demonstrated.

MATERIALS AND METHODS

Preparation of Ribosomes, Ribosomal Proteins, and rRNA

Isolation of E coli K12 (strain PR-C10) 70S ribosomes, zonal separation of 30S and 50S subunits, and purification of 30S ribosomal proteins (r-proteins) were done as described by Held et al [6]. 16S RNA was prepared by phenol extraction of 70S ribosomes according to Traub et al [7]. Then 12.5 mg of 5S, 16S, and 23S rRNA was separated by centrifugation on linear 15–30% sucrose gradients in SSC-EDTA (0.15 M NaCl, 0.015 M Na citrate, 0.01 M ethylenediaminetetracetic acid, pH 7.0) for 2.25 h at 50,000 rpm in a Beckman VTi50 rotor at 4°C. The pooled 16S RNA was precipitated with 2 volumes ethanol at –20°C, dialyzed against 0.03 M Tris (pH 7.4), and stored at –70°C. Salt-washed 30S subunits were prepared according to the method described by Staehelin and Maglott [8]. Alternatively, 70S ribosomes in TMA I (0.03 M NH_4Cl, 0.01 M $MgCl_2$, 0.01 M Tris, pH 7.3, 0.001 M dithiothreitol [DTT]) were brought to 1 M NH_4Cl with 4 M NH_4Cl, then pelleted by centrifugation. Subunits were isolated from these washed ribosomes.

Reconstitution, Isolation, and Analysis of 30S Ribosomal Subunits and Reconstitution Intermediates

Reconstitution of 30S subunits from TP30 or the 21 purified proteins was done according to Held et al [6]. Subribosomal particles were prepared by mixing 1 volume of the desired mixture of pure proteins in TRI buffer (1.0 M KCl, 0.02 M $MgCl_2$, 0.03 M Tris, pH 7.4, 0.001 M DTT) with 2 volumes of 16S RNA in TRO buffer (0.02 M $MgCl_2$, 0.03 M Tris, pH 7.4, 0.001 M DTT) and incubating 1 h at 40°C. Reconstitution was done at a final RNA concentration of 800–1,600 pmoles/ml with a (1.5–2)-fold excess of protein. Reconstituted particles were purified by centrifugation through 15–30% sucrose gradients in TMA I buffer for 75 min at 50,000 rpm in a VTi50 rotor at 4°C. The purified particles were pelleted (10 h at 40,000 rpm, Beckman SW50.1 rotor) from the sucrose-TMA I solution, redissolved in TMA I, and incubated 10 min at 40°C before use. Generally the reconstitution intermediates were prepared just prior to analysis. Poly-U-directed [^{14}C] phenylalanine incorporation activity of 30S subunits was assayed as described by Held and Nomura [9]. Two-dimensional polyacrylamide gel electrophoresis was done by a modification of the method described by Howard and Traut [10]. The size of the first-dimension gel was decreased to 1.5 mm × 7 cm and 80 pmoles of RNase-digested subunits were used per gel. Electrophoresis was for 2.5 h at 80 V in the first dimension and 6 h at 120 V in the second dimension (both at 24°C).

Antisera Production

Antisera to E coli 30S subunit protein S4 were prepared by immunization of New Zealand white rabbits with emulsions prepared by mixing equal volumes of 0.25–1.0 mg/ml S4 in 0.15 M NaCl, 0.5% phenol, and Freund's complete adjuvant (Perrin's modification, Calbiochem). Biweekly injections of 0.25–1.0 mg antigen were administered subcutaneously at 4–8 sites. Blood was collected from ear veins at least seven days after the most recent immunization. Immunization was repeated at varying intervals for up to one year.

Double-Immunodiffusion

Sera were tested for antibody by double-immunodiffusion in 1% agar gels against 4×10^{-6} M antigen, as previously described [11]. Antisera were tested directly and after (5–10)-fold concentration for the presence of antibodies to heterologous 30S ribosomal subunit proteins, as previously described [12].

Preparation of IgG Fractions

The total immunoglobulin G fraction of specific rabbit antisera was partially purified by ion-exchange chromatography on DEAE-Sephadex A50 [13] using a 0–0.5 M NaCl gradient in 0.02 M phosphate buffer (pH 7.2) to elute bound IgG, ammonium sulfate precipitation of all IgG-containing fractions, and gel filtration on Sephacryl S200 (Pharmacia). Alternatively, the IgG fractions were prepared by affinity chromatography on Protein A–Sepharose CL-4B (Pharmacia) and gel filtration. Serum (5 ml) was applied at 4°C to a 0.9-cm × 6-cm Protein A–Sepharose CL-4B column equilibrated with saline-phosphate (SP) buffer (0.15 M NaCl, 0.075 M potassium phosphate, pH 7.15) and recycled over the column for 1–2 h at a flow rate of 0.2–0.3 ml/min. The unbound serum proteins were eluted from the column with SP buffer. Bound IgG was eluted with 0.1 M glycine-HCl (pH 3.0). The neutralized IgG was further purified by gel filtration and concentrated by ammonium sulfate precipitation.

Immunoabsorption Affinity Purification of S4-Specific IgG

S4 (3 mg) in 6 M urea, 0.10 M potassium phosphate buffer pH (7.0) was reacted with a 10-ml slurry of the N-hydroxysuccinimide ester of succinylated aminoalkyl Bio-gel A (Affigel-10, Biorad) for 18 h at 4°C. The coupling reaction was terminated with 1 M methylamine phosphate (pH 9). The gel was washed extensively with cycles of Buffer I (0.20 M KCl, 0.01 M $MgCl_2$, 0.01 M Tris, pH 7.4, at 24°C) and desorption buffer (0.20 M KCl, 0.10 M Na citrate, pH 2.2) ending with Buffer I. Total IgG (5–20 mg) in Buffer I was applied to a 10-ml column of S4-Affigel and cycled over the column for 12 h at 0.2 ml/min and 4°C. Unbound IgG was eluted with Buffer I. Bound, S4-specific IgG was eluted with desorption buffer. IgG-containing fractions were pooled and neutralized with solid Tris (Sigma). Carrier nonimmune IgG was added and the IgG concentrated by precipitation with 50% saturated $(NH_4)_2SO_4$.

Quantitative Immunoprecipitation

Quantitative immunoprecipitation analysis of AS4 IgG was done generally according to the procedure outlined by Mauer [14]. A constant amount (200–400 μg) of purified

IgG was added to a series of tubes containing increasing amounts of S4 (0–10 μg). The reaction was done in reconstitution buffer in 50 to 100-μl volumes. Samples were incubated for 30 min at 37°C followed by 48–96 h at 4°C with occasional vortexing. The precipitates were centrifuged at 4°C in a Beckman Microfuge B for 1.5 min. The supernatants were removed and the precipitates were washed three times with reconstitution buffer, then dissolved in 100 μl of 0.5 N NaOH. Protein was determined by the Lowry assay [15].

Reaction of Antibody With Ribosomal Particles

30S ribosomal subunits or subribosomal particles (16–32 pmoles) were "heat-activated" for 5 min at 40°C in Buffer I or TMA I. Particles were reacted with IgG in 25–100 μl of 0.10 M NH_4Cl, 0.01 M $MgCl_2$, 0.01 M Tris (pH 7.4) for 5 min at 40°C and 40 min at 0°C. The reaction mixtures were layered directly on 5 ml, 15–30% sucrose gradients in TMA I and centrifuged for 2.75 h at 50,000 rpm in a SW50.1 rotor at 4°C. Alternatively, the reaction mixtures were diluted with 1/2 volume of 20% sucrose in reaction buffer and layered between a 5-ml 15–30% sucrose gradient and 250 μl of reaction buffer overlay. Centrifugation was for 20 min (at speed) in a Beckman VTi65 rotor at 50,000 rpm and 4°C in an L5-50 centrifuge. The best gradient separations were obtained by accelerating the rotor at an acceleration rate setting of 1 up to 2,000 rpm, then at an acceleration rate of 10 up to speed. The rotor was braked down to 1,400 rpm, then allowed to coast to a stop. Absorbance at 260 nm was measured with a Gilford density gradient scanner by upward displacement of the gradients with sucrose.

RESULTS

In order to determine the degree of exposure of the antigenic determinants of E coli ribosomal protein S4 in situ, we have characterized the reactivity of S4-specific antibody (AS4) with intermediates of in vitro 30S subunit reconstitution and with the 30S subunit. Since we desired the widest variety of reactions, numerous bleedings were taken from eight rabbits during the course of immunization. Several bleedings from each animal containing high titers of AS4 were further studied. The spectrum of ribosomal protein-specific antibodies in each was determined by concentrating the serum (5–10)-fold and testing it for reaction with each of the 21 small-subunit proteins by double-immunodiffusion. The S4 used as immunogen was judged greater than 98% pure by polyacrylamide gel electrophoresis and the impurities were identified immunochemically as S3 and S7 (data not shown). Several antisera contained antibody to these impurities as well as to S4; others reacted only with S4 (Table I).

A nuclease-free IgG fraction was isolated from the antisera by ion-exchange chromatography or Protein A–Sepharose affinity chromatography. The IgG prepared by each method reacted with S4 in double-immunodiffusion experiments. S4-specific IgG in the IgG fraction was measured by quantitative immunoprecipitation analysis. The AS4 titer of the antibody preparations studied are listed in Table I. The AS4 titer varied from as low as 1.6% up to 16% of the total IgG. We generally saw a rise in the titer over the course of immunization.

To determine if S4 is accessible to antibody on subribosomal particles and to examine any changes in accessibility on completion of the 30S ribosomal subunit, intermediates of

TABLE I. Antibody Content of Rabbit AS4 IgG Preparations

Animal no.	Immunization period (weeks)	Method of IgG purification	Moles S4-specific IgG / 100 moles IgG	Contaminant r-protein antibodies
1	6	A	7	–
1	13	B	16	–
2	25	A	6	–
2	25	B	8	–
2	20–25	C	50–60	–
3	8	A	6	–
3	21	B	9	–
4	8	A	5	–
4	21	B	9	S7
5	24	A	ND	S3,S7
5	32	A	3	S3,S7
5	32	C	ND	–
6	6	A	5	S3,S7
6	6	C	ND	–
7	8	A	2	S3
7	8	C	ND	–
8	7	A	2	–
8	7	C	ND	–

For each rabbit individual IgG preparations are described by number of weeks between initial immunization and bleeding and method of IgG purification. IgG was purified by three methods (indicated in third column): A, DEAE-Sephadex ion-exchange chromatography; B, Protein A–Sepharose affinity chromatography; C, S4-Affigel affinity chromatography. S4-specific antibody in the IgG preparation was measured by quantitative immunoprecipitation (see Methods). Values indicated are percentage of total IgG added that was precipitated by antigen at equivalence (ND; Not Determined.) Contaminating r-protein-specific antibodies were identified by immunodiffusion of concentrated antisera against the 21 pure 30S subunit proteins.

in vitro 30S subunit reconstitution were prepared from purified components according to the following scheme:

$$\text{16S RNA} + \text{S4,S7,S8,S15,S16,S17,S19} \rightarrow \text{RI}[4,7,8,15,16,17,19] \quad (1)$$

$$\text{RI}[4,7,8,15,16,17,19] + \text{S5,S6,S9,S11,S12,S13,S18,S20} \rightarrow \text{RI}^* \quad (2)$$

$$\text{RI}^* + \text{S1,S2,S3,S10,S14,S21} \rightarrow \text{30S ribosomal subunit} \quad (3)$$

The first particle in this scheme is prepared by incubation of 16S RNA with the seven proteins in reconstitution buffer for 1 h at 40°C and purified by zone sedimentation in sucrose gradients (see Methods). The next intermediate in the scheme, the RI* particle, is a complex of 15 proteins and 16S RNA. Twelve of these proteins are required for RI* formation [9]; proteins S6, S13, and S20 were included to stabilize the particle during isolation. The RI* particle was prepared and purified in the same manner as the RI[4,7,8,15, 16,17,19] particle. In early experiments we isolated the particles directly by pelleting through 15% sucrose in reconstitution buffer. We found this method unsatisfactory since it resulted in reduced amounts of some proteins (eg, S5) as a consequence of the high salt in the reconstitution buffer through which they were pelleted.

The isolated RI[4,7,8,15,16,17,19] and RI* particles could be completed by the addition of the missing proteins to form 30S ribosomal subunits that were active in the poly-U translation assay (Table II). The isolated RI[4,7,8,15,16,17,19] particle was completed in two steps. It was first incubated for 1 h at 40°C in reconstitution buffer with the eight proteins needed to form the RI* particle. The remaining six proteins were then added and the incubation was continued for 10 min at 30°C. The isolated RI* particle was completed with this second step only. The results in Table II indicate that the RI[4,7,8,15,16,17,19] and RI* particles are inactive prior to completion. Both particles can be completed upon addition of the missing proteins to yield particles that have 80–100% of the control activity. The two-step reconstitution protocol was used to demonstrate that the RI[4,7,8,15,16,17,19] particle could be converted to the RI* particle and that the RI* particle had undergone the heat-dependent conversion step described by Held and Nomura [9].

Reconstituted 30S ribosomal subunits were prepared either with the 21 pure ribosomal proteins or total 30S protein (TP30). The protein and 16S RNA were incubated for 1 h at 40°C. The reconstituted particles were purified in the same manner as the RI[4,7,8,15,16,17,19] and RI* particles. We found that reconstituted 30S subunits purified in this manner were more active in the poly-U translation assay and had narrower sucrose gradient profiles than subunits isolated directly by pelleting through 15% sucrose in reconstitution buffer.

Analysis of the isolated particles in sucrose gradients indicated that the RI[4,7,8,15,16,17,19] and RI* particles sediment as narrow zones of about 22S and 26S, respectively. The reconstituted 30S subunit sedimented with native 30S ribosomal subunits.

The protein composition of the isolated particles was examined by two-dimensional gel electrophoresis and compared to native 30S subunits prepared from 70S ribosomes. Each of the isolated particles contained the expected proteins in amounts comparable to those found in the native 30S subunit (data not shown).

TABLE II. Completion of Isolated Subribosomal Particles

Particle	Proteins added in completion steps: Step 1	Step 2	Activity (pmoles [^{14}C]phenylalanine incorporated)
16S RNA	A + B + C	–	39
RI[4,7,8,15,16,17,19]	–	–	3
RI[4,7,8,15,16,17,19]	B	C	42
RI*	–	–	2
RI*	–	C	32
ΣSi 30S	–	–	40
Native 30S	–	–	80

30S subunits were reconstituted from 16S RNA or isolated subribosomal particles in two steps by addition of r-protein mixes A (S4,S7,S8,S15,S16,S17,S19), B (S5,S6,S9,S11,S12,S13,S18,S20), and C (S1,S2,S3,S10,S14,S21). 16S RNA or ribosomal particles were prepared in reconstitution buffer. Indicated protein mix(es) or buffer was added and the samples were incubated for 1 h at 40°C in step 1, then chilled on ice. In step 2 incubation was for 10 min at 30°C. As described in Held and Nomura [9], 16 pmoles of the completed particles was assayed directly for poly-U-directed [^{14}C]phenylalanine incorporation activity. ΣSi 30S are isolated 30S subunits reconstituted from 16S RNA and the 21 pure r-proteins. Native 30S subunits were prepared from 70S ribosomes without salt-washing. –; No addition of r-protein.

The AS4 IgG preparations were tested for AS4-dependent aggregation of the reconstitution intermediates and 30S subunits. Antibody was incubated with the ribosomal particles and the reaction mixtures were analyzed by zone sedimentation in sucrose gradients (Fig. 1). Incubation of AS4 IgG with RI[4,7,8,15,16,17,19] particles results in a reduction in the 22S monomer peak area and the formation of faster sedimenting aggregates. The major product was an IgG-linked dimer, although higher aggregates were always seen. Preabsorption of the antibody with S4 at equivalence precipitated the S4-specific IgG and eliminated any reaction of the antibody preparation with the RI[4,7,8,15,16,17,19] particle. The small amount of dimerized particles seen in the absorbed antibody control was also seen with nonimmune IgG and in the absence of IgG and thus represents nonspecific dimers. The production of aggregates larger than dimers indicates that multiple antigenic determinants are accessible on this particle. All of the AS4 IgG preparations reacted with this particle, but the reaction could not be shown to be AS4-specific for those preparations that contained antibody to proteins other than S4. Only a portion of the reactivity could be absorbed with S4 in those IgG preparations.

The IgG preparations also reacted with the RI* particle, but the extent of reaction was less than that seen under identical reaction conditions with the RI[4,7,8,15,16,17,19] particle (Fig. 2). The decrease in AS4 reactivity of the RI* particle may indicate a masking of S4 antigenic determinants on this assembly intermediate as compared to the RI[4,7,8,15,16,17,19]. Alternatively, the conformation of S4 may be altered such that

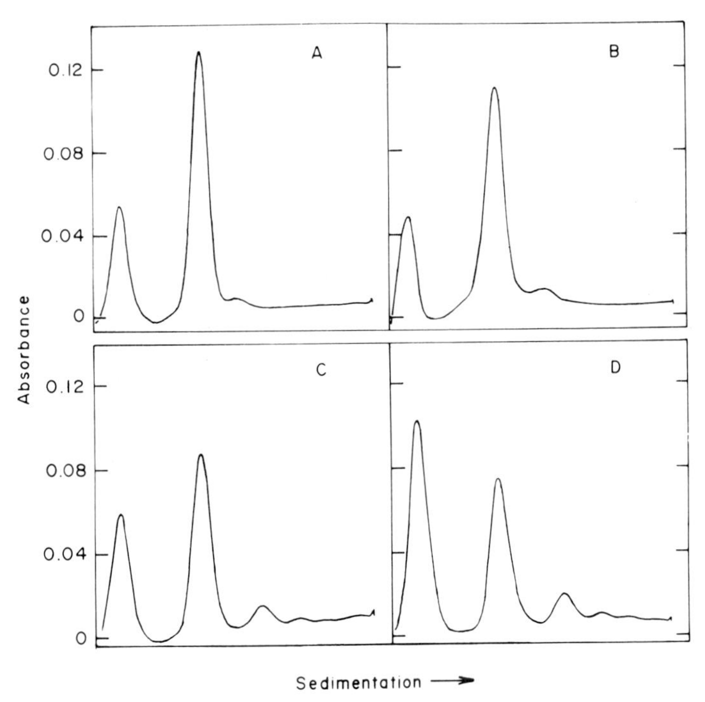

Fig. 1. Sucrose gradient analysis of the reaction of AS4 IgG (1-13B) with RI[4,7,8,15,16,17,19] particles. 24 pmoles of RI[4,7,8,15,16,17,19] was reacted with 140 μg of preimmune IgG (A), 140 μg of S4-absorbed AS4 IgG (B), 140 μg of AS4 IgG (48 pmoles AS4-specific IgG) (C), and 280 μg AS4 IgG (96 pmoles S4-specific IgG) (D). Reaction conditions were those described in Methods. Sedimentation was for 20 min at 50,000 rpm and 4°C in a VTi65 rotor through 15–30% sucrose gradients. Absorbance was monitored at 260 nm in a 2.5-mm pathlength flowcell.

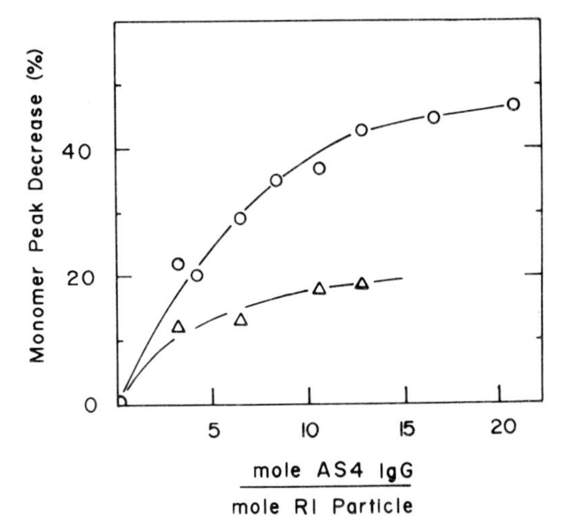

Fig. 2. Reaction of AS4 IgG (1-13B) with RI[4,7,8,15,16,17,19] (○) and RI* (△) particles as measured by antibody-dependent aggregation. The subribosomal particles were reacted with AS4 at various molar ratios and sedimented through sucrose gradient (see Methods and Fig. 1). Aggregation of particles was measured by decrease in monomer peak area from that observed when preimmune IgG was reacted with the same particles. Points shown are the average of 2–4 samples.

it is less strongly bound by antibody. Since the AS4 preparations did react specifically with the RI* particle S4 must be accessible to antibody on this intermediate.

In contrast to the results we observed with the reconstitution intermediates, AS4 IgG from four of the rabbits did not react with native or reconstituted 30S ribosomal subunits (Table III). These antibody preparations were tested over a wide range of antibody-antigen ratios. Reaction conditions such as reaction volume and incubation time were varied. Two methods of IgG purification were utilized and several bleedings over the course of immunization were tested. Glutaraldehyde fixation of the reaction products was attempted [16]. Despite all efforts to favor the reaction of antibody with the 30S ribosomal subunit and to stabilize the products, none of the antibody preparations from four of the eight rabbits reacted with the 30S subunit.

Several methods of 30S subunit preparation were also examined. Salt-washed 30S subunits, subunits isolated from salt-washed 70S ribosomes, reconstituted 30S subunits, low Mg^{2+}-inactivated 30S subunits [17], and 30S subunits from a different strain of E coli (MRE 600) were tested. None of the 30S subunit preparations reacted with these AS4 IgG preparations.

IgG preparations from four rabbits did react with the 30S ribosomal subunits (Table III). However, these preparations contained antibody to other r-proteins. The total IgG was therefore further fractionated by immunoabsorption chromatography on S4-Affigel (see Methods). S4-specific IgG, which bound to the immobilized S4 and was subsequently eluted, did not react with the 30S subunit. Figure 3 shows the results of the reaction of the S4-Affigel passthrough and bound IgG fractions with 30S subunits at various antibody-to-subunit ratios. All of the 30S subunit reactive antibody was found in the passthrough fraction. The bound, S4-specific fraction reacted with S4 and RI particles but not with the 30S ribosomal subunit. This result suggests that the antibody that reacted with the 30S subunit was directed against r-proteins other than S4.

TABLE III. Reactivity of AS4 IgG Preparations With 30S Ribosomal Subunits

IgG preparation[a]	30S subunit reactivity	Inhibition by S4 absorption	AS4 IgG:30S subunit molar ratios tested
1-6A	–		1–23
1-13B	–		1–15
2-25A	–		0.5–4
2-25B	–		0.5–27
2-(20–25)C	–		
3-8A	–		0.5–20
3-21B	–		0.5–19
4-8A	–		0.5–17
5-32A	+	No inhibition	0.2–1
5-32C	–		
6-6A	+	No inhibition	0.5–3
6-6C	–		
7-8A	+	No inhibition	0.5–8
8-7A	+	No inhibition	0.2–5

The 30S subunit reactivity of the IgG preparations was determined by antibody-dependent aggregation measured by zone sedimentation. A reaction is indicated by + and no reaction by –. In the cases where a reaction was detected, the IgG preparations were absorbed at equivalence with pure S4 and the supernatants were tested for reactivity with subunits. "No inhibition" indicates that absorption with S4 did not affect the reaction of the IgG preparation with 30S subunits.
[a]IgG preparations are identified as in Table I by rabbit number, period of immunization, and method of purification (A, DEAE-Sephadex; B, Protein A–Sepharose; C, S4-Affigel).

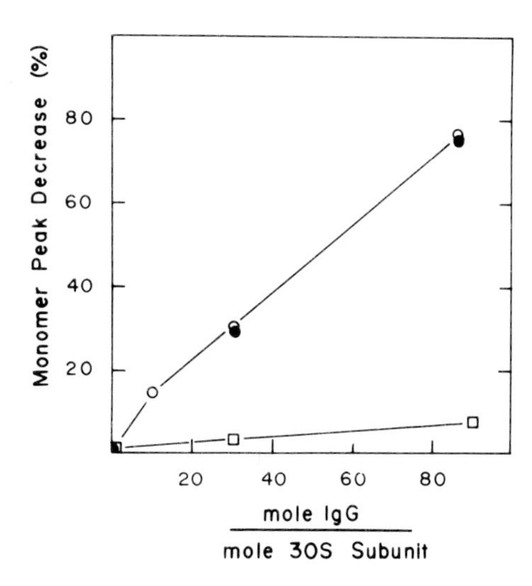

Fig. 3. Reaction of affinity-purified AS4 with 30S subunits. An IgG preparation (6-6A) that reacted with 30S subunits was further fractionated by immunoabsorption on S4-Affigel (as described in Methods). The unfractionated IgG (○), the nonbinding "passthrough" fraction (●), and the bound S4-specific fraction (□) were reacted with 30S subunits. Antibody-dependent aggregation of subunits was analyzed by sucrose gradient centrifugation (20 min at 50,000 rpm in a VTi65 rotor). Aggregation was measured by the decrease in the monomer peak area from that observed with nonimmune IgG. Carrier nonimmune IgG was added to the bound S4-specific fraction to bring the concentration to that of the original preparation.

If the reaction of the IgG preparations with 30S ribosomal subunits was S4-specific, preabsorption of the IgG with S4 should completely eliminate the reaction. However, if the reacting antibodies were directed against some other protein, absorption with S4 should not affect reactivity. Therefore, increasing amounts of S4 were incubated with the IgG preparations that reacted with the 30S subunit and immunoprecipitation was done as described in Methods. The quantitative immunoprecipitation curve for one of the IgG fractions is shown in Figure 4. The S4-absorbed supernatants were assayed for reaction with the 30S subunit and with S4. Despite the complete removal of the S4-specific antibody from the supernatants, there was no inhibition of the reaction of the IgG preparation with the 30S subunit. We concluded that the reaction of this IgG preparation with the 30S ribosomal subunit was due to a contaminant antibody reacting with a protein other than S4. Each of the antibody preparations that reacted with the 30S subunit was similarly tested. None reacted with the 30S subunit in a demonstrably S4-specific manner. Furthermore, in two cases the reaction was inhibited by preabsorption of the antibody with r-protein S7 (animals 5 and 6). These results are summarized in Table III.

Thus, none of the antibody preparations from eight rabbits reacted with the 30S ribosomal subunit in an AS4-specific manner. Since the antibody preparations do recognize S4 in situ on intermediates of 30S ribosome assembly, we have concluded that S4 is inaccessible to antibody on the surface of the 30S ribosomal subunit.

DISCUSSION

The accessibility of E coli ribosomal protein S4 to antibody on the 30S ribosomal subunit and ribosome reconstitution intermediates (RI particles) was studied with antibody preparations from eight rabbits. While the antibody preparations reacted with the RI[4,7,8, 15,16,17,19] and RI* particles in an S4-specific manner, no S4-specific reaction could be detected with complete 30S subunits.

The reaction of the antibody preparations with RI particles was demonstrated to be S4-specific by two methods. When the antibody preparations were passed over a column containing S4 covalently coupled to Bio-gel A, the fraction that bound to the column, the AS4-enriched fraction, reacted with the RI particles while the passthrough fraction showed greatly diminished reactivity. This result is consistent with an S4-specific reaction. However, the enrichment of antibody to contaminants in the S4 might also have occurred. Thus, it cannot be concluded that the antibody preparations react with S4 in reconstitution intermediates without additional supportive evidence.

Absorption of the antibody with S4 was also used to demonstrate the specificity of the reaction of the antibody preparations with RI particles. Immunoprecipitation of the antibody with an equivalent amount of S4 completely inhibited the reaction of some preparations and partially inhibited the reaction of other preparations which were known to be contaminated with antibody to other r-proteins. It is important to note that the immunoprecipitation control experiments used to demonstrate specificity were done at antigen-antibody equivalence. The addition of excesses of antigen may result in absorption of antibody directed against contaminants in the antigen, thus leading to an incorrect appearance of specificity.

We examined two intermediates of 30S subunit reconstitution and found that S4 was accessible on each of these intermediates. We did, however, see a decrease in the extent of reaction of the RI* particle when compared to the RI[4,7,8,15,16,17,19] particle (Fig. 2). The reactions of these two intermediates with several AS4 antibody preparations

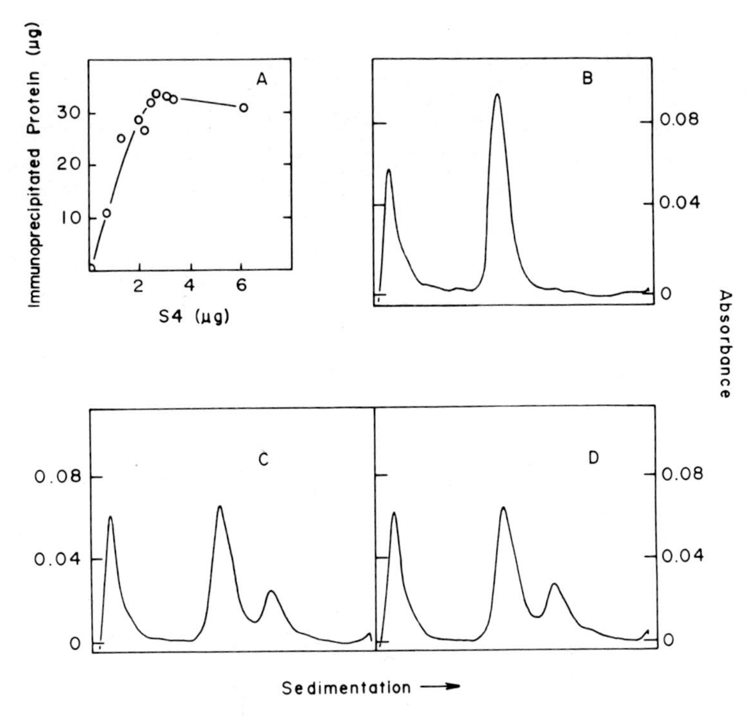

Fig. 4. A: Quantitative immunoprecipitation of IgG preparation 6-6A with S4. Aliquots of IgG (350 μg) were incubated with increasing amounts of pure S4 (0–6 μg) for 72 h in reconstitution buffer. Protein in the immunoprecipitate was measured as described in Methods. B–D: Sedimentation analysis of the reaction of the S4-absorbed IgG with 30S subunits. To measure antibody dependent aggregation of 30S subunits, 16 pmoles of 30S subunits were reacted with 100 μg of IgG and sedimented through sucrose gradients (2.75 h at 50,000 rpm and 4°C in a SW50.1 rotor). B: Preimmune IgG (control); C: "mock"-absorbed IgG (0 μg S4); and D: S4-absorbed IgG from the equivalence zone (3.4 μg S4).

were compared and the RI* was always found to be less reactive (data not shown). The decrease in reactivity may indicate a masking of S4 antigenic determinants by other ribosomal components or a conformational alteration in S4 that affects antibody binding. We have obtained preliminary evidence that S5 and S12 play major roles in the masking of S4 between the RI[4,7,8,15,16,17,19] and the RI* particles [18].

We have been unable to detect any reaction of S4-specific antibody with the 30S subunit. Antibody from eight rabbit antisera and several bleedings from each animal were examined to obtain a broad range of antibody specificities. These preparations have been shown to recognize antigenic determinants in peptide fragments covering residues 1–30, 31–102, 123–174, and 175–203 of S4 and thus recognize determinants widely distributed throughout the protein (data not shown). The antibody preparations do recognize the conformation of S4 in situ, as shown by their reaction with the RI[4,7,8,15,16,17,19] and RI* particles. Thus, we conclude that S4 is inaccessible to antibody on the 30S subunit. This conclusion directly contradicts several reports of S4 accessibility on the 30S subunit (2–5, 19–21). S4 was mapped on the surface of the 30S subunit by Lake et al [4, 19] using an AS4 preparation that was included in the experiments reported here (animal 5). We have now shown that the reaction of this antibody preparation with the 30S subunit was not S4-specific; thus, the reported location of S4 is incorrect.

Stöffler et al [1, 2] used several methods to investigate the accessibility of r-proteins, including S4, to antibody in the 30S subunit. Two of these methods, immunodiffusion and immunoprecipitation of intact 30S subunits, involved incubation of subunits with serum for long periods of time. Such conditions may result in substantial degradation of the subunits by serum nucleases and/or proteases and concomitant exposure of S4.

Sedimentation analysis and inhibition of poly-U translation were also used by Stöffler et al [2] to demonstrate a reaction of antibody with the 30S subunits. However, no test of specificity was reported. Therefore, these experiments did not eliminate the possibility that S4 is completely inaccessible to antibody and that the observed reaction was due to contaminating antibody directed against some other ribosomal protein.

This group has also mapped S4 on the surface of the 30S subunit using the immunoelectron microscopic approach [5, 20, 21]. The report of Tischendorf and Stöffler [5] contains the only data relevant to the specificity of the reaction of AS4 with the 30S subunit. Although it was stated that absorption experiments were done, the conditions and data were not reported. It is possible that contaminating antibodies were absorbed with a contaminant in the S4 used for absorption, particularly if an excess of antigen was used. The possibility of contaminating antibodies must be considered, since two of the antisera used gave multiple precipitin bands when reacted with 30S subunits in immunodiffusion experiments [2]. The explanation given, that these precipitin lines represent the reaction of AS4 antibodies with degraded or heterogeneous ribosomes, seems unlikely since such particles ought to have antigenic determinants of S4 in common and thus give a single precipitin band.

An additional argument for specificity given by Tischendorf and Stöffler [5] was the "enrichment" of the immunoglobulins from sera by affinity chromatography on S4-Sepharose. If the S4 antigen coupled to the column contained any contaminants, reactive antibody against that contaminant would also have been bound. Thus, antibody eluted from such an affinity column should not be assumed to be monospecific. These experiments did not eliminate the possibility that the observed subunit reactivity was due to antibody directed against another r-protein.

Although extensive regions of S4 are not accessible on the 30S subunit, small regions may still be exposed on the ribosome surface. Our findings are in good agreement with other approaches used to determine ribosomal protein accessibility. S4 in the 30S subunit is highly resistant to chemical modification by 2-methoxy-5-nitrotropone [22], trypsin digestion [23], glutaraldehyde modification [16], and reductive methylation with formaldehyde and $NaBH_4$ [24, 25].

Since S4 is inaccessible to antibody in the *intact* 30S subunit, direct immunochemical mapping of S4 determinants on the surface of the subunit will not be possible. However, S4 is exposed on the RI particles and these particles are similar in gross morphology to the intact 30S subunit [26, 27]. Thus, it may be possible to infer the location of S4 from mapping experiments carried out with subribosomal particles. The observation that S4 is accessible to antibody on ribosome assembly intermediates and masked on the 30S subunit also opens the possibility of probing the intermolecular interactions responsible for the masking of S4 antigenic determinants.

ACKNOWLEDGMENTS

We wish to thank M. Nomura, J. Lake, W. A. Strycharz, W. A. Held, and J. Schwarzbauer for helpful discussions, and Catherine Bloomer and Keiko Nomura for

valuable technical assistance. This work was supported by National Institute of Health grants GM22150 and GM07215.

REFERENCES

1. Stöffler G: In Nomura M, Tissières A, Lengyel P (eds): "Ribosomes." Cold Spring Harbor Laboratory, 1974, pp 615.
2. Stöffler G, Hasenbank R, Lütgehaus M, Maschler R, Morrison CA, Zeichhardt H, Garrett RA: Mol Gen Genet 127:89, 1973.
3. Lelong JC, Gros D, Gros F, Bollen A, Maschler R, Stöffler G: Proc Natl Acad Sci USA 71:248, 1974.
4. Lake JA, Pendergast M, Kahan L, Nomura M: Proc Natl Acad Sci USA 71:4688, 1974.
5. Tischendorf GW, Stöffler G: Mol Gen Genet 142:193, 1975.
6. Held WA, Mizushima S, Nomura M: J Biol Chem 248:5720, 1973.
7. Traub P, Mizushima S, Lowry CV, Nomura M: Methods Enzymol 20(Part C):391, 1971.
8. Staehelin T, Maglott DR: Methods Enzymol 20(Part C):499, 1971.
9. Held WA, Nomura M: Biochemistry 12:3273, 1973.
10. Howard GA, Traut RR: FEBS Lett 29:177, 1973.
11. Kahan L, Zengel J, Nomura M, Bollen A, Herzog A: J Mol Biol 76:473, 1973.
12. Kahan L, Held WA, Nomura M: J Mol Biol 88:797, 1974.
13. Dedmon RE, Holmes AW, Deinhardt F: J Bacteriol 89:734, 1965.
14. Mauer PH: Methods Immunol Immunochem 3:40, 1971.
15. Bailey JL: "Techniques in Protein Chemistry." 2nd Ed. Amsterdam: Elsevier, 1967, p 340.
16. Kahan L, Kaltschmidt E: Biochemistry 11:2691, 1972.
17. Zamir A, Miskin R, Elson D: J Mol Biol 60:347, 1971.
18. Winkelmann D, Kahan L: J Supramol Struct (Suppl 3) 8:100, 1979.
19. Lake JA, Pendergast M, Kahan L, Nomura M: J Supramol Struct 2:189, 1974.
20. Tischendorf GW, Zeichhardt H, Stöffler G: Proc Natl Acad Sci USA 72:4820, 1975.
21. Stöffler G, Tischendorf GW: In Drews L, Hahn FE (eds): "Topics in Infectious Diseases." New York: Springer, 1975, vol 1, p 117.
22. Craven G, Gupta V: Proc Natl Acad Sci USA 67:1329, 1970.
23. Crichton RR, Wittmann HG: Mol Gen Genet 114:95, 1971.
24. Moore G, Crichton RR: FEBS Lett 37:74, 1973.
25. Amons R, Möller W, Schiltz E, Reinbolt J: FEBS Lett 41:135, 1974.
26. Vasiliev VD, Koteliansky VE: FEBS Lett 76:125, 1977.
27. Vasiliev VD, Koteliansky VE, Rezapkin GV: FEBS Lett 79:170, 1977.

Journal of Supramolecular Structure 10:457–465 (1979)
Biological Recognition and Assembly 15–23

Structure of the DNA Binding Cleft of the Gene 5 Protein From Bacteriophage fd

Alexander McPherson, Frances Jurnak, Andrew Wang, Ian Molineux, and Alexander Rich

Department of Biological Chemistry, Milton S. Hershey Medical Center of The Pennsylvania State University, Hershey, Pennsylvania 17033 (A.M., F.J.), Massachusetts Institute of Technology, Cambridge, Massachusetts 01239 (A.W., A.R.), and Department of Microbiology, University of Texas, Austin, Texas 78712 (I.M.)

The structure of the gene 5 DNA unwinding protein from bacteriophage fd has been solved to 2.3-Å resolution by X-ray diffraction techniques. The molecule contains an extensive cleft region that we have identified as the DNA binding site on the basis of the residues that comprise its surface. The interior of the groove has a rather large number of basic amino acid residues that serve to draw the polynucleotide backbone into the cleft. Arrayed along the external edges of the groove are a number of aromatic amino acid side groups that are in position to stack upon the bases of the DNA and fix it in place. The structure and binding mechanism as we visualize it appear to be fully consistent with evidence provided by physical-chemical studies of the protein in solution.

Key words: **DNA binding protein, gene 5, fd bacteriophage, X-ray diffraction, protein-nucleic acid interactions**

The gene 5 product of the filamentous bacteriophage fd is a single-strand specific DNA binding protein of 10,000 daltons molecular weight made in approximately 100,000 copies per infected E coli cell [1]. The protein is coded by the phage genome and is elaborated late in infection when the transition from double-stranded replicative form DNA to single-stranded synthesis of the daughter viral genomes occurs [2]. Its primary physiologic role is the stabilization and protection of single-strand DNA daughter virions from duplex formation following replication in the host [3]. Under conditions of low ionic strength in vitro, it will melt double-stranded homopolymers and will reduce the melting temperature of native double-strand calf thymus DNA by 40°C [4].

Received April 16, 1979; accepted April 18, 1979.

0091-7419/79/1004-0457$02.00

Gene 5 protein exists predominantly as a dimer when free in solution [5] and binds, with a stoichiometry of one monomer per four bases [1], to DNA chains running in opposite directions so that it cross-links two strands of a duplex or opposite sides of closed circular single-stranded DNA. The mechanism for DNA unwinding is simply a linear aggregation along the two opposite strands and derives from the highly cooperative nature of the lateral binding interactions [6]. The extensive degree of cooperativity is presumably a product of strong protein-protein forces between adjacent molecules of the gene 5 protein along the DNA strands. On binding to circular, single-stranded fd DNA, the gene 5 protein collapses the circle into a helical rod-like structure containing two anti-parallel strands of DNA [1].

We have solved the structure of the gene 5 protein to 2.3-Å resolution, traced the course of the polypeptide backbone, and constructed a Kendrew model of the molecule that includes all nonhydrogen atoms in the structure. The technical details of the analysis will be presented elsewhere [7], but we will describe here some of the features of the structure directly relevant to the means and interactions by which the single-stranded DNA is bound to the protein. We would like, in addition, to compare some of our findings with those derived from physical and chemical studies of the gene 5 protein in solution.

MATERIALS AND METHODS

The gene 5 protein was isolated from phage lysates and crystallized for X-ray diffraction analysis as described by McPherson et al [8]. Diffractometer data on native crystals plus six isomorphous heavy-atom derivatives were used to calculate to 2.3-Å resolution an electron density map of the gene 5 protein. The phases were deduced using the isomorphous replacement technique of Perutz [9] and refined by the procedures of Dickerson et al [10] by means of the error treatment of Blow and Crick [11]. The sequence of the protein, as determined by Nakashima et al [12], was fitted to the electron density map and a model was then constructed on a scale of 2 cm/Å in a Richards optical comparator [13] using Kendrew model parts. The coordinates from the model were obtained for all atoms using a plumb and line. Graphical presentation and illustration were produced using the program VDW written by Ringle and Hanson and modified by P.M.D. Fitzgerald for the PDP 11/40 computer. Although not yet complete, the atomic coordinates are presently being refined using the constrained least-squares procedures of Konnert and Hendrickson. Many of our deductions regarding the binding of single-stranded DNA to the gene 5 protein are derived from our attempts to fit by hand a model tetranucleotide to the putative binding region of the gene 5 protein Kendrew model. We will pursue this approach using an interactive graphics system when model refinement has been completed.

RESULTS

Figure 1 shows a wooden model of the gene 5 protein at an effective resolution of about 5.0 Å, viewed approximately down the crystallographic 011 direction. The monomer is roughly 45 Å long, 25 Å wide, and 30 Å high. It is essentially globular, with an appendage of density closely approaching the molecular dyad and tightly interlocking with an identical symmetry-related appendage on the second molecule within the dimer. The major portion of the molecular density slants from upper left to lower right in Figure 1, and creates an overhanging ledge of density that serves in part to create an extended shallow groove banding the outside waist of the monomer. In the dimer the two symmetry-related grooves, each about 30 Å in length, run antiparallel courses and are separated by roughly 25 Å.

The course of the polypeptide chain in the gene 5 monomer is shown in Figure 2 as deduced from our 2.3-Å electron density map. The protein is composed entirely of antiparallel β structure with no α-helix whatsoever. This is as expected from spectroscopic measurements [14] and sequence-structure rules [15]. There are three basic elements of secondary structure shown in Figure 3 that comprise the molecule, a three-stranded antiparallel β sheet arising from residues 12–49, a two-stranded antiparallel β ribbon formed by residues 50–70, and a second two-stranded antiparallel β ribbon derived from residues 71–82. It is the first of the two β loops (50–70) that creates the appendage of density near the molecular dyad and maintains the dimer in solution. The second β loop (71–82) forms the top surface of the molecule and we believe it is most involved in producing the neighbor-neighbor interactions responsible for the cooperative protein binding. The central density of the molecule is created by the severely twisted three-stranded β sheet made up of residues 12–49. As a result of the distortion from planarity of these three strands, a distinct concavity is produced on the underside of this sheet. Enhanced in part by density from the β ribbon (50–70) near the dyad, this concavity is extended and deepened to provide the long 30-Å groove.

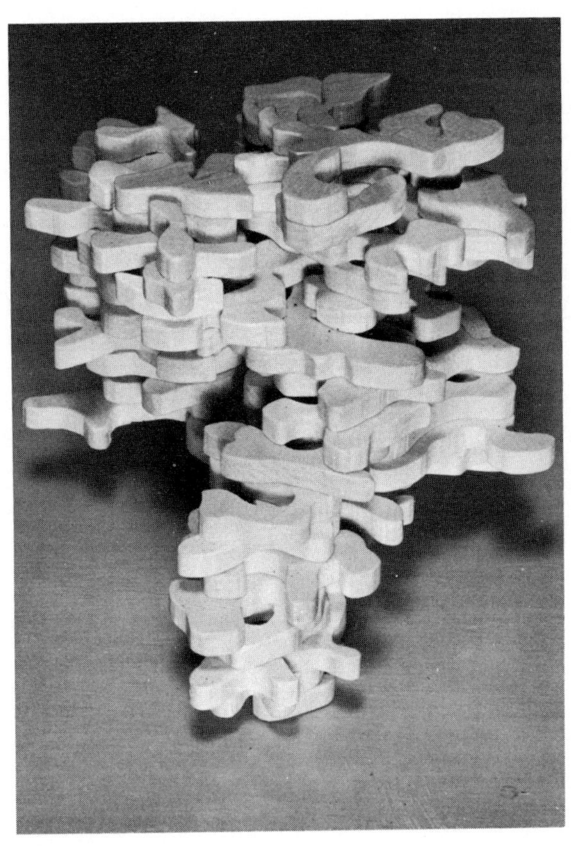

Fig. 1. Representation of the gene 5 protein electron density based on the 2.3-Å Fourier made by cutting appropriate envelopes of density from each section of map and assembling them in the y direction. Model is viewed approximately along the 011 direction and can be seen as an essentially globular mass with a protrusion of density near the molecular dyad.

The long groove beneath the three-stranded sheet by its shape and extent seems to be the single-strand DNA binding interface. There is no other passage through the density that would be consistent with a long polynucleotide binding region. Given that this is the site, then the mode of cross-strand attachment of the gene 5 protein would be that shown in Figure 4. The two monomers within the dimer bind to strands of opposite polarity across the duplex DNA with the molecular dyad roughly perpendicular to the plane of the two bound strands which are separated in the complex by about 25 Å.

Considering that refinement of the structure is not yet complete and that placement of amino acid side chains is still tentative, there still remain a number of interesting features that can be described and that are not likely to be seriously revised. The tetranucleotide binding trough in the gene 5 protein is composed primarily of the amino acid side chains arising from residues 12–49 of the antiparallel β sheet. These strands run more or less parallel with the direction of the DNA strand as it would bind in the trough. The surface of the trough is also comprised in part of residues 50–56 and 66–69, from the interior portions of the two strands forming the β loop near the molecular dyad. A stereo drawing of the gene 5 monomer showing the binding region is shown in Figures 5 and 6.

Aromatic amino acid side chains have been implicated in the binding of DNA to the gene 5 protein by chemical modification and neutron magnetic resonance (NMR) studies. These show that tyrosines 26, 41, and 56 lie near the surface of the protein and are readily nitrated by tetranitromethane, which prevents DNA binding [15]. Conversely, binding of oligonucleotides or DNA prior to reaction prevents nitration of these residues. ^{19}F NMR of the fluorotyrosyl-containing protein confirms these results and further suggests that these tyrosyls intercalate or stack with the bases of the DNA [16]. Similar results, obtained with deuterated protein, implicate at least one phenylalanine residue in a similar fashion [17]. Spectral data lend further support to the idea that the aromatic residues of the protein stack upon or intercalate between the bases [18].

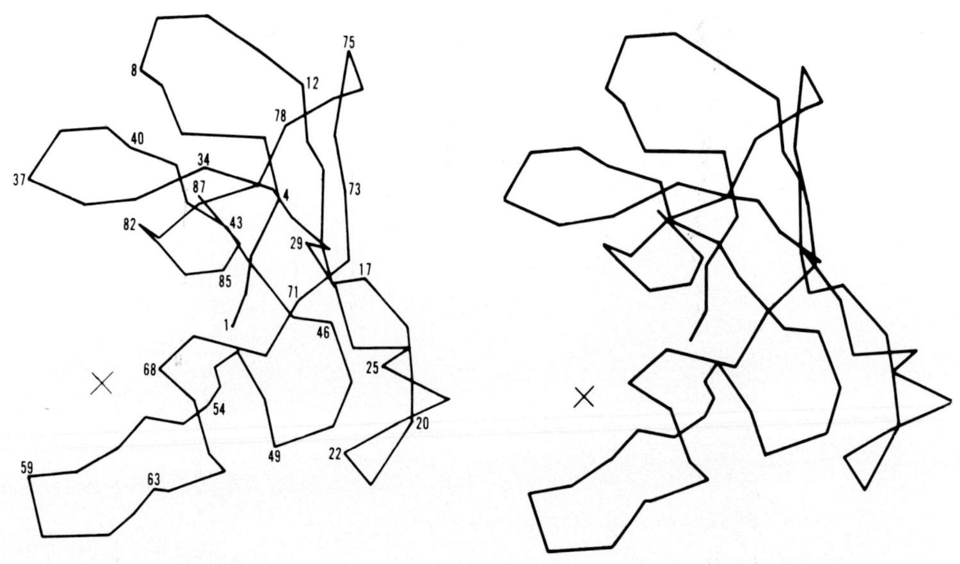

Fig. 2. Stereo drawing of the polypeptide tracing of the gene 5 DNA unwinding protein based on α-carbon coordinates measured from a Kendrew model. The view is roughly along the crystallographic *a* axis.

A number of aromatic residues are arrayed along the binding surface, and these include tyrosines 26, 41, 34, and 56 and phenylalanines 13 and 68. The distribution is not uniform, one end of the trough appearing considerably richer than the other and bearing both phenylalanines as well as tyrosines 34, 41, and 56. The opposite end of the trough, that closest to the viewer in Figure 6, contains only tyrosine 26. The aromatic side chains, with the exception of tyrosine 56 and phenylalanine 68, do not protrude into the binding cleft but are turned away. Each can, however, be brought down into the binding groove by an appropriate rotation about the β carbon. Of particular interest are the side groups of tyrosine 41, tyrosine 34, and phenylalanine 13, which form a triple stack with phe 13 most interior, tyr 41 fully on the outside, and tyr 34 sandwiched in between. The stacking is not precisely one atop the other, but the rings are fanned out like three playing cards. These rings are on the upper edge of the trough; below them on the lower edge and actually positioned in the mouth of the groove is tyrosine 56. Coleman et al [16] note from their NMR data that in the uncomplexed protein a number of tyrosyl proton resonances show upfield shifts, suggesting some ring current effects due to stacking. They hypothesized that the tyrosyl residues involved might be in some organized array such as we observe. These resonances are lost on oligonucleotide binding, suggesting a disruption of the pattern as the residues begin interacting with the bases of the DNA.

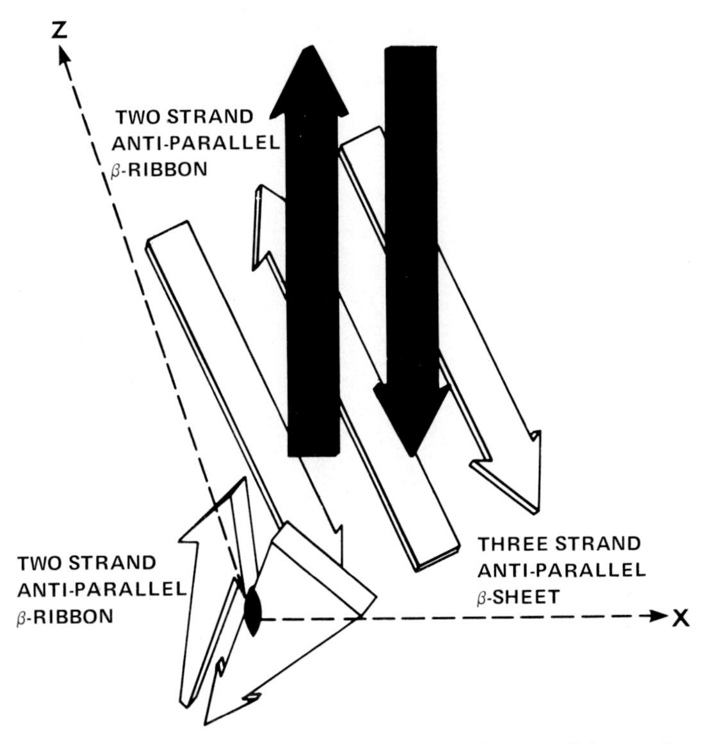

Fig. 3. Schematic drawing showing the secondary structural elements of the gene 5 molecule. Beginning with the N-terminal portion of the sequence, it consists of a three-stranded β sheet which forms the major part of the DNA binding region, a two-stranded β ribbon principally responsible for maintaining the molecule as a dimer in solution, and a second β ribbon that we believe is the primary participant in the lateral interactions from which the cooperativity of protein binding arises.

Tyrosine 26 is near the β bend between strands 1 and 2 of the antiparallel β sheet. This bend appears to be a very flexible elbow of density extending out from the central mass of the molecule and making up one end of the binding region. Even in the crystal, it projects into a large solvent area and seems to be rather mobile and free to move. It is the only tyrosine that we were able to iodinate in the crystal.

We noted that three of the tyrosines in the molecule – 26, 41, and 56 – fall adjacent to, or one removed from, a proline residue. The backbone structure of the protein is engaged in β structure and one would expect that this hydrogen bonding network might restrict the freedom of many bulky side groups. However, by virtue of their proximity to a natural structure-disrupting amino acid, proline, these three tyrosines are endowed with more freedom than they might otherwise enjoy. Because of the proline residues, the tyrosine side chains can rotate from one side of the sheet to the other through the trap door created by the neighbor.

Cysteine 33 is on the inside surface of the binding groove and could certainly interact with the DNA strand. In the conformation that we observe, however, the –SH group is turned up into the interior of the molecule, away from the solvent. It is not in contact with the neighboring tyrosine 34. Although inaccessible to the bulkier Ellmans' reagent, the single cysteine can be reacted with mercuric chloride. Mercuration of cysteine 33 prevents nucleotide or DNA binding to the protein and, conversely, complexation with oligonucleotides prevents reaction with mercury [15]. This is consistent with its placement in the binding groove as is the finding that this –SH group can be photo-cross-linked to thymidine residues of bound nucleic acid [19].

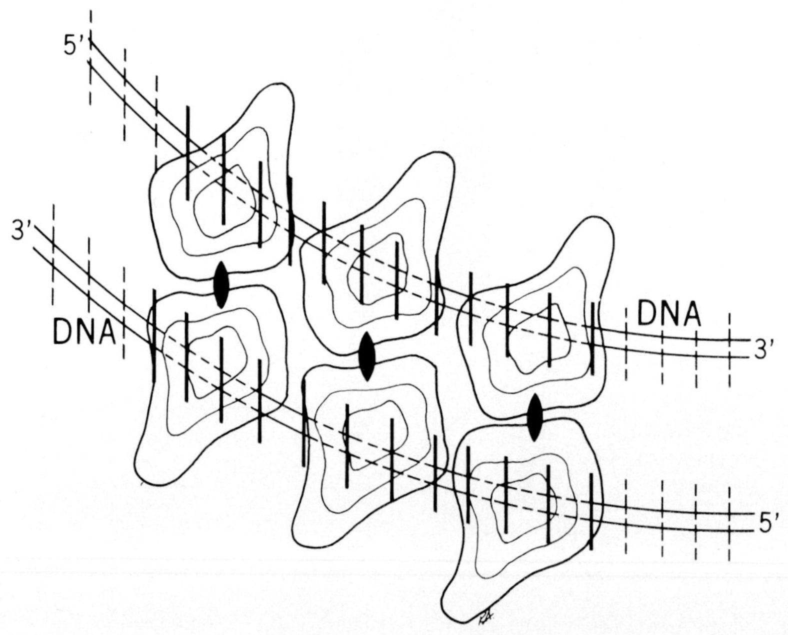

Fig. 4. Schematic representation illustrating the cross-chain binding of the gene 5 dimers to opposing strands of a DNA duplex or opposite sides of a circular single-stranded DNA molecule. The distance between opposing DNA single strands would be about 25 Å.

Acetylation of the ε-amino groups of the seven lysyl residues destroys the binding of gene 5 protein to oligonucleotides and DNA, but these groups are not protected by the presence of DNA from reaction [15]. In addition, NMR spectra show that the ε-amino groups do not undergo chemical shift or line broadening upon complexation and appear to remain highly mobile. This was interpreted as indicating that the ε-amino groups provide a neutralizing charge cloud for the negative phosphate backbone of the nucleotide but do not form highly rigid salt bridges or hydrogen bonds [16]. Resonances from the δ-CH_2 groups of the arginyl residues do undergo chemical shifts and line broadening on DNA complexation, and this could represent direct interaction of the guanidino groups with the phosphate backbone [16].

The DNA binding trough has over its interior surface a fairly large number of basic amino acid side chains which, because of the length and flexibility of these residues, reach into the groove though originating at disparate locations within the molecule. The basic residues most clearly apparent in the cleft are arginines 21, 80, and 82 and lysines 24 and 46. These are all found on the interior surface of the trough, so that the cleft is also something of a postively charged pocket in the protein. It should be noted that other basic amino acids could conceivably approach the binding region but in the conformation we observe in the crystal they are elsewhere. In particular, arginine 16 and lysine 46 are certainly close to the interface, but we see them turned away from the groove rather than toward it.

DISCUSSION

The DNA binding cleft of about 30-Å length and formed principally by the underside of an antiparallel β sheet has been tentatively identified in the gene 5 protein. This assignment is based on the general shape and size of the groove and the distribution of amino acid residues on its surface, which have been implicated in DNA binding by NMR, optical spectra and chemical modification studies in solution.

The binding cleft is very interesting in that the positively charged residues of lysine and arginine are distributed predominantly over the most interior surface, while the aromatic

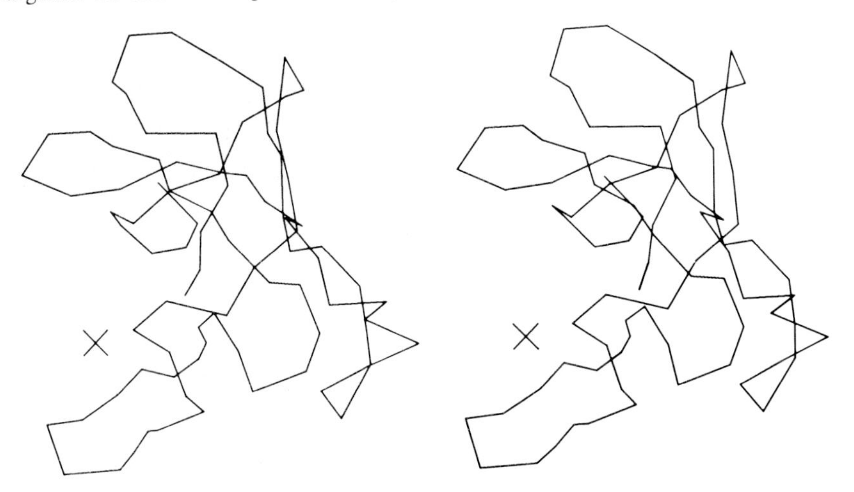

Fig. 5. Stereo representation of the polypeptide backbone of the gene 5 protein rotated so that the view is roughly along the course of the DNA binding groove. This groove is approximately 25 Å in length and runs more or less parallel with the strands of the β sheet.

residues are arrayed primarily along the exterior edges. Thus it appears that the negatively charged polyphosphate backbone of the single-stranded DNA is first recognized by the protein and that it is drawn and fixed to the interior of the groove by the charge interactions. This is followed by rotation of the aromatic groups down and into position to stack upon the bases of the DNA, which are now splayed out toward the exterior of the protein. This is consistent with the finding of Day [14] from micrographic and spectral data that the DNA in the gene 5 complex is completely unstacked and stretched along the filament axis. The cleft then acts as an elongated pair of jaws and the DNA is drawn between them by charge interactions involving the phosphates with the interior lysines and arginines. The jaws then close around the DNA strand through small conformation changes and the rotation of aromatic side chains into position to stack upon the purine and pyrimidine bases. That small, but not gross, conformation changes occur in the protein upon DNA binding is in agreement with the NMR studies by Coleman et al [16] on α-CH and on aliphatic methyl groups which suggest that gene 5 protein must contain a large percentage of fixed structure without large regions of flexible polypeptide chain. Days' spectral evidence also indicates that only small changes occur in overall protein structure on binding [14].

That the interaction between gene 5 protein and DNA is to a great extent electrostatic is clear from the finding that moderate divalent and monovalent cation concentrations cause the complex to dissociate and that binding capacity is lost when the arginines and lysines are chemically modified [15]. The involvement of the aromatic groups, however, is also quite clear from the NMR and spectral data. The minor conformation changes in the gene 5 protein involving other residues and possibly even main-chain atoms are consistent with the physical and chemical studies. Therefore, although our binding mechanism is somewhat speculative, it is to our knowledge entirely consistent with the structure as we visualize it and with the evidence at hand from noncrystallographic analyses.

The binding of the gene 5 DNA unwinding protein to deoxyoligonucleotides is nonspecific in that complexation will occur with oligomers of any sequence. It is found, however, that the gene 5 protein binds oligomers of different sequence with differing affinities and these may vary over two orders of magnitude [16]. Thus the protein does distinguish

Fig. 6. Stereo drawing of the polypeptide backbone of the gene 5 protein oriented as in Figure 5, with each α-carbon represented by a sphere of 3.0-Å radius to give a space-filling effect. The DNA binding region is the pronounced groove running roughly perpendicular to the plane of the drawing.

between different binding possibilities, and this could be the basis for recognition of specific nucleation sites on the fd DNA. Until the structure of a gene 5 protein oligomer complex has been directly visualized we will not be able to confidently establish the interactions that confer the differential binding affinities. The possibility exists, however, that when the interactions of the gene 5 protein with specific sequences of DNA are completely defined, they will suggest how, by only minor structural alterations, a high degree of recognition specificity might be achieved.

ACKNOWLEDGMENTS

This research was supported by grants from the National Institutes of Health, National Science Foundation, National Aeronautics and Space Administration, and the American Cancer Society. A.W. is supported by a grant from the MIT Cancer Center. F.J. is an NIH postdoctoral fellow.

REFERENCES

1. Alberts B, Frey L, Delius H: J Mol Biol 68:139–152, 1972.
2. Oey JL, Knippers R: J Mol Biol 68:125–128, 1972.
3. Mazur BJ, Model P: J Mol Biol 78:285–300, 1973.
4. Salstrom JS, Pratt D: J Mol Biol 61:489–501, 1971.
5. Cavalieri S, Goldthwait DA, Neet K: J Mol Biol 102:713, 1976.
6. Dunker AK, Anderson EA: Biochim Biophys Acta 402:31–34, 1975.
7. McPherson A, Jurnak FA, Wang AHJ, Molineux I, Rich A: J Mol Biol (In press).
8. McPherson A, Molineux I, Rich A: J Mol Biol 106:1077, 1976.
9. Bragg WL, Perutz MF: Proc Roy Soc A225:315, 1954.
10. Dickerson RE, Kendrew JC, Strandberg BE: Acta Crystallogr 14:1188, 1961.
11. Blow DM, Crick FHC: Acta Crystallogr 12:794, 1959.
12. Nakashima Y, Dunker AK, Marion DA, Konigsberg W: FEBS Lett 40:290, 1974.
13. Richards FM: J Mol Biol 37:225, 1968.
14. Day LA: Biochemistry 12:5329, 1973.
15. Anderson RA, Nakashima Y, Coleman JE: Biochemistry 14:907, 1975.
16. Coleman JE, Anderson RA, Ratcliffe RG, Armitage IM: Biochemistry 15:5419, 1976.
17. Coleman JE, Armitage IM: Biochemistry (In press).
18. Pretorius HT, Klein M, Day LA: J Biol Chem 250:9262–9269, 1975.
19. Nakashima Y, Konigsberg W: Paper presented at the International Symposium for Photobiology, Williamsburg, Virginia, 1975.

Journal of Supramolecular Structure 10:467–478 (1979)
Biological Recognition and Assembly 25–36

Reconstituted Low Density Lipoprotein: A Vehicle for the Delivery of Hydrophobic Fluorescent Probes to Cells

Monty Krieger, Louis C. Smith, Richard G. W. Anderson, Joseph L. Goldstein, Yin J. Kao, Henry J. Pownall, Antonio M. Gotto, Jr., and Michael S. Brown

Departments of Molecular Genetics and Cell Biology, University of Texas Health Science Center at Dallas, Dallas, Texas 75235 (M.K., R.G.W.A., J.L.G., M.S.B.); and Department of Medicine, Baylor College of Medicine, Houston, Texas 77030 (L.C.S., Y.J.K., H.J.P., A.M.G.)

Previous studies have shown that the cholesteryl ester core of plasma low density lipoprotein (LDL) can be extracted with heptane and replaced with a variety of hydrophobic molecules. In the present report we use this reconstitution technique to incorporate two fluorescent probes, 3-pyrenemethyl-23,24-*dinor*-5-cholen-22-oate-3β-yl oleate (PMCA oleate) and dioleyl fluorescein, into heptane-extracted LDL. Both fluorescent lipoprotein preparations were shown to be useful probes for visualizing the receptor-mediated endocytosis of LDL in cultured human fibroblasts. When normal fibroblasts were incubated at 37°C with either of the fluorescent LDL preparations, fluorescent granules accumulated in the perinuclear region of the cell. In contrast, fibroblasts from patients with the homozygous form of familial hypercholesterolemia (FH) that lack functional LDL receptors did not accumulate visible fluorescent granules when incubated with the fluorescent reconstituted LDL. A fluorescence-activated cell sorter was used to quantify the fluorescence intensity of individual cells that had been incubated with LDL reconstituted with dioleyl fluorescein. With this technique a population of normal fibroblasts could be distinguished from a population of FH fibroblasts. The current studies demonstrate the feasibility of using fluorescent reconstituted LDL in conjunction with the cell sorter to isolate mutant cells lacking functional LDL receptors.

Key words: **low density lipoprotein, cell surface receptors, receptor-mediated endocytosis, reconstitution of lipoproteins, fluorescent probes, fluorescence-activated cell sorter, familial hypercholesterolemia**

Abbreviations: FH, familial hypercholesterolemia; LDL, low density lipoprotein; PMCA oleate, 3-pyrenemethyl-23,24-*dinor*-5-cholen-22-oate-3β-yl oleate; r-[PMCA oleate]LDL and r-[dioleyl fluorescein]LDL, LDL that had been reconstituted with PMCA oleate and dioleyl fluorescein, respectively.

Receptor-mediated endocytosis is the process by which certain physiologically important molecules such as plasma transport proteins and protein hormones bind to cell surface receptor sites before being internalized by cells. For some receptor-bound molecules, efficient cellular uptake is achieved by the clustering of receptors in specialized regions of plasma membrane called "coated pits," which continually invaginate and pinch off from the cell surface to form coated endocytic vesicles [1–3]. In principle, the process of receptor-mediated endocytosis can be utilized as a system for rapid delivery of drugs to cells, provided that pharmacologically active agents can be attached to transport proteins and hormones that normally enter cells through coated vesicles.

One receptor-mediated uptake system that lends itself to drug delivery is the one that involves low density lipoprotein (LDL), the major cholesterol-transport protein in human plasma [1, 4]. LDL is a large spherical particle that has an average diameter of 220 Å and a molecular weight of approximately 3×10^6 [5]. Each LDL particle consists of an apolar core containing approximately 1,300 molecules of cholesteryl ester that is surrounded by a polar coat composed primarily of phospholipid, small amounts of unesterified cholesterol, and apoprotein B [5]. Cultured cells and body cells derive cholesterol from plasma LDL through receptors that recognize the apoprotein B component of the lipoprotein [4]. Once bound to its receptor in coated pits, LDL is rapidly internalized and delivered to lysosomes where its components undergo hydrolysis. The apoprotein B of LDL is degraded to amino acids, and the cholesteryl esters of LDL are hydrolyzed to yield unesterified cholesterol, which is used by cells for structural and regulatory purposes [4].

The feasibility of using LDL as a vehicle for delivering hydrophobic molecules to cells bearing LDL receptors has recently been established [6–8]. Krieger et al developed a method for removing the cholesteryl ester core of LDL with heptane and reconstituting the particle with exogenous cholesteryl esters [6, 7]. In addition to cholesteryl esters, a wide variety of other hydrophobic molecules can be used to reconstitute heptane-extracted LDL, including compounds that contain esters of long chain *cis*-unsaturated fatty acids (such as triolein and methyl oleate) and compounds that contain polyisoprenoid groups (such as retinyl palmitate and ubiquinone-10) [8]. These reconstituted LDL preparations retain the ability to bind to the LDL receptor and to be taken up by cultured cells [6–8]. Substances introduced into the core of LDL will not enter cells unless the lipoprotein is taken up by the receptor mechanism [6–8].

Previous studies of reconstituted LDL have used radiochemical and biochemical methods to measure cellular uptake. In the present studies we have used the reconstitution technique to incorporate two types of fluorescent molecules into LDL, allowing LDL uptake to be visualized with a fluorescence microscope. The uptake of both preparations of fluorescent reconstituted LDL was found to be dependent on the LDL receptor. With the use of a fluorescence-activated cell sorter, we show that cells that contain LDL receptors can be separated from cells that are genetically deficient in receptors.

MATERIALS AND METHODS

Materials

Sodium [^{125}I]iodide (17 mCi/μg) was purchased from Amersham/Searle. 0.05% Trypsin/0.02% EDTA solution (Cat. No. 610-5300) was obtained from Grand Island Biological Co. 3-Pyrenemethyl-23,24-*dinor*-5-cholen-22-oate-3β-yl oleate (PMCA oleate)

(Fig. 1) was synthesized from 3-pyrenemethyl-23,24-*dinor*-5-cholen-22-oate [9] and oleyl chloride in pyridine, and the compound was purified by silicic acid chromatography. Dioleyl fluorescein (Fig. 2) was synthesized by Dr. T. Y. Shen and associates at Merck Sharp & Dohme Research Laboratories. Other supplies and reagents were obtained from sources as previously reported [6, 10].

Lipoproteins

Human LDL (density 1.019–1.063 g/ml) and lipoprotein-deficient serum (density $>$ 1.215 g/ml) were obtained from the plasma of healthy individuals and prepared by ultracentrifugation [10]. Fetal calf lipoprotein-deficient serum (density $>$ 1.215 g/ml) was prepared by ultracentrifugation [10]. r-[PMCA oleate]LDL and r-[dioleyl fluorescein]-LDL were prepared by a previously described reconstitution method in which the endogenous neutral lipids of LDL were removed by heptane extraction and replaced with PMCA oleate and dioleyl fluorescein, respectively [6, 8]. r-[PMCA oleate]LDL was prepared by incubating 1.9 mg of heptane-extracted LDL-protein with 200 μl of benzene containing 6 mg of PMCA oleate [8]. The mass ratio of PMCA oleate to protein (mg/mg) in the final solubilized r-[PMCA oleate]LDL was 1.24. r-[Dioleyl fluorescein]LDL was prepared by incubating 1.9 mg of heptane-extracted LDL with 200 μl of heptane containing a mixture of 0.6 mg of dioleyl fluorescein and 5.4 mg of triolein [8]. The mass ratio of dioleyl fluorescein to protein (mg/mg) in the final solubilized r-[dioleyl fluorescein]LDL was 0.5. The mass ratio of triolein to protein (mg/mg) in the same preparation of r-[dioleyl fluorescein]LDL was 1.2. The triolein was mixed with dioleyl fluorescein in the reconstitution to dilute the dioleyl fluorescein so that the cellular fluorescence intensity would remain within the dynamic range of the detection system of the cell sorter (see below). The concentrations of all lipoproteins are expressed in terms of their protein content. ^{125}I-LDL was prepared as previously described [11]. Lipoprotein electrophoresis was carried out in agarose gel at pH 8.6 in barbital buffer [12].

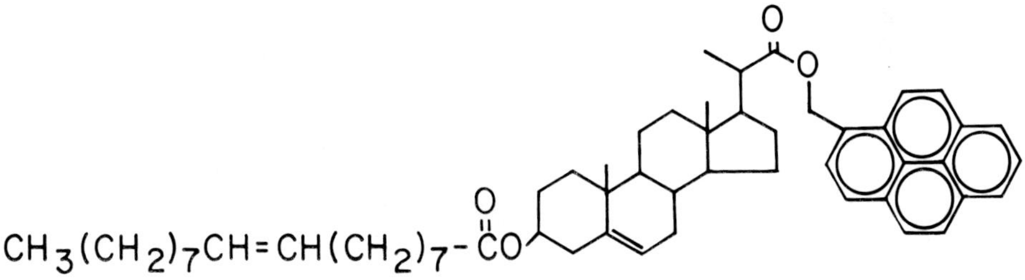

Fig. 1. Structural formula of PMCA oleate.

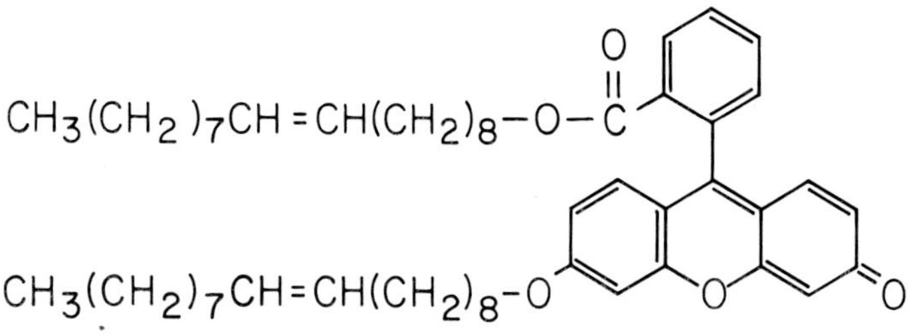

Fig. 2. Structural formula of dioleyl fluorescein.

Cells

Human fibroblasts from normal subjects and from patients with the heterozygous and homozygous forms of familial hypercholesterolemia (FH) were grown in monolayer culture as previously described [13].

Assays

The amounts of surface-bound ^{125}I-LDL (dextran sulfate-releasable ^{125}I-LDL), intracellular ^{125}I-LDL (dextran sulfate-resistant ^{125}I-LDL), and degraded ^{125}I-LDL were measured in intact fibroblast monolayers by previously described methods [9, 13]. The protein content of extracts and lipoproteins was determined by the method of Lowry et al [14], with bovine serum albumin as a standard. The content of PMCA oleate and dioleyl fluorescein in reconstituted LDL was determined by fluorescence spectrophotometry after extracting the lipids from the lipoprotein with chloroform/methanol.

Fluorescence-Activated Cell Sorter

After incubation of cells with r-[dioleyl fluorescein]LDL, the medium was removed and each monolayer was washed 6 times at 4°C with an albumin-containing buffer [13]. The cell monolayers were dissociated by incubation with 1 ml of 0.05% trypsin/0.02% EDTA solution for 1.5 min at 37°C. Nonspecific background fluorescence was reduced by illuminating each cell suspension for 2 min at 4°C with a 75 watt Xenon lamp at a distance of 5 cm. The cells were then kept on ice (3–10 min) until they were analyzed on a Becton-Dickinson FACS III fluorescence-activated cell sorter (laser settings: 488 nm line, 300 milliwatts of power in the light stabilized mode; photomultiplier setting: 500 V; flow rate: 10,000 cells per sample analyzed at approximately 200–300 cells/sec through a 50 μm aperture) [15]. The light scatter gates were set to exclude small nonfluorescent cell debris and large cell aggregates. The observed fluorescence intensity is expressed on a relative scale, which ranged from 1 to 128 units/cell.

RESULTS

Both of the fluorescent compounds used for these studies, PMCA oleate and dioleyl fluorescein, were incorporated into LDL with high yield using the reconstitution procedure [6, 8]. The structures of PMCA oleate (M_r = 825) and dioleyl fluorescein (M_r = 565) are shown in Figures 1 and 2, respectively. Mass analyses of the r-[PMCA oleate]LDL indicate that approximately 750 molecules of PMCA oleate were incorporated into each LDL particle. Similarly, 440 molecules of dioleyl fluorescein and 680 molecules of triolein were incorporated into each particle of r-[dioleyl fluorescein]LDL.

When the r-[PMCA oleate]LDL preparation was subjected to electrophoresis in agarose gel, the fluorescent lipoprotein exhibited the same electrophoretic mobility as native LDL (Fig. 3). When the gel was examined under fluorescent light, the fluorescent material (Fig. 3A) was observed in a position identical to that of the lipid components of the lipoprotein as visualized by fat red 7B staining (Fig. 3B). Similar results were obtained when r-[dioleyl fluorescein]LDL was analyzed by electrophoresis.

Figure 4 shows a series of micrographs of fibroblasts that had been incubated for 24 h with LDL reconstituted either with PMCA-oleate or with dioleyl fluorescein. When incubated with either of these reconstituted lipoproteins, normal fibroblasts accumulated large amounts of intracellular fluorescent material (Fig. 4 A, B, and E). The material appeared as granules that tended to cluster around the nucleus, a region that contains numerous lysosomes [16].

Two observations indicate that the uptake of fluorescent LDL was occurring through the LDL receptor mechanism: 1) Inclusion of a 50-fold excess of native LDL in the culture medium blocked the uptake of both r-[PMCA oleate]LDL (data not shown) and r-[dioleyl fluorescein]LDL (Fig. 4F) by competing for the limited number of LDL receptor sites [4]. 2) Fibroblasts from a patient with the receptor-negative form of homozygous FH, which have a near-total absence of LDL receptor activity [4], failed to accumulate visible quantities of fluorescent LDL when incubated with either r-[PMCA oleate]LDL (Fig. 4 C and D) or r-[dioleyl fluorescein] LDL (data not shown).

The amount of dioleyl fluorescein accumulated by individual fibroblasts incubated with r-[dioleyl fluorescein]LDL was assessed by detaching the cells from the Petri dish with trypsin and passing them through a fluorescence-activated cell sorter of the type de-

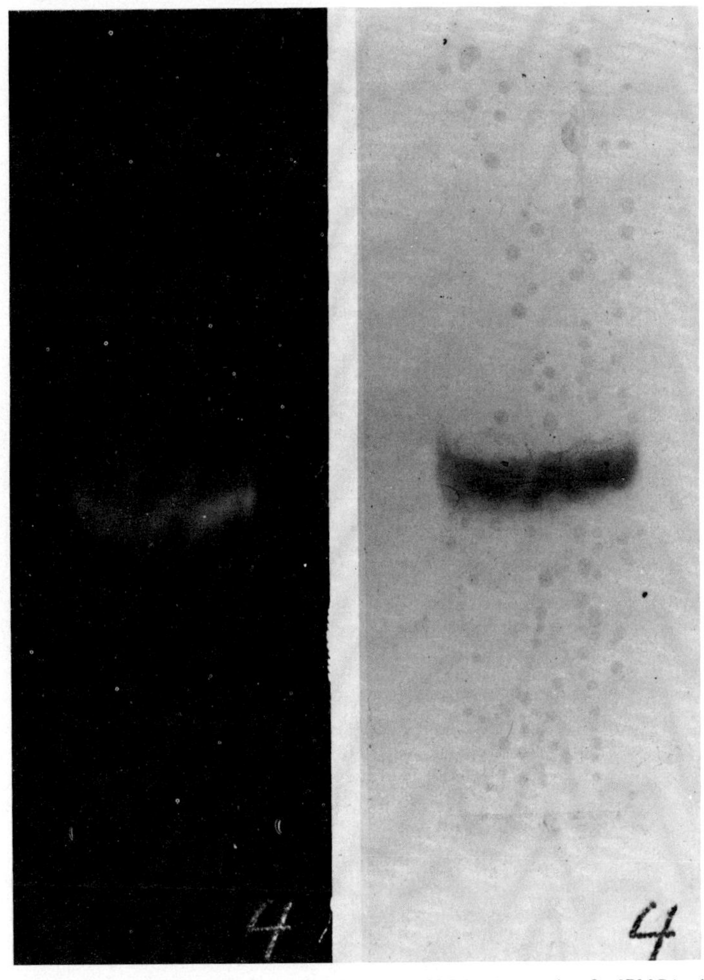

Fig. 3. Electrophoretic mobility of intact r-[PMCA oleate]LDL. A sample of r-[PMCA oleate]LDL (20 μg of protein) was subjected to electrophoresis in agarose gel (7.2 × 2.5 cm) at pH 8.6 (barbital buffer). After electrophoresis, the gel was photographed under ultraviolet light illumination (375 nm) (left), and then fixed and stained with fat red 7B (right). The point of application of the sample is indicated by the well at the bottom of the gel. The electrophoretic mobility of the r-[PMCA oleate]LDL was identical to that of native LDL, which was subjected to electrophoresis in the same study (photograph of native LDL not shown).

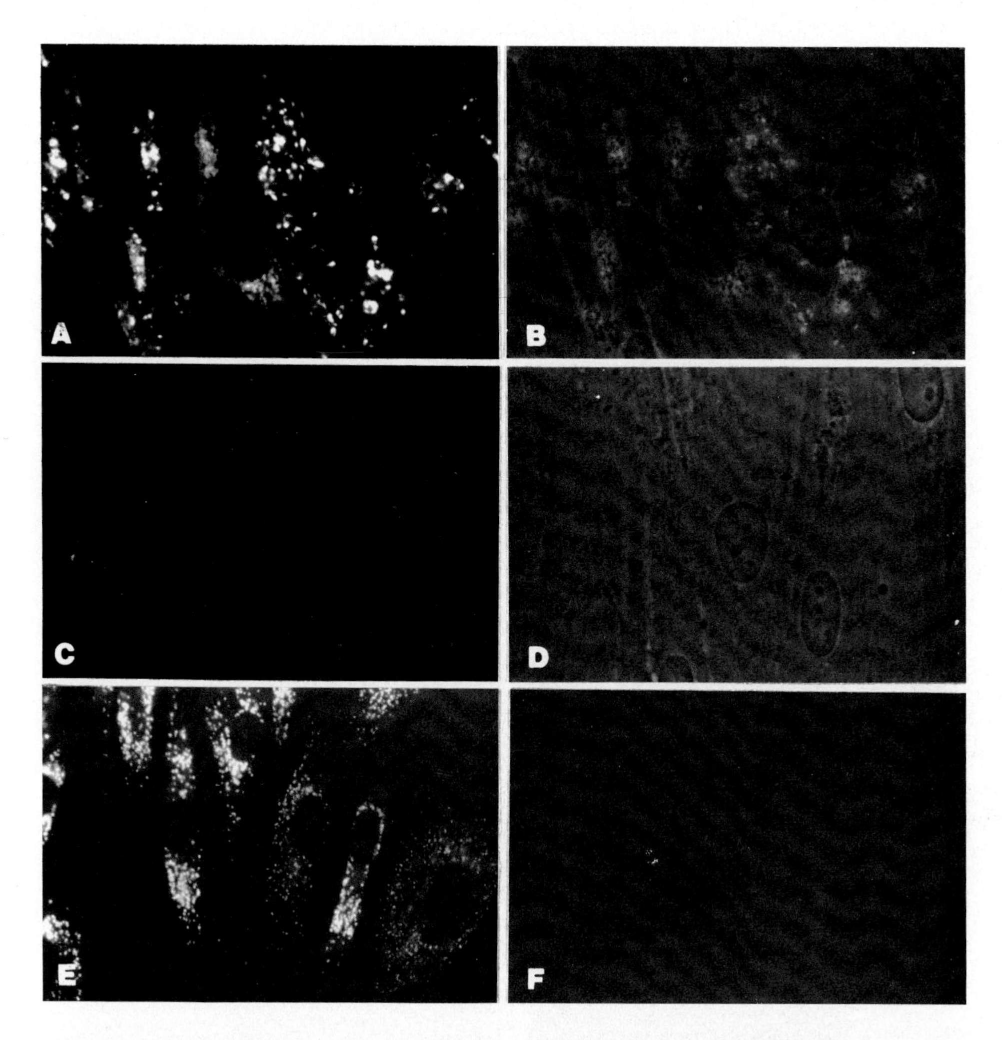

Fig. 4. Fluorescent light micrographs (A, C, E, F) and combined phase contrast-fluorescent light micrographs (B, D) of normal (A, B, E, F) and homozygous FH (C, D) fibroblasts incubated with LDL reconstituted with fluorescent compounds. Cells were seeded (day 0) onto glass coverslips contained within 60-mm Petri dishes at a concentration of 1×10^5 cells per dish in growth medium containing 10% (v/v) fetal calf serum [13]. On day 4 the medium was replaced with medium containing 10% fetal calf lipoprotein-deficient serum. On day 5 the medium was replaced with 2 ml of fresh medium containing 10% fetal calf lipoprotein-deficient serum and either 20 μg protein/ml of r-[PMCA oleate]-LDL (A–D) or r-[dioleyl fluorescein]LDL (E, F) in the absence (A–E) or presence (F) of 1 mg protein/ml of native LDL. After incubation with reconstituted LDL at 37°C for 24 h, the cell monolayers were washed 6 times [12] and fixed with 3% paraformaldehyde in 0.2 M sodium phosphate (pH 7.3) for 10 min at room temperature. The coverslips were removed from the dish, washed with water, mounted with glycerol on glass slides, and viewed in the epifluorescence and phase contrast modes of a Zeiss Photomicroscope III equipped with the appropriate filter packages (Panels A–D: exciter filter, 365/12 nm; chromatic beam splitter, 395 nm; barrier filter, 420 nm. Panels E and F: exciter filters, 455–500 nm; chromatic beam splitter, 510 nm; barrier filter, 528 nm). Panels A and B are photographs of the same cells, as are Panels C and D. Magnifications: A–D, × 600; E and F, × 400.

scribed by Loken and Herzenberg [15]. The results of one such experiment are presented in the dotplots in Figure 5. In this diagram, each dot represents the relative fluorescence (vertical axis) and the relative intensity of light scatter (horizontal axis) of a single cell. In the normal fibroblasts, virtually all cells were highly fluorescent (Fig. 5, left panel). In the FH homozygote cells, much less fluorescence was observed (Fig. 5, center panel). When the two cell populations were mixed together in approximately equal proportions prior to passage through the sorter, the normal cells were clearly distinguished from the FH homozygote cells (Fig. 5, right panel).

Cells from FH heterozygotes have been shown previously by studies using ^{125}I-labeled LDL to express approximately 50% of the normal number of LDL receptors [17, 18]. To determine whether the fluorescence-activated cell sorter could distinguish normal fibroblasts, FH heterozygote fibroblasts, and FH homozygote fibroblasts, cell strains from each of the three genotypes were incubated in the same experiment with both ^{125}I-LDL and r-[dioleoyl fluorescein]LDL at 37°C. In the case of ^{125}I-LDL, the amounts of surface binding, internalization, and degradation of the lipoprotein were measured using standard techniques [4]. Each of these parameters has been shown previously to be a direct reflection of the number of LDL receptors [4]. Fibroblasts from parallel sets of dishes that had been incubated with r-[dioleoyl fluorescein]LDL were harvested with trypsin, the cells from each dish were passed through the cell sorter, and the mean fluorescence intensity per cell was determined.

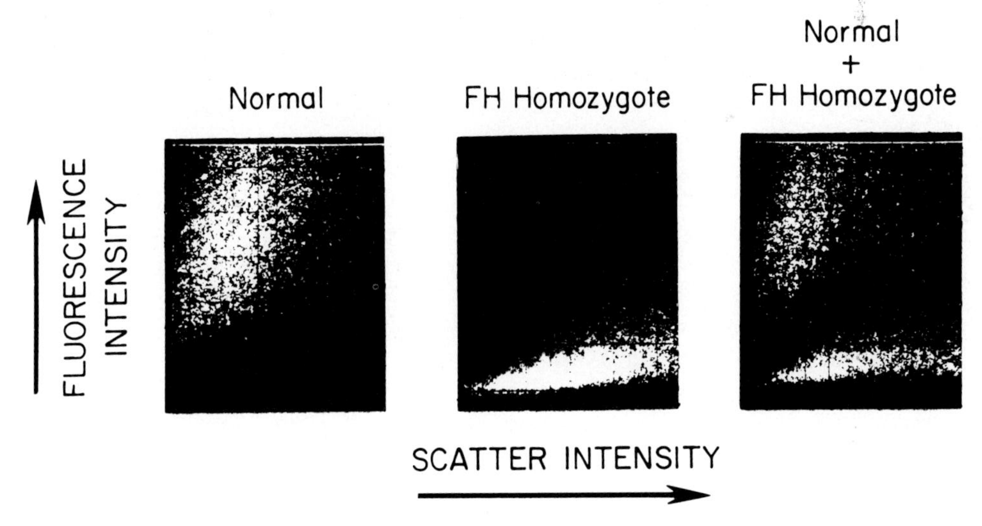

Fig. 5. Dotplots of normal, FH homoxygote, and a mixture of normal and FH homoxygote fibroblasts incubated with r-[dioleyl fluorescein] LDL. Cells were seeded (day 0) in 60-mm Petri dishes at a concentration of 7.5 × 10^4 cells per dish in growth medium containing 10% fetal calf serum [13]. On day 4 fresh medium containing 10% fetal calf serum was added. On day 6 the medium was replaced with 2 ml of fresh medium containing 10% human lipoprotein-deficient serum. On day 7 each dish received 2 ml of fresh medium containing 10% human lipoprotein-deficient serum and 10 μg protein/ml of r-[dioleyl fluorescein]LDL. On day 8, after incubation for 18 h at 37°C, 10,000 normal fibroblast cells (left panel), 10,000 FH homozygote cells (center panel), and a mixture of normal cells and FH homozygote cells (total of 10,000 cells) (right panel) were analyzed in the fluorescence-activated cell sorter as described in Materials and Methods.

Table I shows that LDL receptor activity, as measured by the ^{125}I-LDL techniques, was highest in normal cells, about 50% of normal in the FH heterozygote cells, and virtually undetectable in the FH homozygote cells. Similarly, the average fluorescence intensity was highest in the normal cells (89 and 65 units/cell), about 50% of normal in the FH heterozygote cells (34 and 43 units/cell), and lowest in the FH homozygote cells (17 units/cell). After incubation with r-[dioleyl fluorescein], 54% of the FH homozygote cells (subject M.C.) had fluorescence intensities that were less than 15 units/cell, whereas only 1.7% of the normal cells (subjects D.S. and A.H.) had fluorescence intensities below this value. In normal cells (subject D.S.) that had been subjected to prior incubation with a mixture of 25-hydroxycholesterol and cholesterol to reduce the number of LDL receptors [4], the average fluorescence was reduced from 89 to 22 units/cell (Table I), confirming that the major portion of the fluorescence was attributable to uptake through the LDL receptor. In contrast, prior incubation of the FH homozygote cells (subject M.C.) with 25-hydroxycholesterol plus cholesterol did not significantly reduce the small amount of fluorescence (17 units/cell) that was detected in these cells, confirming that this fluorescence was receptor-independent (Table I).

The reason for the nonspecific fluorescence present in the normal cells incubated with 25-hydroxycholesterol plus cholesterol (22 units/cell) and in FH homozygote cells incubated either in the absence (17 units/cell) or in the presence (14 units/cell) of these sterols is not known. This fluorescence was dependent upon incubation of the cells with r-[dioleyl fluorescein]LDL, but it was independent of the LDL receptor. It may represent transfer of small amounts of dioleyl fluorescein from the surface of LDL to the cell membrane. This low level of nonspecific fluorescence corresponded to a slight diffuse fluorescent glow that was present when cells of all three genotypes were initially observed in the fluorescence microscope after incubation with r-[dioleyl fluorescein]LDL. In cells of all genotypes this glow faded within seconds after exposure to light, whereas the specific fluorescence that was concentrated in the perinuclear region in the normal cells was long-lived (Fig. 4E). Accordingly, in all experiments with r-[dioleyl fluorescein]LDL, the cells were exposed to a bright light prior to being passed through the cell sorter to reduce this nonspecific fluorescence. Such receptor-independent fluorescence was not observed microscopically when cells were incubated with r-[PMCA oleate]LDL.

Heptane-extracted LDL can be reconstituted with a wide variety of nonfluorescent as well as fluorescent hydrophobic molecules [6–8]. Table II lists compounds that have been successfully incorporated into LDL. These molecules fall into several broad classes: 1) lipids that contain esters of long-chain *cis*-unsaturated fatty acids, 2) lipids that contain polyisoprenoid groups, and 3) other lipids, such as cholesteryl ethyl ether, cholesteryl nitrogen mustard, and the aromatic diazo dye fat red 7B. In general, the amount of each hydrophobic molecule that can be incorporated into heptane-extracted LDL is 1–2 mg of neutral lipid per mg of LDL-protein, a value that is approximately equal to the mass of endogenous cholesteryl ester present in native LDL. Inasmuch as most of the lipids listed in Table II do not readily enter mammalian cells in culture, their use as molecular probes has hitherto been limited. The ability to incorporate these compounds into LDL now permits their delivery to cellular lysosomes through the LDL receptor pathway.

DISCUSSION

The results presented here demonstrate that fluorescent reconstituted LDL provides an easily detected visual probe that can be used to determine whether or not a cell in culture expresses LDL receptors. The use of fluorescent LDL should enhance the ease with

TABLE I. LDL Receptor Activity in Fibroblasts From Normal Subjects and Patients With Heterozygous and Homozygous FH: Quantitation with ^{125}I-LDL and r-[dioleyl fluorescein]LDL*

				Metabolism of ^{125}I-LDL			
Clinical phenotype	Age	Sex	Prior treatment of cells	Surface-bound ^{125}I-LDL	Internalized ^{125}I-LDL	Degraded ^{125}I-LDL	Average fluorescence intensity after incubation with r-[dioleyl fluorescein]LDL
					ng/mg protein		units/cell
Normal							
D.S.	Newborn	M	None	150 (140)[a]	920 (840)	3,600 (3,300)	89
D.S.	Newborn	M	25-Hydroxycholesterol + cholesterol	17	75	240	22
A.H.	70	M	None	200 (180)	640 (580)	2,300 (2,200)	65
FH Heterozygote[b]							
G.M.	39	M	None	61 (51)	270 (240)	1,100 (1,000)	34
M.M.	31	F	None	86 (76)	310 (280)	1,200 (1,100)	43
FH Homozygote							
M.C.	8	F	None	5 (1)	21 (16)	37 (33)	17
M.C.	8	F	25-Hydroxycholesterol + cholesterol	4	17	19	14

*Cells were seeded (day 0) in 60-mm Petri dishes at a concentration of 7.5×10^4 cells per dish in growth medium containing 10% fetal calf serum [13]. On day 4 fresh medium containing 10% fetal calf serum was added. On day 6 the medium was replaced with 2 ml of fresh medium containing 10% human lipoprotein-deficient serum and either 2 μl of ethanol or a mixture of 2 μg of 25-hydroxycholesterol plus 24 μg of cholesterol added in 2 μl of ethanol as indicated. On day 7 the dishes were divided into two groups. Cells in Group A were used for studies of ^{125}I-LDL metabolism and cells in Group B were used for fluorescence analysis. Group A: On day 8 each dish received 2 ml of growth medium containing 10% human lipoprotein-deficient serum and 10 μg protein/ml of ^{125}I-LDL (98 cpm/ng protein) in the absence and presence of 400 μg protein/ml of unlabeled LDL. After incubation for 5 h at 37°C the total amounts of surface-bound ^{125}I-LDL, internalized ^{125}I-LDL, and degraded ^{125}I-LDL were measured as described in Materials and Methods. The high affinity values for these processes, which are shown in parentheses, represent the difference between the values observed in the absence and presence of unlabeled LDL. Group B: On day 7 each dish received 2 ml of fresh medium containing 10% human lipoprotein-deficient serum, either ethanol or the mixture of 25-hydroxycholesterol and cholesterol in ethanol as indicated, and 10 μg protein/ml of r-[dioleyl fluorescein]LDL. On day 8, after incubation with r-[dioleyl fluorescein]LDL for 18 h at 37°C, 10,000 cells from each dish were analyzed in the fluorescence-activated cell sorter as described in Materials and Methods.

[a]The values in parenthesis represent high affinity values.

[b]These FH heterozygotes are obligate gene carriers in that each is the parent of an FH homozygote.

which cells are screened for mutations in the LDL uptake pathway, including mutations in patients with FH as well as mutations that are created through in vitro mutagenesis in cultured cells. The ability to separate populations of normal, heterozygous, and homozygous LDL receptor-deficient cells with the fluorescence-activated cell sorter should also facilitate further studies of the somatic cell genetics of the LDL receptor pathway. For example, it should now be possible to mutagenize normal cells in culture and to isolate those that have developed a mutation in a single gene for the LDL receptor. Although one cycle of cell sorting would not be sufficient to separate completely these heterozygous cells from normal cells, it should be possible to subject the mutagenized cells to repeated cycles of growth, incubation with fluorescent LDL, and cell sorting. With each sequential sorting, the percentage of cells that are truly heterozygous should increase progressively. After a clone of cells that is heterozygous for a mutation at the LDL receptor locus has been obtained, the cells could be mutagenized again, and cells homozygous for receptor defects could be selected.

TABLE II. Hydrophobic Molecules That Have Been Incorporated Into Heptane-Extracted LDL to Yield Reconstituted LDL

Lipids containing long-chain *cis*-unsaturated fatty acyl groups and their derivatives
- Cholesteryl esters (monounsaturated fatty acids with chain lengths of 14, 16, 18, 20, 22, and 24 carbon atoms; diunsaturated and triunsaturated fatty acids with chain lengths of 18 carbon atoms; tetraunsaturated fatty acid with chain length of 20 carbon atoms)
- Triacylglycerols (triolein; trilinolein; dioleyl monostearyl glycerate)
- Methyl esters (monounsaturated fatty acids with chain lengths of 14, 16, 18, 20, 22, and 24 carbon atoms; diunsaturated and triunsaturated fatty acids with chain length of 18 carbon atoms)
- Linoleyl alcohol
- 19-Iodocholesteryl oleate
- 4-Methyl umbelliferyl oleate
- 25-Hydroxycholesteryl oleate
- Dioleyl methotrexate
- Dioleyl fluorescein
- PMCA oleate

Lipids containing polyisoprenoid groups
- Vitamin A (retinol)
- Vitamin A palmitate (retinyl palmitate)
- Vitamin E acetate (α-tocopheryl acetate)
- Vitamin K_1
- Coenzyme Q_{10} (ubiquinone-10)
- β-Carotene
- Chlorophyll (a + b)

Other lipids
- Cholestene
- Cholestane
- Cholestan-3-one
- Cholesteryl nitrogen mustard (phenesterine)
- Cholesteryl ethyl ether
- Fat red 7B

A second area of importance involves the use of LDL as a vehicle to deliver hydrophobic molecules of biologic interest to cells that specifically possess LDL receptors. On the basis of this study and other studies previously reported [6–8], it is clear that hydrophobic molecules other than cholesteryl esters can be introduced into LDL. In principle, any molecule (eg, fluorescein) can be incorporated into LDL provided that it can be modified so that it is sufficiently apolar and can be esterified to either an unsaturated long-chain fatty acid (eg, dioleyl fluorescein) or to a polyisoprenoid compound such as phytol (eg, chlorophyll). When taken up by cells in the form of reconstituted LDL, such hydrophobic molecules will be delivered primarily to lysosomes [1, 4, 19, 20]. Some of these probes might prove useful for studying receptor-mediated endocytosis as it relates to lysosome function.

The use of reconstituted LDL as a vehicle for the delivery of a variety of hydrophobic drugs or probes to cells adds a new dimension to the concept of receptor-mediated pharmacotherapy discussed by Neville and Chang [21]. These investigators have suggested that one can devise new approaches to drug delivery by constructing hybrid molecules in which the binding chain of one protein (such as the receptor recognition subunit of a polypeptide hormone) is covalently coupled to the active chain of a different protein that will damage cells as it enters the cytoplasm (such as the toxic subunit of a plant or bacterial toxin) [21]. Indeed, such a disulfide-linked hybrid protein (β-s-s-A) containing the cell-specific β-subunit of human chorionic gonadotropin and the toxic A subunit of ricin has recently been synthesized [22]. Studies in rat Leydig cells have shown that in order for the A subunit of ricin to inhibit protein synthesis it must first bind to cells through the chorionic gonadotropin receptors that are specific for the β subunit of the hybrid molecule. Cells that lack receptors for chorionic gonadotropin, such as mouse L cells, are resistant to the toxic effect of the hybrid molecule [22].

The unique aspect of the LDL reconstitution technique for drug delivery lies in the fact that as many as 1,000 molecules of a hydrophobic compound can be incorporated into a single LDL particle and targeted to cellular lysosomes. One limitation of this approach is that, in order to achieve the desired biologic effects, the drug incorporated into LDL must be able to survive the acidic environment of the lysosome and the action of its multiple hydrolytic enzymes. The types of biologically active molecules that can survive this lysosomal exposure must now be determined.

ACKNOWLEDGMENTS

This research was supported by grants from the National Institutes of Health: HL 20948, HL 15648, and HL 17269. M.K. is a recipient of a U.S. Public Health Service Postdoctoral Fellowship, HL 05657.

REFERENCES

1. Anderson RGW, Brown MS, Goldstein JL: Cell 10:351, 1977.
2. Gorden P, Carpentier J-L, Cohen S, Orci L: Proc Natl Acad Sci USA 75:5025, 1978.
3. Maxfield FR, Willingham MC, Davies PJA, Pastan I: Nature 277:661, 1979.
4. Goldstein JL, Brown MS: Annu Rev Biochem 46:897, 1977.
5. Jackson RL, Morrisett JD, Gotto AM Jr: Physiol Rev 56:259, 1976.
6. Krieger M, Brown MS, Faust JR, Goldstein JL: J Biol Chem 253:4093, 1978.
7. Krieger M, Goldstein JL, Brown MS: Proc Natl Acad Sci USA 75:5052, 1978.
8. Krieger M, McPhaul MJ, Goldstein JL, Brown MS: J Biol Chem 254:3845, 1979.

9. Kao YJ, Charlton SC, Smith LC: Fed Proc 36:936, 1977.
10. Brown MS, Dana SE, Goldstein JL: J Biol Chem 249:789, 1974.
11. Brown MS, Goldstein JL: Proc Natl Acad Sci USA 71:788, 1974.
12. Noble RP: J Lipid Res 9:693, 1968.
13. Goldstein JL, Basu SK, Brunschede GY, Brown MS: Cell 7:85, 1976.
14. Lowry OH, Rosebrough NJ, Farr AL, Randall RJ: J Biol Chem 193:265, 1951.
15. Loken MR, Herzenberg LA: Ann NY Acad Sci 254:163.
16. Poole AR: In Dingle JT (ed): "Lysosomes: A Laboratory Handbook." Amsterdam: North-Holland Publishing Co., 1977, pp 245:312.
17. Goldstein JL, Brown MS, Stone NJ: Cell 12:629, 1977.
18. Bilheimer DW, Ho YK, Brown MS, Anderson RGW, Goldstein JL: J Clin Invest 61:678, 1978.
19. Goldstein JL, Brown MS, Anderson RGW: In Binkley BR, Porter KR (eds): "International Cell Biology 1976–1977." New York: Rockefeller University Press, 1977, pp 639–648.
20. Goldstein JL, Dana SE, Faust JR, Beaudet AL, Brown MS: J Biol Chem 250:8487, 1975.
21. Neville DM, Jr, Chang TA: Current Topics in Membranes and Transport 10:65, 1978.
22. Oeltmann TN, Heath EC: J Biol Chem 254:1028, 1979.

Journal of Supramolecular Structure 10:479–489 (1979)
Biological Recognition and Assembly 37–47

Preliminary Molecular Replacement Results for a Crystalline Gene 5 Protein–Deoxyoligonucleotide Complex

Paula M.D. Fitzgerald, Andrew H.J. Wang, Alexander McPherson, Frances A. Jurnak, Ian Molineux, Frank Kolpak, and Alexander Rich

Department of Biological Chemistry, Milton S. Hershey Medical Center of Pennsylvania State University, Hershey, Pennsylvania 17033 (P.M.D.F., A.M., F.A.J.), Department of Biology, Massachusetts Institute of Technology, Cambridge, Massachusetts 02139 A.H.J.W., F.K., A.R.), and Department of Microbiology, University of Texas, Austin, Texas 78712 (I.M.)

Complexes of the gene 5 protein from bacteriophage fd with a variety of oligodeoxynucleotides, ranging in length from two to eight and comprised of several different sequences, have been formed and crystallized for X-ray diffraction analysis. The crystallographic parameters of four different unit cells, all of which are based on hexagonal packing arrangements, indicate that the fundamental unit of the complex is composed of six gene 5 protein dimers. We believe this aggregate has 622 point group symmetry and is a ring formed by end-to-end closure of a linear array of six dimers. From our results we have proposed a double-helix model for the gene 5 protein–DNA complex in which the protein forms a spindle or core around which the DNA is spooled. Currently 5.0-Å X-ray diffraction data from one of the crystalline complexes is being analyzed by molecular replacement techniques to obtain a direct image of the protein–nucleic acid complex.

Key words: **protein-nucleic acid interactions, X-ray diffraction, gene 5 protein, molecular replacement, DNA, fd bacteriophage**

Determination of the structure of a complex between a DNA binding protein and fragments of nucleic acid by X-ray diffraction analysis promises substantial insight into the means by which these two important macromolecules interact. In addition to delineating the precise atomic interactions by which they recognize and bind to one another, knowledge of such a structure could potentially clarify many of the mechanisms by which the flow of genetic information is controlled. In the case of the DNA unwinding protein, which we discuss here, we believe considerable information may also be gained concerning the assembly and general architectural features of large protein-nucleic acid structures such as are found in chromosomal material and viruses.

Received April 16, 1979; accepted April 18, 1979.

0091-7419/79/1004-0479$02.30

The gene 5 protein from fd bacteriophage has proved to be a particularly useful system for such studies since it can be crystallized as a monomer in the absence of nucleic acid [1]. Its complexes with oligonucleotides and native fd phage DNA can be studied by electron microscopy and a host of other physicochemical techniques. Finally, single crystals of complexes between the gene 5 protein and nucleic acid can be grown. The gene 5 protein is rather small (87 amino acids totaling 9,800 daltons [2]), and its structure in the uncomplexed state is presently known from the single-crystal X-ray diffraction analysis we describe in the accompanying paper. It is our intention to use the structure of the uncomplexed protein to determine the structure in the single crystals of protein-DNA complexes that we have grown and now have characterized by diffraction techniques.

The gene 5 protein–DNA complexes produced in vitro as visualized by electron microscopy are unique in that two protein-covered strands coalesce to yield a helical rod-like structure in which there are 12 gene 5 monomers per turn of the helix. The helix has a width of approximately 100 Å and a longitudinal repeat of about 80 Å [3]. The gene 5 protein–DNA complexes resemble mature filamentous bacteriophage virions though there are clear differences. The mature virus is formed by the displacement of the gene 5 protein at, or in, the host cell membrane by the coat protein, the product of gene 8 [4]. The gene 5 protein is never found in the virion but is returned to the cell for reuse.

In vitro complexes of the gene 5 protein with fd phage DNA have been reported to differ in structure from complexes isolated directly from infected cells. These in vivo complexes were observed to be composed of fibers 40 Å in width that were supercoiled to give an overall width of 160 Å and a longitudinal repeat of 160 Å [3]. More recent electron microscopy studies by Grey [5], however, find the in vivo and in vitro complexes to be identical and to resemble the helical rods described above. One difference has been noted between the two complexes: the stoichiometry of binding from presumably saturated in vitro complexes is one gene 5 monomer per four nucleotides while the in vivo complexes tend to give nonintegral values of approximately 4.6 nucleotides per gene 5 monomer [6].

The gene 5 protein–DNA complexes are considerably different from those of the helix-destabilizing protein from phage T4, the gene 32 protein, when visualized by electron microscopy [2]. These latter complexes, like those of the calf thymus DNA unwinding protein [7], form open ring structures under identical conditions.

There is evidence from cross-linking studies in solution that when gene 5 protein is combined with deoxyoligonucleotides from four to eight in length, high-molecular-weight aggregates containing up to about eight monomers are formed and can be seen on SDS-polyacrylamide gels. It was concluded in these studies that the oligomers gave rise to cross-linked aggregates very similar to those obtained with poly (dA-dT) and that the binding of short stretches of nucleic acid chain appears to induce the association of gene 5 monomers with one another [8].

We have formed complexes of the gene 5 protein with a number of different homogeneous deoxyoligonucleotides in solution, and have crystallized a variety of these complexes in a number of crystal forms. Four of these crystals have been characterized by X-ray diffraction and at least two have been found suitable for a high-resolution structure analysis. At present, we have collected 5.0-Å-resolution diffraction data on one of these crystals, and we describe here our progress in determining its structure using molecular replacement techniques with the monomeric gene 5 protein structure as model.

MATERIALS AND METHODS

The oligonucleotides used in the crystallization experiments were d-pGpC, d-pApT, d-$(Ap)_4$, and d-$(Ap)_8$ from Collaborative Research, Waltham, Massachusetts. The specific-sequence oligomers d-pCpTpTpC and d-$(Tp)_4$ were gifts of Dr. Robert Ratliff and Dr. Lloyd Williams of the University of California at Los Alamos, California; the homopolymers d-$(Cp)_3$ and d-$(Cp)_4$ were gifts of Dr. Gobind Khorana of MIT. d-GGTAAT and its complementary hexamer were supplied us by Dr. Jack Van Boom of the University of Leiden. Gene 5 protein from fd bacteriophage was made by a slight modification of the procedure of Alberts et al [2]. Infected cells were lysed in a French Press in 0.02 M tris (pH 7.6), 0.01 M $MgCl_2$ with no added DNA. The lysate was made 2 M in NaCl and nucleic acids were removed by polyethylene glycol precipitation. The supernatant was dialyzed to low salt by repeated changes of a dialysis buffer containing 0.02 M tris (pH 7.6), 0.05 M NaCl, 0.005 M EDTA, 0.001 M 2-mercaptoethanol and 10% (v/v) glycerol. The protein was isolated using DNA-cellulose and DEAE-cellulose chromatography. Occasionally the gene 5 protein had to be further purified by phosphocellulose chromatography to yield a homogeneous sample.

The complexes were crystallized by the vapor diffusion method in depression plates [9] using polyethylene glycol 6000 (PEG) from Fisher Scientific Co. as the precipitating agent. The methods and conditions employed were virtually the same as those described for crystallization of the native gene 5 protein [1]. The samples contained 5 μl of 14 mg/ml gene 5 protein, 5 μl of the oligonucleotide solution, both buffered at pH 7.5 by 0.01 M tris-HC1, and 10 μl of 12% PEG 6000 in H_2O. The sandwich box reservoirs contained 25 ml of 12% PEG 6000. All operations were carried out at room temperature. A variety of crystal forms were observed; the time required for crystal growth to occur varied from several days to several weeks with the mean at about 20 days. In general, the crystals were stable in their mother liquor for up to several months, although some apparently degraded in response to temperature fluctuations.

For X-ray diffraction analysis, the crystals were mounted by conventional means in sealed quartz X-ray capillaries along with a small aliquot of mother liquor. Precession photographs were recorded on Ilford Industrial G X-ray film, using a crystal-to-film distance of 90 mm and exposure times of about 18 h. Nickel-filtered CuK_α X-radiation was produced by an Elliot rotating anode generator operated at 40 kV and 40 mA with a focal spot size of 200 μm^2.

Three-dimensional X-ray diffraction data were collected to 5.0-Å resolution using the step scan mode [10] on a Picker FACS-1 diffractometer fitted with a 1,600-W Phillips fine-focus X-ray tube. The approximately 8,000 independent reflections were recorded from a single crystal. Standard deviations were estimated from counting statistics [11] and the reflections were merged and Lp-corrected using programs written by Dr. F.L. Suddath.

All computing operations were performed on a PDP 11/40 computer running under the RSX-11M operating system. The rotation function program was that of Crowther [12] as modified by Tanaka [13] for a spherical polar coordinate system and further modified by P.M.D.F. to run on the PDP 11/40 system. The translation function and structure factor calculation programs used were those of Lattmann [14] as modified by P.M.D.F. For rotation function calculations involving self vector searches within the native set, data between 10 Å and 6 Å in resolution and with intensity greater than 4 SD were employed. The maximum length of the vectors included in the search was varied from 20 to 35 Å. For the searches of the complex using the uncomplexed gene 5 protein structure as the search model, we have used in succession only α-carbon atoms, all main-chain atoms, and all main-chain atoms plus tyrosines, phenylalanines, methionines, and cysteine. As above, data between 10 and 6 Å were used and the maximum vector length included was varied between 20 and 35 Å.

RESULTS

At least ten different crystal habits of the gene 5 protein complexed with deoxy-oligonucleotides have been observed in our crystallization trials. The dominant forms are rhombic plates, though triangular and hexagonal prisms and plates as seen in Figure 1 are also frequently encountered. We have commonly observed polymorphism in single samples and transformations between different crystal forms as well. We find that at ratios of oligonucleotide to protein of less than 1.0 only native crystals are grown. At concentrations above this level only the complex crystals appear. We have not observed the co-existence of native crystals with complex crystals. Furthermore, we have never observed the complex crystals in the absence of added oligonucleotide. Most of the crystals grown were multiples, or twinned, or too small, or they presented some other problem to X-ray diffraction analysis. Several, however, grew large enough that we could determine their space groups and cell dimensions.

The unit-cell parameters and symmetry properties of four independent crystal forms of the gene 5 protein–DNA complex are shown in Table I. We noted that three of the crystals are based on hexagonal systems characterized by sixfold symmetry and the fourth, of space group $C222_1$, can be related to the unit cell of the $P6_3$ form if one assumes a pseudohexagonal packing arrangement. In fact, we frequently observe this orthorhombic crystal form growing as a twin or satellite crystal with a crystal of hexagonal habit.

Although we could not measure the density of any of the complex crystals directly, we assumed a volume-to-mass ratio for each that was near the center of the range of crystalline proteins compiled by Matthews [15] and was consistent as well with that measured for the uncomplexed gene 5 protein crystals, $V_m = 2.45$ [1]. Given this, we determined that the most reasonable number of gene 5 monomers in each asymmetric unit was consistently 12 (or six dimers), except for the $P3_1$ form, in which we judged that there are about 24.

The crystals of space group $P3_1$ were the best crystals we examined and X-ray diffraction data to 5.0-Å resolution were collected from this crystal. An example of the diffraction data from these crystals is shown in Figure 2. The volume of the asymmetric unit of this crystal form was twice that of the other forms, but the diffraction patterns of these trigonal crystals show very high 32 pseudosymmetry and this suggested the presence of a nearly crystallographic twofold axis along the 100 or 110 directions in the crystal. Thus the effective asymmetric unit contains 12 gene 5 monomers, the same number determined for the other crystal forms.

The rotation function search for local symmetry in the gene 5 protein–DNA complex crystals revealed three local symmetry axes. All of these can be seen in the section of the rotation function map at kappa = 180°, shown in Figure 3. (The angles psi and phi describe the orientation of a rotation axis; the angle kappa denotes the degree of rotation about that axis. Thus a peak in the map section at kappa = 180° indicates the orientation of a potential local twofold rotation axis.) There is a prominent peak in the kappa = 180° section at psi = 90°, phi = 55°, and an equivalent peak at psi = 90°, phi = 115°. These peaks have magnitude 0.66, expressed as a fraction of the magnitude of the peak corresponding to the crystallographic threefold rotation along *c*. The position of the local peak indicates a twofold rotation axis nearly parallel to the crystallographic a^* or b^* axis and thus confirms our conclusion, based on the extra symmetry seen in the diffraction pattern, that the true space group of $P3_1$ contains molecules packed nearly, but not exactly, with the symmetry of space group $P3_121$, and that the pseudoasymmetric unit contains 12 gene 5 monomers.

A second peak occurs at psi = 0°, phi indeterminate; the magnitude of this peak is 0.71. The interaction of this local twofold parallel to *c* with the crystallographic threefold coincident with *c* will generate a local sixfold axis, and an equivalent peak is indeed found at psi = 0°, phi indeterminate in the section of the map at kappa = 60°. We believe that this twofold/sixfold axis is a manifestation of the packing of the 12-monomer aggregates into the unit cell, but we cannot rule out the possibility that this peak arises from symmetry internal to the 12-monomer aggregate.

The third prominent feature in the map occurs at psi = 90°, phi = 20°, with magnitude of 0.71. We can find no explanation for this local symmetry in the packing of aggregates within the unit cell and therefore we assume that this peak arises from a twofold-symmetry element, parallel to the crystallographic *ab* plane, that is internal to the 12-monomer aggregate.

The rotation function searches in which we seek to determine the 12 orientations of the gene 5 monomer in the pseudoasymmetric unit of the complex structure have so far given inconclusive results. We have found that searches using a model consisting of only main-chain atoms give very nearly the same results as searches using main-chain atoms plus large side chains, and we conclude that the former model is probably adequate for this type of search.

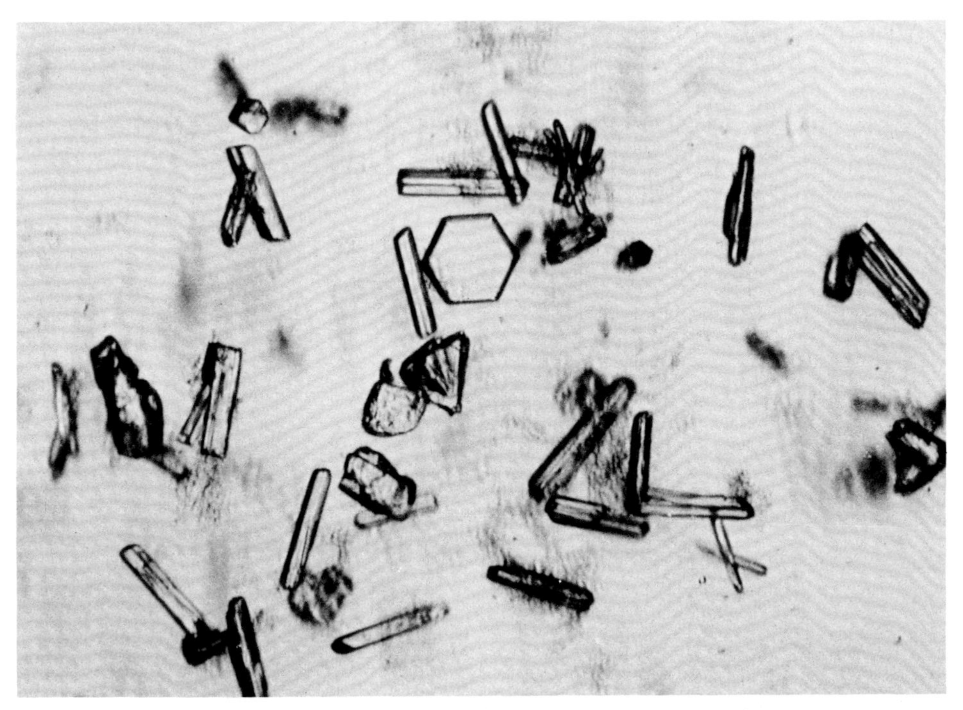

Fig. 1. Low-power light microscope photograph of the hexagonal plate habit of the complex between gene 5 protein and d-pCpTpTpC. These crystals have been observed to exist in the same sample with rhombic plates.

DISCUSSION

In the $P6_3$, $C222_1$, R32, and pseudo $P3_121$ crystals the number of gene 5 monomers per asymmetric unit is observed to be about 12. The repeated occurrence of this number of monomers as the asymmetric unit of the crystals suggests rather strongly a specific aggregate of 12 gene 5 monomers that is formed upon addition of oligonucleotides to the protein. The fact that these aggregates crystallize requires that they be a homogeneous population of identically structured complexes; they must represent some ordered mode of self assembly from the solution species.

There is evidence from solution studies that aggregation of gene 5 protein does occur in the presence of oligonucleotides as well as deoxyribonucleic acid. Rasched and Pohl [8] have found from suberimidate cross-linking and SDS gel electrophoresis of gene 5 protein combined with oligonucleotides that polymeric protein species up to "about eight" are formed. The lack of certainity in their upper limit is due, at least in part, to the anomalous electrophoretic mobility of cross-linked protein aggregates which would be expected to undergo more rapid migration since they are not completely extended polypeptide chains. Thus the size of these aggregates is not inconsistent with 12-monomer aggregate found in our asymmetric unit. In addition, the complex between gene 5 protein and fd phage DNA formed in solution and studied by electron microscopy shows "a helical rodlike structure" in which there are 12 gene 5 monomers per turn of the helix [2]. We believe we are observing crystallographically a structure similar to that observed in solution.

The gene 5 protein binds to DNA in a linear and highly cooperative manner, ie, successive gene 5 molecules tend to bind immediately adjacent to one already bound rather than to an isolated site. This apparently reflects the existence of strong protein-protein interactions between adjacent gene 5 molecules along the DNA strands, and may explain the powerful helix-destabilizing effect exerted by the protein. These strong protein-protein interactions do not occur between gene 5 molecules in solution in the absence of nucleic acid; if they did they would lead to aggregate formation of free molecules and this is not observed. It appears that the potential for forming such interactions is a consequence of conformational changes in the protein molecules induced by binding to DNA. The triggering of conformation change caused by DNA or oligonucleotide binding, the resulting cooperative interaction between protein molecules, and concomitant aggregation of the

TABLE I. Crystal Forms of fd Phage Gene 5 Protein Complexed With Oligodeoxynucleotides

Hexagonal plates	Diamond plates
$a = 107$	$a = 110$
	$b = 180$
$c = 206$	$c = 117$
$P6_3$	$C222_1$
12 * 9,800 daltons per asymmetric unit	12 * 9,800 daltons per asymmetric unit
Rhombohedra	Hexagonal prisms
$a = 140$	$a = 143$
	$c = 83$
$\alpha = 60$	
R32	$P3_1$
12–18 * 9,800 daltons per asymmetric unit	24 * 9,800 daltons per asymmetric unit

protein are likely responsible for the asymmetric unit of 12 monomers that we observe in our complex crystals.

Virtually all protein oligomers and large protein complexes studied so far by X-ray diffraction analysis have demonstrated symmetry relationships, or at least a high degree of quasisymmetry, between the units involved. This seems likely to be the case with the gene 5 protein–oligonucleotide complex as well. We know from crystallographic studies on the free protein that the gene 5 dimers contain perfect dyad axes relating monomers in pairs. The occurrence of six of these dimers in the asymmetric unit of the crystals suggests the likelihood of an aggregate having sixfold symmetry. This is reinforced by the finding that three of the four unit cells encountered are of hexagonal symmetry and that the fourth can be interpreted in terms of hexagonal packing. Although there is no required correlation, objects with hexagonal symmetry do tend to express such symmetry in the crystalline state and the number of hexagonal forms observed in this case argues for such a correlation.

The aggregate occupying the asymmetric unit of the complex crystals, which in all unit cells so far examined has contained 12 gene 5 protein monomers, is most likely a closed arrangement of fixed and determinate size which forms spontaneously in solution only when triggered by the binding of nucleic acid fragments. The simplest model for the asymmetric unit is that of a closed circle or disk having a sixfold axis along its center which is perpendicular to the twofold axes of the dimer units, ie, it possesses 622 point group symmetry.

The shape of the gene 5 protein dimer in the unliganded state is known from X-ray diffraction analysis (see accompanying paper) and has the gross features shown in the model in Figure 4. The structure created when one takes these dimers and arranges them in a

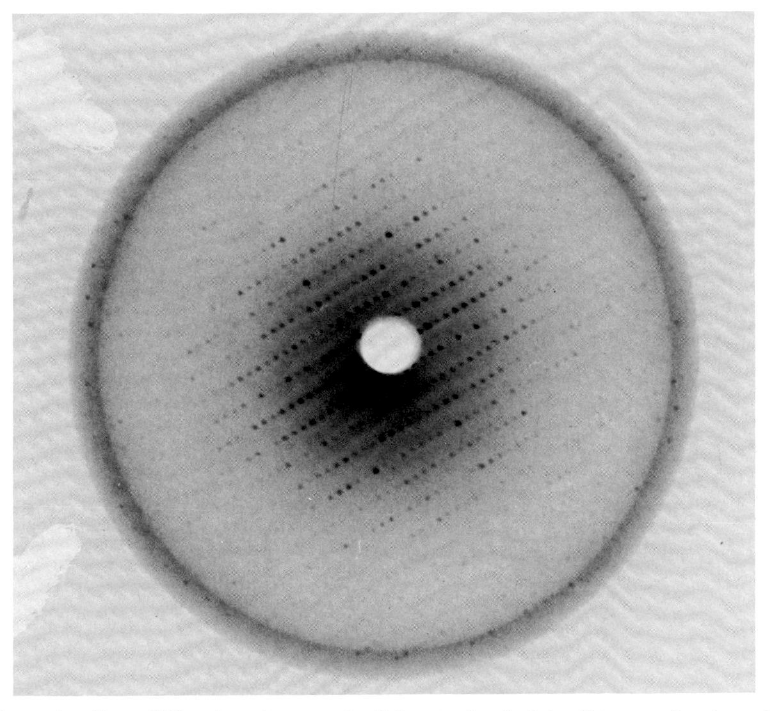

Fig. 2. Precession X-ray diffraction photograph of the zero level of the 0kl zone of reciprocal space of the $P3_1$ crystals of gene 5 protein complexed with d-(GGTAAT). These crystals grow to large size and diffract strongly to at least 3.0 Å in precession photographs.

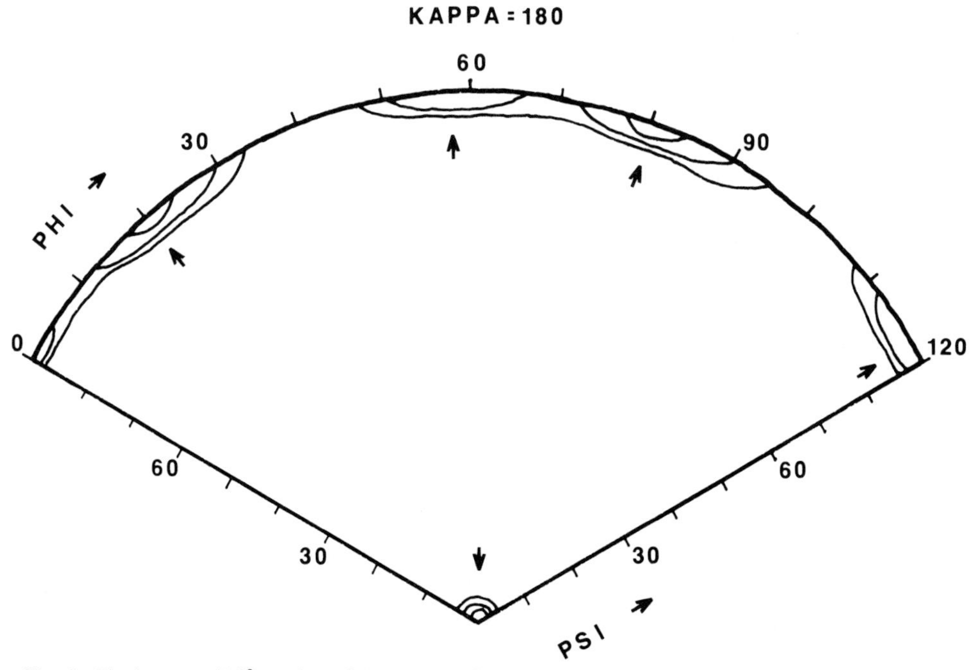

Fig. 3. The kappa = 180° section of the rotation function of the crystals of complexes between the gene 5 protein and d-(GGTAAT). All of the peaks along psi = 90° are of high magnitude with respect to the crystallographic symmetry peak, and all represent local, noncrystallographic symmetry elements present in the crystals.

Fig. 4. A photograph of a three-dimensional model representation of the gene 5 dimer based on the structure derived from X-ray diffraction analysis. The postulated binding regions for the DNA single strands lie beneath the thick mass of density and are separated in the dimer by about 25 Å.

circle such that the dyad axes are perpendicular to a central sixfold axis is shown in Figure 5. Because of the double-wing character of the gene 5 dimer, the aggregate would have a twofold crown shape with a diameter of roughly 100 Å and a thickness of approximately 80 Å. Thus it is not a flat disk shape but that of a squat cylinder. The aggregate we are proposing can be packed without difficulty in each of the unit cells we have characterized.

The aggregation phenomenon we seem to be observing with the gene 5 protein–oligonucleotide complexes is not unprecedented. The tobacco mosaic virus disk [16] is an obvious analogy. Here again nucleic acid binding proteins are stimulated upon nucleotide binding to form a helical rod. Unlike the aggregation seen with the fd gene 5 protein, aggregation of the TMV coat protein can be induced in the absence of nucleic acid by careful selection of the environment. Under these conditions the TMV protein molecules organize into a closed circle or disk having a 17-fold symmetry axis [17].

Another example is the octameric aggregate of histones which forms the core of the chromosomal body [18]. This is clearly different in gross structure since it acts as a spool for superhelically coiled double-stranded DNA, but it may, with regard to the functional organization of the proteins in its center, bear some similarities to the gene 5 nucleoprotein complex.

The aggregate of 12 gene 5 monomers observed in the crystal is not identical in structure to the helical aggregates of the gene 5 protein and DNA observed in the electron microscope. This unit we postulate in Figure 5 is completely closed and does not allow an extended helix to be built up simply by translation along the direction of the sixfold axis. However, the relationship between the two structures may be somewhat analogous to the relationship that exists between the 17-fold TMV closed-disk structure and the TMV helix, which has a 17-fold screw axis. The latter structure arises from the first simply by opening

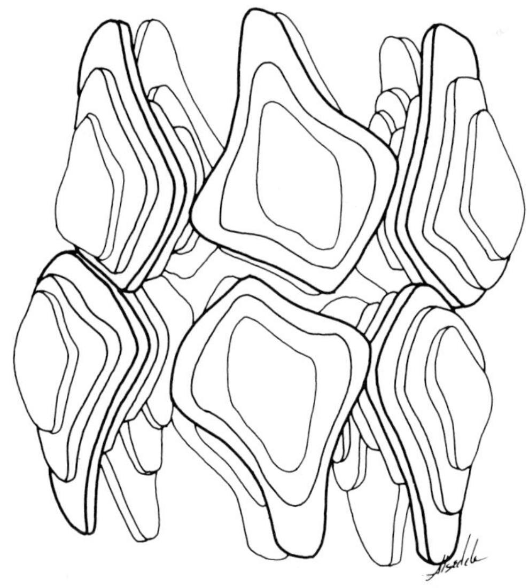

Fig. 5. A proposed model for the asymmetric unit common to the four crystal forms we have so far analyzed. The arrangement is a circle, or disk, having 622 point group symmetry formed by joining the two ends of a linear array of six gene 5 protein dimers, each of which possesses an inherent dyad, to produce closure. The upper hexagon of monomers will bind a single strand of DNA running in one direction and the bottom level a strand running in the opposite direction. The disk has a diameter of about 100 Å and a height of about 80 Å. We believe the DNA binding region of each monomer to be directed toward the outside of the circle.

the disk and displacing the two ends along the direction of the unique axis to produce a "lockwasher" unit. The free ends of these "lockwasher" units are joined as the units are stacked to produce the helix. A model of a helical structure that might be produced by the gene 5 protein binding to two strands of DNA running in opposite directions is shown in Figure 6. This helical structure would contain essentially the same lateral interactions between adjacent protein monomers as occur in the closed disk. This structure is a gene 5 double helix, one chain of which binds a DNA strand running 3′ to 5′ and the other a strand running 5′ to 3′. It has a sixfold screw axis with a linear repeat of about 80–90 Å and a diameter of approximately 100 Å.

The gene 5 protein–DNA aggregate could in principle form with the DNA binding interface to the outside of the ring or to the inside. Based on our analysis of the tertiary structure of the monomer, however, we conclude that the DNA must be to the outside. This suggests that the two DNA single strands are coiled around the gene 5 protein core with a radius of 30–40 Å in the complex. This radius is stereochemically reasonable and is about the same as the radius at which RNA is found in the TMV helical structure. Were the strands of nucleic acid on the inside of the gene 5 protein complex, they would need to be coiled with an unacceptably tight radius. Thus we believe that the DNA strands wind around a core of gene 5 protein which acts as a spool rather than a sheath.

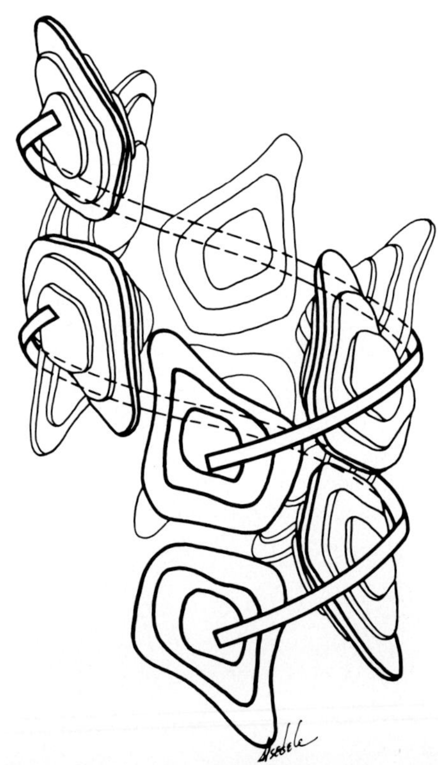

Fig. 6. A proposed model for the structure of one turn of the gene 5 protein–DNA double helix. This arrangement arises by opening the disk structure seen in Figure 5 between any two adjacent dimers and displacing the free ends along the unique axis direction. The stacking of these "lockwasher" units results in a double helix structure having a sixfold screw axis with perpendicular dyads, 12 gene 5 monomers per turn, and dimensions consistent with the helices observed by electron microscopy. The two DNA single strands are spooled around this spindle of gene 5 protein.

A mechanistic advantage of this model is that it provides a simple means for the displacement of the gene 5 protein by the gene 8 coat protein. Since the DNA strands are on the exterior of the gene 5 complex, they are exposed to the approach of the coat protein which also binds to DNA cooperatively but much more tenaciously than the gene 5 protein. Hence, when the gene 5 protein–DNA complex reaches the cell membrane the coat proteins binds very strongly to the DNA strands on the outside of the complex. Since it is of a much smaller size, the coat protein undoubtedly forms a helix with different parameters. This change in the conformation of the DNA produced by coat protein binding would disrupt the gene 5 protein–DNA bonds as well as the gene 5 monomer-monomer interactions. Thus, the gene 5 protein which forms a spindle on the interior of the nucleic acid is exchanged for a sheath of coat protein which must protect the virion after extrusion into the media.

ACKNOWLEDGMENTS

This research was supported by grants from the National Institutes of Health, National Science Foundation, National Aeronautics and Space Administration and the American Cancer Society. A.H.J.W. is supported by a grant from the MIT Cancer Center. F.A.J. and P.M.D.F. are NIH postdoctoral fellows. F.K. is a fellow of the Anna Fuller Fund. We thank Dr. J. H. Van Boom, Dr. Lloyd Williams, Dr. George Ratliff, and Dr. Gabind Khorana for supplying deoxyoligonucleotides.

REFERENCES

1. McPherson A, Molineux I, Rich A: J Mol Biol 106:1077–1081, 1976.
2. Alberts B, Frey L, Delius H: J Mol Biol 68:139–152, 1972.
3. Pratt D, Laws P, Griffith J: J Mol Biol 82:425–439, 1974.
4. Henry TJ, Pratt D: Proc Natl Acad Sci USA 62:800–807, 1969.
5. Grey C: Paper presented at the ICN-UCLA Symposium on Recognition and Assembly in Biological Systems, Keystone, Colorado, 1979.
6. Pretorius HT, Klein M, Day LA: J Biol Chem 250:9262–9269, 1975.
7. Herrick G, Alberts BM: J Biol Chem 251:2124–2146, 1976.
8. Rasched I, Pohl FM: FEBS Lett 46:115–118, 1974.
9. McPherson A: In Glick D (ed): "Methods of Biochemical Analysis." New York: Wiley and Sons, 1976, vol 23, pp 249–345.
10. Wyckoff HW, Doscher M, Tsernoglou D, Inagami T, Johnson LN, Hardman KD, Allewell NM, Kelly DM, Richards FM: J Mol Biol 27:563–578, 1967.
11. Arndt UW, Willis BTM: "Single Crystal Diffractometry." Cambridge, UK: University Printing House, 1966, Chapter 10.
12. Crowther RA: In Rossmann, MG (ed): "The Molecular Replacement Method." New York: Gordon & Breach, 1972, pp 173–178.
13. Tanaka N: Acta Crystallogr A33:191–193, 1977.
14. Lattman EE: PhD thesis, Johns Hopkins University, Baltimore, Maryland, 1969.
15. Matthews BW: J Mol Biol 33:491–497, 1968.
16. Butler PJG: Cold Spring Harbor Symp Quant Biol 36:461–468, 1971.
17. Champness JN, Bloomer AC, Bricogne G, Butler PJG, Klug A: Nature 259:20–24, 1976.
18. Finch JT, Lutter LC, Rhodes D, Brown RS, Rushton B, Levitt M, Klug A: Nature 269:29–36, 1977.

Journal of Supramolecular Structure 11:9–24 (1979)
Biological Recognition and Assembly 49–64

Co-Translational Modification of Nascent Immunoglobulin Heavy and Light Chains

Lawrence W. Bergman and W. Michael Kuehl

Department of Microbiology, University of Virginia, Charlottesville, Virginia 22908

We have investigated the in vivo co-translational covalent modification of nascent immunoglobulin heavy and light chains. Nascent polypeptides were separated from completed polypeptides by ion-exchange chromatography of solubilized ribosomes on QAE-Sephadex. First, we have demonstrated that MPC 11 nascent heavy chains are quantitatively glycosylated very soon after the asparaginyl acceptor site passes through the membrane into the cisterna of the rough endoplasmic reticulum. Nonglycosylated completed heavy chains of various classes cannot be glycosylated after release from the ribosome, due either to rapid intramolecular folding and/or intermolecular assembly, which cause the acceptor site to become unavailable for the glycosylation enzyme. Second, we have shown that the formation of the correct intrachain disulfide loop within the first light chain domain occurs rapidly and quantitatively as soon as the appropriate cysteine residues of the nascent light chain pass through the membrane into the cisterna of the endoplasmic reticulum. The intrachain disulfide loop in the second or constant region domain of the light chain is not formed on nascent chains, because one of the cysteine residues involved in this disulfide bond does not pass through the endoplasmic reticulum membrane prior to chain completion and release from the ribosome. Third, we have demonstrated that some of the initial covalent assembly (formation of interchain disulfide bonds) occurs on nascent heavy chains prior to their release from the ribosome. The results are consistent with the pathway of covalent assembly of the cell line, in that completed light chains are assembled onto nascent heavy chains in MPC 11 cells (IgG_{2b}), where a heavy-light half molecule is the major initial covalent intermediate; and completed heavy chains are assembled onto nascent heavy chains in MOPC 21 cells (IgG_1), where a heavy chain dimer is the major initial disulfide linked intermediate.

Key words: **nascent chains, co-translational modification, glycosylation, polypeptide folding, covalent assembly, heavy and light chains**

Abbreviations: H, heavy chain; L, light chain; F_{CL}, constant region kappa light chain fragment; SDS, sodium dodecyl sulfate; IAA, iodoacetic acid; IAM, iodoacetamide; PBS, phosphate-buffered saline.

Immunoglobulin heavy and light chains undergo several post-translational covalent modifications of their primary sequence prior to attaining the mature form which binds antigens and mediates numerous immunological functions. The immunoglobulin molecule is composed of two disulfide-linked heavy chain glycoproteins, each of which is linked to a light chain by a single disulfide bond [1]. Heavy chains and light chains are composed of four and two sequential domains, respectively. Each 12,500 dalton domain contains a single disulfide loop [2]. A substantial amount of work has been done using the immunoglobulin molecule as a model system for studying these covalent modifications – eg, glycosylation [3–7], intramolecular folding [8–10], and intermolecular assembly [6, 7, 11–17]. However, it is not certain whether such covalent modifications begin during synthesis while the nascent polypeptides are still bound to the polyribosomal complex, or only after the polypeptides have been completed and released from the ribosome into the cisternae of the endoplasmic reticulum.

We have examined the temporal relationship between translation and various post-translational (co-translational) modifications (ie, glycosylation, formation of intrachain disulfide bonds, and intermolecular covalent assembly) by separation of nascent polypeptides from completed polypeptides using ion-exchange chromatography based on the multiple negative charges contributed by the tRNA moiety of the peptidyl-tRNA complex [18].

MATERIALS AND METHODS

Cells

Table I lists the mouse plasmacytoma cell lines used. All cell lines were maintained in suspension in Dulbecco's modified Eagle's medium supplemented with 15% heat-inactivated horse serum, 2 mM glutamine, and nonessential amino acids.

Cell Labeling, Fractionation, and Nascent Chain Isolation

Cell labeling and isolation of membrane-bound ribosomes and nascent chains were essentially as described previously [18, 30]. To prepare [^{3}H]-iodoacetic acid-labeled nascent chains, the nascent chain fraction was dialyzed extensively vs 10 mM NH_4HCO_3 (pH 7.8), lyophilized, resuspended in 1–2 ml 0.1 M Tris-Cl (pH 7.8)–2.0% SDS, and reduced overnight with 100 mM dithiothreitol at 37°C. The sample was then dialyzed for 4–6 h vs 50 volumes 0.1 M Tris-Cl (pH 7.8)–0.1% SDS, and subsequently alkylated for 20 min at 37°C with 1–2.5 mCi [^{3}H]-iodoacetic acid (IAA) (New England Nuclear, 300 mCi/mmole), with unlabeled IAA added to a total concentration of 10 mM IAA. After alkylation the sample was dialyzed extensively vs 0.1 M Tris (pH 7.4)–0.1% SDS and then specifically immunoprecipitated as described below.

Column Chromatography

For purification of L chain and F_{CL} marker proteins and for size separation of nascent chains, the immunoprecipitated samples were resuspended in SDS-sample buffer and analyzed by Sephadex G200 chromatography (1.5 × 95 cm or 1.5 × 190 cm columns) using a buffer system consisting of 0.1 M NH_4HCO_3 (pH 7.8), 0.05% SDS [31].

For tryptic peptide analysis, samples were precipitated with 20% Cl_3CCOOH, collected by centrifugation, and washed successively with 5% Cl_3CCOOH, ethanol: ether (1:1), and ether. The dried pellet was resuspended in 1.0 ml 0.1 M NH_4HCO_3(pH 8.0)

TABLE I. Mouse Myeloma Cells Used in Analysis of Co-Translational Modification*

Cell line	H chain	L chain	Reference
MPC 11 clone 45.6	$\gamma 2b$	κ, F_{CL} [b]	[19]
clone 66.2	–	κ, F_{CL}	[20]
clone NP-2	–	–, F_{CL}	[21]
clone M311	γ_{2b} [a]	κ, F_{CL}	[22–24]
MOPC 21 clone PB00.1	γ_1	κ	[25]
clone NSI	–	κ	[26]
MOPC 104E	μ	λ	[27]
S107	α	κ	[28]
MOPC 46 (tumor)	–	κ	[29]

*The cell lines used in these studies and their immunoglobulin synthetic products are indicated.
[a]The M311 clone synthesizes a variant heavy chain with a carboxy terminal deletion comprising approximately the third constant region domain.
[b]All MPC 11 cell lines synthesize F_{CL}, a constant region kappa light chain gene product with an internal deletion of the variable region; this unusual product is apparently encoded by a separate gene from the gene that encodes the normal kappa light chain expressed in these cells [21, 31].

and digested for 2 h with 250 μg/ml trypsin-TPCK (Worthington) at 37°C. After digestion, the sample was lyophilized, resuspended in 1.0 ml 20% formic acid, and chromatographed on tandem G25-G50 Sephadex columns (1.5 X 96 cm each) equilibrated with 10% acetic acid. Fractions 1.4 ml were collected, dried, and the radiolabeled peptides were located by scintillation counting.

Electrophoresis

Pooled fractions from the Sephadex G25-G50 column were lyophilized, resuspended in a minimal volume of 10% acetic acid, and subjected to high voltage paper electrophoresis in a pH 3.6 pyridine:acetic acid:water (1:10:289 v/v) buffer, for 1.5 h at 3,000 volts. The paper strips were cut into 1.0 cm pieces and counted as described below.

Immunoprecipitated samples were boiled in SDS-sample buffer and electrophoresed in various concentration polyacrylamide gels using a discontinuous SDS-Tris-glycine buffer system essentially as described by Laemmli [32] and Maizel [33]. Cylindrical gels were sliced into 2 mm fractions with a Gilson Aliquogel fractionator and counted as described below. Slab gels were subjected to autoradiography for visualizing the radioactive protein bands.

Immunoprecipitation

Direct immunoprecipitations were performed in antibody excess using appropriate antisera specific for H and L chains and were incubated overnight at 4°C [34]. The immunoprecipitates were collected by centrifugation through 1.0 M sucrose in phosphate-buffered saline (PBS), and washed once with PBS [35].

Radioactivity Measurement

Gel filtration aliquots were dried and counted in a Beckman LS 230 scintillation counter after addition of 0.5 ml of NCS (Amersham/Searle)–H_2O (10:1) and 7.5 ml of toluene containing 0.4% PPO. Crushed polyacrylamide gel samples were counted in a Triton X-100/toluene (1:2) scintillation fluor containing 0.4% PPO and 10% H_2O.

RESULTS

Nascent MPC 11 H Chains Are Quantitatively Glycosylated With a Core Oligosaccharide

Various investigators have shown that the initial glycosylation event involves the transfer of a large molecular weight oligosaccharide (containing N-acetylglucosamine. mannose, and glucose) as a unit from a dolichol lipid intermediate to an asparagine acceptor residue on the polypeptide [44, 45]. Experiments were performed to determine what percentage of nascent MPC 11 H chains are glycosylated prior to completion of H chain translation. The results in Figure 1 indicate that the ratios of [^{3}H]-glucosamine-to-[^{35}S]-Met are very similar for completed H chains (isolated from the rough endoplasmic reticulum) and large nascent H chains, implying that the glycosylation of the nascent H chains is quantitative. The observation that the ratio obtained for the nascent H chains increases from 0.23 to 1.49 in a single fraction (approximately 38,000 daltons) indicates that quanti-

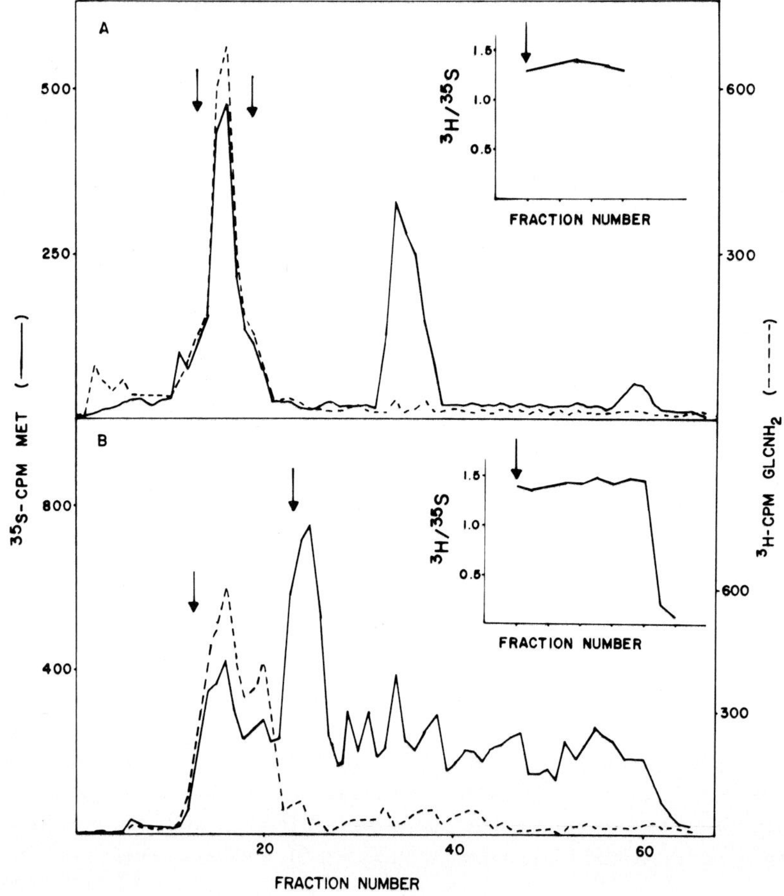

Fig. 1. SDS polyacrylamide gel electrophoresis of MPC 11 completed and nascent immunoglobulin chains. 2 × 10^{9} MPC 11 clone 456 cells were collected and labeled for 5 min with [^{3}H]glucosamine and [^{35}S]Met at isotope concentrations of 100 and 15 μCi/ml, respectively, in the presence of medium lacking glucose but containing all amino acids except Met. The isolation of completed and nascent heavy and light chains was as described previously [18, 30]. The samples were electrophoresed on a 10% polyacrylamide gel in SDS buffer. Migration is from left to right. (A) completed chains; (B) nascent chains. Dashed line, [^{3}H]glucosamine; solid line, [^{35}S]Met. The ratio of [^{3}H]glucosamine to [^{35}S]Met for each fraction marked by the arrows was calculated and plotted (insets) [30].

tative glycosylation of the nascent H chain occurs very soon after the asparaginyl acceptor site (residue 291, 32,000 daltons) passes through the membrane into the cisterna of the rough endoplasmic reticulum. The prominent [^{35}S]Met-labeled peak (fractions 23–27, 34,000–37,000 daltons) seen in the nascent chain sample appears to be due to a translational block related to the glycosylation event (E. Harris, L.W. Bergman, and W.M. Kuehl, unpublished data).

We have determined that the nascent H chains contain at least glucosamine and mannose, but not galactose [18, 36]. After extensive pronase digestion, the [^{3}H]glucosamine-labeled glycopeptide derived from nascent chains is indistinguishable by P-6 chromatography [24] from the [^{14}C]glucosamine-labeled glycopeptide present on completed H chains isolated from the rough endoplasmic reticulum; but both glycopeptides are smaller than either the glycopeptide derived from completed H chains isolated from the smooth membrane/Golgi fraction or the glycopeptide derived from secreted H chains (L.W. Bergman and W.M. Kuehl, unpublished data).

Nonglycosylated Completed H Chains Cannot be Glycosylated

Although the experiment described above demonstrates quantitative glycosylation of the nascent H chains with a core oligosaccharide at a precise time, we attempted to determine whether it is possible for the cell to glycosylate completed H chains of various classes after release from the ribosome. Non-glycosylated completed H chains synthesized in the presence of excess glucosamine (to block glycosylation) cannot be glycosylated during a chase period in the absence of the inhibitor, although the cells regain the ability to glycosylate newly synthesized H chains by 20 min into the chase [30]. Representative results for MOPC 104E (μ) and MPC 11 (γ_{2b}) are seen in Figure 2. Similar results are found for cells synthesizing γ_1 and α H chains [30]. However, Figure 2 also shows that M311 cells, which synthesize an MPC 11 variant H chain having a carboxy-terminal deletion, glycosylate a completed chain to a near normal extent in a similar experiment as described above. Thus, this result provides evidence that the lack of glycosylation of the wild type H chains is a function of the protein itself and is not due either to spatial separation of the H chain from the glycosylation enzymes or to the general inhibitory effect of the glucosamine.

Rapid Folding (Intramolecular Disulfide Bond Formation) Occurs on Nascent Polypeptides

To study the formation of intrachain disulfide bonds on nascent polypeptides we have chosen the MPC 11 L chain as a model system. Figure 3 summarizes the cysteine residues of the L chain and the size and chromatographic properties of the reduced and alkylated Cys-containing tryptic peptides [38]. Our approach has been to isolate the nascent L chains in the presence of excess unlabeled iodoacetamide (IAM) to block any free sulfhydryls and then to selectively label with [^{3}H]-iodoacetic acid (IAA) only those cysteine residues involved in intrachain disulfide bonds. Figure 4 shows an SDS-polyacrylamide gel of [^{3}H]-IAA-labeled immunoprecipitated nascent L chains that were isolated in the presence or absence of excess IAM. The [^{3}H]-IAA-labeled nascent L chains isolated in the absence of IAM (all Cys residues are labeled) show a heterogeneous size distribution of 25,000 daltons (full size) to approximately 10,000 daltons, while the nascent L chains isolated in the presence of IAM (only nascent L chains containing Cys residues in disulfide bonds are labeled) show a size distribution of 25,000–16,000 daltons. Thus only nascent L chains at least 16,000 daltons in size contain disulfide bonds, whereas nascent L chains may be immunoprecipitable as small as approximately 10,000 daltons.

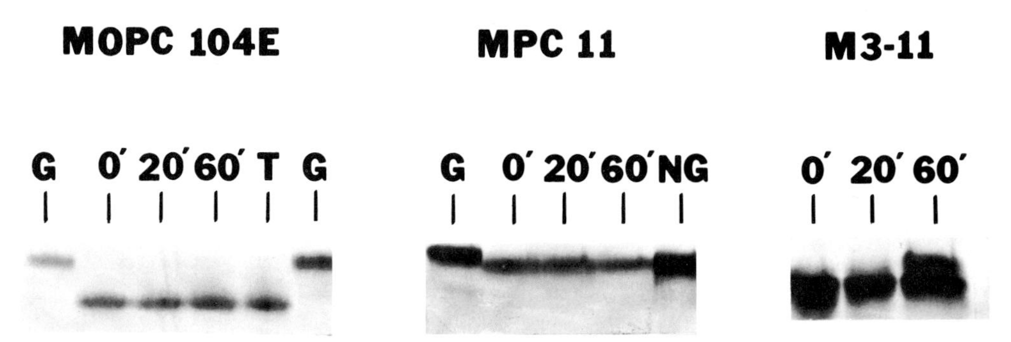

Fig. 2. Effect of glucosamine inhibition on glycosylation of MOPC 104E, MPC 11, and M311 heavy chains. Cells were incubated for 50 min in the presence of 10 mg/ml glucosamine and then labeled for 5 min with [^{14}C]-labeled amino acids in the presence of the inhibitor. Samples were removed at various intervals during a chase in medium containing excess glucose and unlabeled amino acids but lacking glucosamine. The samples were immunoprecipitated and analyzed on 12.5% polyacrylamide slab gels in an SDS buffer as described in Materials and Methods. Column G is glycosylated marker H chain, whereas NG and T are nonglycosylated marker H chains isolated from glucosamine- or tunicamycin-treated cells [37], respectively.

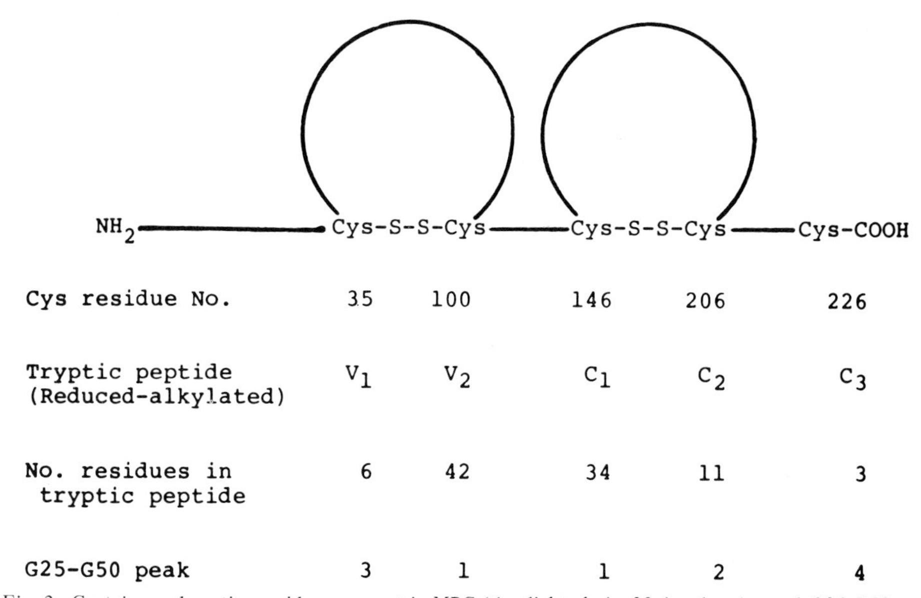

Fig. 3. Cysteine and cystine residues present in MPC 11 κ light chain. Molecular size and G25-G50 Sephadex chromatographic properties of the reduced and alkylated Cys-containing tryptic peptides are indicated.

Only Cys35 and Cys100 Are Involved in a Disulfide Bond on Nascent Chains

To determine which Cys residues were involved in the formation of the nascent intrachain disulfide bond, nascent L chains were isolated in the presence of IAM from cells labeled for 10 min with [^{35}S] Cys. Figure 5 shows the G25-G50 Sephadex chromatography of a tryptic digest of [^{35}S] Cys-labeled nascent L chains in the presence (panel A) or absence (panel B) of reducing agent (vs reduced and alkylated [^{3}H] Cys-labeled marker L chain in each case). In the presence of reducing agent (panel A), peptides V_2 and/or C_1 (peak 1), peptide C_2 (peak 2), and peptide V_1 (peak 3) are identified. As expected for nascent L

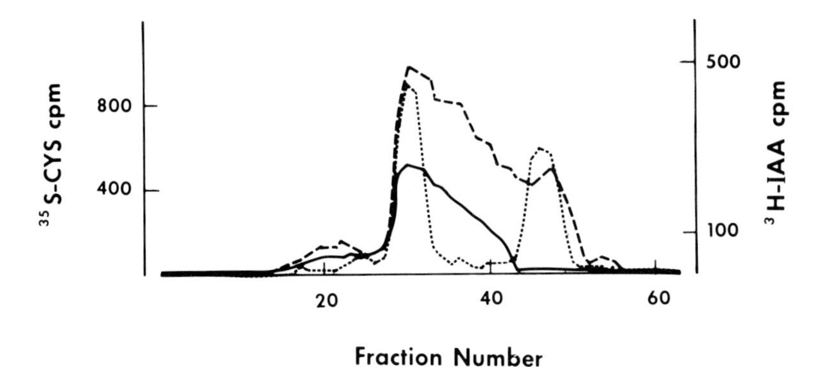

Fig. 4. SDS polyacrylamide gel electrophoresis of nascent L chains. [^{3}H]-IAA-labeled nascent L chains were isolated from MPC 11 clone 662 cells in the presence (——) or absence (– – – –) of excess IAM, as described previously [18, 30] and in Materials and Methods. [^{35}S]-Cys MPC 11 L chain (25,000 daltons) and F_{CL} (11,600 daltons) (. . . .) molecular weight markers were co-electrophoresed with each [^{3}H] cm-Cys sample on 10% polyacrylamide gels in an SDS buffer. Migration is from left to right.

chains, peak 2 (peptide C_2, Cys206) is present in a much lower yield than peak 3 (peptide V_1, Cys35). Peptide C_3 (peak 4), which contains the carboxy-terminal residue of the L chain (Cys226), is completely missing in the nascent chain sample, thereby demonstrating that there is no detectable contamination with completed L chain in the nascent chain sample. In the absence of reducing agent (panel B) there is an additional component (peak X) chromatographing at a larger apparent size than the marker V_2, C_1 peptides (peak 1). Secondly, there is a decreased recovery of peak 3 (peptide V_1) and peak 1 (peptide V_2 and/or C_1), as compared to the sample run in the presence of reducing agent (panel A). The decrease in recoveries of peak 1 and peak 3 in panel B (relative to panel A) may be accounted for quantitatively by the appearance of peak X (peak 1 decreased from 58.7% of the total radioactivity recovered to 36.4%; peak 3 decreased from 32.8% to 9.8% total radioactivity recovered; peak X accounts for 46.0% of the total radioactivity recovered in the absence of reducing agent). Peak X was pooled, reduced, alkylated, and rechromatographed vs [^{3}H] Cys-labeled marker L chain (panel C). It can be seen that reduced peak X contains equal molar quantities of peak 3 (V_1) and peak 1 (V_2 or C_1); in fact, peak 1 isolated from reduced and alkylated peak X contains peptide V_2 and no detectable peptide C_1 (data not shown). Therefore nascent L chains contain some Cys35 and Cys100 in disulfide linkage but no other Cys residues (146, 206, or 226) are present in disulfide bonds (this experiment plus unpublished results).

The First Disulfide Loop (Cys35-S-S-Cys100) Is Formed Rapidly and Quantitatively on Nascent L Chains

To examine the extent and kinetics of disulfide bond formation between Cys35 and Cys100, [^{35}S] Cys-labeled nascent chains were isolated in the presence of excess IAM and subsequently alkylated with IAA to carboxy-methylate only those Cys residues involved in disulfide bonds. The sample was fractionated by G200 Sephadex chromatography and divided into 4 size fractions (25,000– 15,200 daltons). Each fraction was digested with trypsin and chromatographed on a G25-G50 column (profiles are similar to Figure 5). The isolated V_1 peptide was then subjected to high voltage paper electrophoresis at pH 3.6 to resolve the V_1 species that were alkylated with either IAM (Cys present as free sulfhydryl) or IAA (Cys present as disulfide). Table II shows a summary of the data ob-

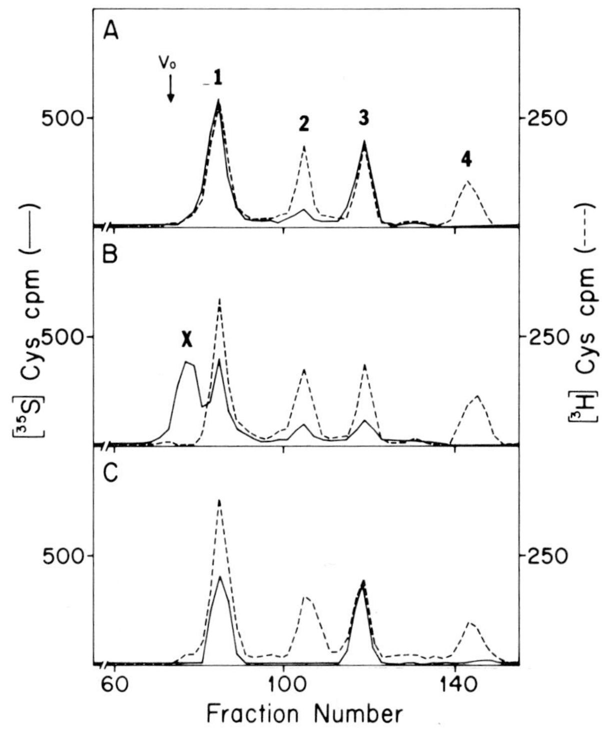

Fig. 5. G25-G50 Sephadex chromatographic profiles of a tryptic digest of [^{35}S]Cys-labeled nascent L chains vs reduced and alkylated [^{3}H]Cys-labeled marker L chain. MPC 11 clone 662 cells were labeled for 10 min with [^{35}S]-Cys. Nascent L chains were isolated, digested with trypsin, and analyzed on tandem G25-G50 Sephadex columns in the presence (panel A) or absence (panel B) of reducing agent. The additional peak (peak X) seen in the absence of reducing agent (panel B) was pooled, reduced, alkylated, and rechromatographed (panel C). Solid line, [^{35}S]Cys-labeled nascent L chains; Dashed line, [^{3}H]Cys-labeled marker L chain. The void volume (Vo) and the tryptic peptide composition of each peak (see Fig. 3) are indicated in panel A.

tained. The data indicate that in fractions 1–3 (25,000–18,000 daltons) essentially all the isolated V_1 peptide co-electrophoreses with the carboxy-methylated V_1 marker (as compared to the amido-methylated V_1 marker). In fraction 4 (18,000–15,200 daltons) 85% of peptide V_1 co-electrophoreses with the carboxy-methylated V_1 marker and 15% with the amido-methylated V_1 marker. This indicates that approximately 85% of the nascent L chains between 15,200 and 18,000 daltons have formed the first intrachain disulfide bond, whereas 100% of larger nascent L chains have formed the first intrachain disulfide bond.

Initial Covalent Assembly (Interchain Disulfide Bond Formation) Also Occurs on Nascent H Chains

We have investigated the question of whether intermolecular covalent assembly (ie, formation of interchain disulfide bonds) begins during synthesis while the nascent polypeptides are still bound to the polyribosomal complex, or only after the polypeptide chains have been completed and released from the ribosome. The major initial disulfide linked intermediate in the assembly pathway is H-L for MPC 11 cells and H-H (H_2) for MOPC 21 cells [12, 13]. After a 30-sec pulse-label with [^{35}S]Met, approximately 49% of the MPC 11 completed H chains and 36% of the MOPC 21 completed H chains are assembled

TABLE II. Kinetics and Extent of Cys35–Cys100 Disulfide Bond Formation*

Sephadex G200 fraction	Estimated size of nascent chains	% Carboxy-methylated $[^{35}S]$-V_1	%Cys 35–Cys100 disulfide
1	25–23.5 × 10^3	100%	100%
2	23.5–21.5 × 10^3	100%	100%
3	21.5–18 × 10^3	100%	100%
4	18–15.2 × 10^3	85%	85%

*Cells were labeled for 10 min with $[^{35}S]$Cys; nascent L chains were isolated in the presence of excess IAM, then reduced with β-mercaptoethanol, and subsequently alkylated with IAA to carboxy-methylate Cys residues involved in disulfide bonds as described in Materials and Methods and Results. The immunoprecipitated nascent L chains were fractionated G200 Sephadex chromatography (1.5 × 195 cm) and divided into 4 size fractions. Each fraction was digested with trypsin, and peptide V_1 was isolated by G25-G50 Sephadex chromatography (see Figs. 3 and 5). Peptide V_1 was subjected to high-voltage paper electrophoresis at pH 3.6 to resolve the V_1 species that were alkylated with either IAM (Cys present as free sylfhydryl) or IAA (Cys present in disulfide linkage).

covalently into various assembly intermediates [39]. Thus most covalent assembly occurs on completed chains.

To examine further the question of whether some of the initial covalent assembly takes place prior to completion of H chain translation, $[^{35}S]$Met-labeled nascent chains were isolated from each cell line and analyzed by immunoprecipitation and SDS-gel electrophoresis for the release of completed chains after reduction of the nascent chain fraction. The results in Figure 6A indicate that for MPC 11 cells completed L chain is specifically released by reduction of the nascent chain fraction, whereas a similar analysis of MOPC 21, seen in Figure 6B, reveals that in this case completed H chain (but no completed L chain) is released by reduction of the nascent chains. Thus in MPC 11 cells completed L chains are covalently assembled onto nascent H chains, whereas in MOPC 21 cells completed H chains are covalently assembled onto nascent H chains.

Minimum Size of Nascent H Chain Required for Formation of Intermolecular Disulfide Bonds

To determine the minimum size of nascent MPC 11 H chains that are covalently bound to completed L chains, we have immunoprecipitated the nascent chain fraction (prior to reduction) with an antiserum directed specifically against κ L chains to precipitate only those nascent H chains that are covalently bound to L chains. The results indicate that completed L chains become covalently bound to some nascent H chains of greater than approximately 38,000 daltons [39].

To determine the minimum size of nascent MOPC 21 H chains that are covalently bound to completed H chains, we have immunoprecipitated nascent H chains prior to reduction. Figure 7A shows the sample analyzed on a 7.5% SDS-polyacrylamide gel in the absence of reducing agent (reduced and alkylated $[^3H]$ Leu-labeled marker H and L chains have been run as markers).

Radiolabeled material is found migrating with an apparent molecular weight range of 112,000–96,000 daltons (fractions 23–31) [slightly more heterogeneous than H chain dimer (H_2) electrophoresed in a parallel gel] and with a very heterogeneous distribution in size characteristic of nascent H chains – ie, from 55,000 daltons (size of completed H chain) to less than 25,000 daltons. The radiolabeled material in fractions 23–31 was sensitive to partial reduction (conditions which reduce interchain disulfide bonds) in that after

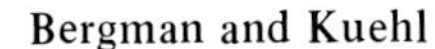

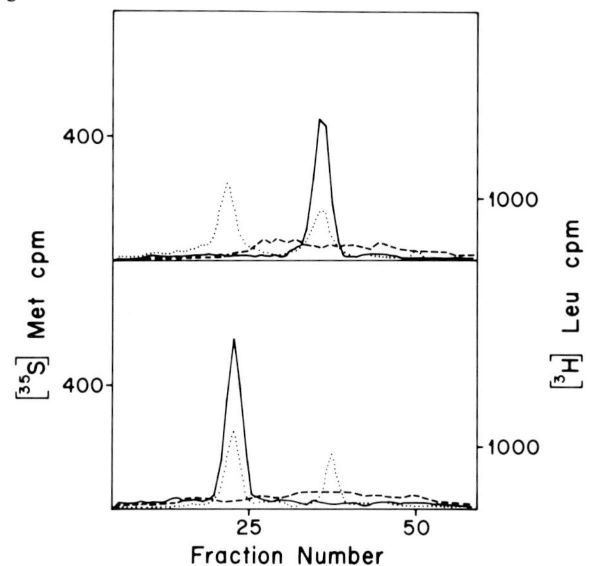

Fig. 6. SDS polyacrylamide gel electrophoresis of completed MPC 11 L chains and completed MOPC 21 H chains released by reduction of nascent chains. Nascent chains were isolated by 2 cycles of QAE-Sephadex chromatography from MPC 11 clone 456 or MOPC 21 cells labeled for 10 min with [^{35}S]Met as described previously [18, 30, 39]. The purified nascent chain fraction was divided into two aliquots, and each aliquot was subjected to a 2 h incubation in elution buffer at 37°C after addition of 0.15 M β-mercaptoethanol to one of the two aliquots. The reduced and nonreduced samples were subjected separately to a third cycle of QAE Sephadex chromatography. The flow-through fraction was collected, dialyzed extensively vs PBS, specifically immunoprecipitated, reduced, and then co-electrophoresed on 10% polyacrylamide gels in an SDS buffer with [^{3}H]Leu-labeled marker H and L chains from the respective cell line. Electrophoretic migration is from left to right. Solid line, immunoprecipitate of flow-through fraction from reduced aliquot; dashed line, immunoprecipitate of flow-through fraction from unreduced aliquot; dotted line [^{3}H]Leu-labeled marker H and L chains. Panel A, MPC 11 cells. Panel B, MOPC 21 cells.

partial reduction no radiolabeled material migrates with an apparent molecular weight larger than the marker H chain (55,000 daltons) (data not shown). The material in fractions 23–31 (Fig. 7A) was pooled and analyzed on a 12.5% SDS-polyacrylamide gel after complete reduction and alkylation (Fig. 7B). Figure 7B reveals that the sample, upon reduction, contains heterogeneous material with an apparent molecular weight range of 55,000–44,000 daltons (ie, contains radiolabeled material co-migrating and migrating slightly faster than the marker H chain). Although approximately 38% of the radioactivity present is due to completed H chains that have covalently assembled to nascent H chains, the heterogeneity present in the reduced sample indicates that nascent H chains of at least 44,000 daltons in size may covalently bind either to a completed H chain or to a second nascent H chain of at least 44,000 daltons in size.

DISCUSSION

Using the mouse plasmacytoma system, we have attempted to determine the temporal relationships between translation and various co-translational modifications. Nascent immunoglobulin heavy and light chains, covalently attached to tRNA molecules, have been separated from completed polypeptides by ion-exchange chromatography based on the multiple negative charges contributed by the tRNA of the peptidyl-tRNA complex. We have examined the glycosylation, folding (formation of intrachain disulfide bonds), and covalent assembly (formation of interchain disulfide bonds) of immunoglobulin polypeptides as they occur on nascent polypeptides.

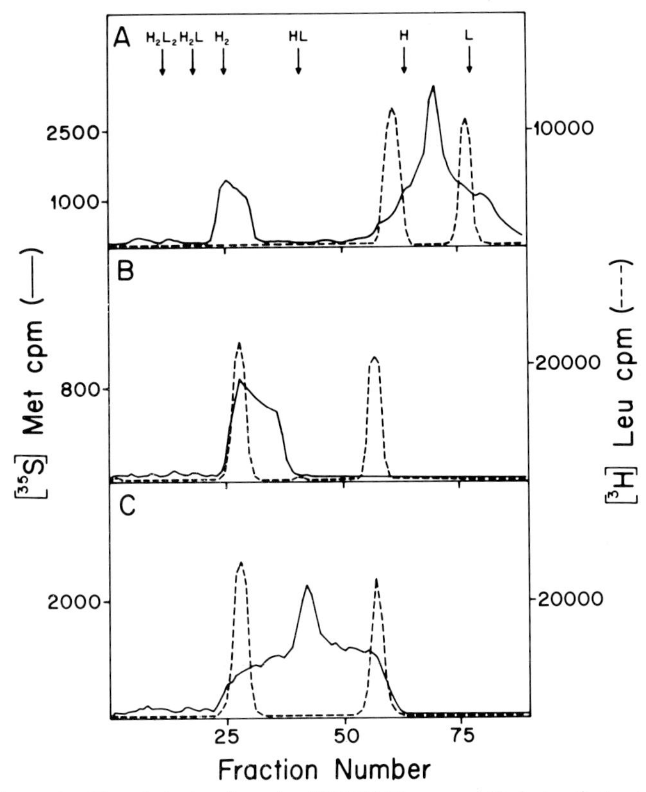

Fig. 7. SDS polyacrylamide gel electrophoresis of MOPC 21 nascent H chains that are covalently attached to completed H chains. MOPC 21 cells were labeled for 15 min with [^{35}S] Met, and nascent chains were isolated by two cycles of QAE-Sephadex chromatography as described previously [39]. The nascent chain fraction was dialyzed extensively vs PBS, subjected to specific immunoprecipitation in the presence of excess unlabeled MOPC 21 κ L chain (prepared from the light chain producer clone NSI), and analyzed by SDS-gel electrophoresis on 7.5% gels in the absence of reducing agent vs [^{3}H] Leu-labeled reduced and alkylated marker H and L chains (panel A). The migration of the covalent assembly intermediates (indicated by arrows) of MOPC 21 immunoglobulin were determined by electrophoresis on a parallel gel. Fractions 23–31 and fractions 61–75 were pooled separately, reduced, and co-electrophoresed on 12.5% polyacrylamide gels in an SDS buffer with [^{3}H] Leu-labeled marker H and L chains as seen in panel B and panel C, respectively. Solid line, [^{35}S] Met-labeled MOPC 21 nascent chains; Dashed line, [^{3}H] Leu-labeled marker H and L chains. Electrophoretic migration is from left to right.

The data presented in Figure 1 and Results indicate that nascent MPC 11 H chains are quantitatively glycosylated with a core oligosaccharide when the chain reaches a size of 38,000 daltons – ie, very soon after the asparaginyl acceptor residue (approximately residue 291, 32,000 daltons) passes through the membrane to the lumen of the rough endoplasmic reticulum and becomes available for glycosylation. Rothman and Lodish [40], using a synchronized wheat germ cell-free synthesizing system in the presence of exogenous pancreatic rough endoplasmic reticulum membranes, have shown a precise temporal sequence of glycosylation and translation of vesicular stomatitis virus G protein. However, in their studies, the observed translation times (23–43 min) are 20–30 times longer than in vivo translation times. Our results provide in vivo evidence that glycosylation occurs quantitatively at a precise interval during translation and that this period may be the only time that glycosylation of the heavy chain can occur (see below).

The possibility that glycosylation can occur on completed H chains after release from the ribosome was investigated by inhibiting the transfer of the core oligosaccharide to the nascent H chain with high concentrations of glucosamine. We have shown that non-glycosylated completed γ_{2b}, γ_1, α, and μ H chain (results for μ and γ_{2b} H chains are shown in Figure 2) synthesized in the presence of the inhibitor cannot be glycosylated during a chase period in the absence of the glucosamine, although the cells have regained the ability to glycosylate newly synthesized H chains by 20 min into the chase [30]. A similar experiment using an MPC 11 variant H chain reveals that in this case glycosylation of the completed chain can occur in that there is the appearance of the glycosylated form of the molecule by 20 min into the chase (experiments with M311 in Figure 2). This result provides evidence that the lack of glycosylation of the wild-type H chains is a function of the protein molecule itself and is not due either to spatial separation of the H chain molecules from the glycosylation enzymes or to the general inhibitor effect of the high levels of glucosamine. The variant M311 H chain includes only 38,000 daltons of amino acids due to a deletion, starting soon after the asparaginyl acceptor site, of the carboxy-terminal region of the molecule. This deletion may affect the conformation of the oligosaccharide acceptor site, thereby allowing glycosylation to occur after release of the completed protein from the ribosome.

Two possibilities are suggested for the inability of the cell to glycosylate completed H chains: 1) intramolecular folding (secondary and tertiary structure) and/or 2) intermolecular assembly (quaternary structure). In support of the first possibility, experiments from the laboratory of Lennarz have demonstrated that the core oligosaccharide can be added in vitro only to denatured ovalbumin and RNase A but not to the native forms of these proteins [41, 42]. Therefore, nascent polypeptides may be providing the asparaginyl acceptor site with little secondary or tertiary structure, allowing glycosylation to occur before the protein folds and the acceptor site becomes inaccessible for the glycosylation enzymes. In support of the second possibility, our experiments with MOPC 46B cells indicate that the completed L chain can be glycosylated if it is in a monomeric form but not if it has been assembled into a dimer [30]. Crystallographic studies by Davies and co-workers have suggested that the oligosaccharide moiety may play a central role as the principal contact between the second constant region domains of the H chains in intact immunoglobulin [43]. At this time, we cannot distinguish these two possibilities for the lack of glycosylation of completed H chains, due to the very rapid intramolecular folding and intermolecular assembly of H chains, as discussed below. However, we propose that the cell has evolved a system for efficiently glycosylating nascent chains as they pass through the membrane, thus circumventing the formation of secondary, tertiary, or quaternary structures in which the asparaginyl acceptor site would be unavailable for glycosylation. Apparently the core oligosaccharide is available for processing on completed chains [44, 45], and the type of processing may be determined by the environment of the core oligosaccharide on the completed chain.

We have demonstrated the formation of the initial intrachain disulfide bond on some nascent MPC 11 L chains (see Figure 3 for summary of L chain Cys residues) by isolating the nascent L chains in the presence of excess IAM to block free sulfhydryl groups and then selectively labeling with [^{3}H] IAA only those Cys residues that have previously formed a disulfide bond (see Fig. 4).

To determine which Cys residues are involved in the disulfide bond present on nascent L chains, we have isolated [^{35}S] Cys-labeled nascent L chains and analyzed the G25-G50 Sephadex chromatographic profile of a tryptic digest chromatographed in the presence or absence of reducing agent. In the absence of reducing agent (Fig. 5B), there is an additional component (peak X) present (which is absent when the tryptic digest is chromatographed in the presence of reducing agent (Fig. 5A). After reduction and alkylation this additional peak yields equal molar quantities of peptide V_1 and peptide V_2 (Fig. 5C; Results). SDS-gel analysis indicates that only nascent L chains of 16,000–25,000 daltons contain Cys residues in disulfide linkage (see Fig. 4). Assuming that approximately 40–50 residues are needed to span the ribosome and membrane of the rough endoplasmic reticulum [46], our results indicate that some nascent L chains have formed the initial intrachain disulfide bond (Cys35-Cys100) as soon as Cys100 enters the cisterna of the rough endoplasmic reticulum. The second intrachain disulfide bond (Cys146-Cys206) present on the MPC 11 L chain is not formed on nascent L chains because Cys206 is only 20 residues from the carboxy-terminus and cannot pass through the membrane before the L chain is completed and released from the ribosome.

The extent and kinetics of formation of the initial disulfide bond on nascent L chains was examined by determining the fraction of Cys35 that was carboxy-methylated with IAA (Cys involved in a disulfide bond) vs the fraction of the Cys residue that was amido-methylated with the excess unlabeled IAM (Cys present as free sulfhydryl) during the isolation of the [^{35}S] Cys-labeled nascent L chains. Table II indicates that for nascent L chains of 25,000–18,000 daltons essentially all the isolated V_1 peptide co-electrophoreses with the carboxy-methylated V_1 marker, indicating that 100% of the nascent L chains in these size fractions have formed the first intrachain disulfide bond. In the smallest size fraction (fraction 4; 18,000–15,200 daltons) approximately 85% of nascent L chains have formed the initial intrachain disulfide bond, using the criteria cited above. Therefore assuming a translation time of 30 sec for L chain [47], these results indicate that the initial disulfide bond is formed quantitatively in less than 4 sec after Cys100 passes through the membrane into the cisterna of the endoplasmic reticulum.

Despite a total lack of corresponding in vivo studies, the kinetics of intrachain disulfide bond formation has been extensively studied in vitro in a number of protein model systems [49–57]. In most of these cases more than one disulfide bond is present in the native protein, and the formation of intermediates with incorrectly paired disulfides occurs during in vitro refolding [58–61]. We have not detected any incorrect disulfide bonds on the nascent L chain. However, it must be mentioned that the Cys residues involved in the intrachain disulfide bonds of L chain are co-linear along the peptide chain. This situation favors formation of the correct disulfide bond because the two Cys residues (Cys35, Cys100) forming this disulfide bond are present in the cisterna of the rough endoplasmic reticulum for a substantial time before other Cys residues (eg, Cys146, Cys206) enter the cisterna. Experiments are now underway to correlate the results presented here with in vitro refolding studies of the MPC 11 L chain, and also to examine the in vivo folding of MPC 11 H chain where the Cys residues involved in interchain disulfide bonds could potentially form an incorrect intrachain disulfide linkage as the protein is being synthesized (see Fig. 8).

Our results provide rigorous in vivo evidence for the independent folding of the domains of a polypeptide, as been proposed from in vitro studies [62–63]. First, although we have provided evidence that the initial intrachain disulfide bond is formed quantitatively on nascent L chains, it seems unlikely that the formation of this disulfide bond is itself the earliest folding event on the nascent chain. Rather, it is likely that a noncovalent nucleation event [64–67] brings Cys35 and Cys100 into close proximity with subsequent formation of the disulfide bond. Additional noncovalent folding may occur after formation of the disulfide bond [56]. Second, the intrachain disulfide loop of the first or variable region domain of L chain is formed quantitatively by the time the molecule is 18,000 daltons in size, which is before a significant part of the second or constant region domain (a domain in immunoglobulin is approximately 12,500 daltons) has been synthesized and passed through the membrane into the cisterna of the rough endoplasmic reticulum. This result indicates that in vivo proteins initiate folding sequentially from the amino terminal end to the carboxy terminal end as growing nascent chains prior to release from the ribosome, although the entire molecule may be required for complete folding of the molecule [68].

In view of the fact that the amino-terminal regions of a nascent polypeptide have a longer time to fold than the more carboxy-terminal regions, it is likely that folding of a denatured completed polypeptide in vitro is not a reliable model of in vivo folding. This deficiency of the in vitro model may, in fact, explain why it is not possible to achieve correct folding of some denatured, completed polypeptides in vitro.

The initial step of intermolecular covalent assembly of immunoglobulin molecules involves formation of H–L or H–H disulfide bonds. From the results presented here we conclude that this initial step of covalent assembly occurs – to a substantial extent – on nascent H chains, as well as on completed H chains, as demonstrated previously by others [12, 13]. Our results demonstrate that in MPC 11 cells (see Fig. 6A) completed L chains are assembled covalently onto nascent H chains since L chains are specifically released by reduction of the isolated nascent chain fraction. This finding is consistent with the H–L

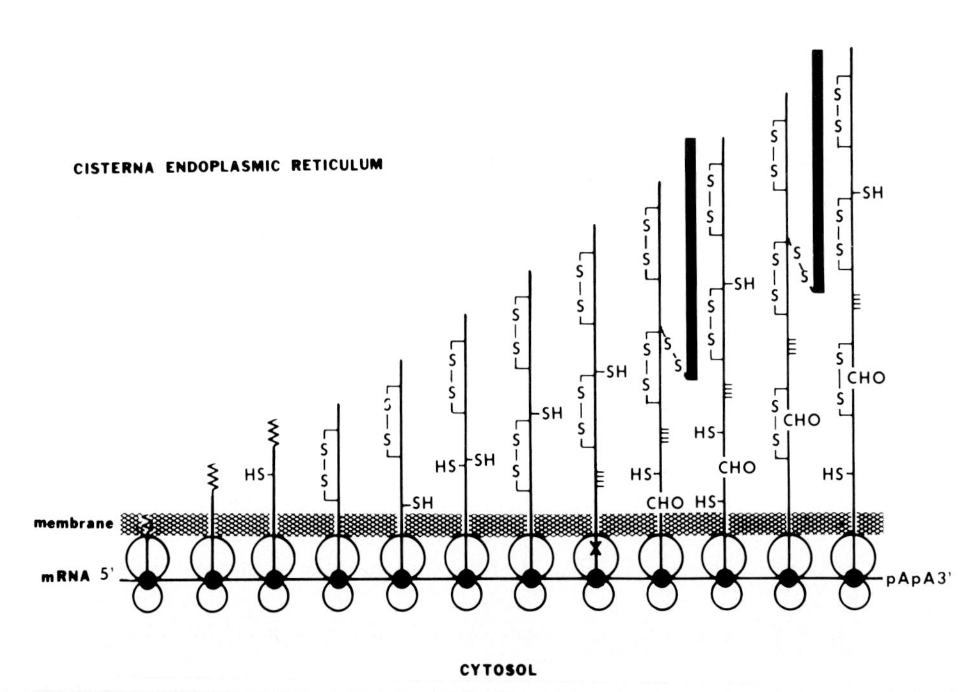

Fig. 8. Model of MPC 11 nascent heavy chain polyribosomal complex showing various co-translational modification events: (1) cleavage of amino-terminal leader or signal sequence (∿∿) ([46]; E. Harris and W.M. Kuehl, unpublished data); (2) transfer of core oligosaccharide (CHO) to asparaginyl acceptor residue (X); (3) formation of intrachain disulfide bonds; and (4) formation of interchain disulfide bond to complete light chain (indicated by thick line). Approximately half of the nascent heavy chains form an interchain disulfide bond with a complete light chain before heavy chain completion and release from the polyribosomal complex.

half molecule being the major intermediate in the assembly of the mature MPC 11 molecule [12]. In contrast, by the criteria stated above, completed H chains are covalently assembled onto nascent H chains in MOPC 21 cells (see Fig. 6B) where an H chain dimer is the predominant intermediate in MOPC 21 covalent assembly [13].

We have determined the minimum size of nascent H chains that are covalently bound to completed L chains for MPC 11 and to completed H chains for MOPC 21. The results indicate that the nascent H chains must be approximately 38,000 daltons before completed L chains become covalently bound in MPC 11 [39], whereas the nascent H chain in MOPC 21 must be approximately 44,000 daltons prior to covalent assembly of a completed H chain (see Fig. 7). The H chain Cys residue involved in the interchain disulfide bond with the L chain is residue 131 of the MPC 11 H chain [69], and the residues involved in H–H disulfide bonds in MOPC 21 are Cys221, 224, and 226 [70]. These residues would be expected to enter the intracisternal space when the nascent H chain has achieved a size of approximately 20,000 daltons and 31,000 daltons for MPC 11 and MOPC 21, respectively (assuming 50 amino acid residues to span the ribosome and rough endoplasmic reticulum membrane [46]). This suggests the possibility that the nascent H chain must first achieve some specific secondary or tertiary structure before intermolecular covalent assembly to a completed L or H chain can occur. However, we are unable to determine when non-covalent association of the completed L or H chain with the nascent H chains occurs. Finally, it should be noted that, although a significant amount of initial covalent assembly (H–L in MPC 11 cells and H_2 in MOPC 21 cells) occurs on nascent chains, some H–L and H_2 assembly occurs on completed chains and all further covalent assembly occurs on completed chains [12, 13].

Figure 8 shows a model of a nascent H chain polyribosomal complex and various co-translational modifications (amino-terminal signal sequence removal, glycosylation, folding, and assembly) that we have shown to occur on nascent immunoglobulin H and L chains in vivo.

ACKNOWLEDGMENTS

Research funds were provided through a grant from the National Institutes of Health (AI 12525-04) to W.M.K.

REFERENCES

1. Edelman GM, Gall WE: Annu Rev Biochem 38:415, 1970.
2. Edelman GM, Cunningham BA, Gall WE, Gottlieb PD, Rutishauser U, Waxdal MJ: Proc Natl Acad Sci USA 63:78, 1969.
3. Uhr JW: Cell Immunol 1:228, 1970.
4. Potter M: Physiol Rev 52:632, 1972.
5. Melchers F, Andersson J: Adv Cytopharmacol 2:225, 1974.
6. Bevan MJ, Parkhouse RME, Williamson AR, Askonas BA: Prog Biophys Mol Biol 25:131, 1972.
7. Kuehl WM: Curr Top Microbiol Immunol 76:1, 1977.
8. Wetlaufer DB, Ristow S: Annu Rev Biochem 42:135, 1973.
9. Anfinsen CB, Scheraga HA: Adv Prot Chem 29:205, 1975.
10. Baldwin RL: Annu Rev Biochem 44:453, 1975.
11. Scharff MD, Laskov R: Prog Allergy 14:37, 1970.
12. Laskov R, Lanzerotti R, Scharff MD: J Mol Biol 56:327, 1971.
13. Baumal R, Potter M, Scharff MD: J Exp Med 134:1316, 1971.
14. Baumal R, Scharff MD: Transplantation Rev 14:163, 1973.
15. Petersen JG, Dorrington KJ: J Biol Chem 249:5633, 1974.

16. Percy JR, Percy ME, Dorrington KJ: J Biol Chem 250:2398, 1975.
17. Sears DW, Mohrer J, Beychok S: Proc Natl Acad Sci USA 72:353, 1975.
18. Bergman LW, Kuehl WM: Biochemistry 16:4490, 1977.
19. Laskov R, Scharff MD: J Exp Med 131:515, 1970.
20. Coffino P, Scharff MD: Proc Natl Acad Sci USA 68:219, 1971.
21. Kuehl WM, Scharff MD: J Mol Biol 89:409, 1974.
22. Birshtein BK, Preud'homme JL, Scharff MD: Proc Natl Acad Sci USA 71:3478, 1974.
23. Weitzman S, Scharff MD: J Mol Biol 102:237, 1976.
24. Weitzman S, Nathanson SG, Scharff MD: Cell 10:679, 1977.
25. Margulies DH, Kuehl WM, Scharff MD: Cell 8:405, 1976.
26. Cowan NJ, Secher DS, Milstein C: J Mol Biol 90:691, 1974.
27. Buxbaum JN, Scharff MD: J Exp Med 138:278, 1973.
28. Bargellesi A, Periman P, Scharff MD: J Immunol 108:126, 1972.
29. Melchers F: Biochemistry 12:1471, 1973.
30. Bergman LW, Kuehl WM: Biochemistry 17:5174, 1978.
31. Rose SM, Kuehl WM, Smith GP: Cell 12:453, 1977.
32. Laemmli UK: Nature 227:680, 1970.
33. Maizel JV Jr: Methods Virol 5:179, 1971.
34. Horwitz M, Scharff MD: In Habel K, Salzman NP (eds): "Fundamental Techniques in Virology." New York: Academic Press, 1969, pp 253–315.
35. Rhoads RE, McKnight GS, Schimke RT: J Biol Chem 248:2031, 1972.
36. Schachter H: Adv Cytopharmacol 2:207, 1974.
37. Hickman S, Kornfeld S: J Immunol 121:990, 1978.
38. Smith GP: Biochem J 171:337, 1978.
39. Bergman LW, Kuehl WM: J Biol Chem 254:5690, 1979.
40. Rothman JE, Lodish HF: Nature 269:775, 1977.
41. Pless DD, Lennarz WJ: Proc Natl Acad Sci USA 74:134, 1977.
42. Struck DK, Lennarz WJ, Brew K: J Biol Chem 253:5786, 1978.
43. Silverton EW, Navia MA, Davies DR: Proc Natl Acad Sci USA 74:5140, 1977.
44. Robbins PW, Hubbard SC, Turco SJ, Wirth DF: Cell 12:893, 1977.
45. Tabas I, Schlesinger S, Kornfeld S: J Biol Chem 253:716, 1978.
46. Blobel G, Dobberstein B: J Cell Biol 67:835, 1975.
47. Shapiro AL, Scharff MD, Maizel JV Jr, Uhr JW: Proc Natl Acad Sci USA 56:216, 1966.
48. Epstein CJ, Goldberger RF: J Biol Chem 238:1380, 1963.
49. Bradshaw RA, Kanareck L, Hill RL: J Biol Chem 242:3789, 1967.
50. Saxena VP, Wetlaufer DB: Biochemistry 9:5015, 1970.
51. Goldberger RF, Epstein CJ, Anfinsen CB: J Biol Chem 239:1406, 1965.
52. Steiner RF, DeLorenzo F, Anfinsen CB: J Biol Chem 240:4648, 1965.
53. Tanford C: Adv Prot Chem 24:1, 1970.
54. Hantgan RR, Hammes GG, Scheraga HA: Biochemistry 13:3421, 1974.
55. Creighton TE: J Mol Biol 87:563, 1974.
56. Isenman DE, Painter RH, Dorrington KJ: Proc Natl Acad Sci USA 72:548, 1975.
57. Teale JM, Benjamin DC: J Biol Chem 251:4603, 1976.
58. Creighton TE: J Mol Biol 87:579, 1974.
59. Creighton TE: J Mol Biol 87:603, 1974.
60. Ristow S, Wetlaufer DB: Biochem Biophys Res Commun 50:544, 1973.
61. Robson B, Pain RH: Biochem J 155:331, 1976.
62. Bjork I, Karlsson FA, Berggard I: Proc Natl Acad Sci USA 68:1707, 1971.
63. Teale JM, Benjamin DC: J Biol Chem 251:4609, 1976.
64. Levinthal C: Scientific Am 214:45, 1966.
65. Lewis PN, Momany FA, Scheraga HA: Proc Natl Acad Sci USA 68:2293, 1971.
66. Anfinsen CB: Biochem J 128:737, 1972.
67. Wetlaufer DB: Proc Natl Acad Sci USA 70:697, 1973.
68. Taniuchi H: J Biol Chem 245:5459, 1970.
69. dePreval C, Pink JRL, Milstein C: Nature 228:930, 1970.
70. Adetugbo K: J Biol Chem 253:6068, 1978.

Journal of Supramolecular Structure 11:139–145 (1979)
Biological Recognition and Assembly 65–71

NMR of fd Coat Protein

T. A. Cross and S. J. Opella

Department of Chemistry, University of Pennsylvania, Philadelphia, Pennsylvania 19104

The conformations of the major coat protein of a filamentous bacteriophage can be described by nuclear magnetic resonance spectroscopy of the protein and the virus. The NMR experiments involve detection of the ^{13}C and ^{1}H nuclei of the coat protein. Both the ^{13}C and ^{1}H nuclear magnetic resonance (NMR) spectra show that regions of the polypeptide chain have substantially more motion than a typical globular protein. The fd coat protein was purified by gel chromatography of the SDS solubilized virus. Natural abundance ^{13}C NMR spectra at 38 MHz resolve all of the nonprotonated aromatic carbons from the three phenylalanines, two tyrosines, and one tryptophan of the coat protein. The α carbons of the coat protein show at least two different classes of relaxation behavior, indicative of substantial variation in the motion of the backbone carbons in contrast to the rigidity of the α carbons of globular proteins. The ^{1}H spectrum at 360 MHz shows all of the aromatic carbons and many of the amide protons. Titration of a ^{1}H spectra gives the pKas for the tyrosines.

Key words: fd coat protein, ^{1}H NMR, ^{13}C NMR

In order to understand the function of a protein it is necessary to correlate the structure of the protein at all levels with its dynamic properties. It is generally more straightforward to determine the primary sequence chemically and the secondary and tertiary structure with scattering techniques than to describe the motions present in the molecule. Nuclear magnetic resonance (NMR) spectroscopy is well suited for studying protein dynamics, because it can focus on individual residues even if more than one of a kind is present in the polypeptide and because relaxation behavior of nuclear resonances can often lead to determinations of rotational diffusion rates [1].

The major coat protein of the E coli bacteriophage fd is the subject of our current NMR studies. This protein exists in several biological situations, including as an intrinsic membrane protein and as the nucleoprotein complex of the virion [2]. The protein has great flexibility in its biological function, yet has only 50 amino acids in a single polypeptide chain. This protein may be an example of molecular dynamics being strongly influenced by all levels of structural organization, including primary structure. The direct influence of sequence on properties of the protein is inferred from the amino acids being segregated into acid, hydrophobic, and basic domains [3]. The NMR results can describe the mobility of the domains as well as individual side chains.

Received May 22, 1979; accepted June 12, 1979.

The coat protein has a high α helix content in a lipid- or detergent-bound state as well as in the assembled virion [4, 5]. In the membrane the protein relies on its hydrophobic midsection to span the hydrocarbon bilayer, with the hydrophilic ends associated with the polar phospholipid headgroups. The 19 amino acid hydrophobic stretch makes this a characteristic integral membrane protein. On the other hand, the collection of basic side chains at one end, may interact with the phosphate groups of DNA in the nucleoprotein complex.

The biology and structure of fd are well characterized [2]. The fd virus is a long filament, 9,000 Å by 90 Å, composed of a single strand circle of DNA encased in 2,700 copies of the major coat protein. The only other component is about 5 copies of the gene 3 product at one end of the virus. Upon binding to an E coli F pilus, the DNA and gene 3 protein enter the cell, while the coat protein is stored in the cell membrane. During viral replication the major coat protein is synthesized as a procoat protein with a leader sequence of 23 amino acid residues that serve to make the protein soluble in the cytoplasm. The procoat protein then is inserted into the membrane where it is trimmed, in a separate step, to the mature coat protein [6]. The major coat protein is stored in the membrane asymmetrically with the carboxyl end on the inside of the cell [7]. The virions are assembled when the DNA is extruded through the membrane, where the DNA is packaged inside the coat protein.

While the procoat protein is soluble in both water and membranes, the coat protein itself is completely insoluble in water. It can only be studied in solution in the presence of lipids or detergents, presumably because of its extensive hydrophobic character. In the virus the coat proteins are arranged in such a way that the hydrophobic amino acids do not interfere with solubility in water. The versatility of the coat protein is illustrated by the dramatic transition from water insolubility during storage in the membrane to high solubility in water when assembled in the virion.

The detergent-solubilized coat protein represents a stable conformation of the molecule. It may be the same structure that exists in the cell membrane after synthesis. On the basis of circular dichroism studies of the protein in sodium dodecylsulfate (SDS) micelles and in phospholipid vesicles, the secondary structure is the same for the two preparations [4, 5]. Proteolytic digests of the detergent-solubilized coat protein show that only the hydrophilic terminal sections are exposed, just as expected for an integral membrane protein spanning a bilayer or a micelle [8]. It is an open question as to how extensive the changes in protein structure are upon virus formation.

Because of the wide range of biological structures that contain the coat protein, we are applying a variety of different NMR techniques to the problem. These include solid-state NMR of the DNA and coat protein in the intact virus, and solution ^{13}C and ^{1}H NMR of the detergent-solubilized major coat protein. This paper describes the ^{13}C and ^{1}H experiments that map out the general features of the protein dynamics.

High-resolution ^{1}H NMR is most suitable for monitoring the conformational states of the protein and the chemical environment of individual residues [9]. In some cases the spectral appearance can provide information on molecular dynamics, and, by observing the presence and exchange of non carbon-bound hydrogens such as amide hydrogens, the overall structural integrity and flexibility of protein folding can be characterized.

Natural abundance ^{13}C NMR is a valuable source of information on structure and dynamics of proteins [1, 10, 11]. Relaxation studies are particularly useful because the well-understood nature of $^{13}C-^{1}H$ interactions makes the motional properties available directly. Small globular proteins in solution have been characterized by ^{13}C NMR.

The detergent-solubilized coat protein of fd is shown to have quite different properties than the previous examples of globular proteins.

MATERIALS AND METHODS

The growth of Escherichia coli F^+ (3,300) on a modified SLBH-rich medium infected in late log phase with fd and the purification of the fd from the growth supernatant by polyethylene glycol precipitation, CsCl block gradient centrifugation, and dialysis will be described in detail elsewhere [12]. Gel chromatography with Sephacryl S-200 SF was used to separate the SDS-solubilized protein from the DNA after the virus was broken up. The coat protein collected from column fractions was dialyzed and lyophilized. Samples for NMR were made by dissolving the protein-detergent powder in a minimum volume of ^2H_2O or H_2O buffered with 40 mM borate at pH 9.0, unless mentioned otherwise. The pH readings on a Beckman model 12 meter with a sodium ion-insensitive electrode are uncorrected for deuterium isotope effects.

^{13}C NMR spectra were obtained on a Nicolet NT-150 NMR spectrometer with an Oxford Instruments magnet. Samples were approximately 12 ml in 20 mm sample tubes with a typical protein concentration of 7 mM. All ^{13}C chemical shifts are referred to external TMS through a dioxane standard.

^1H NMR spectra were obtained on a Bruker WH 360 spectrometer. Samples were about 0.5 ml in 5 mm sample tubes with protein concentrations of 3 mM. Spectra were acquired using rapid-scan correlation spectroscopy. Proton chemical shifts were referenced to internal DSS.

RESULTS

The complete natural abundance ^{13}C NMR spectrum of the fd major coat protein in solution is shown in Figure 1. Despite large peaks from the SDS detergent near 20 ppm and 70 ppm, nearly all types of protein carbon resonances are visible, including many aliphatics above 15 ppm. This paper concentrates on the α carbons in the region 50–70 ppm and the aromatic carbons 100–160 ppm. The carbonyls are in the band around 175 ppm. This spectrum was obtained with relatively slow pulsing to ensure equilibrium signal intensities. Continuous square wave-modulated proton decoupling was used, which means that each carbon resonance is a single line and the intensities have contributions from nuclear Overhauser enhancement (NOE).

The α carbon spectral region (50–70 ppm) has an appearance different from that of any protein previously reported. The α carbons of a native globular protein are a rigid part of the protein backbone. The relaxation parameters of broad linewidth, short T_1, and minimal NOE of α carbons correspond to the rotational correlation time of the entire protein, with no evidence of internal motion. This is not the case for the coat protein, with a number of very sharp resonances superimposed on the more numerous broad α carbon lines. The sharp lines have significantly larger T_1s and NOEs than the broad lines. Clearly, heterogeneity in backbone dynamics is being monitored.

The aromatic region of the ^{13}C spectrum is expanded in Figure 2. The spectrum 2a corresponds to that of Figure 1 and contains resonances from all the aromatic carbons of the 3 Phe, 2 Tyr, and 1 Trp of the coat protein. The spectrum 2b, obtained with weak off-resonance-modulated proton decoupling and refocusing of the ^{13}C magnetization to eliminate the signals from carbons with directly bonded hydrogens, has only the non-

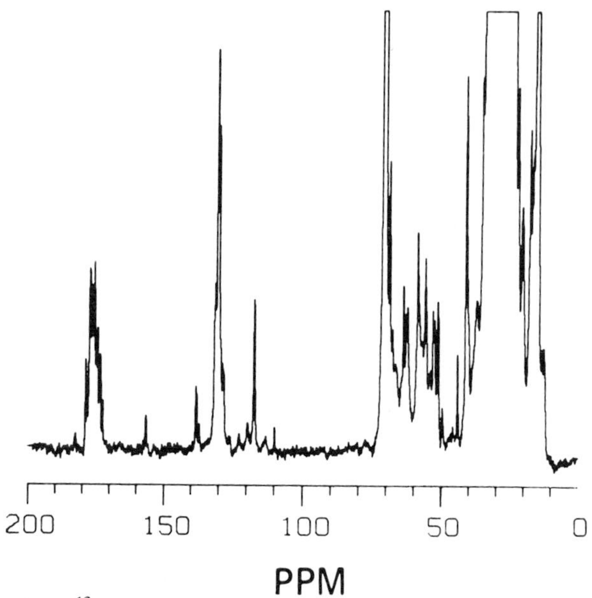

Fig. 1. Natural abundance ^{13}C NMR spectrum of fd major coat protein. 20,000 transients accumulated with 1.5 s recycle delay. Digital resolution of 2.5 Hz with 4 Hz linebroadening added. Referenced to external TMS. ^{13}C frequency of 37.74 MHz.

protonated aromatic carbon resonances. Partial assignments result from the comparison of spectra 2a and 2b.

The C_ζ resonances of the two tyrosines occur at 156.5 ppm at pH 9.0. In the best spectra they have completely separated lines. The 3 phenylalanine C_δ and the $C_{\epsilon 2}$ of the tryptophan come near 138 ppm. The 2 Tyr γ carbons and the δ_2 Trp carbon appear at 129 ppm with the Trp C_γ at 109.5 ppm. These are all narrow resonances, as expected for protein carbons without closeby protons to provide efficient relaxation. And as expected these survive the procedure used for Figure 2b, showing the lines for the nonprotonated carbons.

In a globular protein the other aromatic carbon signals would have broad resonances because of the effect of an attached proton on ^{13}C relaxation. While aromatic group rotation occurs in globular proteins, it is usually limited in extent and is detected through more subtle T_1 and NOE changes rather than drastic line narrowing [11]. A more pronounced effect is seen in the coat protein ^{13}C aromatic spectra. The phenylalanine ring carbons dominate the very sharp signals near 130 ppm. Somewhat broader lines at 118 ppm are from the Tyr C_ϵ. The individual Trp protonated carbons are broader yet and can be seen in the 110–125 ppm region. The finding of such narrow Phe ring carbons and the clear observation of single protonated Trp carbons are remarkable findings. NOE measurements correlate with linewidth measurements, with the sharp lines having high NOE (>2) and the broad lines a minimal amount (1.1).

The downfield region of the ^1H NMR spectrum of the coat protein is shown in Figure 3. Spectrum 3a is run in H_2O, thus resonances are visible from the amide protons of the entire protein and the carbon-bound hydrogens of aromatic groups. In contrast, spectrum 3c in ^2H_2O after hearing, has no amide resonances but only the aromatic hydrogens. The spectrum 3b is an important intermediate case, where the protein is dissolved in ^2H_2O at 20°C after being lyophilized from H_2O and contains 3–5 amide

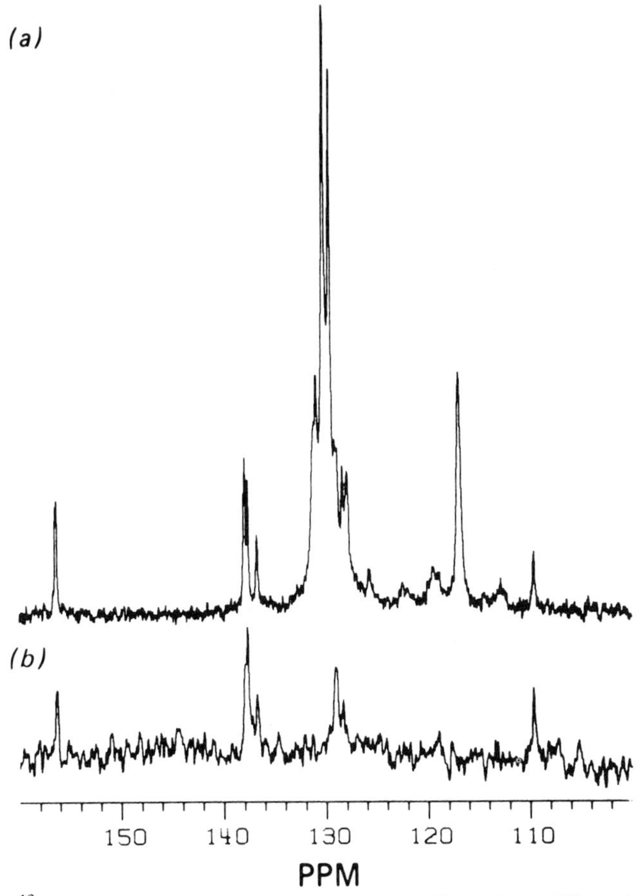

Fig. 2. Aromatic ^{13}C NMR region of fd major coat protein. a) Expansion of Figure 1. b) Non-protonated aromatic carbon spectrum; obtained with off resonance proton decoupling and ^{13}C spin echo formation.

resonances that can be exchanged only by heating. The slow exchange of these few amide protons is evidence for secondary and tertiary structure of the level seen in globular proteins. More sophisticated structural analysis is possible from the individual kinetics of the few amide resonances of spectrum 3b.

The appearance of the aromatic ^1H NMR spectral region is similar to what might be expected from a native protein, especially the upfield two peaks at 6.8 and 6.9 ppm, which can come only from the ϵ protons of the two tyrosines, each resonance being from two equivalent protons on one of the rings. At higher temperatures the doublet structure from coupling to the δ protons is visible, and they titrate upfield with increasing pH. Rapid rotation of the Tyr rings occurs because of the equivalence of the ϵ protons. The corresponding Tyr δ protons are at 7.1 and 7.2 ppm. The phenylalanine ring protons, like the ring carbons, are sharp (7.3 ppm) and show relatively little structure-induced chemical shift nonequivalence. Under some pH and temperature conditions the Trp protons can be discerned among the Phe and Tyr protons. The tyrosine resonances titrate with pKas of 12.5 and 12.6 [12]. The other resonances change little with pH.

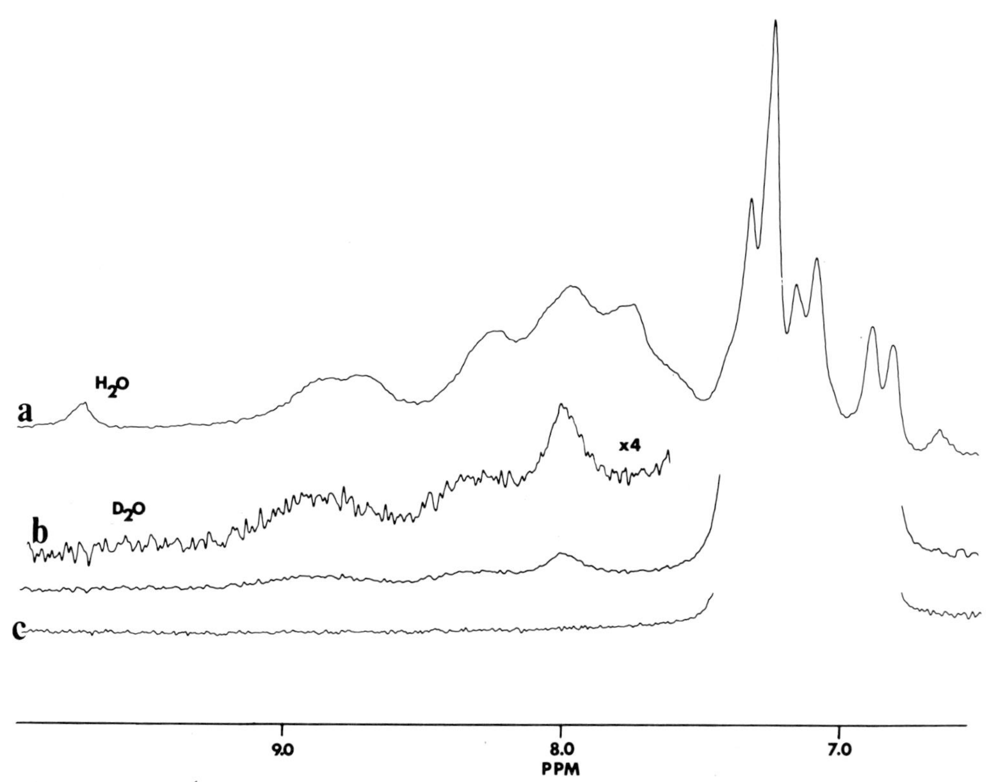

Fig. 3. Downfield ^{1}H NMR region of fd major coat protein. 64 scans of correlation spectroscopy. ^{1}H frequency of 360.06 MHz. a) Protein in H_2O, b) protein in ^2H_2O at 20°C with 4X expansion, c) sample b after heating.

DISCUSSION

Detergent-solubilized fd coat protein has a stable native structure that differs significantly from that of a typical globular protein. A high degree of specific secondary and tertiary structure is reflected in the chemical shift dispersion among identical nuclei from different residues. Examples of this are the Tyr ϵ and δ protons, Tyr ζ carbons, and the Phe γ carbons. Denatured proteins have only a single line for such situations [10]. Other evidence for the coat protein being highly structured comes from the slow exchange of some amide protons and the borad lines of the Trp ring carbons as well as most of the α carbons.

The sequence of amino acids 21–26 is Tyr-Ile-Gly-Tyr-Ala-Trp [3]. These are in the hydrophobic central section of the protein. The very high pKas of the two tyrosines indicate that they are buried in the hydrophobic pocket or that they are involved in a highly specific hydrogen or van der Waals bonding arrangement. One residue away is the Trp, which is a rigidly held residue, as seen in the ^{13}C protonated carbon linewidths. These monitors of the hydrophobic domain of the protein show it to be relatively rigid and highly structured. It is quite likely that many of the broad C_α resonances and the slowly exchanged amide hydrogens will be assigned to this region.

The phenylalanine rings are rapidly rotating. Two Phes are at the basic end, and one is in the acidic region of the protein [3]. There is less organized structure in these regions based on the extensive mobility of these large side chains. Chemical evidence also indicates the exposure of the hydrophilic residues to reagents [8].

The results presented here give a qualitative view of the dynamics of a detergent-solubilized protein. There are clearly different regions of backbone mobility, and there are large differences among side chains in rates of motion. The NMR results to date allow a limited amount of interpretation in terms of specific parts of the protein structure. To the extent that groupings of resonance characteristics with primary structure are possible, the central hydrophobic region seems more highly structured than the hydrophilic end regions. More extensive experiments will give a more complete picture of the protein. NMR of the protein in solution and the intact virus can then give an idea of the extent of conformational change associated with virus assembly.

ACKNOWLEDGMENTS

This work is being supported by grants from NIH (GM-24266) and the American Cancer Society (NP-225). The ^{1}H spectra were taken at the Middle Atlantic NMR Facility. The ^{13}C spectra were taken on a spectrometer obtained with a NSF grant. T.A.C. is supported by a Cell and Molecular Biology Training Grant.

REFERENCES

1. Opella SJ, Nelson DJ, Jardetzky O: In Reising HA, Wade CG (eds): "ACS Symposium Series, No. 34. Magnetic Resonance in Colloid and Interface Science." Washington DC: American Chemical Society, 1976, pp 397–417.
2. For a collection of papers on fd see: Denhardt DT, Dressler D, Ray DS (eds): "The Single Stranded DNA Phages." New York, Cold Spring Harbor Laboratory, 1978: 605–626.
3. Nakashima Y, Konigsberg W: J Mol Biol 88:598, 1974.
4. Williams RW, Dunker AK: J Biol Chem 252:6253, 1977.
5. Nozaki Y, Reynolds JA, Tanford C: Biochemistry 17:1239, 1978.
6. Ito K, Mandel G, Wickner W: Proc Natl Acad Sci USA 76:1199, 1979.
7. Wickner W: Proc Natl Acad Sci USA 72:4749, 1975.
8. Woolford JL Jr, Webster RE: J Biol Chem 250:4333, 1975.
9. Wuthrich K: "NMR in Biological Research." New York: Elsevier, 1976.
10. Allerhand A: Accts Chem Res 11:469, 1978.
11. Opella SJ, Nelson DJ, Jardetzky O: J Am Chem Soc 96:7157, 1974.
12. Cross TA, Opella SJ: (in preparation).

Journal of Supramolecular Structure 11:321–326
Biological Recognition and Assembly 73–78

Fibrous Projections From the Core of a Bacteriophage T7 Procapsid

Philip Serwer

Department of Biochemistry, The University of Texas Health Science Center at San Antonio, San Antonio, Texas 78284

A cylindrical core previously demonstrated in a bacteriophage T7 procapsid (capsid I) has been further examined by electron microscopy. Fibrous extensions of the core have been observed; these fibers appear to connect the core to the capsid I envelope. After infection of a nonpermissive host with bacteriophage T7 amber mutant in any gene coding for a core protein, the resulting lysates contained more noncapsid assemblies of capsid envelope protein than did wild-type lysates; these assemblies had a mass two to at least 500 times greater than the mass of capsid I. This suggests that the internal core and fibers assist the assembly of subunits in the envelope of capsid I.

Key words: bacteriophage T7, procapsid core, electron microscopy, mutant lysates

Protein subunits in the envelopes of the capsids of spherical viruses are arranged in equivalent or quasi-equivalent icosahedral lattices [1, 2]. Interactions between envelope subunits and the viral nucleic acid are sufficient to direct "in vitro" assembly of the envelopes of some of the smaller RNA plant and bacterial viruses (180 envelope subunits) [3–5]. However, nonenvelope proteins are necessary for the efficient "in vivo" assembly of the envelopes of the larger bacteriophages (⩾420 envelope subunits) including λ [6, 7], P22 [8], T4 [9], and T7 [10] (see also Results below). The evidence also indicates that these latter bacteriophages assemble a DNA-free procapsid and that the procapsid subsequently packages DNA (reviewed in Casjens and King [11]). Thus, interactions with DNA probably do not assist procapsid assembly.

The procapsid of bacteriophage T7 (capsid I) has an internal cylinder (core) with an axial hole; the core is attached at one end to the capsid I envelope. This core is also present in bacteriophage T7 [12]. Bacteriophage T7 with an amber mutation in any core protein gene is deficient in capsid I envelope assembly when grown in a nonpermissive host [10] (see also Results) (capsid I is referred to as prohead in Roeder and Sadowski [10]), indicating that the core assists assembly of the capsid I envelope. In the present communication ultrastructural aspects of the core-envelope interaction have been investigated and additional data concerning the phenotypes of T7 core protein-deficient mutants have been obtained.

Received for publication April 6, 1979; accepted August 6, 1979.

METHODS

Bacteriophage and Bacterial Strains

Bacteriophage T7 and T7 amber mutants [13] were received from Dr. F. W. Studier. The following amber mutants were used: gene 8-*am*11; gene 14-*am*140; gene 15-*am*31; gene 16-*am*9. Mutants will be referred to by their gene number followed by *am*. The host for wild-type bacteriophage T7 and the nonpermissive host for T7 amber mutants was Escherichia coli BB/1. The permissive host for amber mutants was E coli 0-11′. Lysates of E coli BB/1 infected with a T7 amber mutant will be referred to by the name of the mutant (ie, an 8 *am*-infected E coli BB/1 lysate will be referred to as an 8 *am* lysate).

Concentration of Protein Aggregates From Lysates and Purification of Capsid I

Capsid I was precipitated from wild-type T7 lysates using carbowax 6000 and was purified as previously described [12]. For the electron microscopic observation of capsids and other protein aggregates in mutant lysates an aerated, log-phase, 6-liter culture of E coli BB/1 (3×10^8 bacteria per milliliter) in M9 medium [12] was infected at 30.0°C with a bacteriophage T7 amber mutant (multiplicity of infection, 15). When lysis occurred (30–45 min after infection), the capsids and other protein aggregates were twice precipitated with Carbowax 6000 and were clarified by low-speed centrifugation as previously described [12].

Electron Microscopy

Support films for electron microscopy were prepared as previously described [12]. A 2- to 5-μl drop of sample was placed on a support film and was incubated at room temperature for 1.0 min. The film was washed with four drops of water and then two drops of sodium methyl phosphotungstate, pH 7.6 [14]. Excess stain was removed from the film with a Kimwipe and the stain was allowed to dry.

Glutaraldehyde Fixation

Capsid I was fixed with glutaraldehyde by diluting three parts of sample into one part of 12.5% glutaraldehyde (TAAB), 0.5 M sodium phosphate, pH 7.4; this mixture was incubated at room temperature for 5 min. Some capsid I samples were immediately dialyzed against 0.2 M NaCl, 0.01 M sodium phosphate, pH 7.4, 0.001 M $MgCl_2$; others were stored at 4°C for times as great as 6 months before dialysis. No significant difference in the properties of capsid I particles (see Results) resulted from the different fixation times. All micrographs shown are of capsid I particles fixed for 6 months at 4°C.

RESULTS

In a previous study electron micrographs of unfixed capsid I were obtained after negative staining with sodium phosphotungstate, pH 7.6 [12]. In an attempt to obtain improved electron micrographs, capsid I particles were fixed with glutaraldehyde and were prepared for electron microscopy using sodium methyl phosphotungstate, pH 7.6, as the negative stain. This stain has better spreading properties [14] and less grain than sodium phosphotungstate. Of 321 capsid I particles observed at random, 71 (22%) appeared to have at least one fiber connecting the internal cylindrical core to the capsid envelope (Fig. 1). Of particles with fibers, 77% were viewed at right angles to the core axis; in the

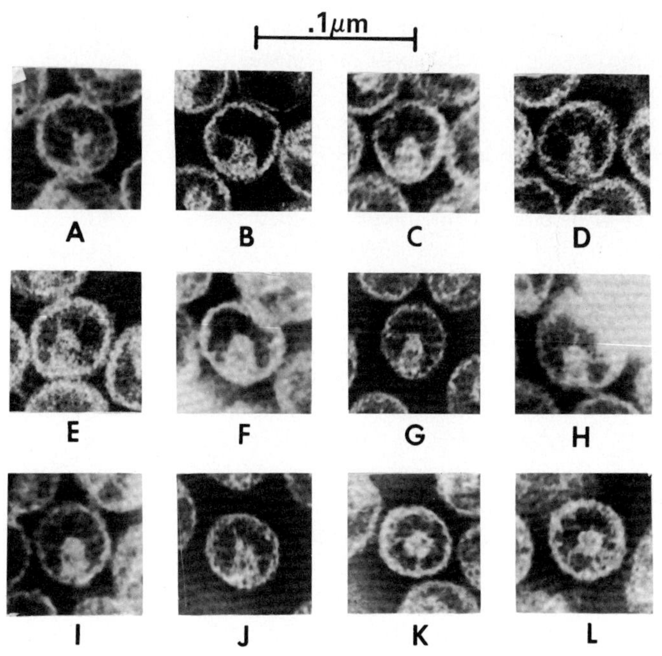

Fig. 1. Capsid I particles with an internal fiber. Capsid I was fixed with glutaraldehyde and prepared for electron microscopy as described in Methods. Particles with an internal fiber are shown.

interior of these particles the following types and numbers of fibers were observed: a) A single fiber was observed in 45% of the particles (Fig. 1A–C); this fiber was usually attached to the end of the core distal to the point of contact with the envelope; b) two to four fibers, with a variable degree of clarity, were observed in the remaining 55% of the particles; some particles had a fiber at the envelope proximal end of the core (Fig. 1D) and 18% had two roughly parallel fibers; these fibers were either parallel or perpendicular to the core axis (Fig. 1E–H, J). Particles viewed parallel to the core axis with fibers connecting the capsid I core to its envelope (23% of the fiber-containing particles) are in Figure 1K–L.

The variation in the number of fibers observed in capsid I particles may be caused by alterations in the fibers induced before or during staining. If it is assumed that capsid I particles have a unique number and configuration of fibers, there must be at least four envelope-distal fibers (two parallel fibers for each direction; at least two directions), and at least one envelope-proximal fiber.

Unfixed capsid I particles in micrographs taken of sodium phosphotungstate-stained specimens [12] were also examined for the presence of internal fibers. Of 280 randomly selected particles, 14 (5%) appeared to have at least one internal fiber. The fibers were not previously noticed in these specimens because of the comparatively small number of particles with fibers and because the fibers were usually not as clearly defined as the fibers in Figure 1.

Mutant Lysates

The proteins associated with the core in capsid I are P8, P14, P15, P16 [10, 12]. (T7 proteins are labeled by P followed by the protein's gene number [15].) In a previous report [10] it was found that 8 *am*, 14 *am*, 15 *am*, and 16 *am* lysates, fractionated by sucrose gradient velocity sedimentation, had more misassembled envelope protein than wild-type lysates. Forty percent of the assemblies of capsid protein were larger and of a different shape than capsids (electron micrographs presented in Roeder and Sadowski [10] suggest that the misassembled envelopes were 2–5 times as massive as the capsid I envelope). However, after sucrose gradient sedimentation of radiolabeled assemblies of capsid proteins from 14 *am*, 15 *am*, or 16 *am* T7 lysates, it has been found that the misassembled capsid protein had a greater tendency to adhere to containers (borosilicate glass) than T7 capsids.* To block this loss of protein in mutant lysates, assembled proteins were concentrated 2,000 × with Carbowax immediately after lysis. Carbowax precipitates, prepared and clarified as described in Methods, were observed by electron microscopy.

Concentrated 8 *am*, 14 *am*, 15 *am*, and 16 *am* lysates were examined. In all of these lysates were observed tubular structures (polycapsids) that were open in at least one location and that had an envelope thickness of 2.0–3.0 nm. This is the same thickness as that of the bacteriophage T7 envelope [16] and a significantly smaller thickness than that of the capsid I envelope (4.0–7.0 nm; from electron micrographs). Some polycapsids had a mass 2–5 times the mass of a T7 capsid, but most of the assembled capsid protein in 8 *am*, 14 *am*, 15 *am*, and 16 *am* lysates was assembled in polycapsids that had a mass at least 50 times the mass of capsid I, and some polycapsids had a mass at least 500 times the mass of capsid I; a polycapsid from an 8 *am* lysate is in Figure 2. The size of these polycapsids is greater by one to two orders of magnitude than the size of polycapsid-like material previously shown [10]. The polycapsids sometimes had pieces of host membrane attached; the membranes varied in size from comparatively small vesicles (white arrows in Fig. 2) to pieces of membrane 10–50 times as large as the vesicles (not shown). Stacks of subunits with roughly the same width as the capsid I core were also observed (black arrows in Fig. 2); these may be partially assembled cores. In negatively stained preparations of concentrated 8 *am*, 14 *am*, 15 *am*, and 16 *am* lysates no less than 95% of the mass of assembled envelope protein was in polycapsids. In wild-type T7 lysates no less than 60% of the mass of assembled envelope protein was in capsids; roughly half of the capsids were capsid I (other T7 capsids observed in lysates are described in Serwer [12]). More precise estimates of polycapsid amounts will be made when high-yield techniques for further purifying polycapsids are developed.

DISCUSSION

The internal fiber of capsid I is probably not made of DNA because DNA synthesis is not required for the assembly of capsid I [15]. Of the possible core proteins (P8, P15, P14, P16) it is unlikely that P8 and P15 are in the fiber because of evidence indicating that P8 and P15 are in the cylindrical part of the core [12]. Remaining are P14 (MW = 18,000) and P16 (MW = 150,000) [15]. Assuming an average amino acid molecular weight of 120 daltons and assuming that the internal radius of the envelope of capsid I is 20.0 nm (determined from electron micrographs), the rise per residue of P14 is 0.13 nm if a single molecule of P14 spans the distance from the center of capsid I to the inner edge of the capsid I envelope. This is 13% less than the rise per residue of an α helix, indicating that the P14 molecule should appear 0.5–0.6 nm in diameter, probably too thin to have been observed by negative-stain electron microscopy. At least three P14 molecules would be necessary to

*A week was sufficient time to adhere 95.0–99.5% of ^{14}C in 80S–200S particles sedimented from the mutant lysates; detectable amounts of T7 capsids were not lost during this time period. The buffer used in the sucrose gradients was 0.15 M NaCl, 0.05 M Tris-HCl, pH 7.4, 0.005 M EDTA, 100 μg/ml gelatin (P. Serwer, unpublished data).

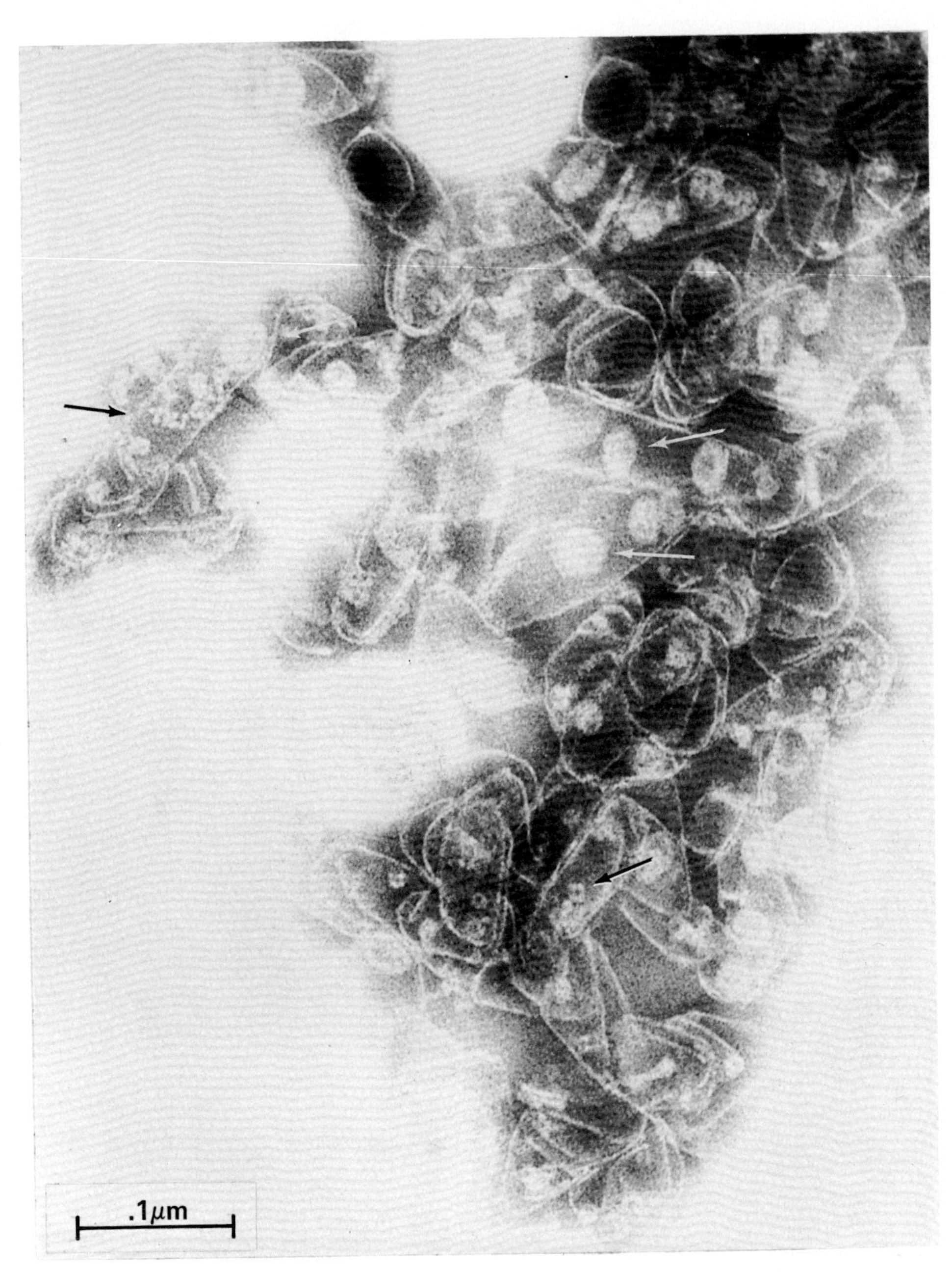

Fig. 2. A T7 polycapsid from an 8 *am* lysate. An 8 *am* lysate was prepared as described in Methods. A clarified carbowax precipitate of the lysate was negatively stained and observed by electron microscopy as described in Methods.

form the fiber observed in the electron microscope. If P16 forms the fiber, the rise per residue is 0.016 nm and a single protein molecule would produce a fiber thick enough to be observed by electron microscopy. A protein fiber with a rise per residue of 0.016 nm would, however, have to be supercoiled or folded.

The occurrence of a deficiency in capsid I assembly when bacteriophage T7 core protein-deficient amber mutants infect a nonpermissive host indicates that the core assists the assembly of the capsid I envelope. Because the cylindrical part of the core contacts the envelope at only one point, parts of the envelope distal to this contact point must interact with the core either a) indirectly, using a signal propagated from the core along the capsid envelope, or b) directly, by contact with the fibrous extensions of the core. The data presented here indicate that possibility (b) occurs; possibility (a) may also occur. The fibers might assist capsid I envelope assembly by binding envelope subunits at a unique distance from some point within the core; this point eventually becomes the envelope center. Testing of this hypothesis, and development of more detailed hypotheses concerning the dynamics of capsid I envelope assembly, require a more extensive knowledge of the structure of the capsid I interior, and the isolation of partially assembled capsid I precursors.

ACKNOWLEDGMENTS

The technical assistance of Miriam H. Mora is acknowledged. For provision of electron microscope facilities, I thank Dr. Edward G. Rennels, Department of Anatomy, The University of Texas Health Science Center, San Antonio. This study was supported by National Institutes of Health grants SO 1 RR 05654 and GM-24365-01.

REFERENCES

1. Caspar DLD, Klug A: Cold Spring Harbor Symp Quan Biol 33:1, 1962.
2. Harrison SC: Trends Biochem Sci 3:3, 1978.
3. Hohn T, Hohn B: Adv Virus Res 16:43, 1970.
4. Bancroft JB: Adv Virus Res 16:99, 1970.
5. Knolle P, Hohn T: In Zinder ND (ed): "RNA Phages." Cold Spring Harbor, New York: Cold Spring Harbor Laboratory, 1975, pp 147–201.
6. Ray P, Murialdo H: Virology 64:247, 1975.
7. Zachary A, Simon LO, Litwin S: Virology 72:429, 1976.
8. Botstein D, Waddell CH, King J: J Mol Biol 80:669, 1973.
9. Steven AC, Aebi U, Showe MK: J Mol Biol 102:373, 1976.
10. Roeder GS, Sadowski PD: Virology 76:263, 1977.
11. Casjens S, King J: Annu Rev Biochem 44:555, 1975.
12. Serwer P: J Mol Biol 107:271, 1976.
13. Studier FW: Virology 39:562, 1969.
14. Oliver RM: In Hirs CHW, Timasheff SN (eds): "Methods in Enzymology." New York: Academic, 1973, vol 27, pp 616–672.
15. Studier FW: Science 176:367, 1972.
16. Serwer P: J Ultrastruct Res 58:235, 1977.

Journal of Supramolecular Structure 11:327–338 (1979)
Biological Recognition and Assembly 79–90

Pyrenesulfonyl Azide as a Fluorescent Label for the Study of Protein-Lipid Boundaries of Acetylcholine Receptors in Membranes

J. M. Gonzalez-Ros, P. Calvo-Fernandez, V. Šator, and M. Martinez-Carrion

Department of Biochemistry, Medical College of Virginia, Virginia Commonwealth University, Richmond, Virginia 23298

Acetylcholine receptor (AcChR) enriched membrane fragments from Torpedo californica electroplax were labeled by in situ photogenerated nitrenes from a hydrophobic fluorescent probe, pyrene-1-sulfonyl azide. Preferential photolabeling of membrane proteins, mainly AcChR, has been achieved and there is a pronounced exposure of the 48,000 and 55,000 molecular weight subunits of AcChR to the lipid environment of the membrane core.

Covalent attachment of the photogenerated fluorescence probe does not perturb the α-neurotoxins' binding properties of membrane-bound AcChR or the desensitization kinetics induced by prolonged exposures to cholinergic agonists. Non-covalent photoproducts can be conveniently removed from labeled membrane preparations by exchange into lipid vesicles prepared from electroplax membrane lipids. Fluorescence features of model pyrene sulfonyl amide derivatives, such as fine vibrational structure of emission spectra or fluorescence lifetimes, are highly sensitive to the solvent milieu. The covalently bound probe shows similar fluorescence properties in situ. PySA photoproducts have great potential to spectroscopically monitor neurotransmitter induced events on selected AcChR subunits exposed to the hydrophobic environment of membranes.

Key words: AcChR-enriched membranes, pyrenesulfonyl azide, fluorescent probes, photolabeling

Assignments of functional role(s) or topographical features of AcChR protein subunits have usually been limited to the 40,000 molecular weight subunit, which is implicated in processes involving ligand interactions with water exposed AcChR segments [1–5]. In addition, the very nature of the *affinity* labels employed in those experiments does not allow the possibility of their use in monitoring possible ligand-induced effects on the receptor protein.

Abbreviations used are: AcChR, acetylcholine receptor; PySA, pyrene-1-sulfonyl azide; α-Bgt, α-bungarotoxin; PySAH, N-(1-pyrenesulfonyl)hexadecylamine; BSA, bovine serum albumin; PMSF, phenylmethylsulfonyl fluoride; carb, carbamyl choline; SDS, sodium dodecyl sulfate.

Received April 3, 1979; accepted June 25, 1979.

0091-7419/79/1103-0327$02.30

On the other hand, an intact membrane environment plays an important role in maintaining the AcChR functional state. Minor perturbations of the membrane components produce AcChR which act as in a pharmacologically desensitized-like high affinity state for cholinergic agonists [6–9]. This lack of manipulative tools has hampered attempts to identify possible structural changes in the receptor assembly which may be implicated in the AcChR function.

We have previously reported [10] the advantages of pyrene as an adequate fluorescent probe that can be introduced, with minimum perturbation, into the lipid phase of AcChR-rich membranes. In this paper, we describe a fluorescent pyrene derivative, PySA, that shows great promise as a non-perturbing probe that can photogenerate covalent labels with the potential to monitor spectroscopically ligand-induced events at the strategic lipid-receptor boundary area.

METHODS

Torpedo californica electroplax was purchased at Pacific Biomarines Supply Co. and AcChR-enriched membrane fragments in Ca^{++} free-Ringer solution were obtained according to the procedures of Duguid and Raftery [11] and Lee et al [12].

Specific activities of the preparations (between 1 and 2 nmoles of α-Bgt binding sites/mg protein) were determined by a DEAE-cellulose filter disc assay procedure [13] using $[^{125}I]$ α-Bgt prepared from α-Bgt purified from Bungarus multicinctus venom (Sigma Chemical Co.) [14] and iodinated by the solid phase Enzymobead (glucose-oxidase and lactoperoxidase) method (Bio Rad Laboratories). Protein concentrations were determined by the method of Lowry et al [15]. PySA and PySAH were purchased from Molecular Probes and $[^3H]$PySA was obtained by titration by the Whilzbach method (by ICN Corp.) and purified by TLC. The purified material shows no presence of impurities, as revealed by TLC using mixtures of chloroform/methanol (10:1, 5:1, and 1:1, by vol) or n-hexane/diethylether/acetic acid/methanol (60:40:1:1, by vol) as solvent systems. The specific activity of $[^3H]$PySA was 8 Ci/mole. All handling of PySA was performed in the dark and glassware covered with aluminum foil.

Cobratoxin was purified from Naja naja siamensis venom (Miami Serpentarium) [16] and used to prepare cobratoxin-sepharose affinity gel [17].

Absorption spectra were measured in a Cary model 15 spectrophotometer. Emission spectra and fluorescence lifetimes were measured at 20°C in an SLM model MC 320 subnanosecond fluorimeter after deaeration of the samples by bubbling through pure argon during 10 minutes.

Membrane Fragments Labeling Procedure

AcChR-enriched membrane fragments were suspended in Ca^{++} free-Torpedo Ringer solution (5 mM Tris buffer, pH 7.4, NaCl 262 mM, KCl 5 mM, NaN_3 0.02% and PMSF 0.1 mM) to give a final concentration of 8 mg protein/ml and transferred into a tube containing PySA-coated glass beads. The mixture was stirred until the radioactivity per ml of suspension remained constant (∿ 1 h). Final concentration of PySA in the range of 10^{-3} M. The suspension was flushed with nitrogen during 10 minutes, placed in a thermostated cell holder (20°C) and irradiated with UV light (Mineral light UVS 58, Ultraviolet Products Inc.) during 13 minutes under continuous stirring. After irradiation, the labeled AcChR membrane preparation was sedimented by centrifugation at 27,000 × g during 10 minutes and the supernatant was discarded. The pellet was resuspended twice in Ca^{++} free-Ringer buffer containing 10 mg/ml BSA and recentrifuged. BSA remaining in the suspension was removed by washing twice with Ca^{++} free-Ringer buffer and, finally, resuspended to give a protein concentration of about 1 mg/ml. A portion of this membrane preparation was disrupted in 10 mM sodium phosphate buffer, pH 7.4, 3 mM EDTA, 0.1 mM PMSF, 0.02% NaN_3, containing 1% Triton X-100, to obtain labeled purified receptor by cobratoxin-affinity chromatography. The AcChR was displaced from the cobratoxin-sepharose 4B affinity gel by overnight incubation with 1 M carb solution in 10 mM sodium phosphate buffer, pH 7.4, containing 0.1 M NaCl, 1 mM EDTA, 0.01 mM PMSF, 0.02% NaN_3 and 0.1% Triton X-100. The specific activity of the labeled purified receptor proceeding from disrupted membranes was 1.5–2.0 $\times$ 10^5 cpm/mg protein. Distribution of the radioactive label on the AcChR subunits was measured by scintillation counting after SDS polyacrylamide disc gel (7.5%) electrophoresis [18] using N,N-diallyltartardiamide as polyacrylamide cross-linking agent [19] to allow for easy dissolution of gel slices (1 mm thick) in 2% periodic acid, and efficient counting. The position of protein bands in the gels was elucidated after staining with Coomassie blue. The molecular weights were determined by comparison of electrophoretic mobility with protein standards.

Time Courses of [^{125}I] α-Bgt Binding to AcChR-Enriched Membrane Fragments

Time dependence of [^{125}I] α-Bgt binding to AcChR-rich membranes was followed basically according to Quast et al [20]. In all assays the concentration of AcChR was 10^{-7} M and [^{125}I] α-Bgt concentration was 2.5 $\times$ 10^{-7} M. Exposures of 30 minutes to carb (10^{-6} M) prior to the addition of [^{125}I] α-Bgt were performed. Upon 40-fold dilution of carb containing sample, the rate of α-Bgt binding reverts, within a period of 15 minutes, to that observed in the absence of carb pretreatment.

Lipid Extraction and Fractionation

Total lipids from PySA-labeled or native AcChR-rich membranes were extracted by the Bligh and Dyer procedure [21]. Lipid classes were fractionated by TLC on 0.25 mm thick layers of silica gel G using n-hexane/diethylether/acetic acid/methanol (60:40:1:1, by vol) as solvent systems. Lipid spots were visualized with iodine vapors and identified by comparison with standards. Relevant gel zones were scraped off the plate and radioactivity was determined.

Removal of Non-Covalently Bound PySA Photoproducts From Labeled Membrane Preparations

Non-covalently bound photoproducts contained in labeled AcChR-rich membrane fragments were partially removed by exchange with lipid vesicles prepared from AcChR-rich membrane fragments total lipid extracts. Total lipids were previously dissolved in benzene and evaporated to dryness. Deaerated Ca^{++} free-Ringer buffer, containing 2 mM MgCl, was added to give a final proportion of 8 mg total lipids/ml Ringer and the mixture was sonicated at 20°C, under nitrogen, in a Sonifier Cell Disruptor model W 140 (Heat Systems-Ultrasonics Inc.) for 6 periods of 3 minutes. The lipid suspension was then centrifuged at 800 $\times$ g for 5 minutes to eliminate possible titanium particles from the sonicator probe. A certain volume of lipid vesicles was added to the labeled AcChR-rich membranes preparation in such a way that the ratio between lipids in the vesicles/lipids in the membranes preparation was approximately 2:1. The mixture was incubated at room temperature for 2 h, with gentle stirring, layered on top of 10–20 ml 5 mM Tris buffer, pH 7.4, containing 0.5 M sucrose and centrifuged 1 h at 35,000 $\times$ g. The pellet was resuspended in Ca^{++} free-Ringer buffer and radioactivity contents, protein concentration, and α-Bgt binding activity were assayed in this fraction and in the supernatant.

RESULTS

Properties of PySA and Derivatives

PySA is a hydrophobic fluorescence compound, soluble in organic solvents and with poor water solubility (10^{-5} M).

Irradiation of PySA with long wavelength (>300 nm) UV light in solutions containing BSA (10 mM sodium phosphate buffer, pH 7.4, 0.03% Triton X-100, 0.02% NaN_3 and BSA 3 mg/ml) produces reactive nitrenes. Photoproducts' appearance is easily followed by measuring absorbance at 347 nm, which is 4-fold increased during the process (Fig. 1). On the other hand, when PySA is incorporated into membrane fragments, absorbance changes associated to photolysis are hard to follow because of the high degree of light scattering. For this reason, the optimum irradiation time is monitored using an AcChR rich membrane fragments suspension in front of the sample containing the BSA solution to filter light in a manner similar to that when PySA incorporated into membrane fragments is directly irradiated (Fig. 1B). Under these conditions, the optimum irradiation time is about 15

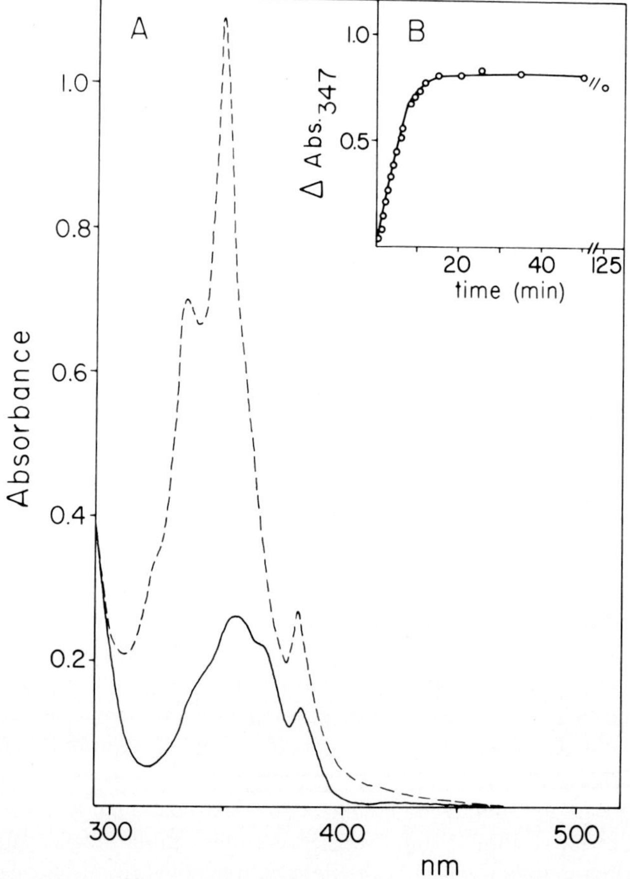

Fig. 1A) Absorption spectra of PySA before (——) and after (· · ·) irradiation. PySA (3×10^{-5} M) in 10 mM sodium phosphate buffer, pH 7.4, containing 0.03% Triton X-100, 0.02% NaN_3 and 3 mg/ml BSA, was irradiated at 20°C with long wavelength ultraviolet light. B) Time dependence of PySA photolysis when it was irradiated through a 1 pathlength quartz cuvette containing a suspension of AcChR-enriched membrane fragments (1.6 mg protein/ml). Experimental conditions as in Figure 1A.

minutes and the resulting photoproducts appear to be photostable up to 2 h of additional irradiation.

Figure 2A shows the emission fluorescence spectrum of PySA photoproducts resulting by photolysis when incorporated into AcChR-enriched membrane fragments. Upon excitation at 346 nm, the compound, probably a blend of pyrenesulfonyl amide derivatives, shows fluorescence maxima at 379, 398, and 416 nm. Fluorescence of the parent compound cannot be observed because of the rapid formation of photoproducts under the light source of the fluorimeter.

Figure 2B shows the solvent dependence of the emission fluorescence spectra of a pyrenesulfonylamide derivative, PySAH. Upon excitation at 346 (the absorption maximum of PySAH in the solvents assayed), the fluorescence maxima are at similar wavelengths to those observed for emission maxima of PySA photoproducts in the membrane preparations. On the other hand, the fluorescence emission intensities depend on the nature of the solvent environment (Figs. 2 and 3). In addition, other fluorescence parameters, such as the lifetime of the excited state, undergo dramatic changes dependent on the solvent (Table I).

PySA Labeling of the Components of AcChR-Enriched Membrane Fragments

The removal of possible non-membrane incorporated PySA photoproducts was performed by washing with Ca^{++} free-Ringer buffer or Ca^{++} free-Ringer buffer containing BSA as a scavenger. About 3–4% of the total radioactivity was present in the first Ringer buffer

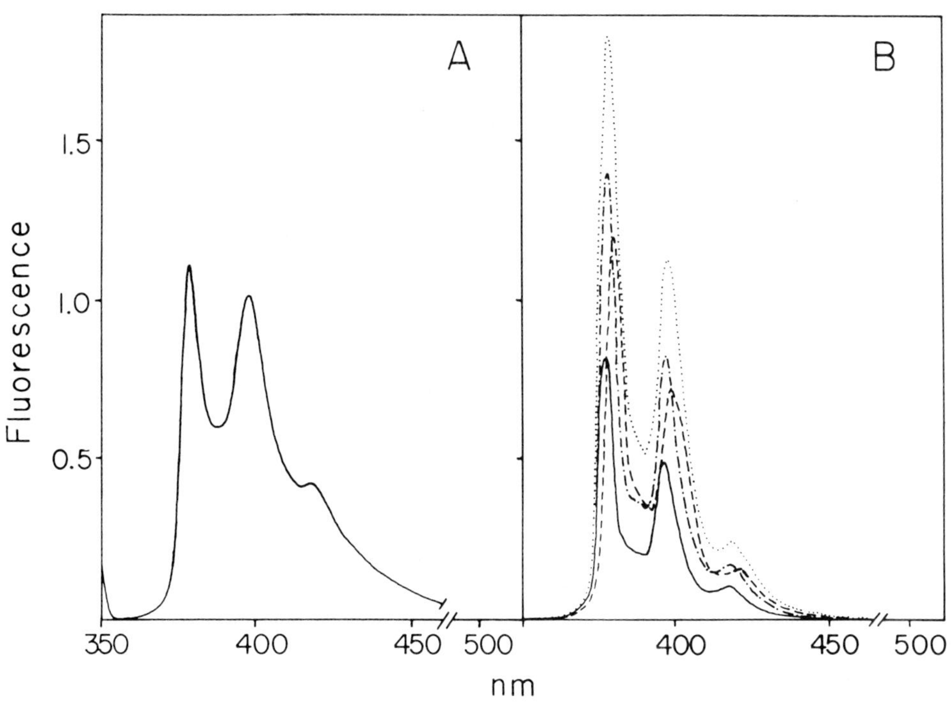

Fig. 2A) Fluorescence emission spectrum of PySA (10^6 cpm/mg protein) in AcChR-enriched membrane fragments (1.94 mg protein/ml) after irradiation with long wavelength ultraviolet light for 13 minutes. 100 μl of the membrane suspension was diluted to 1.5 ml in 5 mM Tris buffer, pH 7.4, containing 0.5 M sucrose and 0.02% NaN_3. Samples were purged with argon and fluorescence measured. Excitation at 346 nm. B) Fluorescence emission spectra of PySAH (2×10^{-6} M) in spectra grade hexane (——), chloroform (- - -), cyclohexane (-·-·-), and methanol (· · ·). Excitation at 346 nm.

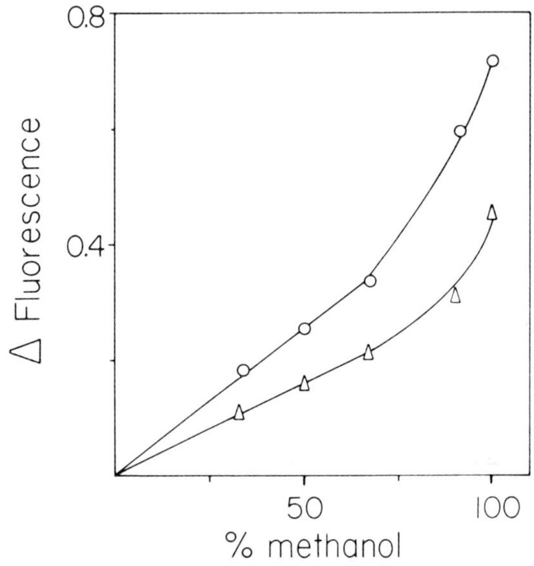

Fig. 3. Variation of fluorescence intensity of PySAH (2×10^{-6} M) at 337 (○) and 396 (△) nm in chloroform/methanol mixtures at different proportions.

TABLE I. Fluorescence Lifetimes Values of PySAH (2×10^{-6} M) in Different Solvents

	Florescence lifetime (nsec)[a]
Methanol	14.54 ± 0.03[b]
Chloroform/methanol (1:9, v/v)	14.15 ± 0.05
Chloroform/methanol (1:2, v/v)	13.21 ± 0.09
Chloroform/methanol (1:1, v/v)	12.15 ± 0.04
Chloroform/methanol (2:1, v/v)	10.91 ± 0.07
Chloroform	9.28 ± 0.06
Hexane	12.36 ± 0.09
Cyclohexane	18.84 ± 0.15

[a]Emission filters: Corning 4-96 and 3-144.
[b]Standard deviations from 4 different determinations.

supernatant and amounts no greater than 10–12% were removed in the 3 BSA washes. The extent of the removal was not altered if BSA was present during the incorporation of PySA into membranes or if the membranes were washed with Ca^{++} free-Ringer buffer containing BSA before or after the irradiation, which indicates the lack of partition of PySA or its photoproducts into water solvent and membranes.

About 90% of the membrane incorporated PySA photoproducts are extracted with the lipid components of the membrane fragments. This association, however, is by non-covalent interaction of the PySA photoproducts with the lipid core, since about 95% of these lipid-associated photoproducts are not bound to any of the lipid classes extracted. Comparison of the radioactivity distribution in TLC plates of lipid extracts from labeled AcChR-enriched membranes and a standard of PySA photolyzed in water shows that non-bound PySA photoproducts produce several spots on the TLC plates. Only a small amount

(∿5%) of the total PySA photoproducts present in the lipid extracts shows any appearance of being lipid-bound PySA photoproducts as they comigrate with cholesterol esters.

About 10% of the membrane-contained PySA photoproducts are detected in the protein fraction that can be precipitated by treatment of the Triton X-100 solubilized membrane fragments with trichloroacetic acid. AcChR protein was isolated by cobratoxin-affinity chromatography from Triton X-100 solubilized labeled membrane fragments and it contains (in terms of specific activity per milligram protein) most of the protein-bound PySA photoproducts. The molar ratio of bound PySA photoproducts to AcChR under the conditions described in the Methods section is about 27 ± 7.

Figure 4 shows the distribution of PySA photolabeling on the AcChR protein subunits determined after their separation by SDS-gel electrophoresis. The majority of the label is associated with the 48,000 and 55,000 molecular weight subunits, whereas the 40,000 and 68,000 subunits show practically no PySA photoproducts attachment. Different initial concentrations of PySA were assayed (10^{-5} to 5×10^{-3} M) without any substantial modification of the qualitative labeling pattern. In addition, control experiments in which non-irradiated or previously photolyzed (in buffer) PySA was used showed no radioactivity associated with protein bands on the SDS gels; instead, the radioactivity traveled with the dye front. The electrophoretic pattern of AcChR subunits (protein stain) is not modified upon photolabeling, which indicates that no apparent loss of polypeptide material occurred as a result of cross-linking or any other kind of polymerization or hydrolysis.

Removal of Non-Covalently Bound PySA Photoproducts

Removal of non-covalently bound PySA photoproducts contained in labeled AcChR-enriched membrane preparations was attempted in order to increase the relative proportion of AcChR covalently bound label to free PySA photoproduct in the membrane. Figure 5 shows that up to 70% of the radioactivity due to the presence of non-protein-bound tritiated PySA photoproducts can be removed from the electroplax PySA-labeled membranes. Total lipid extracts from AcChR-enriched membrane fragments, instead of commercially available lipid mixtures, as well as short periods of incubation time, were used in order to avoid the possibility of lipid composition modification by exchange between the labeled membrane fragments and the "acceptor" lipid suspensions. Under these conditions (Fig. 5) the resulting "clean" membranes retain most of the initially present total amount of proteins as well as both the specific activity (in molar terms of α-Bgt binding sites) and susceptibility to desensitization by a cholinergic agonist.

Binding Properties of PySA-Labeled AcChR-Enriched Membrane Fragments

The covalent attachment of PySA photoproducts to AcChR by photolysis within AcChR-enriched membrane fragments does not affect the extent of [^{125}I] α-Bgt binding; therefore, the specific activity of the membrane preparations in molar terms of α-Bgt binding sites remains substantially unchanged.

In addition, the kinetics of transitions from low to high carb affinity states of AcChR-enriched membrane fragments [20, 22, 23] are not modified by covalent labeling with PySA nitrene. In this regard the modified membranes display a kinetic behavior similar to the well-known "desensitization" kinetics of freshly prepared (control) membrane preparations.

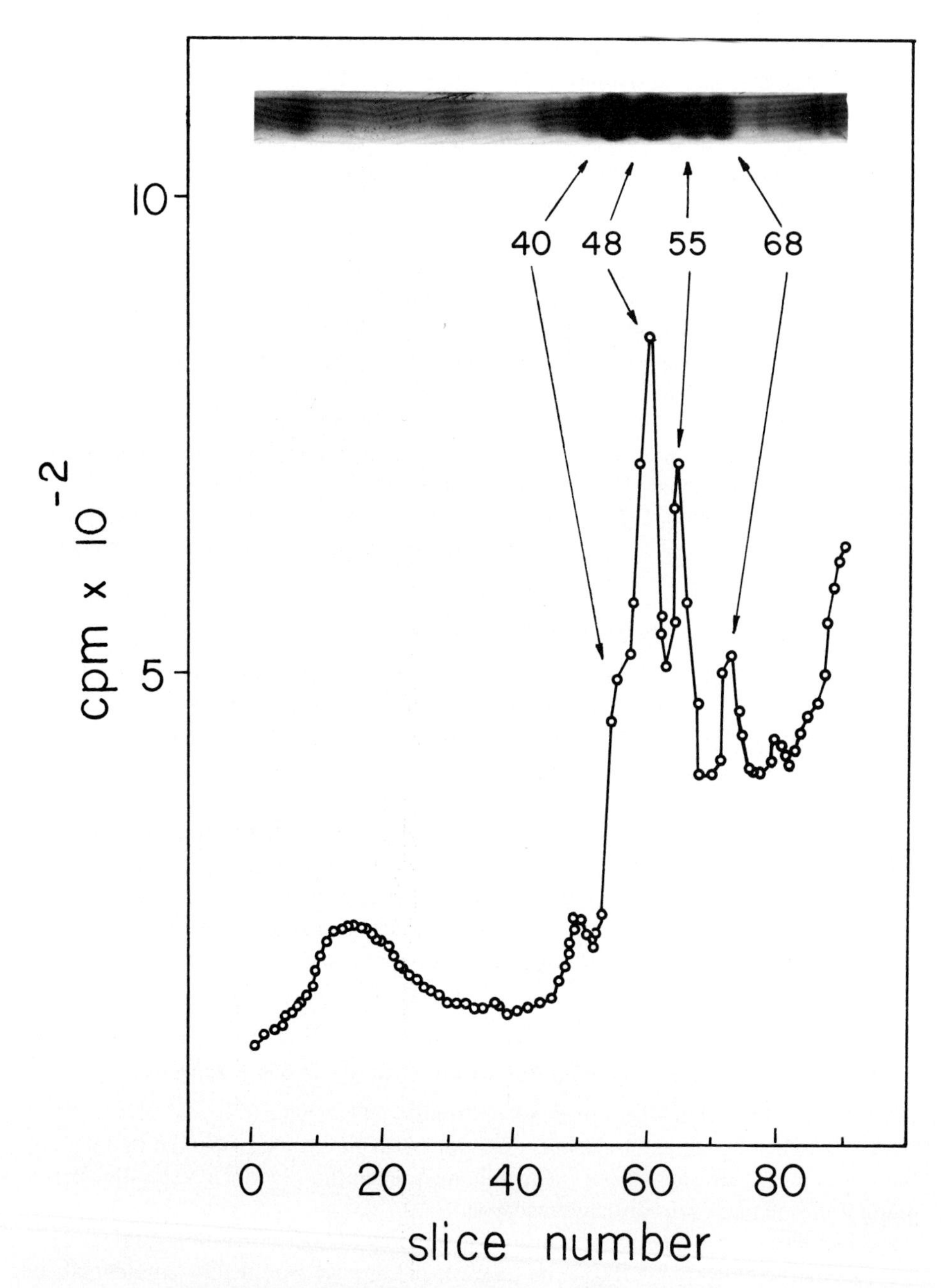

Fig. 4. Distribution of radioactivity in AcChR purified from PySA labeled AcChR-enriched membrane fragments (see Methods). The protein electrophoretic pattern is shown in the upper part of the figure. Numbers within the figure indicate molecular weight values in thousands.

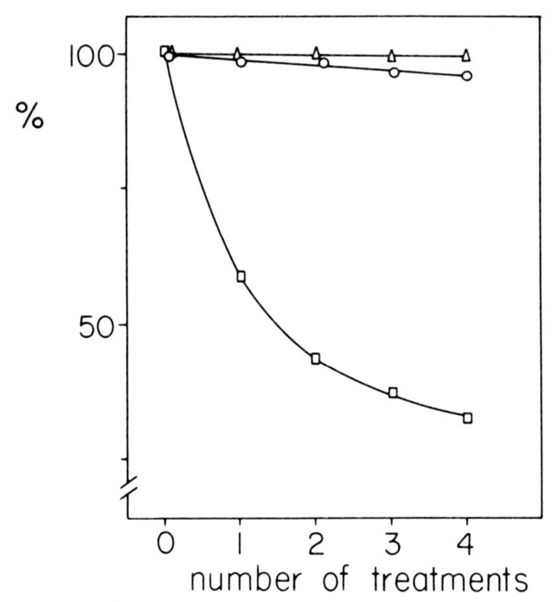

Fig. 5. Removal of unbound PySA photoproducts from labeled AcChR-enriched membrane fragments. Membranes were treated with lipid vesicles as many times as indicated in the figure (see Methods) and protein concentration (○), specific activity of the preparation in molar terms of α-Bgt binding sites per milligram of protein (△), and radioactivity contents (cpm/mg of protein) due to the presence of radioactive PySA within the membranes (□), were determined after every treatment. Data expressed as percentages of initial values.

DISCUSSION

Four different types of polypeptide chains have been previously detected in solubilized and purified AcChR and in highly enriched membrane preparations from Torpedo californica electroplax [1, 2, 24–27]. Nevertheless, the functional roles and structural arrangement of these protein subunits in the quaternary assembly of the AcChR within the membrane core is poorly understood. Only a partial assignment of the functions regarding the binding of some ligands to the 40,000 molecular weight subunit located in the external side of the membrane has been possible [3–5, 28]. Furthermore, selective iodination by the lactoperoxidase method [29] shows that the 40,000, 50,000, and 60,000 molecular weight subunits are easily iodinated from the outside membrane surface.

In our attempt to determine the portions of this AcChR macromolecule exposed to the lipid core of AcChR-enriched membrane fragments, we utilize PySA, a photogenerating hydrophobic covalent probe without specific AcChR binding sites. This compound, like pyrene [10], apparently, and almost quantitatively, partitions into the hydrocarbon regions of membrane lipids. Once incorporated into the membrane, extensive washing with excess of BSA does not remove the incorporated probe, before or after conversion of PySA into its photoproducts. In addition, tetracaine, which is known to be a membrane perturbant with great accessibility to hydrocarbon regions of membranes [30, 31], and which quenches significantly the fluorescence of pyrene incorporated into the AcChR-rich membrane fragments [10], appears to be an equally efficient quencher of fluorescence of membrane incorporated PySA photoproducts [Gonzalez-Ros, in preparation].

In the covalent binding to AcChR-enriched membrane fragments, the light-generated nitrene seems to have a preference for proteins. In general, different hydrophobic photolabels such as those generating arylnitrenes or carbenes have shown variable efficiency by labeling both protein and lipid components of different membrane components [32–36]. In our case, although only a relatively small fraction (∿10%) inserts into proteins, it is the predominant, if not the only, covalent interaction with the membrane components, a fact that can be partially explained by the peculiar composition of these membrane fragments in which protein is the main component (60%). Furthermore, it is now apparent [37] that AcChR is the major, integral protein present in AcChR membrane fragments isolated from Torpedo electroplax. The most likely interaction of water-insoluble nitrene with membrane proteins is from the lipid core, where it is presumably photogenerated. Nitrenes, produced in the lipid core in the vicinity of (or on) the protein hydrophobic segments in contact with lipids, rapidly react with protein components. Therefore, the extensive labeling of 48,000 and 55,000 molecular weight subunits in the membrane fragments should be expected if these two subunits come into contact, or expand, the lipid bilayer. These two polypeptide chains are known to be the only ones devoid of carbohydrates within the AcChR molecule [38], which probably indicates their "buried" character within the lipid bilayer. On the other hand, the absence of PySA labeling on the 40,000 subunits agrees with the results obtained by others [3–5, 28] using water-soluble (as distinct from hydrophobic) affinity labels.

A troublesome consequence of experimental tampering with electroplax membrane is the frequent final isolation of membranes in a "desensitized" state as an unfortunate consequence of handling these AcChR-rich membranes in the presence of membrane perturbants [6–9, 23]. From this work, it is apparent that not all perturbations of membrane architecture necessarily lead to loss of cholinergic ligand-induced, low to high, affinity state transitions of the receptor. As in the case of introduction of pyrene into the membrane lipid core [10], in situ, covalent labeling, with PySA, of the AcChR subunits in contact with the lipid phase does not affect the affinity state transitions either. The question of why pyrene compounds do not affect the sensitization-desensitization affinity states while other perturbations of the membrane do remains unclear.

These non-perturbing effects of PySA labeling on the activity of AcChR-bound membranes in conjunction with the ability to remove unbound photoproducts from the membrane preparations by exchange with lipid suspensions, allows us to obtain a functionally "unperturbed" receptor membrane preparation enriched in AcChR with a covalently bound probe at the protein-lipid boundary area.

Ideally, spectroscopic probes of molecular events in a protein must a) be identified with at least general regions of the protein; b) be non-perturbing as to the biological functions of the protein; and c) possess highly sensitive spectral properties which will reflect on the nature of its microenvironment. From the above discussion, it is apparent that certain hydrophobic domains of AcChR can be covalently labeled by photoproducts of PySA meeting criteria a) and b). Our studies with the model compound PySAH are a good indication that similar compounds, to be created as covalent photoproducts, should show equivalent fluorescent behaviors. The sensitivity of these compounds' fluorescence emission and lifetime of the excited singlet state to solvent polarity are very promising as to their potential for use in the monitoring of the probes' environment. Of particular interest is their potential for the detection (through physical means) of the transformation of protein-lipid boundary area in the AcChR from the sensitized to desensitized states.

Ligand binding induced conformational transition(s) of AcChR has been proposed [39]. Bonner et al [40] and Barrantes [41] monitored slight changes in intrinsic fluorescence of the receptor during its association with various agonists, and Witzemann and Raftery [4] have observed an increase of the AcChR labeling with a water-soluble probe, ethidium azide, in the presende of cholinergic ligands. Postulated conformational transitions may not only involve water-exposed regions of the receptor molecule but may also encompass the subunits in contact with the hydrocarbon core of the membrane. PySA appears advantageous as a generator of a non-perturbing covalent fluorescence tool to monitor events at this most interesting boundary.

In conclusion, the nitrene, photogenerated from the lipophilic PySA, appears to be a useful probe to determine the exposure of receptor subunits to the hydrophobic core of AcChR-enriched membrane fragments, contributing to a better understanding of the arrangement of the receptor within the membrane environment. Furthermore, because of its non-perturbing effects and fluorescence properties, PySA shows promise as a spectroscopic tool to detect events at the strategic area of receptor-lipid interfaces on AcChR-enriched membrane fragments.

REFERENCES

1. Karlin A, Weill CL, McNamee MG, Valderrama R: Cold Spring Harbor Symp Quant Biol 40:193, 1975.
2. Hucho F, Layer P, Kiefer HR, Bandini G: Proc Natl Acad Sci USA 73:2624, 1976.
3. Witzemann V, Raftery MA: Biochemistry 16:5862, 1977.
4. Witzemann V, Raftery MA: Biochemistry 17:3599, 1978.
5. Hsu HPM, Raftery MA: Biochemistry 18:1862, 1979.
6. Brisson AD, Scandella CJ, Bienvenue A, Devaux PF, Cohen JB, Changeux JP: Proc Natl Acad Sci USA 72:1087, 1975.
7. Weiland G, Georgia B, Lappi S, Chignell CF, Taylor P: J Biol Chem 252:7648, 1977.
8. Andreasen TJ, McNamee MG: Biochem Biophys Res Commun 76:958, 1977.
9. Young AP, Brown FF, Halsey MJ, Sigman DS: Proc Natl Acad Sci USA 75:4563, 1978.
10. Sator V, Thomas JK, Raftery MA, Martinez-Carrion M: Arch Biochem Biophys 192:250, 1979.
11. Duguid JR, Raftery MA: Biochemistry 12:3693, 1973.
12. Lee T, Witzemann V, Schimerlik M, Raftery MA: Arch Biochem Biophys 183:57, 1977.
13. Schmidt J, Raftery MA: Anal Biochem 52:349, 1973.
14. Clark DG, Macmurchie DD, Elliot E, Wolcott RJ, Laudel AM, Raftery MA: Biochemistry 11:1963, 1972.
15. Lowry OH, Rosebrough NJ, Farr AL, Randall RJ: J Biol Chem 193:265, 1951.
16. Ong DE, Brady RN: Biochemistry 13:2822, 1974.
17. Moore WM, Brady RN: Biochim Biophys Acta 498:331, 1977.
18. Osborn M, Weber K: J Biol Chem 244:4406, 1969.
19. Anker HS: FEBS Lett 7:293, 1970.
20. Quast V, Schimerlik M, Lee T, Witzemann V, Blanchard S, Raftery MA: Biochemistry 17:2405, 1978.
21. Bligh EG, Dyer WJ: Can J Biochem Physiol 39:911, 1959.
22. Weber M, David-Pfeuty T, Changeux JP: Proc Natl Acad Sci USA 72:3443, 1975.
23. Weiland G, Georgia B, Wee VT, Chignell CF, Taylor P: Mol Pharmacol 12:1091, 1976.
24. Chang HW, Bock E: Biochemistry 16:4513, 1977.
25. Flanagan SD, Barondes SH, Taylor P: J Biol Chem 251:858, 1976.
26. Nickel E, Potter LT: Brain Res 57:508, 1973.
27. Raftery MA, Vandlen RL, Michaelson D, Bode J, Moody T, Chao Y, Reed K, Deutsch J, Duguid J: J Supramol Struct 2:582, 1974.
28. Weill CL, MacNamee MG, Karlin A: Biochem Biophys Res Commun 61:997, 1974.

29. Hartig PR, Raftery MA: Biochem Biophys Res Commun 78:16, 1977.
30. Koblin DD, Pace WD, Wang HH: Arch Biochem Biophys 171:176, 1975.
31. Martinez-Carrion M, Thomas JK, Raftery MA, Sator V: J Supramol Struct 4:373, 1975.
32. Nieva-Gomez D, Gennis RN: Proc Natl Acad Sci USA 74:1811, 1977.
33. Bayley H, Knowles JR: Biochemistry 17:2414, 1978.
34. Bayley H, Knowles JR: Biochemistry 17:2420, 1978.
35. Klip A, Gitler C: Biochem Biophys Res Commun 60:1155, 1974.
36. Bercovici T, Gitler C: Biochemistry 17:1484, 1978.
37. Neubig RR, Krodel EK, Boyd ND, Cohen JB, Proc Natl Acad Sci USA 76:690, 1979.
38. Vandlen RL, Wu WCS, Eisenach JC, Raftery MA: Biochemistry 18:1845, 1979.
39. Nachmansohn D: "Harvey Lectures, 1953/1954." New York: Academic, 1955, vol 49, p 57.
40. Bonner R, Barrantes FJ, Lovin TM: Nature 263:429, 1976.
41. Barrantes FJ: Biochem Biophys Res Commun 72:479, 1976.

Journal of Supramolecular Structure 11:339–347 (1979)
Biological Recognition and Assembly 91–99

Effect of Vanadate on Gill Cilia: Switching Mechanism in Ciliary Beat

Jacobo Wais-Steider and Peter Satir

Albert Einstein College of Medicine, Bronx, New York 10461

Lateral (L) cilia of freshwater mussel (Margaritana margaritifera and Elliptio complanatus) gills can be arrested in one of two unique positions. When treated with 12.5 mM $CaCl_2$ and 10^{-5} M A23187 they arrest in a "hands up" position, ie, pointing frontally. When treated with approximately 10 mM vanadate (V) they arrest in a "hands down" position, ie, pointing abfrontally. L-cilia treated with 12.5 mM $CaCl_2$ and 1 mM NaN_3 also arrest in a "hands down" position; substitution of 20 mM KCl and 1 mM NaN_3 causes cilia to move rapidly and simultaneously to a "hands up" position.

The observations suggest that there are two switching mechanisms for activation of active sliding in ciliary beat one at the end of the recovery stroke and the other at the end of the effective stroke; the first is inhibited by calcium and the second by vanadate or azide. This is consistent with a model of ciliary beating where microtubule doublet numbers 1, 2, 3, and 4 are active during the effective stroke while microtubule doublets numbers 6, 7, 8, and 9 are passive, and the converse occurs during the recovery stroke.

Key words: switch hypothesis, cilia, motility, vanadate, calcium, dynein

Following the identification by Cantley et al [1] of vanadium in the +5 oxidation state (vanadate) as a potent (Na, K)-ATPase inhibitor that does not affect sarcoplasmic reticulum Ca^{2+}-activated ATPase, mitochondrial ATPase (F1), or actomyosin ATPase [2], Gibbons et al [3] and others [4, 5] have shown that micromolar concentrations of vanadate inhibit the Mg^{2+}-activated dynein ATPase of cilia. Particularly, Gibbons et al [3] showed that vanadate inhibits the motility of detergent-treated sea urchin sperm reactivated with ATP. However, at much higher concentrations vanadate is ineffective in arresting untreated sperm, presumably because it does not penetrate the intact sperm cell membrane. Sale and Gibbons [6] showed that vanadate also inhibits the sliding of microtubule doublets of trypsin-treated axonemes.

In this paper we report on the effects of vanadate and other solutions on the motility of in vivo L-cilia of freshwater mussel gills. There are three main types of cilia on the frontal face of freshwater mussel gills, distinguished by their length and arrangement: the frontal (F), laterofrontal (LF), and lateral (L) cilia. There are about 200 cilia per L-cell;

Received May 23, 1979; accepted August 29, 1979.

0091-7419/79/1103-0339$02.00

these cilia beat with metachronal rhythm when perfused with 12.5 mM $CaCl_2$, producing the well-known metachronal waves (MW) with wavelengths of approximately 10 μm traveling along the epithelium at rates of several hundred micrometers per second.

We have found that millimolar concentrations of vanadate inhibit the motility of in vivo L-cilia of freshwater mussel gills. These cilia arrest in a "hands down" position, ie, pointing abfrontally. This is different from the position of L-cilia arrested with calcium and ionophore [7]. Satir found [7] that L-cilia treated with calcium and ionophore arrest in a "hands up" position, ie, pointing frontally. This arrest has been shown by Walter and Satir [8] to be due to an increase of Ca^{2+} around the ciliary axoneme. L-cilia also arrest in a "hands up" position in response to various types of physical and chemical stimulation [9, 10].

Based on the observation that L-cilia can be arrested in one of two unique positions, we suggest that there are two switch points for activation of active sliding in ciliary beat.

MATERIALS AND METHODS

Materials

Sodium orthovanadate (V) – probable formula, $Na_3VO_4 \cdot 16H_2O$ – and ionophore A23187 were obtained from Fisher Scientific Co. and Eli Lilly and Co., Indianapolis, respectively. Vanadate solutions were prepared in reliance on the probable formula weight and are approximate; the ionophore was prepared as a 10^{-3} M stock solution in 0.25% dimethyl formamide and 0.75% ethanol. Freshwater mussels (Margaritana margaritifera and Elliptio complanatus) were obtained from Connecticut Valley Biological Supply Co., Inc. Triton X-100, ATP disodium salt, and DL-dithiothreitol (DTT) were obtained from Sigma Chemical Co.

Methods

Gills of freshwater mussels were stripped while still in the animal; then they were excised and placed in 12.5 mM $CaCl_2$ and 1 mM Tris-HCl, pH 7.4. For experimentation, pieces of stripped gill were placed on microscope slides, where they were gently stretched and held in place. The slide with the specimen was then placed in a Petri dish containing the experimental solution. All solutions used here contained 1 mM Tris-HCl, buffered at pH 7.4. The pH remained unaltered during experimentation. Observations were made with a Zeiss microscope adjusted for bright-field illumination. For detailed examination of the specimens, pieces of gill were quick-fixed [11], embedded in Epon, and sectioned with glass knives to about 2 μm in thickness. The sections were observed and photographed in appropriate orientation under phase-contrast microscopy. The reactivation experiments utilized the following protocols: Incubation of gills in 35 mM KCl and 5 mM phosphate buffer (pH 7.0) at room temperature (25°C) for 3 h produced an en masse exfoliation of motile ciliated cells [8]. The cells were washed and resuspended in 50 mM KCl, 2 mM EGTA, 4 mM $MgSO_4$, 1 mM DTT, and 30 mM HEPES, pH 7.0 (wash solution), and stored in crushed ice. A 20-μl aliquot of cell suspension was mixed with 10 μl of detergent solution (wash solution containing 0.04% w/v Triton X-100 and stored at 0°C) and incubated at room temperature for 15–45 sec. Triton treatment was terminated by the addition of 200 μl of wash solution. This treatment rendered the ciliated cells motionless; perfusion of reactivation solution (wash solution containing 2 mM ATP) across these cells caused them to resume beating. Inhibitors were added to the reactivation solution as indicated.

RESULTS

L-cilia beat with metachronal rhythm with a mean beat frequency of about 17 Hz. Figure 1a shows an instantaneously fixed preparation that preserves the MW on both sides of the gill filaments. Satir [11] has analyzed such preparations and has indicated correspondence between the preserved wavelength and the ciliary beat. Each wavelength contains cilia captured in a variety of stroke positions; these repeat from wavelength to wavelength. Figure 2a is a transverse section through a gill filament where L-cilia were fixed in MW positions. Note that the appearances of the L-cilia on the two sides of the filament are different; actually, portions of several differently positioned cilia are sectioned in each half of the filament. In addition, in such a transverse orientation, different beat positions are also seen from filament to filament. LF and F cilia are also seen in Figure 2a. This appearance is in contrast with the position of L-cilia arrested with 12.5 mM $CaCl_2$ and 10^{-5} M A23187 [7]. In this case all L-cilia stop in a specific position. Figure 1b shows a random longitudinal section of an instantaneously fixed preparation of a gill filament treated with calcium and ionophore. The portions of the L-cilia shown in this figure form a dense mat, in which no MW positional variation can be seen. The section in Figure 1b is taken frontally to the position of the L-cells, suggesting that the cilia point in a frontal direction. This is better seen in Figure 2b, which is a transverse section of the same preparation as Figure 1b. The L-cilia in both sides of the filament are all in the same stroke position, a "hands up" position, corresponding to the end of the recovery stroke.

We have found that within 10 min following treatment of the excised gill epithelium with about 20 mM V, the MW also disappears and the L-cilia again appear to have stopped in one specific position. Figure 1c shows a random longitudinal section of an instantaneously fixed preparation from such an experiment. The L-cilia on both sides of the filament form a dense mat, indicating that the cilia have been stopped. However, this section is taken abfrontally to the L-cells, suggesting that the cilia point in an abfrontal direction, ie, opposite to Figure 1b. This is better seen in Figure 2c, where a random transverse section of a preparation of the same experiment shows that all L-cilia are arrested in the same position, a "hands down" position, corresponding to the end of the effective stroke.

We have been able to switch arrested L-cilia from one arrest position to another without restarting beat. Treatment of L-cilia with 5 mM $CaCl_2$ and 10^{-5} M A23187 causes the cilia to arrest in the "hands up" position, as shown in Figure 3a. Changing this solution to 5 mM $CaCl_2$, 10^{-5} M A23187, and 10 mM V causes the cilia to move and arrest in a neutral position, where the cilia are approximately straight, as shown in Figure 3b. Further replacement by 10 mM V alone causes the cilia to move and arrest in the "hands down" position, as shown in Figure 3c. Switching the arrested position of the L-cilia can also be accomplished in the reverse direction. Treatment of L-cilia with 10 mM V causes the cilia to arrest in a "hands down" position (Fig. 4a). Addition of Ca^{2+} and ionophore causes the cilia to move and arrest at a neutral position (Fig. 4b), where they are approximately straight, and complete substitution causes most of the L-cilia to move to a "hands up" position (Fig. 4c). Not all cilia move to a "hands up" position, presumably because the effect of the vanadate on L-cilia in vivo is not readily reversible when treatment with vanadate is continued for periods of time longer than 10 min.

We wanted to determine whether the movement of L-cilia from one arrested position to the other is energy-dependent. We attempted to do this by treating the gill with solutions containing azide, and the results were somewhat unexpected. Treatment of a gill with 12.5 mM $CaCl_2$ and 1 mM NaN_3 causes L-cilia to arrest in the "hands down" position

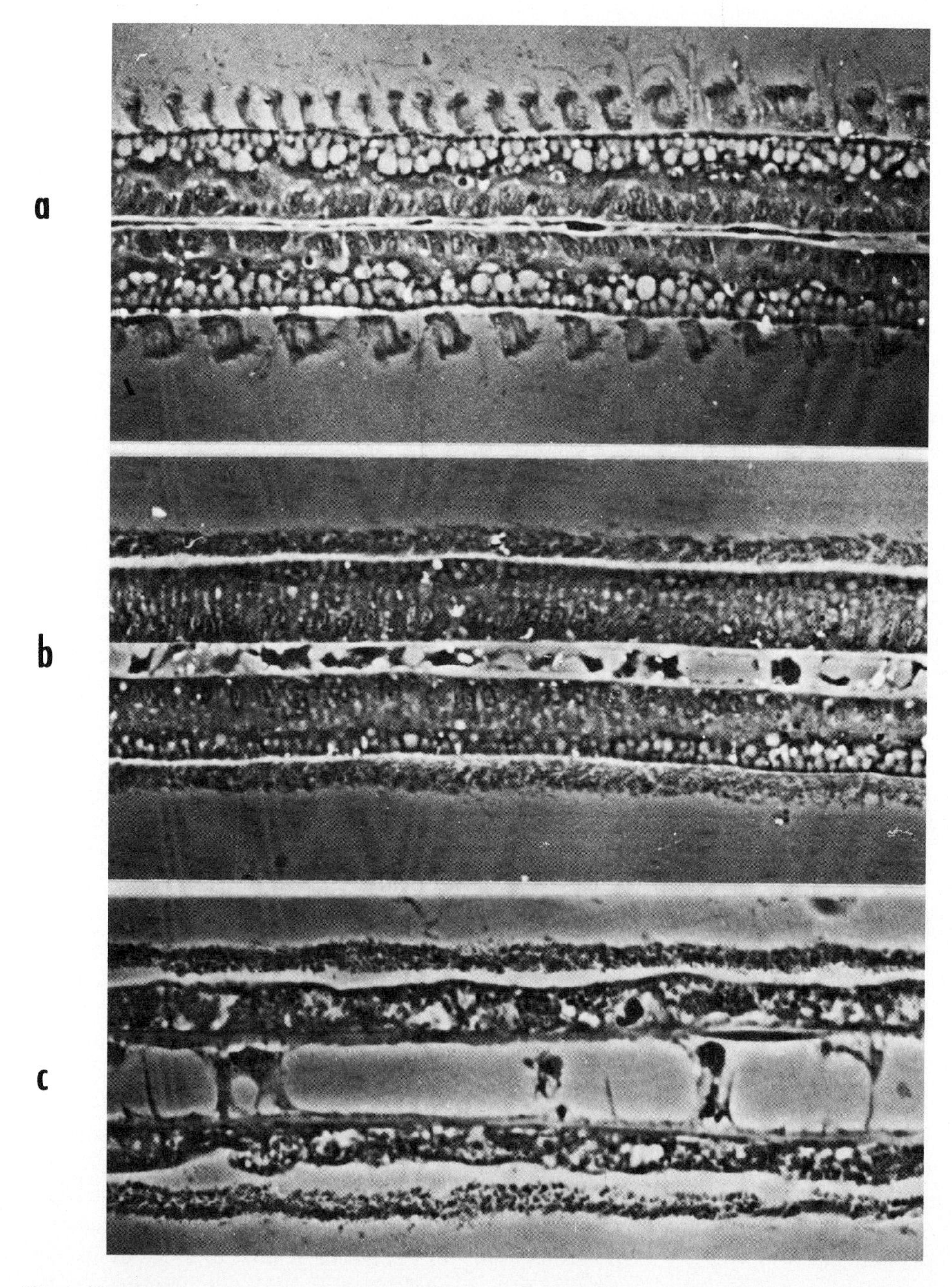

Fig. 1. Phase-contrast micrographs of longitudinal sections of gill filaments after quick fixation. a: Section through the metachronal wave showing L-cilia captured in a wide variety of beat positions; these repeat from wavelength to wavelength. b: Section through a gill filament treated with 12.5 mM $CaCl_2$ and 10^{-5}M A23187; L-cilia arrest and form a dense mat. This section is taken frontally to the position of the L-cells, indicating that the L-cilia point frontally. c: Section through a gill filament treated with 20 mM vanadate; L-cilia also arrest and form a dense mat, but this section is taken abfrontally to the L-cells, indicating that the cilia point abfrontally. × 670.

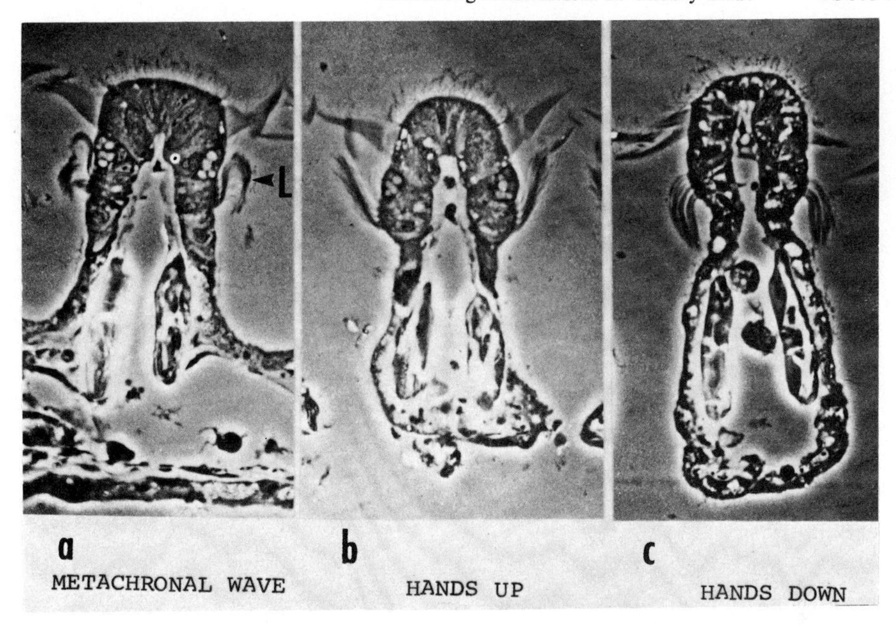

Fig. 2. Phase-contrast micrographs of transverse sections through gill filaments showing L-, LF-, and F-cilia after quick fixation. a: L-cilia fixed in metachronal wave position. b: L-cilia treated with 12.5 mM $CaCl_2$ and 10^{-5} M A23187; cilia arrest in a "hands up" position. c: L-cilia treated with 20 mM vanadate; cilia arrest in a "hands down" position. × 525.

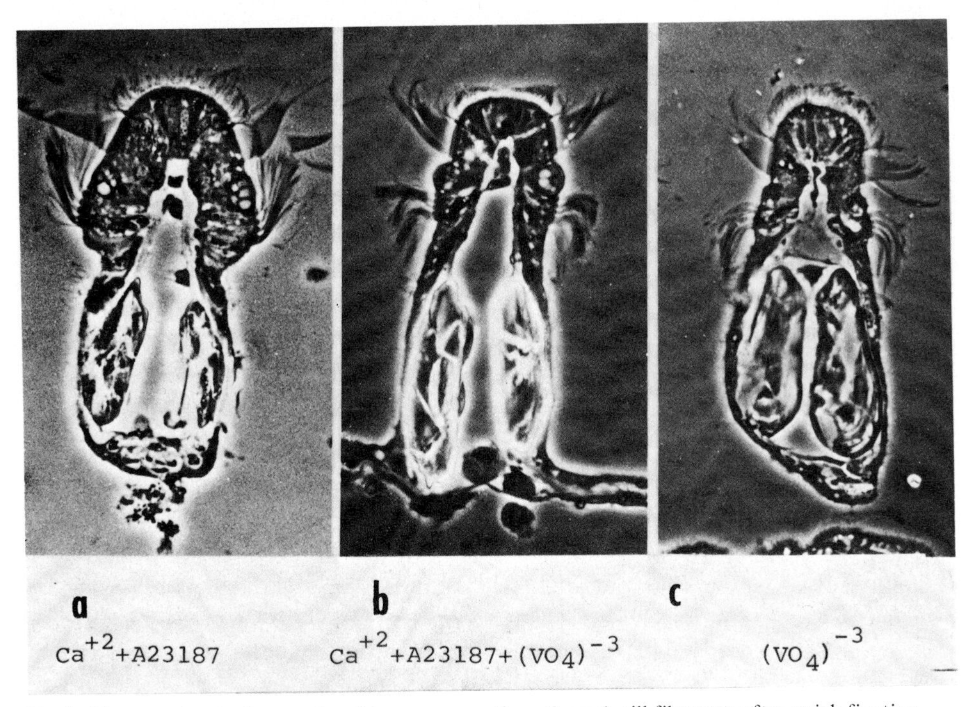

Fig. 3. Phase-contrast micrographs of transverse sections through gill filaments after quick fixation. a: L-cilia treated with 5 mM $CaCl_2$ and 10^{-5} M A23187 arrest in the "hands up" position. b: Addition of 10 mM vanadate to this solution causes cilia to move and arrest at a neutral position. c: Further replacement by 10 mM vanadate alone causes cilia to move and arrest in the "hands down" position. × 525.

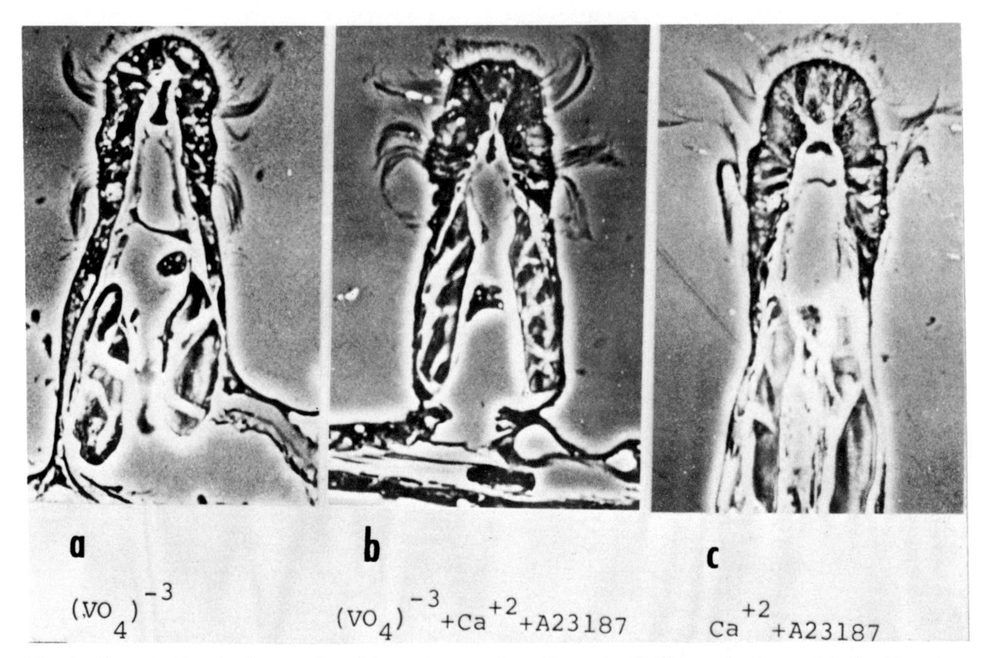

Fig. 4. Phase-contrast micrographs of transverse sections through gill filaments after quick fixation. a: L-cilia treated with 10 mM vanadate arrest in the "hands down" position. b: Addition of Ca^{2+} and ionophore to (a) causes the cilia to move and arrest at a neutral position. c: Complete substitution causes most of the cilia to move and arrest in the "hands up" position. × 525.

as shown in Figure 5a (ie, similarly to vanadate-induced arrest). This suggests that calcium has not entered the cilia. However, within 5 sec of a change of this solution to 20 mM KCl and 1 mM NaN_3 the cilia flick synchronously from the "hands down" position to the "hands up" position, as shown in Figure 5b. After 15 min in KCl and azide or 5 min after return to 12.5 mM $CaCl_2$ and 1 mM NaN_3, the cilia revert to a "hands down" position (Fig. 5c).

If, in the experiment described above, either 20 mM $MgSO_4$, 20 mM NaCl, 25 mM $CaCl_2$, 6 mM $CaCl_2$, or 40 mM sucrose is substituted for 20 mM KCl, essentially no flicking of cilia occurs. When a gill is treated with 1 mM NaN_3 solutions containing 20 mM $MgSO_4$, 20 mM NaCl, or 40 mM sucrose, instead of 12.5 mM $CaCl_2$, arrest is also "hands down," but no flicking of cilia occurs following substitution of these solutions by 20 mM KCl and 1 mM NaN_3. Thus it seems that azide is the agent responsible for "hands down" arrest in these experiments and that a rise in internal Ca^{2+} is needed to cause flicking of cilia. The latter assumption was tested by arresting cilia in a "hands down" position using 12.5 mM $CaCl_2$ and 1 mM NaN_3 and then adding 10^{-5} M A23187. In this experiment the L-cilia moved from a "hands down" position to a "hands up" position in an asynchronous manner and in a longer time (within 15 min) than when treated with 20 mM KCl. The addition of 1 mM $LaCl_3$, a known calcium channel blocking agent, to azide solutions containing Ca^{2+} and K^+ also abolishes flicking of the L-cilia.

We also studied the effect of vanadate and azide on reactivated L-cilia. While 10 mM V was necessary to arrest L-cilia of gills in vivo, reactivated (and presumably demembranated) L-cilia were arrested with 5 μM V. This arrest was readily reversible by washing in reactivation solution free of vanadate. We found that treatment of reactivated L-cilia directly with 1 mM NaN_3 had no effect on their motility. This suggests that the site of action of azide in the in vivo situation is not directly on the axoneme, but presumably is mitochondrial [12], such that respiratory ATP production is blocked or greatly reduced.

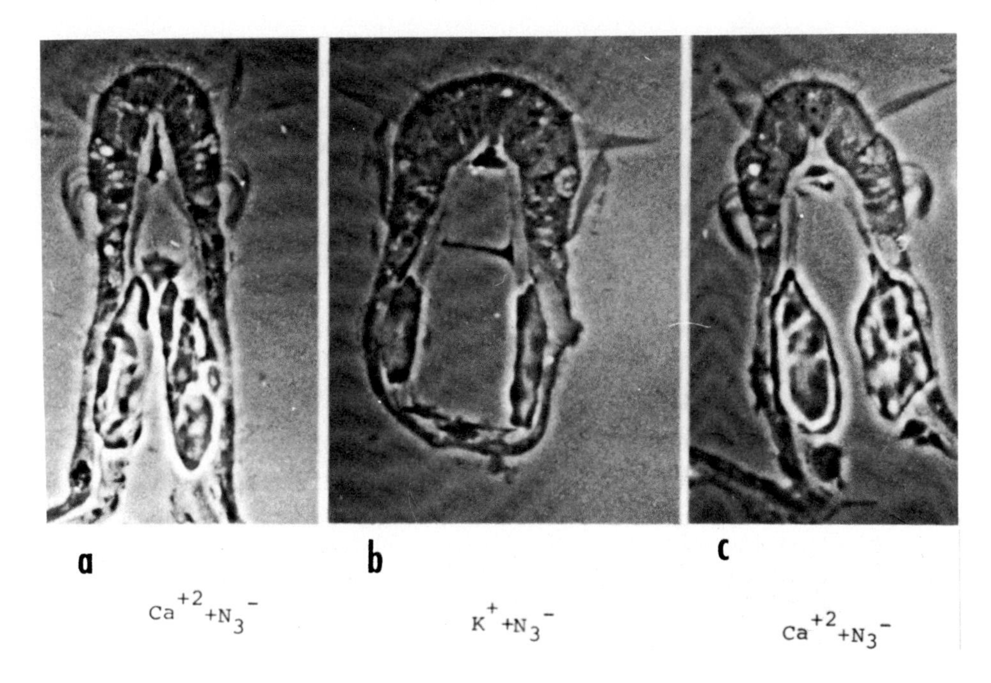

Fig. 5. Phase-contrast micrograph of transverse sections through gill filaments after quick fixation. a: L-cilia treated with 12.5 mM $CaCl_2$ with 1 mM NaN_3 arrest in a "hands down" position. b: Changing of solution to 20 mM KCl and 1 mM NaN_3 causes cilia to flick synchronously to a "hands up" position. c: Returning the gill filament to 12.5 mM $CaCl_2$ and 1 mM NaN_3 causes cilia to revert to a "hands down" position. $\times$ 525.

DISCUSSION

L-cilia move through a wide variety of stroke positions during a beat cycle. The observation that these cilia can be made to arrest using the appropriate solution at one of two unique positions, ie, a "hands up" and a "hands down" position, suggests that there are two metastable switch points in a beat cycle; one at the end of the recovery stroke and the other at the end of the effective stroke.

The first arrest position has been described previously [7] and is apparently dependent upon an increase of axonemal free Ca^{2+} to $> 8 \times 10^{-7}$ M [8]. In vivo, such an increase in cytoplasmic Ca^{2+} presumably depends upon depolarization of the cell membrane. This can also be experimentally produced by a divalent cation ionophore (A23187) and high external Ca^{2+}.

The second arrest position has not been described in the recent literature. We have succeeded in producing this arrest either with high external vanadate or with azide. The effects of vanadate and calcium on the motility of L-cilia are directly upon the axoneme. These substances inhibit when applied directly to reactivated cilia in low concentration. Applied externally, the substances must be present in thousand-fold or higher excess, presumably to reach the axonemal concentrations required for inhibition.

Gibbons et al [3] reported that micromolar concentrations of vanadate inhibited the motility of reactivated sea urchin sperm flagella and their Mg^{2+}-activated dynein ATPase. We have found that 5 μm V produces a reversible arrest of reactivated L-cilia. It is then reasonable to assume that the effect of vanadate on the gill cilia is due to its inhibitory effect on the dynein ATPase.

Arrest is produced by 1 mM azide independently of ionic content of the solutions tested, and even in the presence of high external calcium, this arrest is in the "hands down" position. It does not inhibit axonemal reactivation even at relatively high concentrations and therefore does not act directly on the axoneme. Insofar as azide interferes with ATP production it may also indirectly affect the activity of dynein ATPase, lowering sliding velocity to zero. This suggests that the "hands down" position is the passive (low-energy-requiring) ciliary arrest position. In addition, azide alone must be insufficient to open the cell membrane to Ca^{2+}. The presence of K^{+}, which presumably depolarizes the membrane, or the presence of A23187 produces a change in position (flicking) because Ca^{2+} can enter the cell. In the presence of $LaCl_3$, a Ca^{2+} channel blocking agent, or in the absence of sufficient external Ca^{2+}, the K^{+}-induced flicking is abolished.

In earlier experiments it was shown that Ca^{2+}-induced arrest was reversible [7]. Here, we have shown that both Ca^{2+}- and V-induced arrest depends on the continued presence of the specific inhibitors and that the arrest positions can readily be switched without restarting beat by substituting one inhibitor for the other. The vanadate-induced arrest is not as easily reversed, presumably because the cells cannot readily pump vanadate out of the axoneme. The presence of two ciliary arrest positions that may be interchanged without restarting beat may provide important clues about the mechanism of doublet microtubule coordination within the axoneme.

Sale and Satir [13] have provided evidence that in trypsin-treated axonemes the microtubule doublets slide actively past each other in only one direction. If this single polarity of active sliding applies to untreated axonemes, as seems likely, it must imply that not all doublets are active at the same time during ciliary beat. The geometry of ciliary bend indicates that during bend generation some of the doublets are sliding towards the tip and others towards the base: Thus, applying the conclusion of Sale and Satir [13] regarding the direction of sliding, we find that only doublets that slide towards the base have active dynein arms, while the others move passively. An analysis of the displacements of adjacent doublets at the end of the recovery and effective strokes by Satir and Sale [14] indicates that doublets numbers 1, 2, 3, and 4 are active while doublets numbers 6, 7, 8, and 9 are passive during the effective stroke. The converse must occur during the recovery stroke. This is illustrated in Figure 6. In L-cilia, doublet No. 5 is always seen attached to doublet 6 and thus is presumably passive at any time. This model predicts that there must be a point during the effective stroke where doublets 1, 2, 3, and 4 are switched off and doublets 6, 7, 8, and 9 are switched on. The converse happens at the end of the recovery stroke.

Our results indicate that the switch point at the end of the effective stroke is vanadate- or azide-sensitive and the switch at the end of the recovery stroke is calcium-sensitive. In the presence of azide or vanadate, L-cilia will not be able to switch from the effective stroke to the recovery stroke and will remain in a "hands down" position. Addition of calcium and ionophore to vanadate-arrested L-cilia causes the cilia to move from the "hands down" position to a neutral position. This suggests that calcium may antagonize the effect of vanadate and allow some of the microtubules that are "on" to be turned "off" and some other microtubules that are "off" to be turned "on." The resulting combination of partially active microtubules moves the cilium. Similarly, in the presence of calcium the L-cilia will not be able to switch from the recovery to the effective stroke and will remain in a "hands up" position. Addition of vanadate to calcium-arrested L-cilia again causes the cilia to move to a neutral position.

Azide, vanadate, and calcium have different effects on ciliary motility. The specificities of these substances suggest that the switch point mechanism at the end of the effective stroke is dependent on the dynein cycle, while the second switch is dependent on

Fig. 6. Diagram of a cross section of an L-cilium axoneme illustrating the mode of operation of doublet microtubules during ciliary beating, as proposed in this paper. In these axonemes doublet 5 is always seen attached to doublet 6 and is thus presumably passive at any time. The stippled doublets are active during the effective stroke, while those not stippled are passive. The activity pattern reverses during the recovery stroke.

a different mechanism, sensitive to calcium. Warner and Satir [15] have proposed that the active sliding of microtubule doublets is converted into local bending by the cyclic attachment and detachment of radial spokes extending from the A microtubule to the projections of the central sheath. This interaction may be the calcium-sensitive mechanism where the second switch point mechanism resides.

ACKNOWLEDGMENT

This work was supported by grant HL-22560 from the US Public Health Service.

REFERENCES

1. Cantley LC, Josephson L, Warner R, Yanagisawa M, Lechene C, Guidotti G: J Biol Chem 252: 7421, 1977.
2. Josephson L, Cantley LC: Biochemistry 16:4572, 1977.
3. Gibbons IR, Cosson MP, Evans JA, Gibbons BH, Houck B, Martinson KH, Sale WS, Tang WY: Proc Natl Acad Sci USA 75:2220, 1978.
4. Kobayashi T, Martensen T, Nath J, Flavin M: Biochem Biophys Res Commun 81:1313, 1978.
5. Martensen T, Kobayashi T, Nath J, Flavin M: Fed Proc, vol 37, p 2864, 1978 (Abstract).
6. Sale SW, Gibbons IR: J Cell Biol 82:291, 1979.
7. Satir P: Science 190:586, 1975.
8. Walter MF, Satir P: J Cell Biol 79:110, 1978.
9. Murakami M, Takahashi K: Nature 257:48, 1975.
10. Satir P, Reed W, Wolf DI: Nature 263:520, 1976.
11. Satir P: J Cell Biol 18:366, 1963.
12. Hochster RM, Quastel JH: "Metabolic Inhibitors." New York: Academic, 1963, p 395.
13. Sale WS, Satir P: Proc Natl Acad Sci USA 74:2045, 1977.
14. Satir P, Sale WS: J Protozool 24:498, 1977.
15. Warner FD, Satir P: J Cell Biol 63:35, 1974.

Journal of Supramolecular Structure 12:35–46 (1979)
Biological Recognition and Assembly 101–112

The Errors in Assembly of MuLV in Interferon Treated Cells

Paula M. Pitha, Bruce Fernie, Frank Maldarelli, and Nelson A. Wivel

Laboratory of Biochemical Virology, Johns Hopkins University Oncology Center, Baltimore, Maryland 21205 (P.M.P., B.F., F.M.), and the Laboratory of Cell Biology, National Cancer Institute, Bethesda, Maryland 20205 (N.A.W.)

Interferon treatment of JLSV-6 cells chronically infected with Rauscher MuLV leads to the formation of noninfectious particles (interferon virions) containing the structural proteins of env and gag genes as well as additional viral polypeptides. In the control virions the major glycoprotein detected is gp71, interferon virions contain in addition to gp71 and 85k dalton (gp85) glucosamine-containing, fucose-deficient glycoprotein which is recognized by antiserum to MuLV but not by the gp71 antiserum. The surface iodination of the intact virions indicates that both gp71 and gp85 are the major components of the external virions envelope. However, unlike the control virions in which gp71 associates with p15E (gp90), the gp71-p15E complex was not detected in interferon virions. The analysis of the iodinated proteins of the disrupted interferon virions revealed the presence of 85k and 65k dalton polypeptides preciptable with antiserum against MuLV, which are not present in the control virions. The difference in the polypeptide pattern of virions produced in the presence of interferon does not seem to be a consequence of the slowdown in the synthesis of viral proteins or their processing in the interferon-treated cells. Both the structural proteins of env and gag genes seem to be synthesized and processed at a comparable rate in the interferon-treated and -untreated cells. These results indicate an alteration of virus assembly in the presence of interferon.

Key words: **MuLV, uninfectious particles, interferon, virus assembly**

The several mechanisms by which interferon inhibits virus replication in an infected cell is through the impairment of mRNA translation [1–3]. In murine leukemia virus (MuLV) inhibition of viral replication occurs after the synthesis of viral RNA [4–7] and viral proteins [7–10] and seems to be mediated through changes in the cellular membrane [11, 12] and consequent interference with virus assembly and maturation [13–16]. Er-

Received for publication May 3, 1979; accepted August 16, 1979.

0091-7419/79/1201-0035$02.30

roneous assembly is reflected in some systems by the formation of noninfectious MuLV particles [5, 7, 17], and the virus particles accumulated on the cell surface of interferon-treated cells were shown to be less thermostable than the control virus [16]. In the AKR virus system, no noticeable morphologic difference was observed between the appearance of the particles from interferon-treated cells and controls [16], while interferon treatment of Friend erythroleukemia cells led to the formation of aberrant virus particles [18]. Virus particles produced in the presence or absence of interferon contain comparable amounts of 70S RNA [16], which indicates that the decrease in infectivity is not caused by the absence of the viral genome.

We now report that in the presence of interferon, MuLV is produced with abnormal structural proteins. By a combination of a high-resolution sodium dodecyl sulfate polyacrylamide gel electrophoresis (SDS PAGE) and immunoprecipitation with antisera against viral proteins of the gag and env genes, we identified distinct groups of viral polypeptides of both high and low molecular weight in virions produced in the presence of interferon, but not in virions produced in the absence of interferon. Under these conditions, however, the synthesis of viral proteins in the interferon-treated cells was not altered.

METHODS

Cells and Viruses

Rauscher leukemia virus was obtained from chronically infected JLSV-6 cells (derived from BALB/c mouse bone) at titers of 2×10^6 plaque-forming units (PFU) per milliliter, as assayed by the UV-XC test in SC-1 cells [19]. Cells were grown in minimal essential medium (MEM) with Earle's salts containing 10% fetal bovine serum (FBS) and antibiotics. Treatment of the cells with 150 units/ml of interferon reduced the amount of infectious virus 20-fold to 50-fold, while the amount of virus particles in the medium (quantitated by reverse transcriptase activity or by uridine labeling) decreased five-fold.

Radioisotopic Labeling of Virus Particles and Cells

Several different labeling procedures were utilized depending on subsequent analysis of the proteins. Confluent cells chronically infected with MuLV were labeled with ^{3}H leucine (15 h) in leucine-free medium containing 10 μCi/ml of ^{3}H leucine in the presence of 5% dialized serum. To label intracellular viral proteins cells were labeled with interferon for 24 h prior to labeling and then incubated in amino acid-free medium for 30 min, then labeled for 20 min with ^{3}H amino acid mixture (100 μCi/ml) in MEM. Radioactivity was removed, and cells were washed and incubated in MEM with 5% FBS and interferon for an additional 3 or 6 h. Cells were harvested by scraping, and radioactive proteins were analyzed before and after immunoprecipitation on SDS gels.

Virus Purification

The labeled virus was purified from the medium by a two-step gradient technique described previously [16]. In principle the harvested medium was clarified at 10,000g for 10 min and the supernatant was filtrated through a 0.4-μ Millipore filter to remove cell debris and banded at the interface of 20% (w/v) sucrose and 40% (w/v) potassium tartrate in standard buffer. The banded virus was then centrifuged to equilibrium on a continuous sucrose gradient (24–48%, w/v). Fractions containing the virus were diluted with standard buffer and virus was pelleted at 340,000g for 1 h.

The unlabeled virus used for subsequent iodination was purified by the Sepharose C14B chromatographic method of McGrath et al [21]. In this method gp71 is preserved in the virus particle to a higher degree than by centrifugation. This seems to be especially important in the purification of virus assembled in the presence of interferon, which is fragile and easily loses gp71 during the centrifugation (Pitha, unpublished). Briefly, 12-h collections of culture fluids were clarified, and cell debris was removed by filtration on 0.4-μ filters and concentrated with immersible molecular separators (Millipore Corp.), mixed with tracer amounts of ^{3}H uridine-labeled virus, and chromatographed on a column of Sepharose C14B (Pharmacia) at 4°C in TEN buffer (20 m M Tris, 1 m M EDTA, 0.1 M NaCl pH 7.5). Virus appeared in this void column; fractions containing the virus peak were monitored by optical density at both 260 and 280 nm. The OD_{260}/OD_{280} ratio of the purified virus ranged from 1.22 to 1.27.

Antisera

The antisera were kindly provided by R. Wilsnack, Huntington Laboratories, Brooklandville, Maryland. The Rauscher MuLV antiserum was obtained from a goat immunized with purified virus particles disrupted by treatment with Tweenether. Goat antisera to gp71 and p30 were prepared by immunization of the animals with purified gp71 and p30 from Rauscher MuLV. The titers measured by radioimmunoassay (50% binding) were: gp71 antiserum 10.2×10^4 when gp71 was used as an antigen and lower than 50 for p15E and p30; p30 antiserum 9.5×10^4 for p30 and lower than 50 for p10, p12, and gp71; RLV antiserum 9.6×10^4 measured for gp71 and 2×10^2 for p12. Antiserum to FBS was a gift from Dr. Stephen Kennel, Oak Ridge National Laboratories.

Immunoprecipitation

Labeled virions or cells were disrupted and immunoprecipitated in the presence of Nonidet P40 (1%) and sodium deoxycholate (0.5%) in 25 mM Tris-HCl buffer, Ph 8.0, and 50 mM NaCl. Volumes were adjusted to 200 μl with the same buffer, to which 1 μl of the indicated serum was added. Serum used was preabsorbed on a monolayer of the uninfected cells fixed with methanol. The mixture was incubated for 30 min at 37°C, then for 2 h at 4°C. The immune complexes were precipitated by adding 50 μl of a 10% suspension of Staphylococcus aureus, Cowan strain (American Type Culture Collection) and incubated for 30 min at room temperature.

The precipitates were pelleted in a Brinkmann microfuge and washed twice with 50 mM Tris, pH 7.5, 0.1 M NaCl, 0.5% NP-40, 2.4 M KCl, and once with 50 mM Tris, pH 7.5, 0.10 M NaCl, 0.1% Trition X-100, and 5 mM EDTA. After the final wash, 45 μl of a mixture of 0.0625 M Tris, pH 6.8%, SDS, and 10% glycerol was added to the pellet. Samples assayed under reducing conditions then received 5 μl of 0.5 M iodoacetamide (0.05 M final concentration). All samples were placed in a boiling water bath for 10 min prior to electrophoresis.

Separation and Identification of the Proteins

The labeled viral proteins were separated by polyacrylamide (10% or 13%) gel electrophoresis in the presence of 0.1% SDS [23] and stained with Coomassie Blue R 250. Gels containing ^{3}H-labeled proteins were impregnated with scintillator and dried, and radioactive bands were detected by scintillation autoradiography using Kodak RP Xomat film. Dried gels containing ^{125}I-labeled proteins were exposed to Kodak RP Xomat film in the presence of a fast tungstate intensifying screen.

Interferon

Interferon was a generous gift from Dr. E. Knight. It was produced in L cells by induction with MM virus and purified as described previously [22]. The specific activity of the interferon was 5×10^6 units/mg protein. The antiviral activity of interferon was standardized against the reference mouse interferon (NIH) that had an assigned activity of 12×10^3 units/ml. The amount of interferon used throughout this study was 150 units/ml.

RESULTS

Effect of Interferon on Structural Proteins of MuLV

The polypeptide patterns of virions produced in the presence of interferon and purified on discontinuous and continuous sucrose gradients were examined by SDS polyacrylamide gel electrophoresis, followed by autoradiography, and compared to those of control virus. Labeling with ^{3}H amino acids for 15 h in the presence and absence of interferon gave patterns characteristic of MuLV [23] (Fig. 1). The virions assembled in the

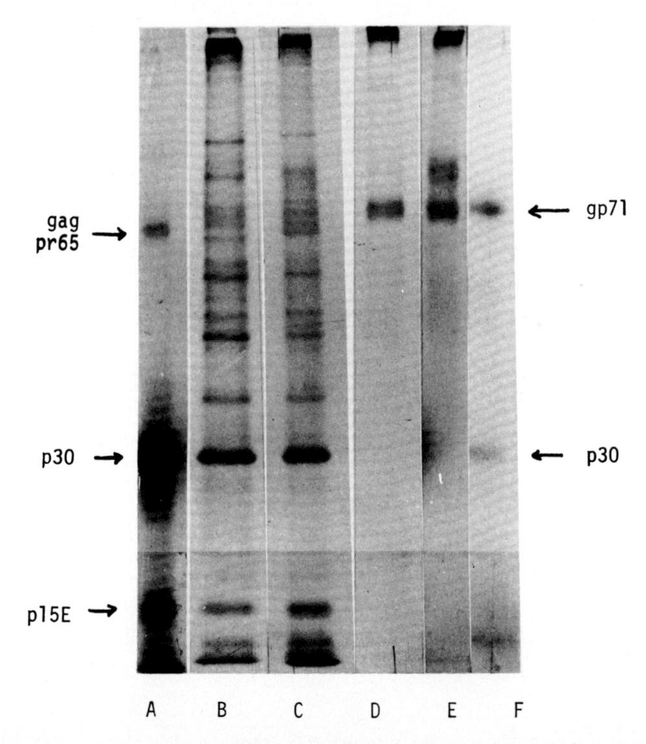

Fig. 1. Polypeptide pattern of MuLV synthesized in the presence and absence of interferon. JLSV-6 cells chronically infected with MuLV were grown without or with interferon (150 units/ml from 0 h to 40 h) and labeled (22–40 h) with 10 μCi/ml of ^{3}H amino acid mix. To label the virion glycoproteins with ^{3}H glucosamine, cells were incubated in glucose-free medium with 100 μCi/ml of ^{3}H glucosamine for 8 h (24–32 h) with or without interferon. Virus was purified by sucrose gradient centrifugation, disrupted, and analyzed on SDS PAGE as described in Methods. Approximately the same number of cpm of each sample were applied to the gel. ^{125}I-labeled gag pr65, p30, and p15 (A), ^{3}H amino acid-labeled proteins of control (B) and interferon (C) virions, ^{3}H glucosamine-labeled glycoproteins of control (D) and interferon (E) virions, and ^{125}I-labeled gp71 and p30 (F) are presented.

presence of interferon contained higher amounts of ^{3}H-labeled peptides which correspond in mobility to p15E and gag pr65 than the controls, while the amount of labeled p30 in the two types of virions was comparable. The interferon virions also have additional ^{3}H-labeled polydisperse bands present in the 72–90K, 40–50K, and 30–35K regions.

To determine whether the proteins detected in the 78–90K regions of the virions assembled in the presence of interferon represent the viral glycoproteins, MuLV was labeled with ^{3}H glucosamine for 24 h, and the glycopeptide patterns of the virions released both in the presence and absence of interferon were compared. SDS gels (Fig. 1) show a difference between the glycopeptide pattern of virions assembled in the presence and absence of interferon. In the control virions the main ^{3}H glucosamine-labeled band was gp71 (gp45 and gp32 were detected as minor bands only after much longer exposure), while the interferon virions contained an additional ^{3}H glucosamine containing 85K protein. This glycoprotein was not detected in virions labeled with fucose (data not shown). The gp45 and gp32 were detected only as minor bands seen after the longer exposure.

The new high-molecular-weight glycoprotein present in the virions from interferon-treated cells was precipitated with antiserum to disrupted MuLV but not with the anti-gp71 serum. Furthermore the precipitation with anti-MuLV serum revealed the presence of the high-molecular ^{3}H glucosamine-labeled glycoproteins not detected in unprecipitated, disrupted virions. These glycoproteins seem to be highly reactive with the MuLV serum. The presence of the high-molecular-weight glycoproteins was not affected by the degree of purification, and thus it is unlikely that these polypeptides represent cellular contamination.

While this work was in progress it was reported that MuLV produced in interferon-treated cells contains a 90K protein containing glucosamine; whether this is a viral or cellular protein has not been established [24].

Components of MuLV Reactive With Antiserum to MuLV, p30, and gp71

To determine the antigenic specificity of the proteins present in the interferon virions, MuLV produced in the presence of interferon was purified and disrupted with Triton X-100, and viral proteins were iodinated, analyzed, and compared to proteins present in the control virions.

The viral proteins were first precipitated with the gp71 antiserum to remove the envelope glycoprotein, and the rest of the ^{125}I-labeled protein was precipitated with MuLV antiserum, the precipitates being analyzed by SDS PAGE electrophoresis and autoradiography (Fig. 2). The antiserum to gp71 removed iodinated gp71 from the protein mixture of both control and interferon virions. In the control virus, the sequential precipitation with MuLV antiserum precipitated only the ^{125}I-labeled structural proteins of the gag gene (p30, p15, p10) and p15E. In interferon virions the MuLV antiserum precipitated, in addition, 85K and 65K proteins as major ^{125}I-labeled bands. These data indicate that the interferon virions contain at least two high-molecular-weight proteins (85K and 65K) which are not detected in the control virions. The 85K protein may be identical to gp85 detected in ^{3}H glucosamine-labeled interferon virions.

Differential Localization of Viral Glycoproteins in the Virion Membranes

It was shown previously [24] that lactoperoxidase treatment of intact MuLV labels predominantly the 78K viral protein, which is antigenically related to gp71. To examine how the presence of gp85 in the interferon virions affects their membrane topology, virions produced in the presence and absence of interferon were purified on a Sephadex C14B

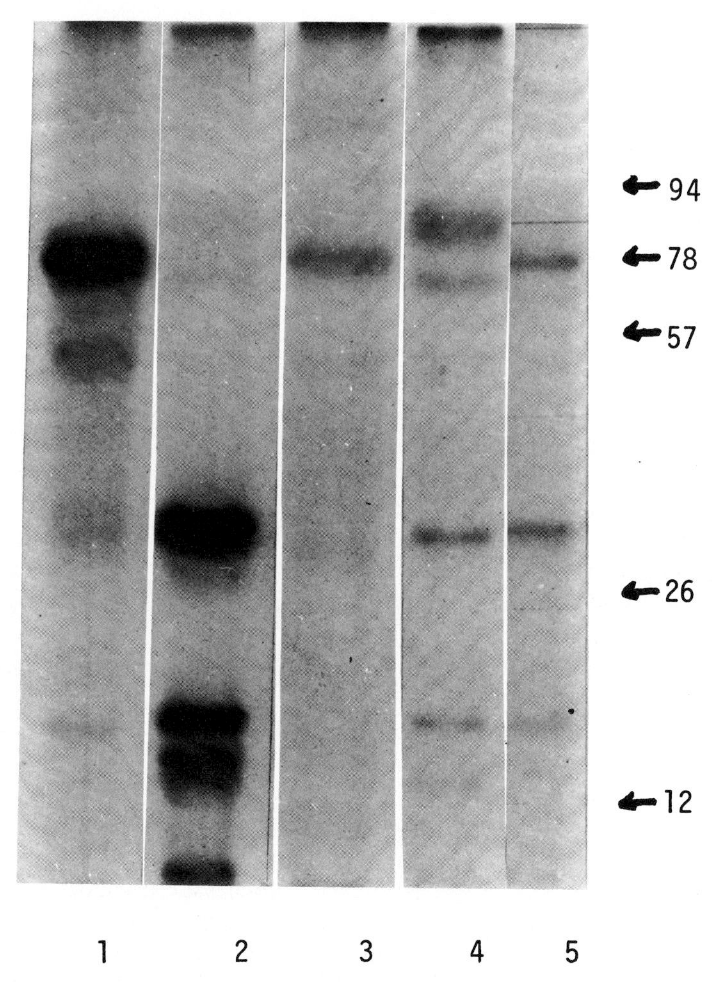

Fig. 2. Sequential immunoprecipitation and SDS PAGE gel electrophoresis of iodinated MuLV proteins detected in virions produced in the presence and absence of interferon. Virus produced in the presence and absence of interferon was harvested, purified, iodinated, and analyzed as described in Methods. Iodinated proteins (2×10^6 cpm) were precipitated first with anti-gp71 serum [control (1) and interferon (3)] and the supernatant was precipitated with anti-MuLV serum [control (2) and interferon (4)]. ^{125}I-labeled gp71, p30 and p15 (5).

column, iodinated by the solid-state lactoperoxidase method [20], disrupted, and treated with respective antiserum, and the precipitates were analyzed by polyacrylamide gel electrophoresis followed by autoradiography.

Surface iodination of the control virions revealed the presence of one major surface protein with the mobility of gp71 (Fig. 3), when assayed in the presence of mercaptoethanol. To confirm its relationship, immunoprecipitation was done with the gp71 antiserum. When the precipitate was analyzed under reducing conditions gp71 was detected as a major labeled band with minor bands in the gp45 and gp32 regions. The antiserum to p15(E) precipitated 90K but not gp71 (data not shown). The fact that no radioactive p15(E) was detected indicates that in the intact virions p15(E) is not accessible for surface iodination. No iodinated viral proteins were precipitated with p30 antiserum.

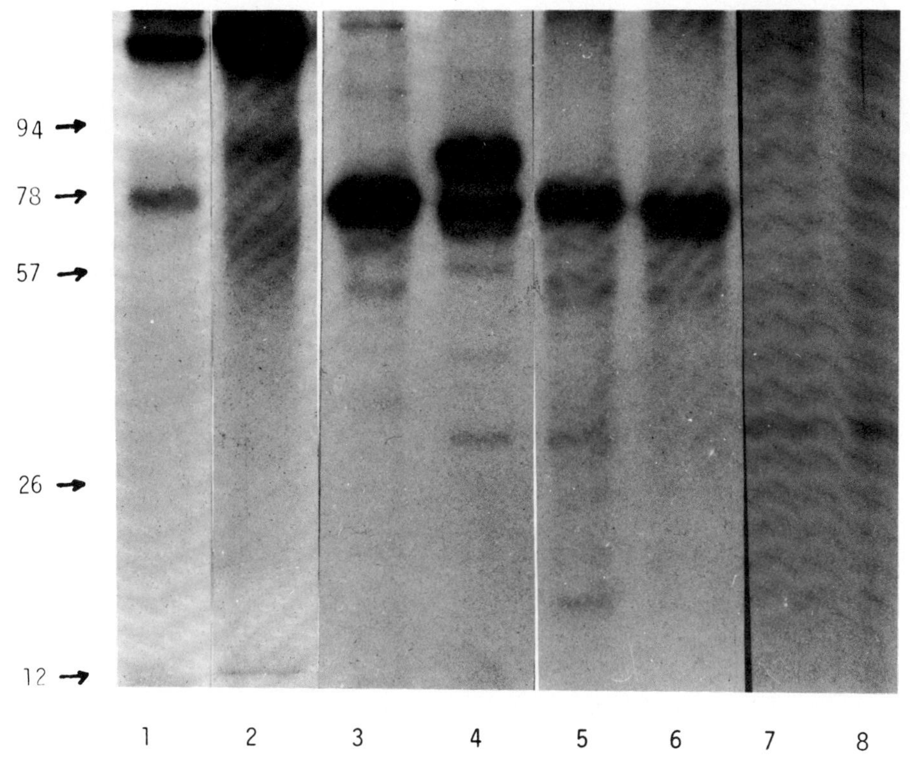

Fig. 3. Accessibility of viral glycoproteins on the surface of virions produced in the presence of interferon. Virions produced in interferon-treated cells (150 units/ml, interferon present 0–36 h, virions harvested 24–36 h) were purified, labeled, and analyzed as described in Methods. Unprecipitated samples (0.1 μg of protein) control virions (1) and interferon (2) virions. Immunoprecipitation of iodinated proteins (2×10^6 cpm); gp71 antiserum:controls reducing (3) nonreducing (4) conditions; interferon reducing (5) nonreducing (6) conditions; p30 antiserum:controls (7) interferon (8).

Surface iodination of the intact interferon virions revealed patterns different from those of the control virions (Fig. 3). Analysis of the iodinated proteins after precipitation with gp71 antiserum under both nonreducing and reducing conditions revealed only the presence of a gp71. Antiserum to p30 did not precipitate any labeled viral proteins. No precipitation of ^{125}I-labeled proteins was detected with the normal goat serum (data not shown). These data provide evidence that the mature MuLV produced in the presence and absence of interferon have different surface morphology.

Effect of Interferon on the Rate of Synthesis of Viral Proteins

We have shown previously that the synthesis of p30 and gp71 is not inhibited by the interferon treatment in AKR-2B cells chronically infected with MuLV [10]. These experiments, however, did not eliminate the possibility that in interferon-treated cells the rate of viral protein synthesis or the cleavage of the precursors of viral proteins is slowed down. To determine whether the observed lack of fidelity in the virus assembly could be explained by the slowdown in the rate of viral protein synthesis or its processing, the interferon-treated cells were pulsed for short periods of time with ^{3}H amino acids, and then the radioactivity was chased for 3 and 6 h in the presence of interferon. The ^{3}H-labeled proteins

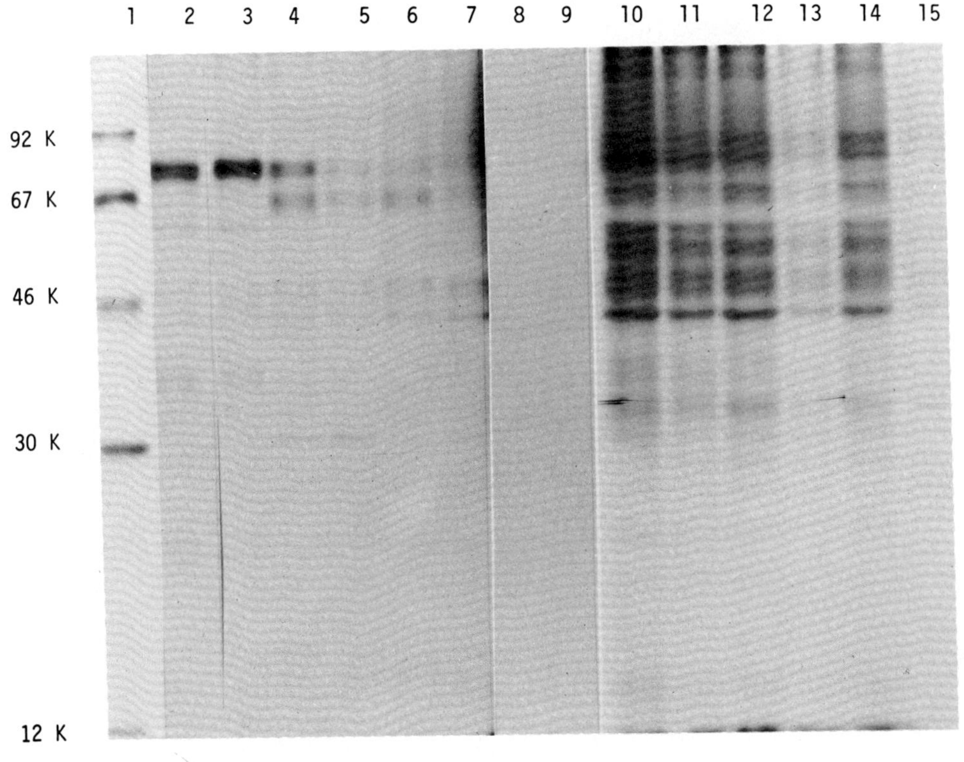

Fig. 4. Synthesis of viral proteins in the cells in the presence and absence of interferon. The cells were treated with interferon (for 24 hrs) and pulsed with ^{3}H amino acid (100 μCi/ml) for 20 min; the radioactivity was chased for 3 and 6 h. The ^{3}H-labeled proteins were analyzed before precipitation (10–15) (0.1 μg protein per well) and after precipitation (2–7) with anti-MuLV serum (10^5 cpm/well). Controls: pulse (2, 10), 3-h chase (4, 12), and 6-h chase (6, 14). Interferon-treated cells: pulse (3, 11), 3-h chase (5, 13), and 6-h chase (7, 15). ^{125}I-labeled gp71 and p30 (1). Non immune serum pulse controls (8), interferon (9).

were analyzed on SDS PAGE before and after immunoprecipitation with anti-MuLV, gp71, and p30 serum. The analysis of the labeled proteins (Fig. 4, lanes 10–15) on acrylamide gels indicated that in interferon-treated cells the amount of proteins labeled in a 20-min pulse was lower than in the controls. This difference became much more obvious when the proteins were analyzed after a 3- and a 6-h chase, indicating faster turnover of labeled proteins in MuLV-infected, interferon-treated cells. No qualitative difference, however, was observed after immunoprecipitation (Fig. 4, lanes 2–7). In the cells labeled for a short period of time with ^{3}H amino acid in the presence and absence of interferon, the antiserum against MuLV precipitated env pr 85 and gag pr 65, and these were further processed into gp71 and p30, respectively, both in the interferon-treated cells and controls. We have shown that the virions from interferon-treated cells contain gp85, which is antigenically unrelated to gp71 but is recognized by the anti-MuLV serum. In order to determine whether the env pr 85 and gag pr 65 synthesized in the cells show antigenic similarity to gp71 and p30, the ^{3}H-labeled proteins from interferon-treated cells and controls were precipitated with anti-gp71 and anti-p30 serum and analyzed on SDS PAGE. The results, in Figure 5, indicate that the env pr 85 present in the interferon-treated cells is antigenically related to gp71 and there-

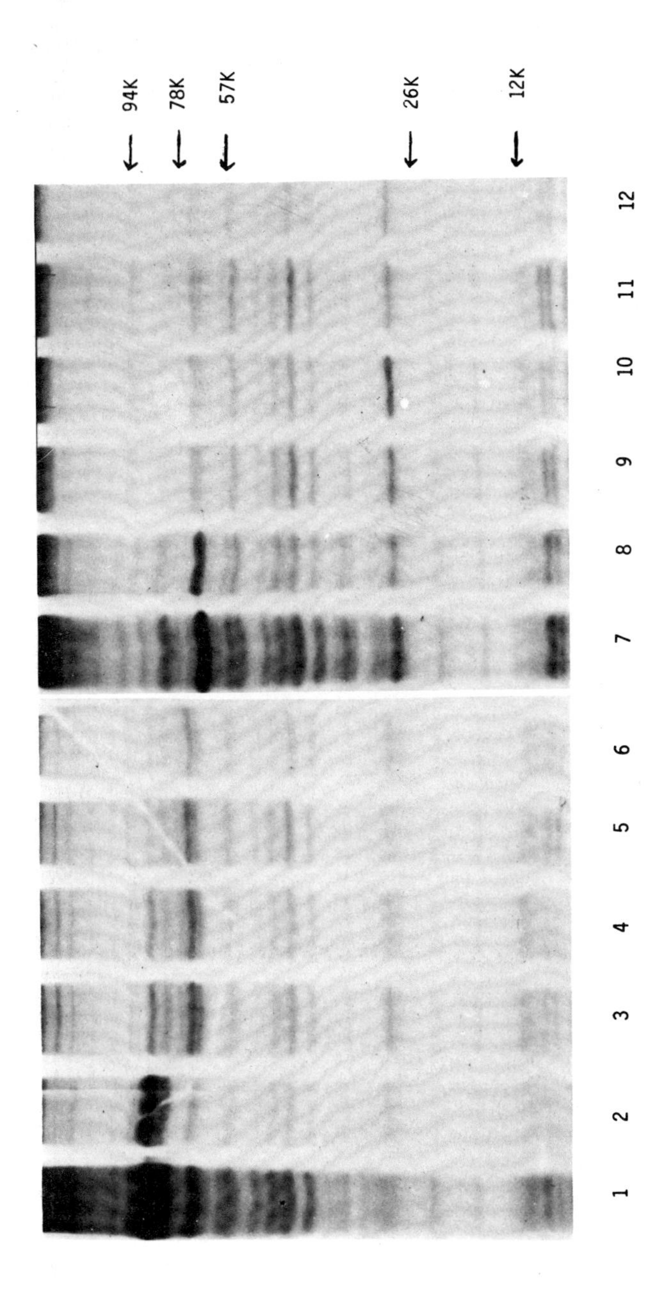

Fig. 5. Synthesis and processing of viral proteins in interferon-treated cells and controls. Cells were labeled as described in Figure 4 and viral proteins were precipitated with anti-gp71 serum (1–6) and anti-p30 serum (7–12). Controls: pulse (1, 7), 3-h chase (3, 9), and 6-h chase (5, 11). Interferon-treated cells: pulse (2, 8), 3-h chase (4, 10), and 6-h chase (6, 12).

fore probably represents a different protein from that present in the virions. Similarly, the gag pr 65 present in the interferon-treated cells has antigenic similarity to p30.

These results indicate that the rates of synthesis and processing of viral proteins in interferon-treated cells are comparable to those in the control cells.

Finally, we examined the possibility that the interferon treatment leads to a redistribution of the viral proteins in the plasma membrane and thus to erroneous assembly. The surface membrane proteins of the interferon-treated cells and controls were selectively labeled with ^{125}I iodine, the cells were lysed, and viral proteins were analyzed after precipitation with anti-gp71 serum. The results, in Figure 6, indicate that both in the interferon-treated and in the untreated cells the major protein labeled was gp71, and that this protein survived under nonreducing conditions as gp90. Thus, in the interferon-treated cells as well as in the control the gp71 glycoprotein is a major viral surface protein.

DISCUSSION

The results presented indicate that interferon treatment of JLSV-6 cells persistently infected with MuLV leads to the formation of virus particles containing several novel polypeptides not detected in the control virions.

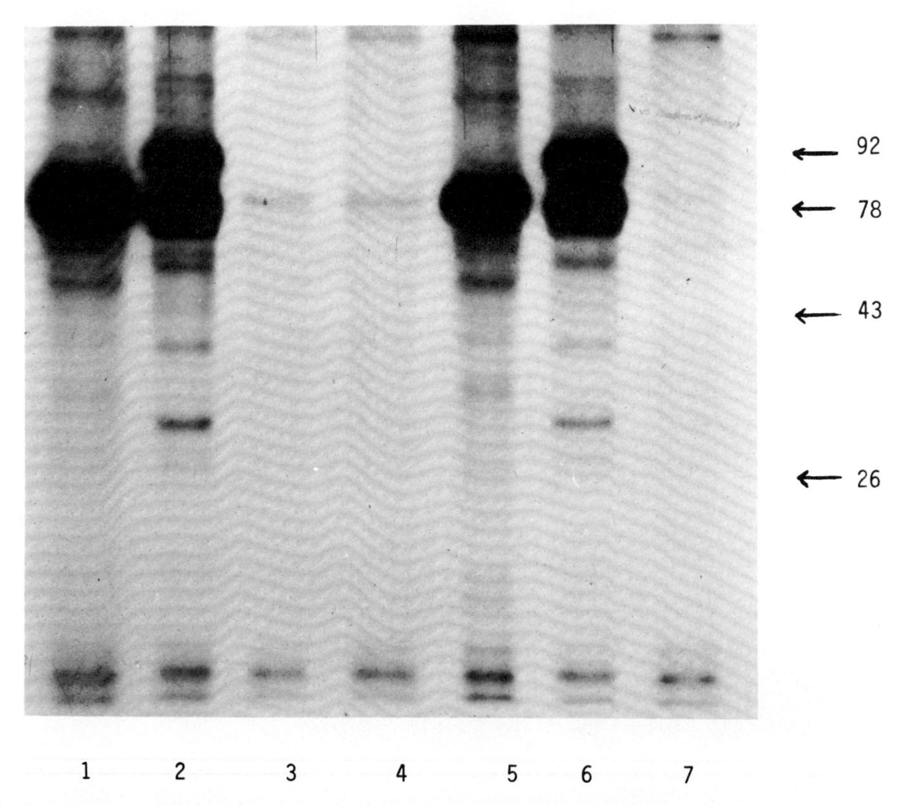

Fig. 6. The surface ^{125}I-labeling of infected cells grown in the presence and absence of interferon. The surface of the cell was labeled as described in Methods. Viral proteins were precipitated with anti-gp71 serum (1, 2, 5, 6) and anti-p30 serum (3, 7) and analyzed on SDS PAGE in reducing (1, 3, 5, 7) and nonreducing (2, 6) conditions. Controls:anti-gp71 serum (1, 2); anti-p30 serum (3); normal goat serum (4); interferon-treated cells; anti-gp71 serum (5, 6); anti-p30 serum (7).

Surface iodination of intact virions revealed that in interferon-treated virions both 85k(gp85) and gp71 are major components of the virions envelope. Unlike the control virus, in which gp71 (a major membrane protein) can associate through a disulfide bond linkage to p15E [25], the gp71-p15E complex was not detected in the virions assembled in the presence of interferon.

The analysis of the virion proteins iodinated after virus disruption indicated that the interferon-treated virion contain, in addition to gp85, a 65k protein which was not seen in the control virus. It is particularly interesting that these two peptides in interferon-treated virions are precipitable with antiserum against MuLV; however, these proteins are not efficiently recognized by the gp71 and p30 antiserum. The question whether the new proteins in the interferon virions are coded by the env and gag genes needs further investigation. The post transcriptional modifications of the gene products may be altered in interferon-treated cells and this may change the antigenicity of the products. Further identification of these proteins by fingerprinting or partial digest analysis should be able to determine their relationship to the env pr85 and gag pr65. The possibility that the new proteins are the cellular proteins or viral nonvirion proteins also cannot be completely eliminated. It has been shown recently that vesicular stomatitis virus assembles several cellular glycoproteins into the virions (Lodish, personal communication). If a similar situation happens with MuLV, then the antiserum to disrupted MuLV would contain the antibodies to cellular glycoproteins as well. Since the activity of the antiserum to MuLV to precipitate these proteins was not removed by preadsorption with fixed uninfected cells, this would indicate that these proteins detected in the interferon virions are not present on the membranes of the uninfected cells.

Retrovirus assembly and maturation are membrane-associated events. It is assumed that a cleavage of the precursors of the gag proteins occurs at the time of virus budding or core formation [26]. Previous work [27, 28] indicates that the precursors of the gag proteins are detectable in extracellular virus particles; this may indicate that only a partial cleavage of gag pr65 is required for virus release and that a final maturation of the core is coupled with further cleavage. Proteolytic activity associated with Rauscher MuLV has been shown to cleave gag pr 65 into the lower-molecular-weight structural proteins [26]. Accordingly, the amount of p30 detected in the virions released from interferon-treated cells seems to be time-dependent (data not shown).

An alternative explanation would be that interferon affects post transcriptional modification and processing of one or several primary gene products. Failure of the processing of env and gag precursors in interferon-treated cells could lead to insertion of these proteins into virions together with the fully processed viral proteins. However, the intracellular synthesis and processing of precursor by viral proteins does not seem to be affected by interferon treatment.

Thus the molecular basis of the erroneous virus assembly in interferon-treated cells remains unknown. However, the finding that interferon virions contain significantly more high-molecular-weight proteins than the controls and the fact that in interferon-treated cells mature noninfectious virus particles accumulate on the cell surface [9, 14–16, 30]. indicate that the erroneous assembly leads both to the inhibition of virus release and a decrease in its infectivity.

ACKNOWLEDGMENTS

We wish to thank Drs. M. Strand and W. P. Rowe for helpful suggestions and advice during the course of this study, Dr. E Knight for the gift of purified interferon, and Dr. R.

Wilsnack for the antisera. Paula M. Pitha is a scholar of the Leukemia Society of America. This research was supported by Public Health Service grant CA 10961 from the National Cancer Institute.

REFERENCES

1. Joklik WK, Merigan TC: Proc Natl Acad Sci USA 56:558–565, 1966.
2. Jungwirth CJ, Horak G, Bodo J, Lindner J, Schultz B: Virology 48:59–70, 1972.
3. Billiau A, Edy VG, Sobis H, De Somer P: Int J Cancer 14:335–340, 1974.
4. Aboud M, Shoor R, Salzberg S: Virology 84:134–141, 1978.
5. Billiau A, Heremans H, Allen PT, Baron S, De Somer P: Archiv Virol 57:205–220, 1978.
6. Bolognesi DP, Montelaro RC, Frank H, Schafer W: Science 199:183–186, 1978.
7. Bonner WM, Laskey RA: Eur J Biochem 46:83–88, 1974.
8. Friedman RM, Chang EH, Ramseru JM, Myers MW: J Virol 16:569–574, 1975.
9. Friedman RM, Ramseur JM: Proc Natl Acad Sci USA 71:3542–3544, 1974.
10. Pitha PM, Rowe WP, Oxman MN: Virology 70:324–338, 1976.
11. Duesberg LH, Robinson HL, Robins WS, Heubner RJ, Turner HC: Virology 36:73–86, 1968.
12. Pitha PM, Rowe WP: In Chirigos MA (ed): "Control of Neoplasia by Modulation of the Immune System." New York: Raven Press, 1977.
13. Metz DH, Esteban J: Nature 238:385–388, 1972.
14. Chang EH, Friedman RM: Biochem Biophys Res Commun 77:392–397, 1977.
15. Chang EH, Mims SJ, Triche TJ, Friedman RM: J Gen Virol 34:363–368, 1977.
16. Pitha PM, Wivel NA, Fernie BF, Harper HP: J Gen Virol 42:467–480, 1979.
17. Wong PKY, Yuen PH, MacLeod R, Chang EH, Myers MW, Friedman RM: Cell 10:245–252, 1977.
18. Luftig RB, Conscience JF, Skoultchi A, McMillian P, Revel M, Ruddle F: J Virol 23:799–810, 1977.
19. Hartley JW, Rowe WP: Virology 65:128–134, 1975.
20. Kennel SJ, Lerner RA: J Mol Biol 76:485–502, 1973.
21. McGrath M, Witte O, Pincus T, Weissman IL: J Virol 25:923–927, 1978.
22. Knight Jr E: Biol Chem 250:4139–4144, 1975.
23. Laemmli UK: Nature 227:680–685, 1970.
24. Lengyel P, Leary P, Gresser I: Proc Natl Acad Sci USA 70:2785–2788, 1973.
25. Pinter A, Fleissner E: Virology 83:1417–1422, 1977.
26. Yeger H, Kalnins VI, Stephenson JR: Virology 89:34–44, 1978.
27. Jamjoom G, Karshin WL, Naso RN, Arcement LJ, Arlinghaus RB: Virology 68:135–145, 1975.
28. Bolognesi DP, Leis JJ, Moennig V, Schafer W, Atkinson PH: J Virol 16:1453–1463, 1975.

Journal of Supramolecular Structure 12:47–61 (1979)
Biological Recognition and Assembly 113–127

Membrane Dynamics of Differentiating Cultured Embryonic Chick Skeletal Muscle Cells by Fluorescence Microscopy Techniques

Hannah Friedman Elson and Juan Yguerabide

Department of Biology, C-016, University of California, San Diego, La Jolla, CA 92093

Changes in membrane fluidity during myogenesis have been studied by fluorescence microscopy of individual cells growing in monolayer cultures of embryonic chick skeletal muscle cells. Membrane fluidity was determined by the techniques of fluorescence photobleaching recovery (FPR), with the use of a lipid-soluble carbocyanine dye, and by fluorescence depolarization (FD), with perylene used as the lipid probe. The fluidity of myoblast plasma membranes, as determined from FPR measurements in membrane areas above nuclei, increased during the period of myoblast fusion and then returned to its initial level. The membrane fluidity of fibroblasts, also found in these primary cultures, remained constant. The fluidity in specific regions along the length of the myoblast membrane was studied by FD, and it was observed that the extended arms of the myoblast have the highest fluidity on the cell and that the tips at the ends of the arms had the lowest fluidity. However, since the perylene probe used in the FD experiments appeared to label cytoplasmic components, changes in fluidity measured with this probe reflect changes in membrane fluidity as well as in cytoplasmic fluidity. The relative change in each of these compartments cannot yet be ascertained. Tips have specialized surface structures, filopodia and lamellipodia, which may be accompanied by a more immobile membrane as well as a more rigid cytoplasm. Rounded cells, which may also have a more convoluted surface structure, show a lower apparent membrane fluidity than extended cells.

Key words: **plasma membrane, fluidity, skeletal muscle, myogenesis, laser, fluorescence photobleaching recovery, fluorescence depolarization, carbocyanine, perylene, fluorescence anisotropy, microviscosity**

The evolution of the concept of the cell membrane from a static interface to a dynamic fluid, lipid bilayer containing proteins with varying degrees of mobility [1–3] has been accompanied by the development of sensitive techniques for measuring membrane fluidity.

Received April 20, 1979; accepted August 31, 1979.

0091-7419/79/1201-0047$02.90

Many membrane phenomena may involve fluidity changes, such as formation of specialized surface structures, hormonal modulation, endocytosis, and some aspects of cellular differentiation. These membrane functions may be clarified when the dynamic changes underlying them are illuminated.

Skeletal muscle cells provide a good system for study of the involvement of membrane recognition and fluidity in differentiation. During myogenesis, mononuclear myoblasts fuse to form myotubes, or multinucleate skeletal muscle fibers. Monolayer cultures of dissociated skeletal muscle cells also differentiate to produce striated, multinucleate muscle cells synthesizing the normal complement of muscle-specific proteins [4–6]. Fusion is accompanied by specific cell–cell recognition. When cells of different tissue type are cocultured, fusion of the plasma membranes occurs between skeletal muscle cells only, although skeletal muscle cells of different species will recognize each other and fuse [7]. The mechanisms for cell recognition, as well as for fusion, are still unclear. Transmission electron micrographs reveal a row of vesicles along the two apposed and fusing cell membranes [8–10]. One report, however, indicates that fusion is initiated at a single site, with a single, widening, cytoplasmic bridge forming between the two cells [11].

In molecular terms, cell recognition and fusion could be envisioned as requiring rigid complementary protein structures, much like the protein rosette targets to which secretory vesicles fuse [12]. Such assemblies could be sufficient for fusion. On the other hand, a transient increase in membrane fluidity could be required during fusion. Fusion in model systems is enhanced by agents that increase lipid fluidity. For example, fusion of bilayers will occur only when the temperature is above the liquid-gel transition temperature [13]. Agents that perturb the bilayer structure, such as lysolecithin [14], enhance fusion, whereas agents that decrease fluidity, such as cholesterol, inhibit fusion [13].

In the present study, one of these possibilities was explored by the measurement of lipid bilayer fluidity during myogenesis in monolayer cultures of embryonic chick skeletal muscle cells. In a previous study [15], it was reported that an increase in membrane fluidity preceded fusion. The method involved measuring rotational diffusion of a fluorescent probe in muscle cultures brought into suspension by trypsin digestion. There are, however, certain limitations in the use of cell suspensions, such as heterogeneity of the cell population and possible membrane damage caused by trypsin. In addition, the cellular location of the fluorescent probe must be defined to be certain that only the plasmalemma is being monitored. Finally, myoblasts are anatomically polarized cells, having a central ellipsoidal nuclear region, two long thin arms or extensions on opposite sides, and tips, with terminal foot pads. The fluidity of the membrane may not be uniform along the entire cell, and any changes during differentiation may be restricted to one of these regions.

To resolve these problems we undertook this investigation of muscle membrane fluidity by making measurements under the fluorescence microscope. The use of the microscope allows us to determine the location of the probe in the labeled cells and to choose specific regions of the cell for measurements of fluidity.

Fluidity of specific regions of myoblasts and fibroblasts was studied by two methods. Lateral mobility was measured by the technique of fluorescence photobleaching recovery (FPR), with a carbocyanine dye used as the fluorescent lipid probe. Rotational mobility was measured by a fluorescence depolarization (FD) technique, with perylene used as the probe [16].

MATERIALS AND METHODS

Cultures

Eleven-day chick embryo breast muscle tissue was dissociated by trypsin [17] and plated on 18 mm diameter collagenized coverslips (Calbiochem calf skin collagen or Sigma Type IV collagen). About 2–3 × 10^6 cells were plated on the coverslips in a 10 cm diameter Petri dish (Falcon) containing 5 ml medium 8:1:0.25 (Eagle's minimum essential medium with Earle's salts supplemented with 10% horse serum, 2.5% embryo extract, 0.2 mM glutamine, and 125 units/ml of penicillin and streptomycin). Medium components were obtained from Gibco. Material that can adsorb to collagen and produce a fluorescent background was removed from the medium by preincubation in a collagen-coated Petri dish at room temperature overnight. The cells were maintained in a humidified incubator with 5% CO_2: 95% air at 37°C, and the growth medium was changed daily.

Lateral Mobility

Lateral mobility of membrane lipids was measured using the lipid probe N,N′-di(octadecyl)oxacarbocyanine (K-1) (Fig. 1a) generously supplied by A. S. Waggoner [18]. Fluorescence photobleaching recovery was performed in a manner similar to that previously described [19, 20]. A coverslip to which the cells were attached was washed twice with Dulbecco's phosphate-buffered saline (PBS), labeled in 0.3 ml PBS containing 13 μM K-1 and 1% ethanol at 22°C for 4 min, rinsed once in PBS plus 0.2% bovine serum albumin, and finally rinsed in PBS. The coverslip was secured face down to a glass slide by a silicone rubber gasket. One coverslip was used for measurements for no longer than one hour after removal from the incubator.

The apparatus for photobleaching was similar to that illustrated elsewhere [19], except that a helium/cadmium laser (Liconix model 4110) was employed, having an output of 10 mwatt and wavelength of 442 nm. The laser beam was focused onto a specific spot (2–3 μm diameter) on a cell through a 40X Pol objective on a Zeiss microscope under epi-illumination. Fluorescence was monitored by using the laser beam attenuated by three orders of magnitude by a neutral density filter. For bleaching, the attenuator was removed for 0.2–0.4 sec. The recovery of fluorescence into the bleached spot was detected and recorded with a photon counting system made from Ortec components consisting of a photomultiplier (model 9201), D-A convertor (9325), photon counter (9315), amplifier discriminator (9302), and strip-chart recorder.

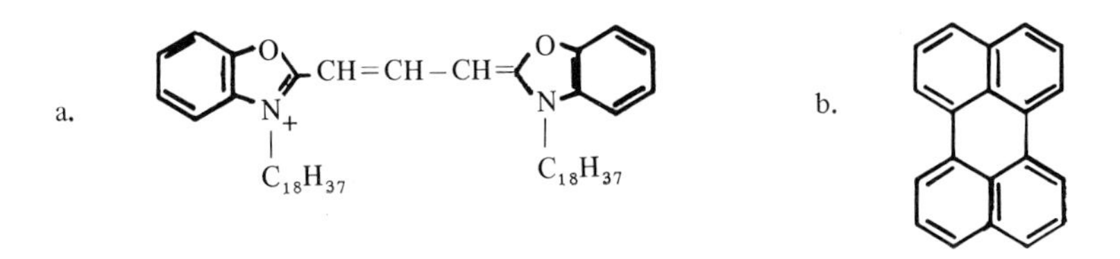

Fig. 1. Fluorescent probes: (a) N,N′-di(octadecyl)oxacarbocyanine (K-1); (b) Perylene.

Rotational Mobility

The laser beam is highly polarized and hence can be used to measure fluorescence depolarization, or anisotropy, at a single point on a cell through a microscope. The perylene probe used here is weakly oriented when embedded in a lipid bilayer and is thus suitable for measurement of microviscosity by fluorescence depolarization [21]. Cells were processed as above, except that they were labeled for 15 min at 37°C in PBS containing 10 μM perylene and 1% ethanol. A rotatable polarizer was placed in the path of the emitted beam, and the intensity was recorded with the polarizer parallel and perpendicular to the direction of polarization of the exciting light beam. Fluorescence anisotropies were corrected for light scattering, background fluorescence, and distortions introduced by 1) differences in transmission efficiencies through the microscope for light polarized parallel ($I_{\parallel}$) and perpendicular ($I_{\perp}$) to the direction of polarization of the incident beam and 2) depolarization introduced by the finite collection angle of the microscope objective. Light scattering and background fluorescence corrections were made by subtracting background fluorescence as determined from parallel and perpendicular intensities adjacent to cells under study. A correction factor α for transmission and collection angle artifacts was obtained by two different methods. In one, a thin film of perylene in a viscous oil was formed between a cover glass and microscope slide, and the directly measured, uncorrected ratio $(I_{\perp}/I_{\parallel})_{OM}$ was determined under the microscope under conditions identical to those used in our cell experiments. The correct value of the polarization ratio, $(I_{\perp}/I_{\parallel})_{OS}$, for perylene in the viscous oil had been previously determined in a steady-state spectrofluorimeter [22]. The corrrection factor is given by

$$\alpha = \left[\frac{I_{\perp}}{I_{\parallel}}\right]_{OS}\left[\frac{I_{\parallel}}{I_{\perp}}\right]_{OM} \tag{1}$$

The corrected ratio $(I_{\perp}/I_{\parallel})^c$ for the cell experiments was obtained from experimental values of $I_{\perp}$ and $I_{\parallel}$ by the equation

$$R = \left[\frac{I_{\perp}}{I_{\parallel}}\right]^c = \alpha\left[\frac{I_{\perp} - I_{\perp}{}^b}{I_{\parallel} - I_{\parallel}{}^b}\right] \tag{2}$$

where $I_{\perp}{}^b$ and $I_{\parallel}{}^b$ are the perpendicular and parallel background intensities, respectively. Anistropy was calculated with the equation

$$A = \frac{1 - (I_{\perp}/I_{\parallel})^c}{1 + 2\,(I_{\perp}/I_{\parallel})^c} = \frac{1 - R}{1 + 2R} \tag{3}$$

In a second method for evaluating α, human red cell ghosts were labeled with perylene, and the value of $(I_{\perp}/I_{\parallel})_{GS}$ of the ghost membranes was measured in a steady-state spectrofluorimeter using a membrane suspension. The $(I_{\perp}/I_{\parallel})_{GM}$ was then measured under the microscope, corrected for background, and α was calculated with the expression

$$\alpha = \left[\frac{I_{\perp}}{I_{\parallel}}\right]_{GS}\left[\frac{I_{\parallel}}{I_{\perp}}\right]_{GM} \tag{4}$$

This procedure assumes that the weak orientation of perylene in the lipid bilayer does not significantly influence the values of the polarized intensities. Close agreement was obtained between the values of α determined by the two procedures described here. Furthermore, by calibrating in the steady-state spectrofluorimeter a series of solutions of varying microviscosities formed by mixing two solvents of different viscosities, we determined that α did not change in the range of anisotropies of interest here. The correction factor was found to be $\alpha = 1.50 \pm 0.04$. All measurements were made at 21–22°C.

RESULTS

Cell Cultures

Monolayer cultures of embryonic chick skeletal muscle cells grown under the conditions described here differentiate, as shown in Figure 2. Cells proliferate for the first 2 days. Myoblasts enlarge, elongate, and overlap in chains at about 30 h after plating. Fusion of many cells has occurred at about 36 h and is complete in the culture at about 60–70 h after plating. The myotubes enlarge, become striated, and develop spontaneous contractions after 5 days in culture.

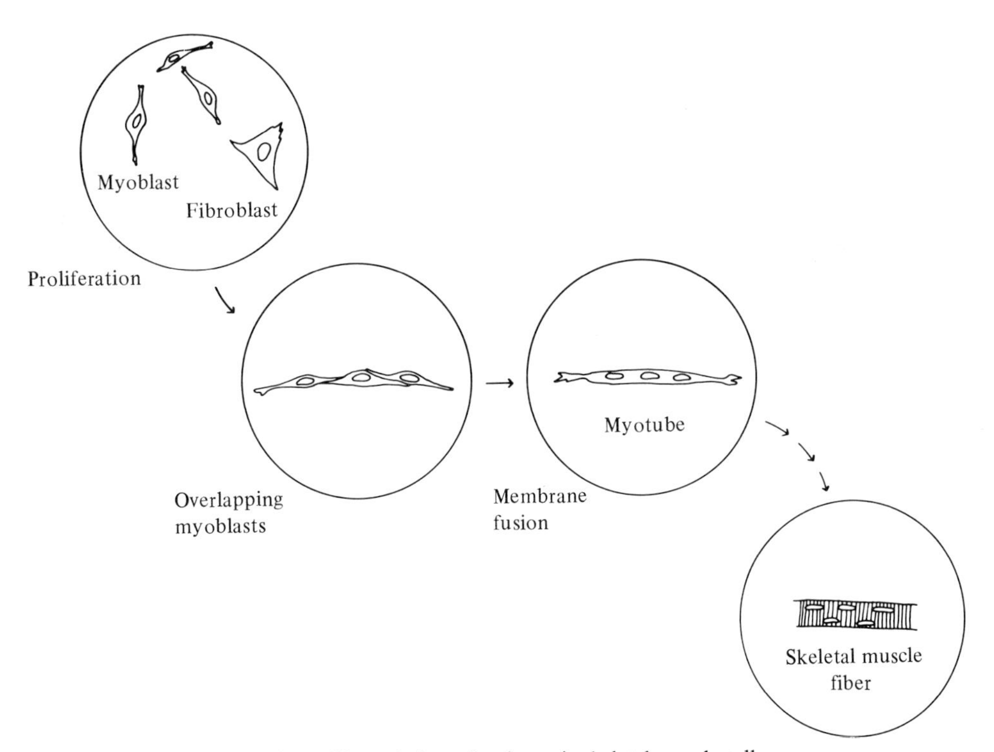

Fig. 2. Differentiation of embryonic skeletal muscle cells

Labeling of Cells With Fluorescent Probes

The fluorophore K-1 did not penetrate the plasma membrane of intact cells under our incubation conditions, as evidenced by distinct rim-labeling around the edges of the cells and a lack of labeling of internal membranous structures (Fig. 3). A striking observation was that the tips of the myoblast displayed much weaker fluorescence than the rest of the plasmalemma.

Two probes were tested for fluorescence depolarization (FD) measurements under the microscope. The probe 1,6-diphenyl-1,3,5-hexatriene (DPH) is widely used for studies of microviscosity in cell suspension and has been used in a previous study of suspended myoblasts [15]. Our observations under the microscope, however, indicate that DPH accumulates in very large spherical vacuoles in both myoblasts and fibroblasts. The fluorescence from these internal vacuoles represented a significant fraction of the total DPH fluorescence intensity. In addition, the cells deteriorated rapidly after labeling with DPH, probably due to the tetrahydrofuran solvent used to solubilize this fluorophore. We therefore examined another weakly oriented lipid probe, perylene, and found that this probe was also quickly internalized, but that the labeling procedure, which uses ethanol instead of tetrahydrofuran, did not damage the cells. With perylene, intense fluorescence was observed from small lipid granules around the nucleus, especially on the sides of the nucleus adjacent to the two cell extensions or arms (Fig. 3). Areas above the nuclei and in the arms and tips were more weakly fluorescent. In mature myotubes, fluorescence intensity displayed a striated pattern along the cell, probably due to internal labeling of the sarcoplasmic reticulum and transverse tubules. As with K-1, fluorescence intensity along the cell showed a marked decrease at the flattened, widening tips of the cells. In general, the lowest fluorescence intensity was seen at the tips of the cells and directly over the nuclei. The lower fluorescence intensity in the nuclear region seems to be due to the paucity of cytoplasm in this region, which decreases the fluorescence intensity from internalized probe. The intensity in this region is probably due mostly to fluorescence from perylene in the plasmalemma and nuclear membranes. The reason for the decreased fluorescence at the tips of the cells is not clear. By focusing the 2–3 μm illuminating light beam on specific areas above the nuclei or in the regions of the arms or tips, one can avoid exciting fluorescence from discernible lipid granules in adjoining areas.

Lateral Mobility

For measuring fluidity of membranes by photobleaching, the probe K-1 was incorporated into the plasma membranes of the cells. Figure 4 shows typical recordings from a single spot over the nucleus of a fibroblast successively bleached and allowed to recover fluorescence three times. The measured light intensity on the ordinate (counts per 0.1 sec) varied from cell to cell from 10^4 to 10^5 photons per 0.1 sec. The recovery of initial intensity was better than 85%. The 10% to 15% incomplete recovery of fluorescence intensity after a bleaching pulse could be due to internalization and labeling of slowly moving pinocytotic vesicles immediately below the membrane or to a small percentage of probe molecules immobilized by interaction with proteins. Our present experiments cannot distinguish between these two possibilities. Photo-cross-linking of probe molecules does not seem likely, since the percent fluorescence recovery was not dependent on the percent of bleach. For the recording in Figure 4, the bleached spot diameter was 2.5 μm. For most of the measurements, the spot diameter was 3.3 μm, to slow down the recovery times.

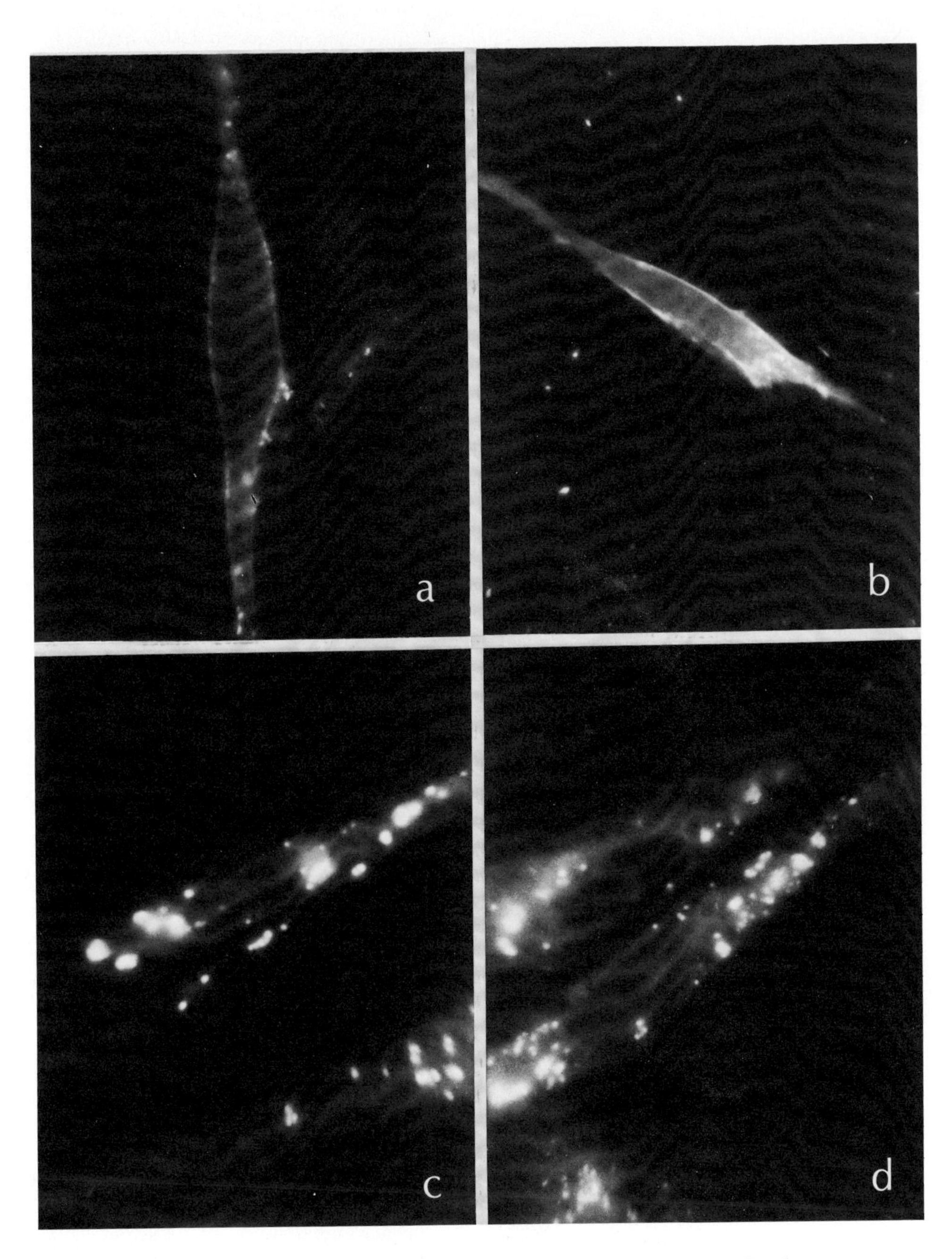

Fig. 3. Fluorescence micrographs of cells labeled with: (a,b) K-1; (c,d) Perylene.

The results of FPR measurements made at several intervals after plating the cells are shown in Figure 5. The $t_{1/2}$ ordinate represents the time after bleaching for half-maximal recovery of fluorescence and is plotted against age of the culture. These results indicate that the plasmalemma above the nuclear region of myoblasts underwent a 40% decline in $t_{1/2}$ over the first 3 days, or a 40% increase in the apparent rate of lateral diffusion of the lipid probe over this time. The $t_{1/2}$ value for a flat membrane is related to the diffusion coefficient of the fluorophore by the expression

$$t_{1/2} = \frac{\beta r^2}{4D} \tag{5}$$

where D is the diffusion coefficient and r is the effective ($1/e^2$) radius of the bleaching beam [23]. For a laser bleaching beam with a Gaussian distribution of intensity along its diameter, β is a parameter that depends on the percent of bleached fluorophore. Its value ranges from 1.0 for a very low bleaching to about 1.7 for 90% bleaching. Values of $t_{1/2}$ are thus not directly comparable in general unless reduced to the same extent of bleaching. However, theoretical calculations indicate that for the range of bleaching intensities and precision in our experiments, a value of $\beta \sim 1$ is justified in all experiments. In this case, values of $t_{1/2}$ can be directly compared without corrections for percent bleach. Newly formed myotubes showed the same higher membrane fluidity as the unfused myoblasts. After a week in culture, the $t_{1/2}$ returned to its initial level for myoblasts that never fused, as well as for myotubes. It appears that fusion was not required for the changes that made the membrane more fluid, and that fusion was also not required to return the membrane to its initial state. Thus, the membrane changes are not coupled to fusion, although they may be required to make fusion possible.

The fibroblasts present in the culture maintained a constant plasma membrane fluidity in a region over their nuclei throughout the period examined. Their membranes appeared to be more fluid than those of myoblasts. A diffusion coefficient for the fibroblasts of 3×10^{-9} cm^2/sec was calculated using the relationship given above with $\beta = 1$.

The mechanism for the apparent increase in myoblast membrane fluidity during the period of active fusion is not clear. One trivial possibility is a smoothing out of the myoblast membranes following trypsinization. A very convoluted membrane with many microvilli, blebs, and filopodia along it would have a larger surface area (and larger effective value of r in equation 5) within the bleached spot and hence a slower recovery time and lower apparent fluidity than a smooth surface. It is known that trypsin causes myoblast cells to round up, and scanning electron micrographs show that rounded cells produced by trypsinization or during mitosis have very irregular surfaces [24, 25]. Indeed, our direct comparisons of stretched out and rounded myoblasts or fibroblasts by FPR indicate an increase in $t_{1/2}$ as large as twofold for rounded cells. In the studies summarized in Figure 5, however, only extended cells were used. Scanning electron micrographs of cultured primary muscle cells have revealed relatively smooth surfaces on myoblasts and myotubes [26, 27]. Interestingly, primary chick fibroblasts have been shown to have very irregular surfaces [27], yet a higher rate of diffusion of K-1 was obtained on the fibroblast plasma membrane than on the smoother myoblast membrane. Furthermore, the fibroblasts in the culture were also trypsinized and yet did not exhibit the same large $t_{1/2}$ at early times after plating that the myoblasts did. Thus, changes in $t_{1/2}$ measured with K-1 in the region of a nucleus do not appear to be due to changes in surface topography but seem to reflect real changes in membrane fluidity.

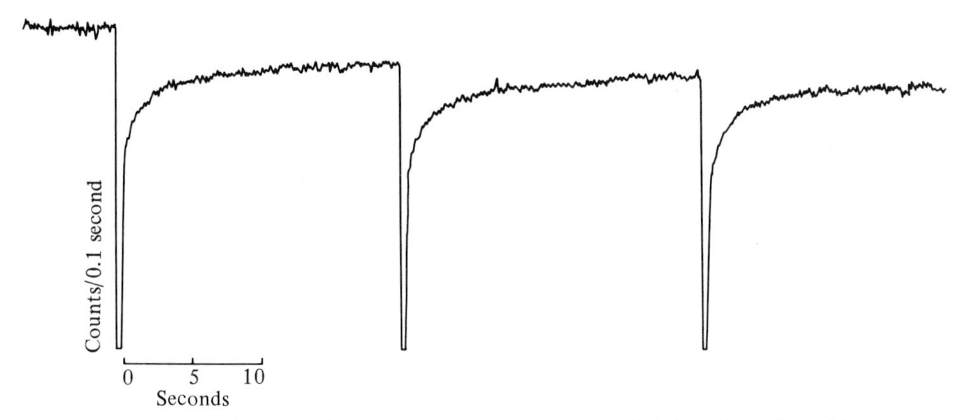

Fig. 4. Fluorescence photobleaching recovery record. A fibroblast labeled with K-1 was bleached on a 2.5 μm in diameter spot for 0.4 sec. Fluorescence recovery into the spot was monitored for three successive bleaches.

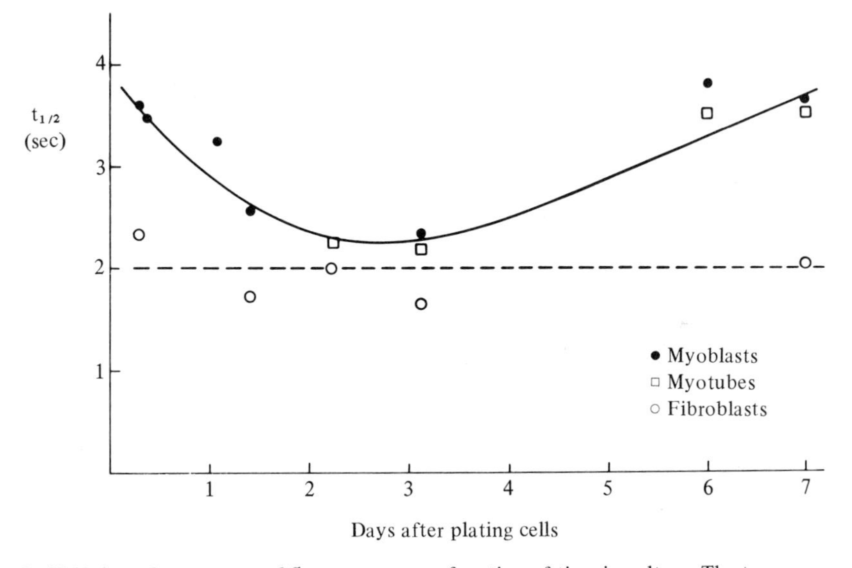

Fig. 5. Half-times for recovery of fluorescence as a function of time in culture. The $t_{1/2}$ were measured as described in Methods. All data from many experiments were averaged for these points. Between 9 and 38 data were averaged per point, with an average of 18 data per point. Standard deviations of the mean were less than ± 10%.

Rotational Mobility

As discussed above, the interpretation in terms of membrane fluidity of $t_{1/2}$ values obtained by the FPR technique requires some knowledge of the microtopography of the cell membrane. Moreover, FPR results obtained from geometrically narrow and irregular cell regions where the bleaching beam diameter is larger than the illuminated cell structure, such as tips of myoblasts, are difficult to interpret. These questions of surface and cell geometry led us to attempt to follow the time course of membrane fluidity by a technique of fluorescence depolarization (FD), which does not depend on such geometry. The technique measures the amount of rotational mobility displayed by a fluorescent probe during its lifetime, which is related to the membrane's fluidity, or microviscosity [21, 28]. Results obtained with weakly oriented probes such as perylene or DPH should be independent of the cell and surface geometry of the plasma membrane around or within the spot illuminated by the laser beam. Perylene was chosen rather than the commonly used DPH because it can be excited in the visible region, which minimizes autofluorescence, and because the labeling procedure does not damage cells, and it does not bleach as readily. It has the problem of not being as sensitive as DPH to changes in viscosity.

Time course of microviscosity above nuclear region. Fluorescence anisotropy was measured by FD over the central nuclear region of myoblasts and fibroblasts. The anisotropy of the perylene emission was found to remain fairly constant on both cell types throughout the culture period examined. The ratio of anisotropies of myoblasts to fibroblasts ranged from 1.0 to 0.95 at 30 h after plating, with no clear trend of a change thereafter. The FPR results of increasing fluidity over the first 3 days and a subsequent decrease would have led to the prediction that the ratio of anisotropies of myoblasts to fibroblasts would have decreased and then increased, yet this was not clearly observed for the nuclear region. The results of the fluorescence anisotropy measurements, however, do not necessarily negate a similar change in fluidity, since calculations that we have made indicate that a 40% change in fluidity, as suggested by the FPR results, would produce only a 5% change in the ratio of anisotropies. Such a slight change would be difficult to detect in our experiments with the experimental precision which we have been able to achieve so far under the microscope with perylene. Moreover, the perylene probe, in contrast to K-1, is internalized, and some of the fluorescence registered in our experiments probably arises from perylene in the nuclear membrane as well as from the plasma membrane.

Fluorescence from the nuclear membrane would tend to obscure possible fluidity changes in the plasma membrane. To see if similar results are obtained at different locations along the length of the cells, fluorescence anisotropy at different regions of a myoblast were measured.

Microviscosity along length of myoblast. Figure 6 is a sketch of a myoblast and the letters a (tip), b (arm), and c (nuclear region) indicate the points on the cell at which fluorescence anisotropy measurements were made. The results for cells that had been cultured for about 50 h are given in Table I.

The arm of the myoblast appeared to have a lower fluorescence anisotropy, or higher rotational mobility (assuming a constant lifetime), than the nuclear region of the cell. The ratio of anisotropies was 0.41, suggesting that perylene in the arm region is in a relatively more fluid environment. However, since perylene penetrates into the cytoplasm, it is not possible at present to state to what extent changes in anisotropy in the arm region reflect changes in membrane versus cytoplasmic fluidity.

In all cases examined, the flattened, widened, or bulbous tips at the ends of the myoblasts had a higher fluorescence anisotropy, or lower rotational mobility, than the rest of the cell. The anisotropy of the tip is twice that of the nuclear region. It is possible that the specialized structures at the leading edge of migrating cells, the filopodia and lamellipodia, represent extensions of membrane that has somehow become more rigid. We have also done FPR measurements at the tips of myoblasts with K-1, which does not label the cytoplasm, and found a manyfold increase in $t_{1/2}$. The FPR results at the tips, however, are difficult to interpret because of the small size of these structures and the resulting uncertainty in the bleaching geometry.

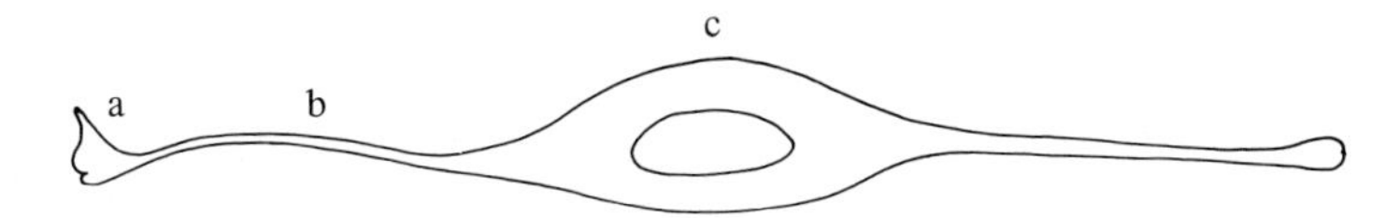

Fig. 6. An elongated myoblast. Letters indicate the positions assayed by the fluorescence depolarization technique: a. tip; b. arm; and c. nuclear region.

TABLE I. Fluorescence Anistropy Along the Length of a Myoblast

Location	$I_{\perp}/I_{\parallel}$[a]	Anisotropy	Ratio to nuclear region
a. Tip	0.78 ± 0.05 (9)	0.086	1.7
b. Arm	0.94 ± 0.02 (20)	0.021	0.41
c. Nuclear region	0.86 ± 0.02 (13)	0.051	

[a]$I_{\perp}/I_{\parallel}$ were corrected and anisotropy was calculated as described in the text. One standard deviation of the mean is given, as well as the number of measurements averaged (in parenthesis).

TABLE II. Fluorescence Anisotropy of Rounded vs Extended Myoblasts

	$I_{\perp}/I_{\parallel}$[a]	Anisotropy
Extended	0.92 ± 0.02	0.029
Rounded	0.78 ± 0.02	0.088
Ratio (extended/rounded)	1.2	0.32

[a]$I_{\perp}/I_{\parallel}$ as well as anisotropies for rounded and extended myoblasts were averaged from three experiments. One standard deviation of the mean is given.

Rounded vs extended cells. During mitosis the normally extended cells become rounded and develop many filopodia and microvilli [24]. Such cells are found at random in the tissue culture during the first 2 or 3 days. It was of general interest to study the microviscosity of these cells. Results with the FD technique are shown in Table II. Rounded cells had about a 20% lower ratio of $I_{\perp}/I_{\parallel}$, or a three times higher fluorescence anisotropy than extended cells. Hence, round cells seemed to show less rotational freedom for the lipid probe perylene. These findings may be related to the relatively lower rotational mobility found at the tips of myoblasts. Filopodia, and possibly other surface projections, may require relative immobilization of some structures in the region of the projections. In the case of membranes, one way this could be brought about is by increasing the protein concentration. This possibility could also account for our observation of lower fluorescence intensity at the tips of cells with the surface-specific lipid probe K-1, suggesting a lower percentage of lipids at these tips.

FPR also showed that rounded cells had a longer recovery time for fluorescence intensity after bleaching than extended cells, suggesting a lower fluidity. However, as mentioned above, this may be due in part to geometrical effects.

DISCUSSION

The FPR and FD techniques are potentially powerful methods of unraveling the role of membrane fluidity and lateral mobility of membrane components in cell functions. The FPR technique can measure the lateral mobility of both fluorescent labeled lipids and proteins, while the FD technique can measure rotational mobility of lipids. Our extension of the FD technique to measurements through a microscope with a polarized laser beam makes it possible to use FD to measure fluidity in specific regions of anatomically polarized cells, and allows one to avoid exciting fluorescence from internal structures which can be discerned under the microscope. Insofar as mobility of lipids is concerned, the FPR and FD techniques complement or reinforce each other. With probes which are only weakly oriented in the membrane [29–31], the FD technique gives information on the average local microviscosity independent of the local cell geometry or topography of the cell membrane, including the presence of microvilli, filopodia, blebs, or lamellipodia, and aids in the interpretation of $t_{1/2}$ values of the FPR technique, which are sensitive to these parameters. In addition, the intense bleaching beam used in the FPR technique could alter membrane properties and yield artifactual results. The much lower intensities used in the FD measurements are not anticipated to alter membrane properties, and thus can be used to support or question results obtained by the FPR technique. The FD technique, on the other hand, is sensitive not only to microviscosity but also to the lifetime of the fluorophore. The lifetime of perylene is not highly sensitive to environmental changes but, in general, it must be ascertained that a change in polarized fluorescence is not due to changes in lifetime before it can be definitely interpreted in terms of fluidity changes.

The proper use of the FPR and FD techniques in the study of the lipid bilayer requires the availability of suitable fluorescent probes. K-1 labeled exclusively the plasmalemma of cultured muscle cells, unless they were damaged, in which case interior structures were labeled. It was thus an ideal lipid probe for FPR experiments with viable cells. However, we could not use it to study rotational mobility by polarized emission, because its short lifetime and relatively slow rotational motion in its surface location make its polarized emission insensitive to fluidity changes [32]. For FD experiments, DPH is a sensitive microviscosity probe, but its accumulation in large vacuoles produced in our cultures during the labeling steps precluded its use. Perylene is less sensitive to microviscosity, but labeling did not damage the cells. This fluorophore, however, is also internalized, which can complicate results of polarized fluorescence measurements. The source of the internal fluorescence can be the nuclear membrane, cytoplasmic membranes (mitochondria, sarcoplasmic reticulum, Golgi bodies, vesicles, etc), as well as components in the cytosol too small to be seen under the microscope (lipid droplets and lipophilic proteins, for example). Under the microscope, one can avoid areas where the major part of the fluorescence is clearly due to discernible internal organelles.

Although perylene proved to be the more useful, neither perylene nor DPH is an ideal probe for polarized fluorescence measurements under the microscope. An ideal probe should 1) not be internalized, 2) have a polarized emission that is highly sensitive to fluidity, 3) be only weakly oriented in a bilayer, 4) absorb in the visible region of the spectrum, 5) not be highly sensitive to bleaching, and 6) label without cell damage. To avoid internalization one must use an amphipathic probe [32]. This would seem to contradict the requirement of only weak orientation. However, covalent attachment of a fluorophore to the terminal region of a lipid hydrocarbon chain results in an amphipathic probe where the fluorescent moeity is weakly oriented [32]. Such probes may eventually allow much better use of the FD technique under the microscope than is possible with perylene and DPH.

With these caveats in mind, we have found that the fluidity of muscle plasma membranes appears to change during myogenesis, as revealed by FPR measurements in regions above nuclei. The fluidity in myoblasts apparently increases as cells become competent to fuse and returns to its initial level after all fusion is essentially complete.

In a previous study on membrane fluidity of myoblasts, Prives and Shinitzky [15] observed a 50% increase in membrane fluidity preceding fusion of muscle cells by following the rotational diffusion of the lipid probe DPH. However, their studies were done on cell suspensions obtained by trypsinization of cultured cells at various times after plating. They thus measured a population of cells that was heterogeneous and whose membranes may have been altered by trypsinization. In addition, our observations under the microscope indicate that the DPH probe used in their experiments is highly internalized and that the fluorescence intensity in a cell suspension may be dominated by structures other than the plasmalemma. However, as mentioned, our results with the surface-specific probe K-1, and using a microscope to assay the three cell types (myoblasts, myotubes, fibroblasts) individually while they were attached to their substratum, indicate that fluidity changes do indeed occur in the plasma membrane of the muscle cell population.

Edidin and Fambrough [33] used fluorescent antibody fragments of anti-muscle plasma membrane antibody to look at muscle membrane fluidity. Antibodies micropipetted onto cells formed a small spot whose rate of enlargement gave a rough diffusion coefficient on the order of $1-3 \times 10^{-9}$ cm^2/sec. Schlessinger et al [34] measured both lipid and protein fluidity by FPR on an established rat muscle cell line, L-6. Their lipid probe was another carbocyanine dye (diI) and gave a diffusion coefficient of $9 \pm 4 \times 10^{-9}$ cm^2/sec. This is a somewhat higher membrane fluidity than observed in our studies on primary chick cells. However, the small differences may be due to differences between 1) the 2 species, 2) established cell lines vs primary cell cultures, or 3) particular laboratory protocols.

Another study implicating an involvement of lipids in myoblast fusion was one in which treatment of cultures with phospholipase C was shown to inhibit fusion [35]. The lipid composition of plasma membranes can affect the rate of growth and fusion of myogenic cells [15, 36], and such changes may produce small changes in lipid mobility as measured by FPR [37]. In addition, tetrameric conconavalin A, which can patch certain glycoproteins and glycolipids, will inhibit fusion at high concentrations [38], and it has been suggested that mobility of some of these sugar-containing molecules is required. Sha'afi et al have measured the microviscosity of muscle sarcolemma preparations from 20-day-old chicks using FD with perylene. They obtained a membrane microviscosity of 61 ± 8 cP [39]. It is not clear, however, whether extraction can alter the membrane properties.

We also observed distinct changes when different locations on a myoblast were assayed by FD using perylene. The tip of the cell was found to be the least fluorescent and least fluid. Fibroblasts in these cultures, when examined by FD using perylene, were also found to have anisotropies at their tips and ruffling membranes that were twice the value measured over their nuclear regions. Although these results indicate interesting changes in fluidity, we cannot at present, using perylene, separate changes in plasma membrane fluidity from changes in cytoplasmic fluidity. Since the same lower intensity was found with the surface-specific probe K-1, the results seem to reflect, at least in part, membrane properties. On the other hand a variety of observations from other laboratories indicate that the cytosol in the tips may be in a state of low fluidity. Thus it has been reported that actin filaments in motile cells form a network at the cell edges and within lamellipodia [40–42]. It has been proposed that the network is in the form of a gel containing actin as well as gelation proteins [43]. If such a gel network should indeed exist in the myoblast tips, it is quite possible that it could be sensed by internalized probe. It should finally be noted that changes in plasma membrane and cytoplasm microviscosities are not mutually exclusive but may, in certain instances, go hand in hand; that is, changes in microviscosity in the cytoplasm might be reflected in similar changes in the adjacent plasma membrane. To determine whether indeed such a correlation exists in the myoblast tips, we are examining various fluorophores as probes of cytosol microviscosity [32].

The arms of the myoblasts showed the highest levels of fluidity as determined from FD measurements. The interpretation here is again limited by the unknown contribution of fluorescence from internal membranes and the cytoplasm. If these contributions are not significant, then it is interesting that the region of higher fluidity coincides with the region where fusion first occurs [11, Fig. 9]. The intermediate level of fluidity over the nuclear region measured with perylene is probably an average between that of the nuclear membrane and the sarcolemma, although their relative contributions are not known.

The observations presented here are that myoblast fusion is accompanied by a higher membrane fluidity, as revealed by FPR measurements with the membrane-specific probe K-1. The membrane changes, however, do not seem to require fusion since they occur in the entire population of muscle cells, even those that do not fuse. In addition, the tips of the cells appear to be more rigid than the other regions of the cell as revealed by FD measurements with the penetrating probe perylene, but it is not clear whether these reflect changes in the plasma membrane, the cytoplasm, or both. The advantages of fluorescence microscopy methods are that single cells can be assayed and that specific locations on the membrane can be selected for assay with a finely focused laser beam.

ACKNOWLEDGMENTS

We wish to acknowledge helpful discussions with M. C. Foster and the assistance of E. Yguerabide.

This work was supported by the Muscular Dystrophy Association (H.F.E.) and NSF PCM75-19594 (J.Y.).

REFERENCES

1. Frye LD, Edidin M: J Cell Sci 7:319, 1970.
2. Hubbell WL, McConnell HM: J Am Chem Soc 93:314, 1971.
3. Singer SJ, Nicolson GL: Science 175:720, 1972.
4. Holtzer H, Abbott J, Lash J: Anat Rec 131:567, 1958.
5. Cooper WG, Konigsberg IR: Anat Rec 133:368, 1959.
6. Shainberg A, Yagil G, Yaffe D: Dev Biol 25:1, 1971.
7. Yaffe D, Feldman M: Dev Biol 11:300, 1965.
8. Hay ED: Z Zellforsch 59:6, 1963.
9. Shimada Y: J Cell Biol 48:128, 1971.
10. Rash JE, Fambrough D: Dev Biol 30:166, 1973.
11. Lipton BH, Konigsberg IR: J Cell Biol 53:348, 1972.
12. Beisson J, Lefort-Tran M, Pouphile M, Rossignol M, Satir B: J Cell Biol 69:126, 1976.
13. Papahadjopoulos D, Poste G, Schaeffer BE, Vail WJ: Biochim Biophys Acta 352:10, 1974.
14. Poole AR, Howell JI, Lucy JA: Nature 227:810, 1970.
15. Prives J, Shinitzky M: Nature 268:761, 1977.
16. Elson HF, Yguerabide J: J Supramol Struct 9:116, 1979.
17. Bischoff R, Holtzer H: J Cell Biol 44:134, 1970.
18. Yguerabide J, Stryer L: Proc Natl Acad Sci USA 68:1217, 1971.
19. Yguerabide J: Presented at Biophysical Society Meeting, 1971.

20. Koppel DE, Axelrod D, Schlessinger J, Elson EL, Webb WW: Biophys J 16:1315, 1976.
21. Shinitzky M, Dianoux A-C, Gitler C, Weber G: Biochemistry 10:2106, 1971.
22. Kehry M, Yguerabide J, Singer SJ: Science 195:486, 1977.
23. Wode B, Yguerabide J, Feldman J (in preparation).
24. Porter K, Prescott D, Frye J: J Cell Biol 57:815, 1973.
25. Follett EAC, Goldman RD: Exp Cell Res 59:124, 1970.
26. Shimada Y: Dev Biol 29:227, 1972.
27. Papadimitriou JM, Dawkins RL: Cytobios 8:227, 1973.
28. Perrin F: J Phys Radium 7:390, 1926.
29. Kawato S, Kinosita K Jr, Ikegami A: Biochemistry 16:2319, 1977.
30. Chen LA, Dale RE, Roth S, Brand L: J Biol Chem 252:2163, 1977.
31. Yguerabide J, Foster MC: In Grell E (ed): "Membrane Spectroscopy." New York: Springer-Verlag (in press).
32. Yguerabide E, Yguerabide J: unpublished results.
33. Edidin M, Fambrough D: J Cell Biol 57:27, 1973.
34. Schlessinger J, Axelrod D, Koppel DE, Webb WW, Elson EL: Science 195:307, 1977.
35. Nameroff M, Trotter JA, Keller JM, Munar E: J Cell Biol 58:107, 1973.
36. Horwitz AF, Wight A, Ludwig P, Cornell R: J Cell Biol 77:334, 1978.
37. Axelrod D, Wight A, Webb W, Horwitz A: Biochemistry 17:3604, 1978.
38. Sandra A, Leon MA, Przybylski RJ: J Cell Sci 28:251, 1977.
39. Sha'afi RI, Rodan SB, Hintz RL, Fernandez SM, Rodan GA: Nature 254:525, 1975.
40. Abercrombie M, Heaysman JEM, Pegrum SM: Exp Cell Res 67:359, 1971.
41. Lazarides E: J Cell Biol 68:202, 1976.
42. Lazarides E: J Supramol Struct 5:531, 1976.
43. Brotschi EA, Hartwig JH, Stossel TP: J Biol Chem 253:8988, 1978.

Journal of Supramolecular Structure 12:165–175 (1979)
Biological Recognition and Assembly 129–139

Evidence of Microfilament-Associated Mitochondrial Movement

Timothy J. Bradley and Peter Satir

Department of Anatomy, Albert Einstein College of Medicine, Bronx, New York 10461

The mitochondria in the lower Malpighian tubule of the insect Rhodnius prolixus can be stimulated by feeding in vivo and by 5-hydroxytryptamine in vitro, to move from a position below the cell cortex to one inside the apical microvilli. During and following their movement into the microvilli, the mitochondria are intimately associated with the microfilaments of the cell cortex and microvillar core bundle. Bridges approximately 14 nm in length and 4 nm in diameter are observed connecting the microvillar microfilaments to the outer mitochondrial membrane and microvillar plasma membrane. Depolymerization of all visible microtubules with colchicine does not inhibit 5-HT-stimulated mitochondrial movement. On the other hand, treatment with cytochalasin B does block mitochondrial movement, suggesting that microfilaments play a role in the mitochondrial motility. We have labeled the microvillar microfilaments, which are 6 nm in diameter, with heavy meromyosin, which supports the contention that they contain actin. A model of the mechanism of mitochondrial movement is presented in which mitochondria slide into position in the microvilli along actin-containing microfilaments in a manner analogous to the sliding actin-myosin model of skeletal muscle.

Key words: microvilli, Malpighian tubule, cytoskeleton, actin, cell motility

The Malpighian tubules of insects are the site of primary urine formation [1]. In the bloodsucking insect Rhodnius prolixus the Malpighian tubules are divided into two functional units: the upper tubule, which is the site of blood filtration by means of an osmotic gradient, and the lower tubule, where KCl resorption and urine acidification occur [2]. The periodic intake of a liquid meal in this insect imposes a cyclic nature on diuresis in the Malpighian tubules.

Dr. Bradley is now at the Dept. of Developmental and Cell Biology, Univ. of California, Irvine, CA 92717.

Received and accepted August 27, 1979.

0091-7419/79/1202-0165$02.30

We have shown that substantial ultrastructural modifications occur in the lower tubule as ion transport is initiated, ie, in response to feeding in vivo or to 5-HT stimulation in vitro [3, 4]. These modifications include microvillar growth, changes in mitochondrial conformation, and the movement of mitochondria from a position inside the cell below the cell cortex to one inside the microvilli. Because both the physiological and morphological changes can be induced rapidly (in 10 min) in vitro by 5-HT, this system has proved very useful.

The microvilli of cells of the lower Malpighian tubule, which form a striated border, differ from more conventionally studied microvilli such as vertebrate intestine in that, under certain conditions, they may contain mitochondria or endoplasmic reticulum (ER) [5]. They share with all microvilli the characteristic that they are apical cell projections enclosed by a unit membrane surrounding a core of microfilaments that extend down out of the base of the organelle into a terminal web.

Based on ultrastructural examination of 5-HT-stimulated tubules, we have suggested that mitochondrial movement in the lower tubule is dependent upon interactions with the microvillar microfilaments [5]. This evidence includes the close and regular association of microvillar core microfilaments with extended mitochondria and the movement of mitochondria into microvilli in which no microtubules are visible.

In this paper we provide further experimental support for this hypothesis. We demonstrate for the first time that the core microfilaments of insect cell microvilli, like vertebrate cell microvilli, contain actin. In addition, we show that drugs such as cytochalasin B, which interferes with microfilament-associated movement, affect mitochondrial extension, whereas drugs such as colchicine do not.

MATERIALS AND METHODS

Animals were reared and dissected as described elsewhere [6]. Dissected tubules were maintained in vitro in normal Rhodnius Ringer [6] in small drops suspended under mineral oil. Tubules were fixed for 1 h in 4% glutaraldehyde in 0.5 M cacodylate buffer (pH 6.9), followed by further fixation in 1% O_5O_4 in 0.5 M cacodylate containing 0.1 M sucrose, 10 min en block staining in 1% uranyl acetate in 70% ethanol, ethanol gradient dehydration, and embedding in Epon 812. Sections were cut on a Reichert ultramicrotome, stained with uranyl acetate in 50% ethanol followed by Reynolds lead stain, and observed in a Siemens 101 electron microscope.

The effect of cytochalasin B on mitochondrial movement was examined by treating cells for 15 min with 10 μm cytochalasin B (Sigma) in dimethylsulfoxide (DMSO) solvent dissolved in Rhodnius Ringer (final DMSO concentration, 1%). Control preparations were placed for an equal time in Ringer's containing 1% DMSO. Treatment with colchicine was carried out at room temperature for 30 min, using 10 mM colchicine. Tubules treated with colchicine were either poststimulated with 5-HT or the two agents were added together. Parallel samples from the same animal were run as controls. Fixation and preparation for EM were as described above.

Heavy meromyosin (HMM) was prepared from rabbit skeletal muscle following the procedure of Kielly and Harrington [7] with only minor modifications. All procedures up to fixation were carried out at 0°C. Tubules were made permeable by a series of glycerine solutions (25%, 10 min; 33%, 10 min; 50%, 1 h; 25%, 10 min) made up in solution A (100 mM KC1, 5 mM $MgCl_2$, pH adjusted to 7.0 with 7 mM sodium phosphate buffer). After glycerine treatment, the tubules were treated with 10 mM sodium pyrophosphate for 10 min, washed 3 times in 25% glycerol in solution A, and placed overnight in solution A with 12.5% glycerol and 2.8 mg/ml HMM. After 3 additional rinses with solution A, fixation and preparation of the tubules for EM was as described above, with the exception that 1% tannic acid was added to the glutaraldehyde and osmium fixation was carried out at pH 6.0 for 20 min.

RESULTS

Colchicine Experiments

The morphology of the lower Malpighian tubule has been described in detail elsewhere [3–5]. Treatment of the lower Malpighian tubules with 10 mM colchicine does not result in gross nonspecific changes (Fig. 1), so that, for example, the basal infoldings of the cell membrane, basal mitochondria, nuclei, and ER appear normal. However, microtubules in the cells are substantially reduced in frequency within 10 min of exposure to colchicine. After 30 min, 75–100% of the microtubules formerly present have been depolymerized, with many cells, upon EM observation, containing no visible microtubules. Probably the most striking effect of colchicine is the gradual disappearance of the microtubule-containing axopods that lie among the apical microvilli [6]. Microtubules also depolymerize in other microtubule-rich regions, such as the apical subcortical area, the periphery of the nucelus, and between the basal infolds.

Treatment with colchicine does not block 5-HT-stimulated mitochondrial movement. This is true whether the tubules are 1) treated with colchicine for 30 min and then stimulated or 2) treated with colchicine and 5-HT simultaneously for 30 min. In the latter case, mitochondrial movement occurs within 10 min after stimulation and is unaffected by the continued presence of colchicine. The tubule shown in Figure 1 is of particular interest, because this section and adjacent sections were found to contain no microtubules when exhaustively searched under the electron microscope. Yet, when stimulated with 5-HT subsequent to the colchicine treatment, mitochondrial movement into the microvilli was normal (Fig. 1). Similar results with many other tubules demonstrate that mitochondrial movement is not dependent on intact arrays of cytoplasmic microtubules and can proceed even when no microtubules are observed near the apical microvilli.

Evidence for Microfilament Involvement

Treatment of the tubules with cytochalasin B has no effect either on gross cell morphology or on the appearance of microfilaments in the apical microvilli. Nevertheless, upon subsequent stimulation with 5-HT, mitochondrial movement is entirely blocked (Fig. 2). In control tubules treated only with DMSO, 5-HT-stimulated mitochondrial movement is normal (Fig. 3).

In nonstimulated tubules, microvillar core microfilaments are evenly spaced throughout the microvillus (Fig. 4a). Following stimulation with 5-HT and mitochondrial movement into the microvilli, there is a redistribution of the microvillar microfilaments such that a sheath of microfilaments is observed surrounding the mitochondria. In microvillar cross-sections this is visualized as a ring of points around the mitochondrion (Fig. 4b). In DMSO-treated tubules, where mitochondrial movement is normal, bridges are observed between the microvillar microfilaments and the outer mitochondrial membrane as well as between microfilaments and the microvillar plasma membrane (Fig. 4c). The length of the bridges average around 14 nm (13.7 ± 1; n = 16). In images where the bridges are free of surrounding stained material (white arrows, Fig. 4c) the thickness of the bridges can be seen to be thinner (∿4 nm) than the core microfilaments (∿ 6 nm).

In nonstimulated tubules the mitochondria are associated with microtubules in the subcortical region of the cell. When tubules are treated with cytochalasin B before stimulation, the mitochondria remain in that region (Fig. 5), suggesting that the mechanism whereby the mitochondria are drawn away from the microtubules and into the microvilli has been inhibited.

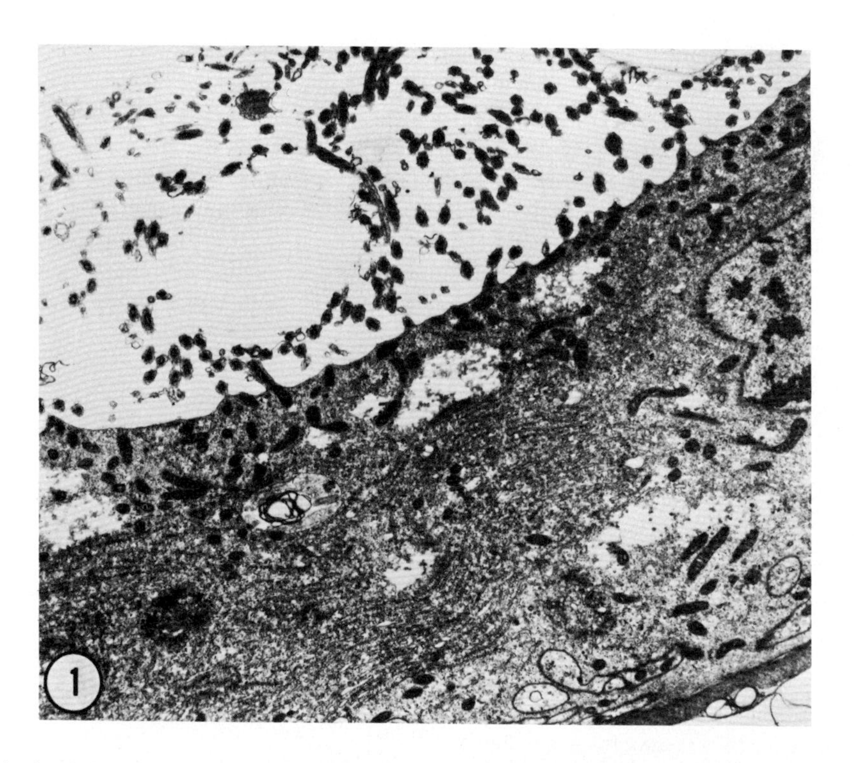

Fig. 1. A tubule treated for 30 min with 10 mM colchicine followed by 5-HT stimulation. Notice that mitochondrial movement into the microvilli is not blocked. Several sections of the tubule shown were examined at high magnification, and no microtubules were observed. Although some microtubules may have been overlooked, one can conservatively state that over 95% of the microtubules previously in the cell were depolymerized by the colchicine treatment; yet mitochondrial movement is unaffected (magnification, × 8,200).

Fig. 2. A tubule treated with 10 μm cytochalasin B in 1% DMSO for 15 min, followed by stimulation with 5-HT. Mitochondria have not moved into the microvilli but remain in the subcortical region as in nonstimulated cells (magnification, × 11,400).

Fig. 3. A tubule treated with 1% DMSO followed by 5-HT stimulation. This is a solvent control for the cytochalasin experiments. Mitochondrial movement into microvilli is normal (magnification, × 12,300).

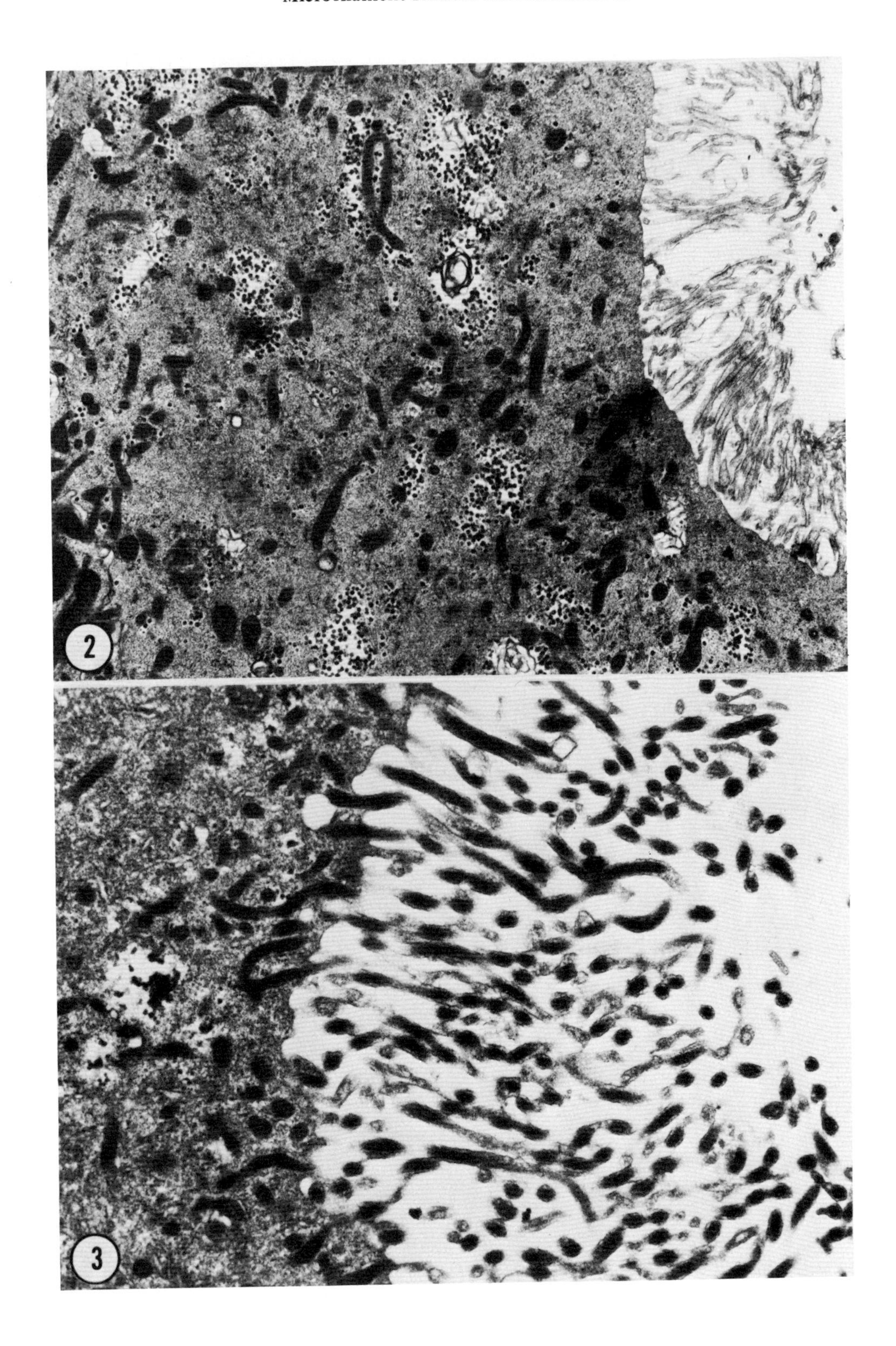
2
3

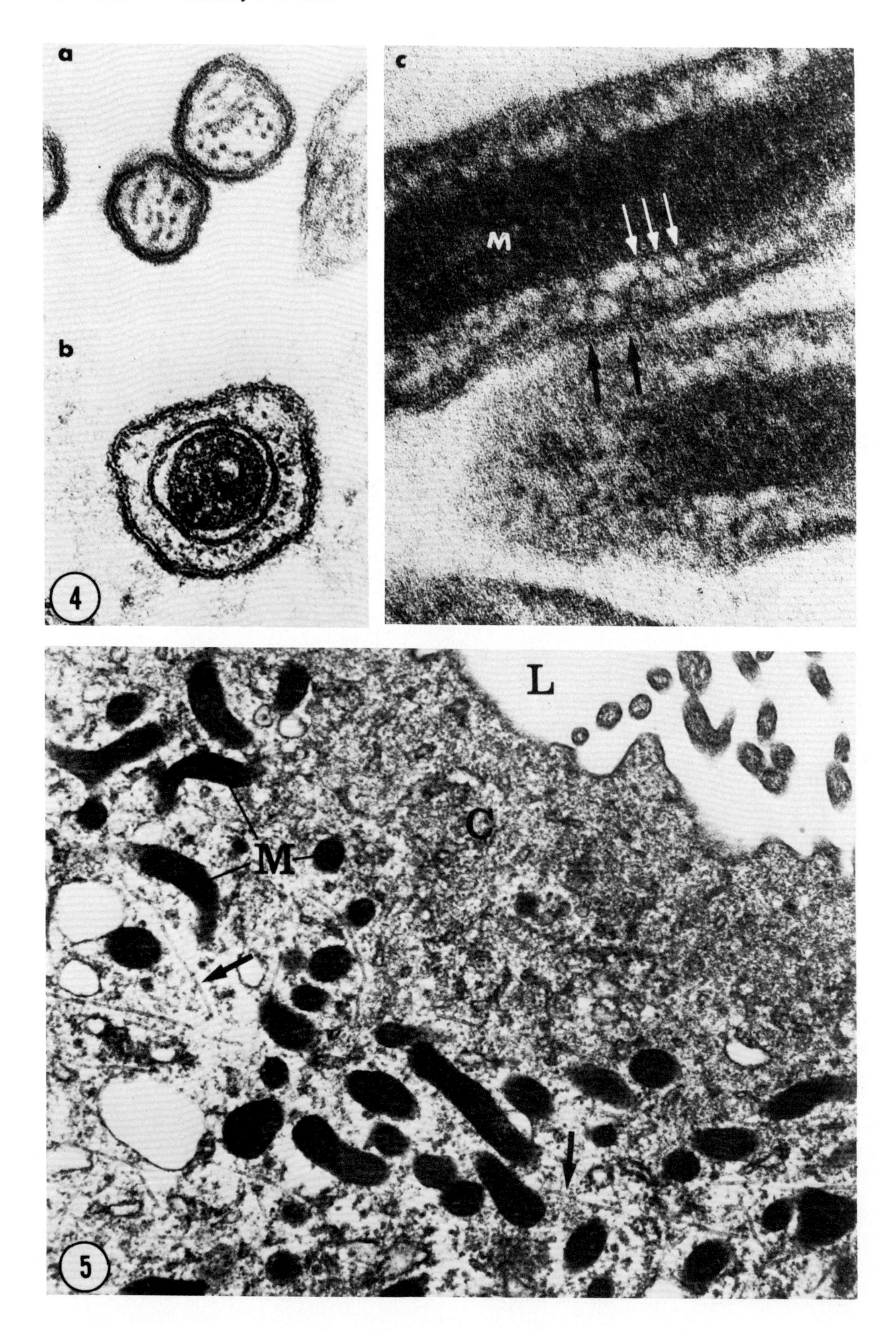
a
b
c
M
4
L
C
M
5

HMM Labeling

We have succeeded in decorating microfilaments in the lower tubule with rabbit skeletal muscle HMM to form arrowheads (Figs. 6 and 7). This finding confirms our hypothesis that the microfilaments in the central cell region and in the microvillar core bundle contain actin. Although the direction of some of the arrowheads in the microvillar core bundles is clearly toward the cell, we are unable to state definitely that all the microfilaments have identical polarity in these preparations.

DISCUSSION

Movements of cell organelles have repeatedly been shown to be associated either with microtubules [8–13] or with microfilaments [14–17]. Based on morphological associations, it has been suggested that mitochondrial movement in the Malpighian tubules of insects is associated with microfilaments [4, 18]. Our work reported here with Rhodnius Malpighian tubules is the first to demonstrate experimentally the role of microfilaments in mitochondrial movement.

5-HT-stimulated mitochondrial movement in the lower tubule clearly is not microtubule dependent, since mitochondria can enter microvilli where no microtubules are visible and depolymerization of approximately 95% of the cellular microtubules with colchicine does not inhibit subsequent mitochondrial movement. This clearly differentiates the mechanism of mitochondrial movement into microvilli in the lower tubule from that postulated for mitochondrial movement in nerve axons, where microtubules are implicated on morphological grounds [19, 20] and on the basis of the blockage of movement by colchicine and vinblastine [21]. It is not clear, however, if a component in positioning and holding the mitochondria below the microvilli in the subcortical region involves microtubules and might be comparable to movement along the axon.

Instead, 5-HT-stimulated mitochondrial movement into the microvilli in the lower tubule evidently involves an interaction with actin-containing microfilaments: 1) We have shown that when mitochondria enter the microvilli in response to 5-HT, there is a reorganization of the microvillar core bundles into a sheath of microfilaments which surrounds the mitochondrion along its entire length within the microvillus. This close association of microfilaments with the mitochondria is observed wherever extended mitochondria are seen, both in the microvilli and in the axopods [5]. Bridges 14 nm in

---

Fig. 4. a. Cross-section of a microvillus from a nonstimulated tubule. The microvillar core microfilaments are evenly spaced (magnification, × 131,250). b. Cross-section of a microvillus from a tubule stimulated with 5-HT. The core microfilaments are organized in a sheath around the mitochondrion. In cross-section the microfilaments of the sheath are seen as a ring of points (magnification, × 137,500). c. Bridges are observed connecting the microfilament 1) to the mitochondrion (white arrows) and 2) to the microvillar plasma membrane (black arrows) in tubules stimulated in 1% DMSO (mitochondrion, M; magnification, × 204,000).

Fig. 5. In tubules stimulated following cytochalasin B treatment the mitochondria (M) remain below the cell cortex (C) in close association with microtubules (arrows) (tubule lumen, L; magnification, × 28,250).

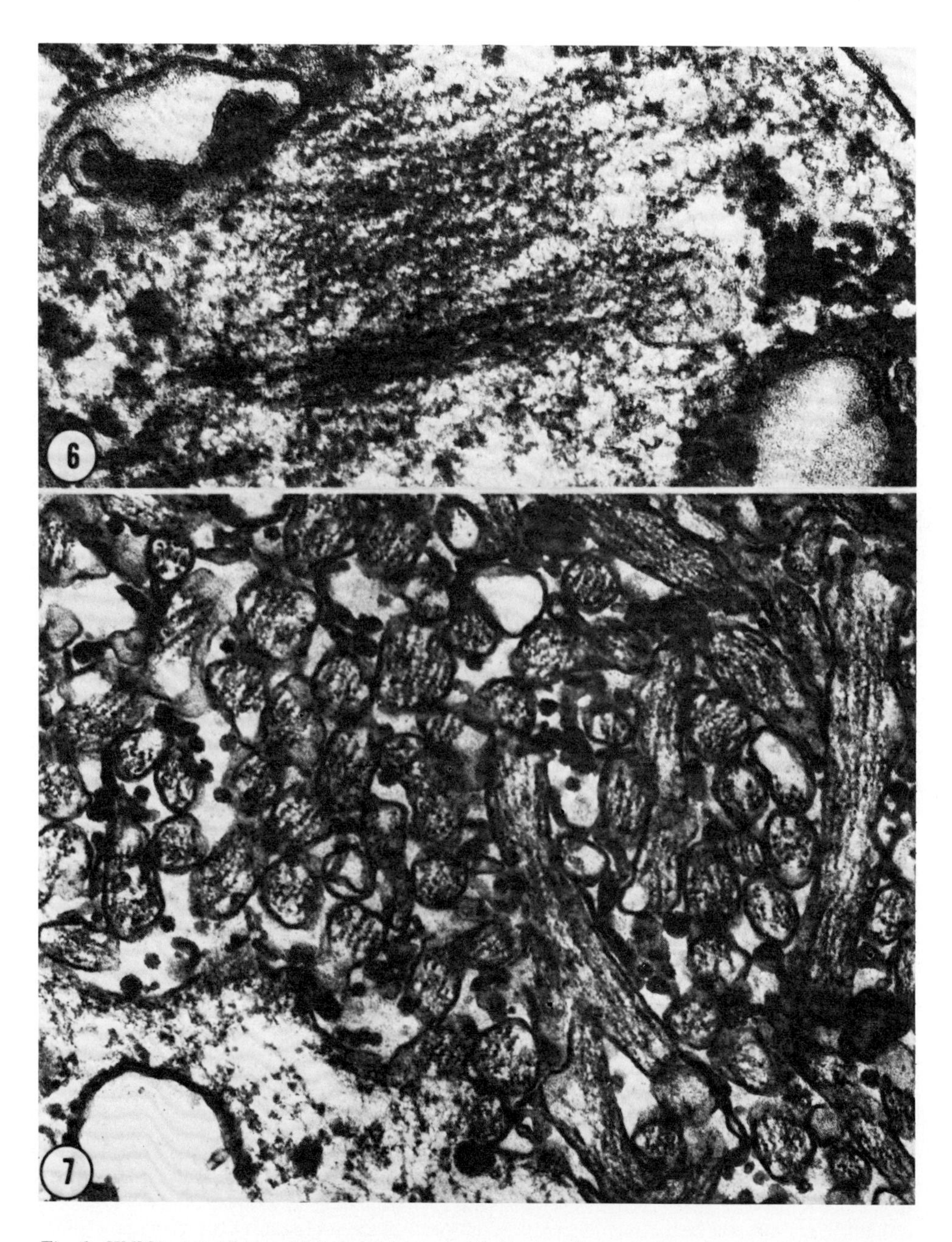

Fig. 6. HMM isolated from rabbit skeletal muscle labels microfilaments throughout the lower tubule cells with an "arrowhead" pattern (magnification, × 87,500).

Fig. 7. The core microfilaments of the microvilli in the lower tubule label with HMM, suggesting that they contain actin (magnification, × 47,500).

length connect the microfilaments to the mitochondrial outer membrane and to the microvillar plasma membrane. The appearance of these bridges is not an artifact induced by DMSO treatment, since they can also be observed, albeit with less contrast, following fixation in the presence of tannic acid where no DMSO is used. The increased clarity of the bridges following DMSO treatment may be due to traces of DMSO remaining during fixation with glutaraldehyde, thereby facilitating the rapid entry of fixative. 2) Treatment of lower tubules with cytochalasin B completely blocks mitochondrial movement upon subsequent stimulation with 5-HT. Other workers have shown that cytochalasin B, although it is not completely specific, blocks microfilament-associated cell and cell organelle motility in vivo [22] and inhibits gelation in vitro of purified actin [23, 24]. 3) We are able to decorate microfilaments throughout the lower Malpighian tubule cell with HMM. This demonstration of actin in the microvillar core microfilaments is the first using insect microvilli. In the microvilli of chicken intestine [25] and sea urchin egg [26, 27] the core microfilaments all label identically with HMM such that the "arrowheads" point toward the base of the microvillus. This orientation has important implications for assembly and function. We are unable to confirm this point in our material at present. We are, however, carefully investigating the orientation of the core microfilaments in the microvilli of the lower tubule, since mitochondria not only move into the microvilli upon stimulation but also move out when stimulation ceases (unpublished observation). We have no information about the presence or location of myosin in these cells.

We summarize our results in a model for 5-HT-stimulated mitochondrial movement in the lower tubule (Fig. 8). In the nonstimulated cell, the mitochondria are located below the cell cortex in close association with microtubules. Occasional bridges are observed between microtubules and mitochondria (Fig. 8a), and we suggest that microtubules may be important in positioning the mitochondria in the subcortical region. At this time the microvillar core microfilaments are evenly spaced within the microvillus, perhaps structurally connected to each other and to the plasma membrane, as is also the case in intestinal microvilli in vertebrates [25, 28].

Upon stimulation (Fig. 8b) the mitochondria partially detach from the microtubules to become associated with the actin-containing microfilaments extending down from the microvillar core bundles. We suggest that this lateral association of the mitochondria with the microvillar core bundles allows the mitochondria to enter the microvilli (Fig. 8c) via an actin-myosin sliding mechanism. During this movement there is a fundamental rearrangement of the microvillar microfilaments from a microvillar core pattern to a regularly spaced single row surrounding the mitochondria. Bridges form links from the outer mitochondrial membrane to the plasma membrane via the microfilaments.

Intestinal microvilli do not contain myosin [29], and it has been proposed that the connections between microfilaments in the core bundle are composed of another protein [28]. Because the location of myosin in the lower tubule has not been determined, it is not clear whether the bridges observed perform a myosin-like function in moving the mitochondria or whether they are composed of other proteins like those in vertebrate microvilli. Further refinements or modifications of our model should therefore result from our present investigations regarding the presence and localization of myosin in the microvilli of the lower tubule and the mechanism by which stimulation associated with 5-HT binding at the basal cell surface is communicated across the cell to influence cytoskeletal events in the apical region.

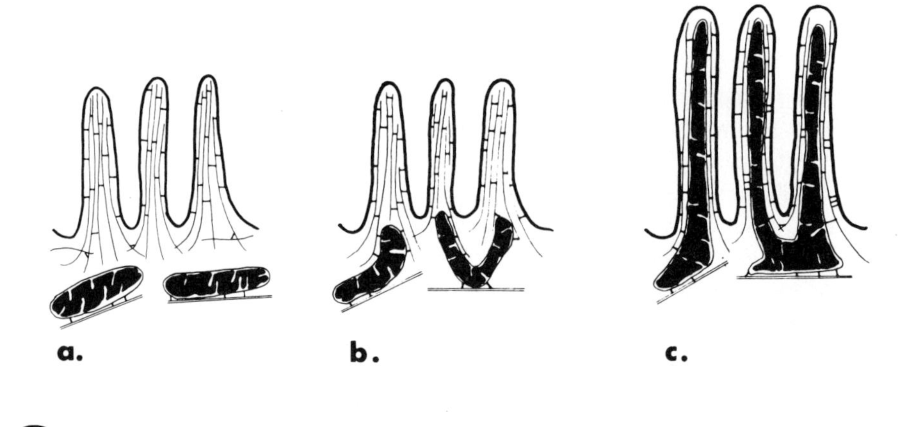

Fig. 8. This diagram presents our conception of the events associated with 5-HT-stimulated mitochondrial movement in the lower tubule. In nonstimulated tubules (a) the mitochondria are located in the subcortical region attached to anchoring microtubules. The core microfilaments, which are evenly spaced in the microvilli, may be positioned by bridges to each other and to the microvillar membrane. Following stimulation (b), the mitochondria become laterally associated with microfilaments which extend down from the microvillus. As mitochondrial movement into the microvillus proceeds (c), the core microfilaments reorganize into a sheath surrounding the mitochondrion. Bridges are observed attaching the microfilaments to the mitochondrial outer membrane and microvillar plasma membrane. Mitochondrial extension proceeds by an actin-based sliding mechanism (see text). Basal regions of the mitochondria remain anchored to the subcortical microtubules.

ACKNOWLEDGMENTS

We would like to thank Drs. John Condeelis and Jeffrey Salisbury for their help in the production of HMM. We thank Ms Rosica Ramer for typing the manuscript, and Mr. Steven Lebduska for photographic assistance.

This work was supported by USPHS grant HL 22560 and NIH postdoctoral fellowship AM 05499.

REFERENCES

1. Maddrell SHP: Adv Insect Physiol 8:199, 1971.
2. Maddrell SHP, Phillips JE: J Exp Biol 62:671, 1975.
3. Wigglesworth VB, Salpeter MM: J Insect Physiol 8:299, 1962.
4. Bradley TJ, Satir P: J Cell Biol 75:255a, 1977.
5. Bradley TJ, Satir P: In preparation.
6. Bradley TJ, Satir P: J Cell Sci 35:165, 1979.
7. Kielly WW, Harrington WF: Biochim Biophys Acta 41:401, 1960.
8. Porter K: In Wolstenholme GEW (ed): "Principles of Biomolecular Organization." Boston: Little, Brown and Co., 1961.
9. Lacy PE, Holwell DA, Young DA, Fink CJ: Nature 219:1177, 1968.
10. Smith DS: Phil Trans R Soc London Biol Sci 261:395, 1971.
11. Bardele CF: Z Zellforsch Mikrosk Anat 126:116, 1971.

12. Satir P: In Sleigh MA (ed): "Cilia and Flagella." New York: Academic Press, 1974.
13. Murphy DB, Tilney LG: J Cell Biol 61:757, 1974.
14. Burnside B: J Supramol Struct 5:257, 1976.
15. McGuire J, Moellmann G, McKeon F: J Cell Biol 52:754, 1972.
16. Palevitz B, Hepler PK: J Cell Biol 65:29, 1978.
17. Goldman R, Pollard T, Rosenbaum J: "Cell Motility." Cold Spring Harbor. New York: Cold Spring Harbor Laboratory, 1976.
18. Ryerse JS: Proc Microscop Soc Can 4:48, 1977.
19. Smith DS, Jarlfors U, Cameron BF: Ann NY Acad Sci 253:472, 1975.
20. Smith DS, Jarlfors U, Cayer ML: J Cell Sci 27:255, 1977.
21. Friede RL, Ho K: J Physiol 255:507, 1977.
22. Wessels NK, Spooner BS, Ash JF, Bradley MO, Luduena MA, Taylor EL, Wrenn JT, Yamada KM: Science 171:135, 1971.
23. Hartwig JH, Stossel TP: J Cell Biol 79:271a, 1978.
24. MacLean S, Griffith LM, Pollard TD: J Cell Biol 79:267a, 1978.
25. Mooseker MS, Tilney LG: J Cell Biol 67:725, 1975.
26. Burgess DR, Schroeder TE: J Cell Biol 74:1032, 1977.
27. Begg DA, Rodewald R, Rebhun LI: J Cell Biol 79:846, 1978.
28. Bretscher A, Weber K: J Cell Biol 79:839, 1978.
29. Mooseker MS, Pollard TD, Fujiwara K: J Cell Biol 79:444, 1978.

Journal of Supramolecular Structure 12:177–183 (1979)
Biological Recognition and Assembly 141–147

Structural Comparisons of the Aggregates of Tobacco Mosaic Virus Protein

Gerald Stubbs, Stephen Warren, and Eckhard Mandelkow

Rosenstiel Center, Brandeis University, Waltham, Massachusetts 02254 (G.S., S.W.) and Abteilung Biophysik, Max-Planck-Institut für Medizinische Forschung, Jahnstrasse 29, 69 Heidelberg, Federal Republic of Germany (E.M.)

The coat protein of tobacco mosaic virus forms numerous aggregates, including the small A-protein, the disk, and two helical forms. The structures of the disk, the helical protein forms, and the virus are compared. Most of the differences are in the conformation of the chain between residues 89 and 113, which lies in the region of protein at the center of the virus, inside the RNA. It is disordered in the disk, but has a fixed conformation in the virus and the protein helices. The differences between the virus and the two helical protein forms are largely in the conformations of arginines and carboxylic acids in this region.

Key words: tobacco mosaic virus protein, X-ray diffraction, protein structure

INTRODUCTION

Tobacco mosaic virus (TMV) is a rod-shaped plant virus, 3,000 Å long and 180 Å in diameter. The coat protein subunits (MW 17,500) form a helix of 49 subunits in three turns. A single strand of RNA follows the basic helix at a radius of about 40 Å, with three nucleotides bound to each protein subunit. Early work on structure and assembly of TMV has been reviewed by Caspar [1].

The coat protein of TMV forms a large number of aggregates, which fall into three main groups: A-protein, disk, and helical forms. The interconversions between aggregates have been summarized by Durham and Klug [2]. The A-protein is a mixture of oligomers (3–12 subunits) present at low ionic strength and high pH. Lowering the pH to near neutral or raising the ionic strength produces the disk, which consists of two layers of 17 subunits each [3]. The layers face the same way, but further polymerization is hindered by a pairing interaction between them [1, 4]. At pH below neutral the protein forms long helices [5] with a morphology similar to that of the virus [6]. There are two forms of helical aggregate (as well as the virus itself): The A form has 16-1/3 subunits per turn, like the virus, whereas the B form has 17-1/3 subunits per turn [7]. The existence of two separate forms was found by X-ray diffraction from oriented gels, and the technique for preparing such gels [8] is such that it has not been possible to correlate the conditions of formation with the type of helix formed.

Received April 9, 1979; accepted September 9, 1979.

X-ray structural studies have been made of the intact virus [9], the disk [10], and both forms of protein helix. Some features of the protein helix structures will be described in this paper, but full details will be given elsewhere (Mandelkow, Stubbs, and Warren, in preparation). The virus structure was determined at a resolution of 4 Å by X-ray diffraction from oriented gels. That part of the structure containing the RNA binding site and some of the features which are important in virus assembly are shown in Figure 1. A more complete illustration of the virus structure is given in Stubbs et al [9: Fig. 2a]. The virus contains four approximately radial α-helices, termed [11] the left and right radial (LR and RR) and the left and right slew (LS and RS). LS and RS lie on top of LR and RR, at angles of 10–20°, as in a coiled-coil of α-helices [10]. LR and RR are connected at the inside wall of the protein (the hole down the center of the virus) by a short length of chain called V [9]. V contains two or three turns of rather irregular α-helix. The protein at higher radius than these five helices has not yet been well described for the virus, because the resolution of the map falls to about 5.5 Å at radii between 60 Å and 80 Å. At a radius of 40 Å the RNA is bound to the protein. The three phosphate groups form ion pairs with arginines 90, 92, and either 41 or 113. The bases lie flat against the hydrophobic surface of the LR helix, forming a shape like a saddle. In the region between the RNA and V, there are six carboxylic acids, which have been called the "carboxyl cage" [9]. Four of these (Glu 95, Glu 97, Glu 106, and Asp 109) are near a site which binds divalent cations, and could be a physiologically significant metal-binding site, perhaps binding calcium [9, 12]. The other two (Asp 115 and Asp 116) are about 10 Å away, nearer the RNA. This arrangement is reminiscent of the double metal-binding sites of thermolysin [13] and concanavilin A [14, 15], although the two sites in TMV are further apart.

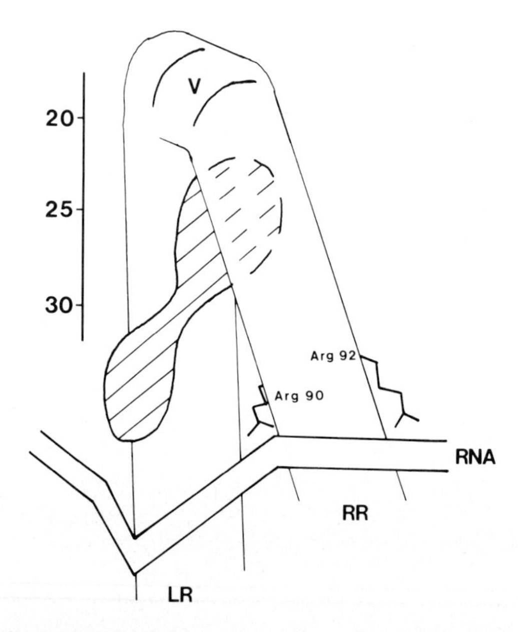

Fig. 1. RNA binding site in TMV (schematic), viewed in the direction parallel to the helix axis. The bar at the side runs in a radial direction, and indicates distance from the axis. The shaded region represents the carboxyl cage. V and parts of LR and RR are shown, together with the backbone of the RNA. Bases of the RNA are not shown, but if they were, the bases of the RNA strand *behind* the one shown would bind to LR.

The disk crystallizes [3], and its molecular structure has been determined to 2.8 Å resolution [10]. Although the intersubunit arrangement is different, the internal structure of the disk subunit [11: Fig. 1] is very similar to that of the virus. The major difference is that the loop of protein containing V, residues 89–113, is disordered in the disk crystal [11] and has been shown to be very flexible in solution [16]. A feature which was not evident in the virus map because of the lower resolution but is presumably common to both structures is the "hydrophobic girdle," a concentration of aromatic and proline residues about 75 Å from the center of the disk [10].

It is evident that the morphology and general fold of the protein chain are well understood for all the aggregates. However, the differences in molecular structure between the aggregates are less clear. This paper addresses these differences and attempts to describe some of them.

METHODS

The methods used to calculate the electron density maps of the protein helices will be described briefly here and more fully elsewhere. Samples of repolymerized TMV protein were prepared as described by Mandelkow et al [7] and made into oriented gels [8]. These gels were of as high a quality as those used to determine the virus structure [9]. Diffraction patterns were recorded on film, measured on a computer-controlled flat-bed scanner and corrected for geometric factors [17]. These diffraction patterns were very similar to those of the virus but fell into two classes, A and B [7]. Electron density maps of both forms were calculated to a resolution of 4 Å using intensities from the protein helices and phases from the virus. The most useful maps were $(2|F_{obs}|-|F_c|)\exp(i\alpha_c)$ Fourier-Bessel syntheses. $|F_{obs}|$ was obtained from the diffraction patterns of the protein helices. $|F_c|$ and α_c were usually obtained from the virus map in the following way: A subunit of TMV in the map was defined by an envelope and transformed to calculate $|F_c|$ and α_c. Various envelopes were used, deleting the RNA or regions of protein which were of particular interest. Deleting such a protein region from the calculated structure factors ensures that the map will not be biased towards identity with the virus. For comparison, one map of the A form was calculated using the measured virus structure factors as $|F_c|$ and the virus isomorphous replacement phases [9] as α_c. This was possible because the A form is isomorphous with the virus.

The protein helices were compared with the virus by careful sheet-by-sheet comparisons of the electron density maps. The disk was compared with the virus using the published descriptions of the disk structure [10, 11].

RESULTS

The principal differences between the various aggregates are shown schematically in Figure 2. Many of the differences between the disk and the virus have been described by Champness et al [11]. The RNA is not part of the disk structure. In its place there is a large open area, formed by the pairing interaction between the layers, which tilts the layers apart at low radius. It has been described as a pair of jaws ready to accomodate the RNA. From the RNA binding site to the center of the disk there are no structural features in the map, so that the hole in the center of the disk appears to be 80 Å across, compared with 40 Å in the virus. Part of the hole must be occupied by a disordered chain from residue 89 to residue 113. By contrast, this chain is fully ordered in the virus: Two

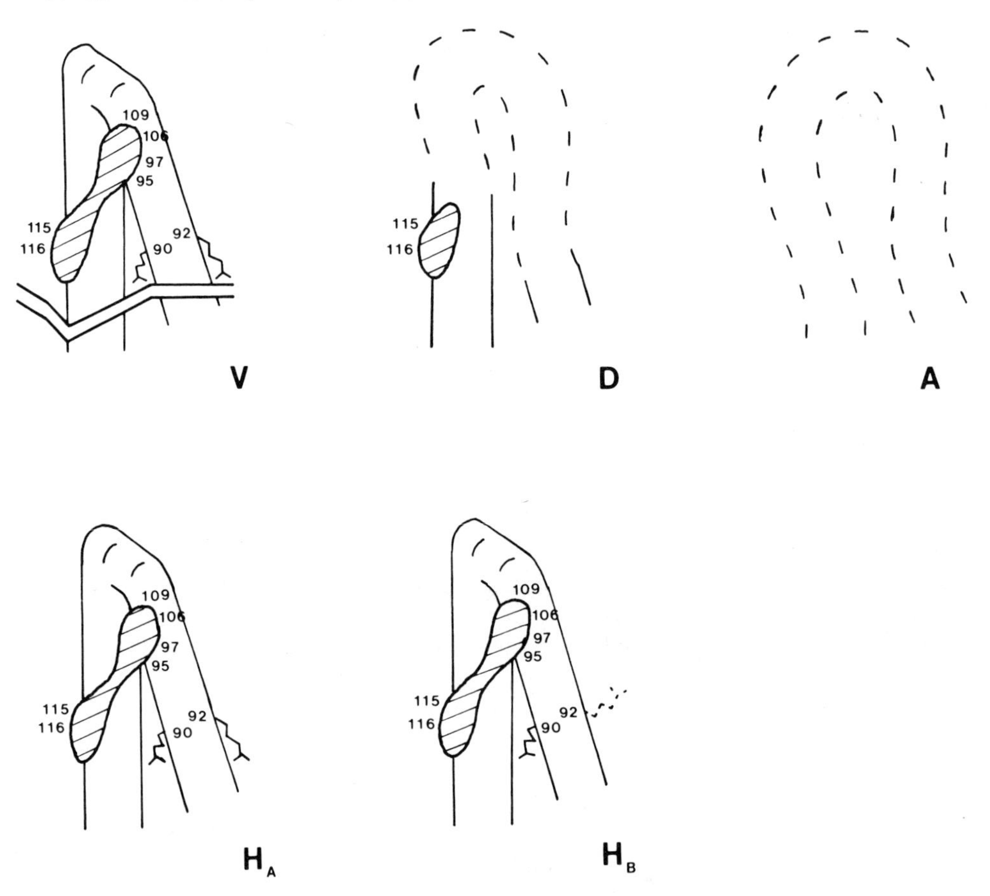

Fig. 2. Schematic representation of the central loop (residues 85–120) in the TMV aggregates. Letters indicate aggregate: Virus, disk, A-protein (speculative), helix A, helix B. Numbers indicate approximate residue locations. Solid lines: Ordered structure. Broken lines: Disordered structure. Shading: "carboxyl cage."

of the radial helices (LR and RR) of the virus extend in beyond the RNA, and are connected by a length of chain which is also partly α-helical (V) [9]. V forms a wall which lines the central hole of the cirus, shielding the RNA from the solvent. It has been suggested [11, 16, 18] that this part of the protein is flexible in the disk in order to allow the RNA to reach its binding site from the central hole.

The packing of the subunits is different in the disk and the virus. The side-to-side interactions are very similar: Two polar regions alternate with two hydrophobic regions [10]. A salt bridge between Arg 122 and Asp 88 has been seen in both structures [9, 10]. However, the interaction in the axial direction (between the rings) of the disk is quite different from the interaction between successive turns of the virus helix [9, 11]. In the virus, the four approximately radial α-helices are closely packed, even between subunits, with the LR helix above and between the RS helix from one subunit and the LS helix from another [11: Fig. 3]. In the disk, one layer is displaced laterally about half a subunit, so that LR is now approximately between LS and RS of the *same* subunit below, although intersubunit packing of helices is now much looser. The molecular details of this packing have been established [10] : There is a complex network of salt bridges and hydrogen bonds between the subunits. Looking at the 4 Å map of the virus, it appears possible that the inner end of this network (involving part of RS and part of LR) may be maintained by changing the conformation of the side chains, but the rest of the network must be completely different. More details must await a higher resolution map of the virus.

Comparing the protein helices with the virus, we find much more similarity than exists between the disk and the virus. All the significant changes which we can see at this resolution are in or near the RNA binding site. Once again the RNA is absent, but at least one of the three phosphate groups of the RNA is replaced by an anion, probably forming an ion pair with Arg 90. The hydrophobic base binding site on the LR helix is exposed to the solvent, as it is in the disk. The structure of the inner protein, which is disordered in the disk, is intact in both forms of protein helix. This region contains most of the carboxyl cage, and while there are some changes around these carboxyl groups, it must be borne in mind that the virus structure was determined at pH 8, well above the pKs of all the carboxyl groups, whereas the protein helix structures were determined at pH 5.5, where some of the groups would be protonated [1].

Only one significant difference has been found between the A form of the protein helix and the B form. There is a peak of electron density near the inner end of RR which is present in A, but absent in B. This difference was observed regardless of whether this peak was included in the calculation of phases from the virus map. In the virus model [9] the peak has been tentatively attributed to the guanidinium group of Arg 92, which forms an ion pair with an RNA phosphate. We might thus suggest that this arginine retains its conformation in the virus-like A form, but is disordered or substantially moved in the B form. It could bind to an anion, or interact with a nearby carboxyl group – the maps do not allow us to make a definite statement.

DISCUSSION

The transitions between forms of TMV protein all appear to involve conformational changes of the charged groups in the low radius region: The six carboxyls, two of the phosphate-binding arginines (90 and 92), and possibly the other arginines in this region: 41, 112, and 113. The importance of these charged groups is not surprising, since the main parameters affecting aggregation state are pH and ionic strength [2]. Arg 41 is the only arginine in the inner part of the protein for which the main chain is ordered in all forms so far examined. It is possible, however, that the side chain of this residue is more disordered in the protein helical form than it is in the virus, and it may have different conformations in the virus and the disk. The maps are difficult to interpret in this region. Arg 112 and Arg 113 are near the carboxyl cage [9] and probably form ion pairs with two of the carboxylic acids. It is not possible to say whether they change conformation between the virus helix and the protein helix forms. They are part of the flexible loop in the disk, and are therefore disordered in this aggregate, as are Arg 90 and Arg 92. Arg 90 has the same conformation in the virus and both protein forms, binding a phosphate of the RNA in the virus, and an anion in the protein forms. Arg 92 has an ordered conformation in the virus and the A form of the protein helix, binding RNA in the virus and perhaps an anion in the phosphate site in the protein. However, it has a different conformation in the B protein helix. This is the only major difference so far observed between the A and B protein helix forms and may well be the cause of the difference between the forms. Arg 92 is in the side-to-side intersubunit boundary, and a conformational difference which affected the intersubunit spacing at this point would change the number of subunits per turn in the helix.

The carboxyl residues (the carboxyl cage) are the source of the anomalous pK values near 7 observed by Caspar [1]. Such pKs arise when carboxyl groups are forced into close proximity by the protein structure. The resulting structure can serve as a switch, active under physiological conditions, between different states of the protein [9, 19]. The virus and protein helices have two or three anomalous pKs [1, 12], and in these structures all six carboxyl residues are held in fixed conformations. In the disk, only residues 115 and 116 of the carboxyl cage retain any trace of rigid conformation [10]. This is because, in the absence of the binding energy of the RNA (as in the virus) or the neutralization of some of the carboxyl groups at lower pH (as in the protein helices), electrostatic repulsion within the flexible loop is too great to allow it to fold [9]. This is consistent with the observation [12] that there is only one anomalous pK value in the disk form, presumably arising from these two residues. Since the A-protein (the small oligomer mixture) has no anomalous pK values [12], we might expect the extent of disorder to be even greater in this form. Asp 115 and Asp 116 are just beyond the end of the flexible loop, at the start of the LR helix in the disk, but LR extends further back in the virus. It may unwind even more in the A-protein than it does in the disk, allowing these two residues to lose their fixed conformation.

Comparing all the major forms of TMV protein so far examined, we see that almost all the differences occur in the RNA binding site and the protein region inside this site. This region, containing the carboxyl cage and the V helix, is where control of assembly is believed to reside [9]. The stability of its secondary and.tertiary structure appears to depend on the stability of the carboxyl cage: Four of the carboxyls (95, 97, 106, and 109) are actually part of the flexible loop. Since it is ordered in all helical forms, and probably not in any other form, it seems likely that the formation of this central ordered protein region is an essential step in the transition from 17-fold (disk) symmetry to helical symmetry. V helices from neighboring subunits are only 9 Å apart, that is, they are close-packed, so the interactions between neighboring V helices could well determine the aggregation state. Thus we see that if the carboxyl cage acts as a switch [19] between the disk and helical forms, the transition is probably mediated through the formation of the ordered central protein region.

ACKNOWLEDGMENTS

The authors would like to thank Drs. D.L.D. Caspar and K.C. Holmes for valuable discussions during this work, and Drs. L. Makoswski and C.V. Stauffacher for helpful criticisms of the manuscript.

REFERENCES

1. Caspar DLD: Adv Protein Chem 18:37, 1963.
2. Durham ACH, Klug A: J Mol Biol 67:315, 1972.
3. Finch JT, Leberman R, Chang Y-S, Klug A: Nature 212:349, 1966.
4. Finch JT, Klug A: Phil Trans Roy Soc B261:211, 1971.
5. Schramm G: Z Naturforsch 2b:249, 1947.
6. Franklin RE: Nature 177:928, 1956.
7. Mandelkow E, Holmes KC, Gallwitz U: J Mol Biol 102:265, 1976.

8. Gregory J, Holmes KC: J Mol Biol 13:796, 1965.
9. Stubbs GJ, Warren SG, Holmes KC: Nature 267:216, 1977.
10. Bloomer AC, Champness JN, Bricogne G, Staden R, Klug A: Nature 276:362, 1978.
11. Champness JN, Bloomer AC, Bricogne G, Butler PJG, Klug A: Nature 259:20, 1976.
12. Durham ACH, Vogel D, deMarcillac GD: Eur J Biochem 79:151, 1979.
13. Matthews BW, Colman PM, Jansonius JN, Titani K, Walsh KA, Neurath H: Nature New Biol 238:41, 1972.
14. Hardman KD, Ainsworth CF: Biochemistry 11:4910, 1972.
15. Edelman GM, Cunningham BA, Reeke GN, Becker JW, Waxdal MJ, Wang JL: Proc Natl Acad Sci USA 69:2580, 1972.
16. Jardetzky O, Akasaka K, Vogel D, Morris S, Holmes KC: Nature 273:564, 1978.
17. Holmes KC, Stubbs GJ, Mandelkow E, Gallwitz U: Nature 254:192, 1975.
18. Butler PJG, Finch JT, Zimmern D: Nature 265:217, 1977.
19. Caspar DLD: In Markham R, Horne RW (eds): "Proceedings of the Third John Innes Symposium." Amsterdam: North-Holland, 1976, p 85.

Journal of Supramolecular Structure 12:299–304 (1979)
Biological Recognition and Assembly 149–154

Gliding Mycoplasmas Are Inhibited by Cytochalasin B and Contain a Polymerizable Protein Fraction

Jack Maniloff and Utpal Chaudhuri

Department of Microbiology, University of Rochester Medical Center, Rochester, New York 14642

Studies are presented on the effect of cytochalasin B (CB) on the growth of five Mycoplasma species, three Acholeplasma species, and one Spiroplasma species. The three gliding mycoplasma species (M gallisepticum, M pneumoniae and M pulmonis are the only mycoplasmas inhibited by CB. These are the only prokaryotes reported to be inhibited by CB. This suggested that these three mycoplasmas might have some sort of cytoskeletal structure. A protein fraction has been isolated from M gallisepticum which polymerizes in 0.6 M KCl and depolymerizes when KCl is removed. This fraction contains a major 58,000-dalton protein, a 46,000-dalton protein, and a minor 87,000-dalton protein.

Key words: mycoplasma, cytochalasin B, actin-like protein, cytoskeleton

The mycoplasmas are a group of prokaryotes which lack cell walls; each cell is bounded by a single lipoprotein cell membrane (see, for example, Maniloff and Morowitz [1] and Razin [2]). These are the smallest free-living cells, with genomes of only (0.5–1.0) $\times 10^9$ daltons of DNA. From a comparison of mycoplasma and bacterial 16S rRNA oligonucleotides (see Maniloff et al [3]; C. R. Woese, J. Maniloff, and L. B. Zablen, manuscript in preparation), it has been concluded that the mycoplasmas are close to one of the clostridia sublines. Therefore, mycoplasmas are not related to some simpler prokaryote which was ancestral to eubacteria; instead, mycoplasma arose from the gram-positive eubacteria. The only reported exception is a thermophilic acidophilic mycoplasma which belongs to the archaebacteria, not the eubacteria.

Among the more than 50 reported mycoplasma species, three species are different because their cells have polar subcellular structures and exhibit gliding motility [4]. These three species, all of which are respiratory pathogens in different animal hosts, are Mycoplasma gallisepticum, M pneumoniae, and M pulmonis.

Utpal Chaudhuri is now at the Department of Physics, Calcutta University, Calcutta.

Received April 10, 1979; accepted August 24, 1979.

0091-7419/79/1203-0299$01.40

Studies on the subcellular polar structures and molecular events in the cell cycle have focused on M gallisepticum. These cells are pear-shaped with a polar bleb structure, and prior to division cells are elongated with bleb structures at both ends [5]. Some micrographs show a bleb area in the cytoplasm near a polar bleb structure [6], indicating that each new bleb might be assembled at a preexisting one. The polar bleb can be separated from the cell by sonication, isolated, and characterized [6–9]. These studies showed that the blebs contain the sites of DNA replication and that the chromosome origin remains attached to a bleb throughout the cell cycle. Studies with synchronized cells showed that the M gallisepticum chromosome is replicated once per cell generation and that there is a gap in DNA synthesis during cell division [8].

These data led us to suggest that the M gallisepticum bleb structures might function as a primitive "mitotic-like" apparatus for chromosome organization and segregation [9]. If one of the nascent chromosomes were attached to a new bleb (made at a preexisting one and, hence, near the DNA origin attachment site), then bleb migration to form the opposite pole of the cell for division would assure chromosome segregation. Since each cell may only contain a single genome copy [10], this cell cycle could have evolved to assure segregation of the two daughter chromosomes before cell constriction and division.

Such a cell cycle requires a mechanism for the intracellular movement of subcellular structures. To investigate this, the effect of several drugs on M gallisepticum was examined [11]. Colchicine and vinblastine had no effect on cell growth, but cytochalasin B (CB) inhibited M gallisepticum cell division. The data indicated that CB blocks at two points in the cell cycle: at the time of bleb structure formation and at the time of cell division [11].

We report here studies to see what other mycoplasma species might be inhibited by CB and the isolation of a polymerizable protein fraction from M gallisepticum.

CYTOCHALASIN B STUDIES

The effect of CB on the growth of five Mycoplasma species, three Acholeplasma species, and one Spiroplasma species was examined. Typical dose-response curves are shown in Figure 1. Some species were slightly inhibited at higher ethanol concentrations. However, only M gallisepticum, M pneumoniae, and M pulmonis were inhibited by CB. The inhibition of M gallisepticum, but not of A laidlawii or M capricolum, confirms recently reported results [11], although different A laidlawii and M capricolum strains were used in this and the previous study. Hence, the three mycoplasma species having polar subcellular structures are inhibited by CB.

CHARACTERIZATION OF POLYMERIZING PROTEIN FRACTION

The isolation protocol was based on those used to isolate actin-like proteins (see, for example, Pollard and Weihing [19]). However, n-butanol was used instead of acetone, since n-butanol has been reported to solubilize mycoplasma membranes more effectively than acetone [20]. Preliminary experiments indicated that a better yield of mycoplasma polymerizable protein was obtained with n-butanol than with acetone (data not shown).

M gallisepticum cells from a 10- to 12-liter culture were harvested by centrifugation (10,000g for 10 min at 4°C) and washed once with 200 ml 0.01 M Tris–0.001 M EDTA buffer (pH 7.5). The cell pellet was resuspended in 50 ml depolymerizing buffer (0.2 mM ATP, 0.5 mM β-mercaptoethanol, 0.5 mM Tris, pH 7.6) and freeze-thawed (–196°C to

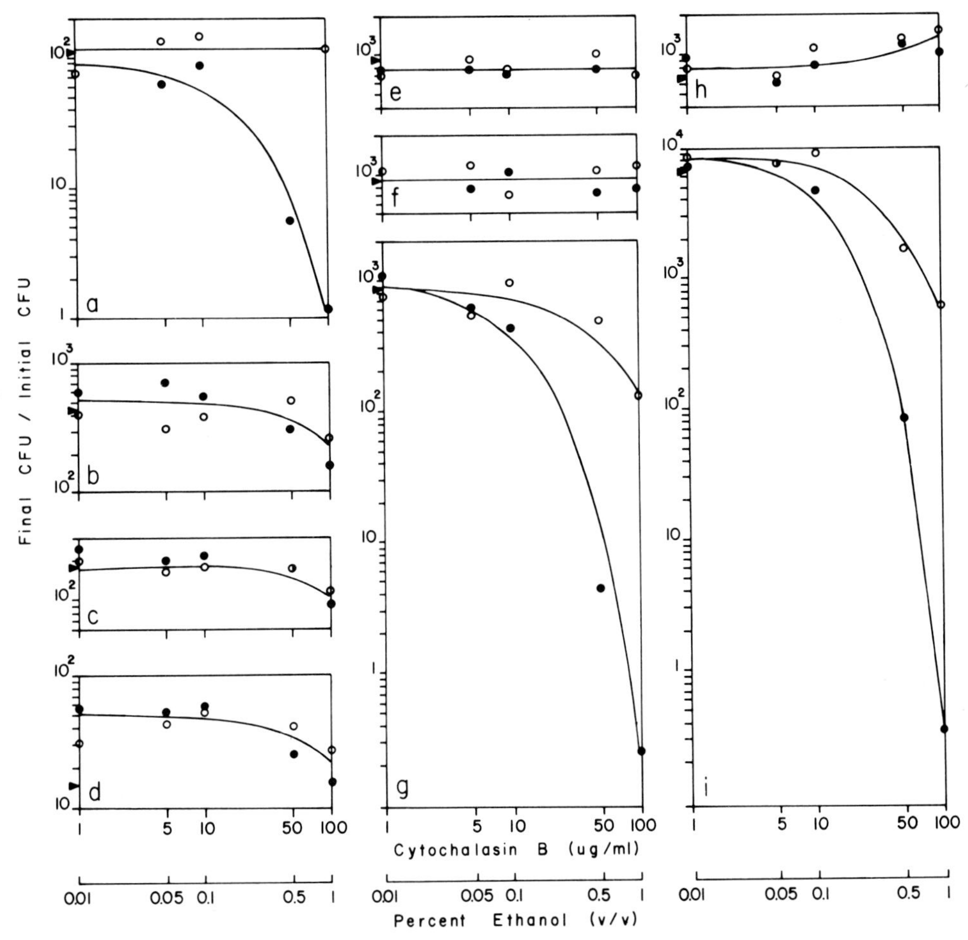

Fig. 1. Effect of CB on mycoplasma growth. CB was dissolved in ethanol and diluted in growth media. Mycoplasmas were added to tubes containing medium plus CB (●) and, as controls, to tubes containing medium plus ethanol concentrations equivalent to those in the CB-containing tubes (○). At each concentration growth is expressed as the number of colony-forming units (CFU) assayed after a 2- 4-day incubation, divided by the number of CFU assayed at the start of the experiment. The arrow on each ordinate shows growth in a control tube containing no CB or ethanol. a) M gallisepticum strain A5969 [12]; b) A modicum strain PG-49; c) A axanthum strain S-743; d) S citri strain Maroc; e) M capricolum strain Kid [13]; f) M arginini strain G230; g) M pneumoniae strain FH; h) A laidlawii strain JA1 [14]; i) M pulmonis strain PG-34. The strains without literature citations were supplied by Dr. J.G. Tully (NIAID, Bethesda, Maryland), except S citri, which was supplied by Dr. R.F. Whitcomb (US Department of Agriculture, Beltsville, Maryland. b,h) Grown on tryptose broth medium [15]; g) grown on a mycoplasma broth base medium [16]; d) grown in a PPLO broth medium [17, 18]. All others grown in a mycoplasma broth base medium [10].

40°C) twice. An equal volume of n-butanol was added and the suspension was stirred for 4–6 h at 4°C. The suspension was filtered through Whatman No. 3 paper and dried. The butanol powder was resuspended in 20 ml depolymerizing buffer. The solution was stirred for 16 h at 4°C and centrifuged (48,000g for 2 h at 4°C) to remove cellular debris. The supernatant was removed and centrifuged again (120,000g for 2 h at 4°C) to remove other slower-sedimenting material. Protein in this supernatant was polymerized by the addition of KCl to a final concentration of 0.6 M. The solution was stirred gently for 16 h at 4°C and then centrifuged (120,000g for 2 h at 4°C). The pellet was resuspended in 3 ml polymerizing buffer (depolymerizing buffer containing 0.6 M KCl, pH 7.6) and allowed to disperse overnight at 4°C. The solution was dialyzed for 3 days at 4°C against 1 liter depolymerizing buffer. The protein was then polymerized again by the addition of KCl to a final concentration of 0.6 M.

The final solubilized protein after two cycles of polymerization-depolymerization was termed polymerizable protein fraction. The yield from 10–12 liters of culture was 300–600 μg protein. SDS-polyacrylamide gel electrophoresis in 10% acrylamide slab gels [21] of this protein revealed three bands: a major band at 58,000 daltons, a band with a little less material at 46,000 daltons, and a faint band at 87,000 daltons (data not shown).

The polymerization-depolymerization kinetics of the protein fraction were followed by viscometry. The addition of KCl (to a final concentration of 0.6 M) to the protein fraction in depolymerization buffer caused an increase in the specific viscosity from 0.05 to 0.13 in about 4 h (Fig. 2a). The viscosity increase was less if the protein fraction was mixed with both KCl (0.6 M final concentration) and $MgSO_4$ (0.001 M final concentration) (Fig. 2b). No increase in viscosity was seen with $MgSO_4$ alone (data not shown).

Dialysis of the KCl polymerized protein against depolymerizing buffer, to reduce the KCl concentration, caused a decrease in viscosity (Fig. 2c). However, after 72 h dialysis the specific viscosity had not returned to its original depolymerized value of 0.05. Addition of KCl, to 0.6 M final concentration, caused this material to polymerize again (Fig. 2d).

DISCUSSION

The three mycoplasma species, which have terminal structures and exhibit gliding mobility, are the only mycoplasmas found to be inhibited by cytochalasin B. These three (M gallisepticum, M pneumoniae, and M pulmonis) are the only prokaryotes reported to be inhibited by CB. Previous studies have shown that the effect on M gallisepticum is not due to an inhibition of the uptake of glucose or macromolecule precursors [11], indicating that CB may be affecting processes dependent on some sort of microfilament structure.

With regard to the antibacterial action of the cytochalasins, cytochalasins B, D, and E have been reported to have no effect on gram-positive and gram-negative bacteria [22–24]. Cytochalasin A (5–25 μg/ml) has been reported to inhibit the growth of gram-positive bacteria (probably by inhibiting solute transport), but not to inhibit gram-negative bacteria [25]; The latter observation conflicts with a report that cytochalasin A (25 μg/ml) causes the lysis of Escherichia coli [23].

An actin-like protein has been reported to have been isolated from M pneumonia [26]. However, the M pneumoniae results need to be reevaluated in terms of the more recent data that the actin isolation procedure applied to E coli leads to the isolation of the translation elongation factor EF-Tu, and that EF-Tu has properties similar to actin [27–29]. Also, the M pneumoniae experiments [26] followed an isolation protocol described for platelet actin, which leads to a protein fraction that polymerizes in low salt and depolymerizes in high salt [30]. Opposite properties have been reported for muscle actin and all other nonmuscle actins (see, for example, Pollard and Weihing [19] and Korn [31]). Recently, it has been shown that the polymerization behavior of platelet actin can be explained by the salt dependence of its critical concentration [32]; when this is taken into account, it was found that platelet actin polymerization was similar to that of other nonmuscle actins. Hence, the polymerization behavior of the M pneumoniae protein needs reexamination as well.

The studies described here were begun to investigate possible mechanisms for the movement of M gallisepticum polar structures during the cell cycle and for DNA segregation. The CB inhibition results were consistent with the idea that there might be some sort of microfilament structure in M gallisepticum and led to experiments to isolate these structures. A protein fraction has been isolated from M gallisepticum which polymerizes in 0.6 M KCl and depolymerizes when KCl is removed. This fraction contains a major 58,000-dalton protein, a 46,000-dalton protein, and a minor 87,000-dalton protein. Experiments are in progress to further characterize this protein fraction.

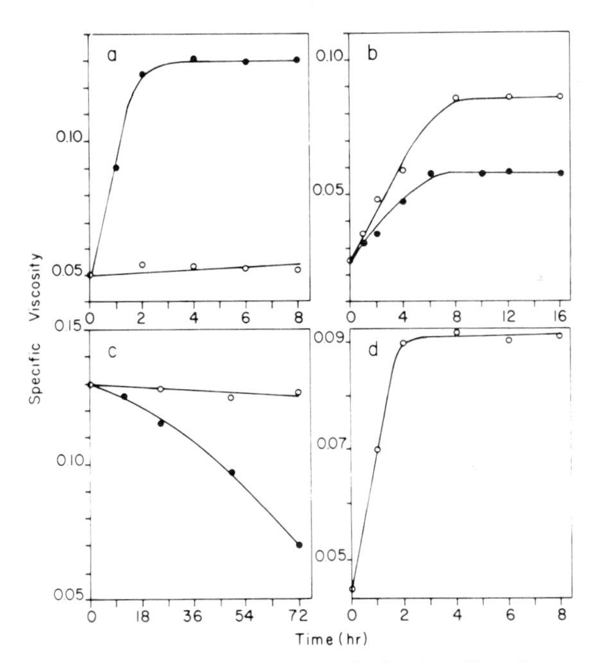

Fig. 2. Viscosity of M gallisepticum polymerizable protein fraction. Viscosity was measured at room temperature (22°C) using an Ostwald-type viscometer, having a minimum volume of 2 ml and a flow time of about 90 sec for buffer solutions. The specific viscosity is the flow time of the protein divided by the flow time of the buffer, minus unity. a) Protein (300 μg/ml final concentration) in depolymerizing buffer was mixed with KCl (0.6 M final concentration) at room temperature and viscosity was measured as a function of time (•). A control protein sample received no KCl (○). b) The protocol was as for (a), except that one sample received 0.6 M KCl (○) and the other received 0.6 M KCl plus 0.001 M $MgSO_4$ (•). The final protein concentration was 200 μM/ml. c) Protein (300 μm/ml) in polymerizing buffer was dialyzed against depolymerizing buffer at 4°C and viscosity was measured as a function of dialysis time (•). A control polymerized protein sample was dialyzed against polymerizing buffer (○). d) Protein depolymerized as described in (c) was mixed with KCl; final concentrations were 0.6 M KCl and 200 μg protein/ml. Viscosity was measured as a function of time after KCl addition.

ACKNOWLEDGMENTS

We thank Mr. David Gerling for his technical assistance in these studies and Dr. Klaus Haberer for many helpful discussions. This work was supported by US Public Health Service grant AI07939 from the National Institute of Allergy and Infectious Diseases.

REFERENCES

1. Maniloff J, Morowitz HJ: Bacteriol Rev 36:263, 1972.
2. Razin S: Microbiol Rev 42:414, 1978.
3. Maniloff J, Magrum L, Zablen LB, Woese CR: Zentralbl Bakteriol Parasitenkd Infektionskr Hyg Abt 1:Orig, Reihe A 241:171 (1978).
4. Bredt W: Colloq Inst Natl Sante Rech Med, Paris 33:47, 1974.
5. Morowitz HJ, Maniloff J: J Bacteriol 91:1638, 1966.
6. Maniloff J, Quanlan DC: Ann NY Acad Sci 225:181, 1973.
7. Quinlan DC, Maniloff J: J Bacteriol 112:1375, 1972.
8. Quanlan DC, Maniloff J: J Bacteriol 115:117, 1973.
9. Maniloff J, Quinlan DC: J Bacteriol 120:495, 1974.
10. Ghosh A, Das J, Maniloff J: J Mol Biol 116:337, 1977.
11. Ghosh A, Maniloff J, Gerling DA: Cell 13:57, 1978.
12. Tourtellotte ME, Jacobs RE: Ann NY Acad Sci 79:521, 1960.
13. Ryan JL, Morowitz HJ: Proc Natl Acad Sci USA 63:1282, 1969.
14. Liss A, Maniloff J: Virology 55:118, 1973.
15. Maniloff J: Microbios 1:125, 1969.
16. Barile MF: In Fogh J (ed): "Contamination in Tissue Culture." New York: Academic Press, 1973, p 131.
17. Liao CH, Chen TA: Proc Am Phytopathol Soc 2:100, 1975.
18. Igwegbe ECK: Appl Environ Microbiol 35:146, 1978.
19. Pollard TD, Weihing, RR: CRC Crit Rev Biochem 2:1, 1974.
20. Razin S: Adv Microbial Physiol 10:1, 1973.
21. Weber K, Osborn M: J Biol Chem 244:4406, 1969.
22. Wessels NK, Spooner BS, Ash JF, Bradley MO, Luduena MA, Taylor EL, Wrenn JT, Yamada KM: Science 171:135, 1971.
23. Betina V, Micekova D, Nemec P: J Gen Microbiol 71:343, 1972.
24. Demain AL, Hunt NA, Malik V, Kobbe B, Hawkins H, Matsuo K, Wogan GN: Appl Environ Microbiol 31:138, 1976.
25. Cunningham D, Schafer D, Tanenbaum SW, Flashner M: J Bacteriol 137:925, 1979.
26. Neimark HC: Proc Natl Acad Sci USA 74:4041, 1977.
27. Rosenbusch JP, Jacobson GR, Jaton J-C: J Supramol Struct 5:391, 1976.
28. Beck BD, Arscott PG, Jacobson A: Proc Natl Acad Sci USA 75:1250, 1978.
29. Wurtz M, Jacobson GR, Steven AC, Rosenbusch JP: Eur J Biochem 88:593, 1978.
30. Abramowitz JW, Stracher A, Detwiler TC: Arch Biochem Biophys 167:230, 1975.
31. Korn ED: Proc Natl Acad Sci USA 75:588, 1978.
32. Gordon DJ, Boyer JL, Korn ED: J Biol Chem 252:8300, 1977.

Journal of Supramolecular Structure 12:305–320 (1979)
Biological Recognition and Assembly 155–170

Protein-RNA Interactions During TMV Assembly

K. C. Holmes

Max Planck Institut für medizinische Forschung, Heidelberg, Germany

A review of the structural studies of tobacco mosaic virus (TMV) is given. TMV is essentially a flat helical microcrystal with 16 1/3 subunits per turn. A single strand of RNA runs along the helix and is deeply embedded in the protein. The virus particles form oriented gels from which high-resolution X-ray fiber diffraction data can be obtained. This may be interpreted by the use of six heavy-atom derivatives to give an electron density map at 0.4 nm resolution from which the RNA configuration and the form of the inner part of the protein subunit may be determined. In addition, the protein subunits form a stable 17-fold two-layered disk which is involved in virus assembly and which crystallizes. By the use of noncrystallographic symmetry and a single heavy-atom derivative, it has been possible to solve the structure of the double disk to 0.28 nm resolution. In this structure one sees that an important structural role is played by four alpha-helices, one of which (the LR helix) appears to form the main binding site for the RNA. The main components of the binding site appear to be hydrophobic interactions with the bases, hydrogen bonds between aspartate groups and the sugars, and arginine salt bridges to the phosphate groups. The binding site is between two turns of the virus helix or between the turns of the double disk. In the disk, the region proximal to the RNA binding site is in a random coil until the RNA binds, whereupon the 24 residues involved build a well-defined structure, thereby encapsulating the RNA.

Key words: tobacco mosaic virus, structure, RNA-binding site, assembly, protein-nucleic acid interactions

There are still very few systems from which we can obtain detailed information about the geometry of the interactions between proteins and nucleic acids. One such is tobacco mosaic virus (TMV). TMV particles are rod-like, 300 nm long, and 18 nm in diameter. TMV consists of 2,140 protein subunits, each of molecular weight 17,420 daltons (158 residues), arranged on a helix of pitch 2.3 nm with 16 1/3 subunits per turn. Winding through this helix is a single strand of RNA 6,400 nucleotides long (Fig. 1) [1, 2].

Structural studies on TMV have a long history. In 1936 Bernal obtained the first X-ray diffraction patterns from orientated gels of TMV and realized the implication of the existence of layer-lines in this pattern, namely, that the particle is built up from a regular arrangement of repeating subunits [3]. At the time, this conclusion did not find immediate

Received May 4, 1979; accepted August 10, 1979.

0091-7419/79/1203-0305$02.90

acceptance. Furthermore, the detailed interpretation of the diffraction pattern eluded Bernal and Fankuchen. Only when the necessary theoretical development, the elucidation of the structure factor of a helix, had been carried out in 1952 [4] was it possible to proceed with the analysis of the diffraction of TMV gels. Using the theory of Cochran, Crick, and Vand, J. D. Watson [5] was able to show that the structure of TMV was helical with 3n + 1 subunits every 6.9 nm. Watson's best value for n was 10. Independently of each other, R. E. Franklin [6] and D. L. D. Caspar [7] took up the study of the structure of the orientated gels. Franklin was soon to be joined by A. Klug. After Franklin's untimely death in 1958 Klug took over the leadership of the group which was at that time located in Birkbeck College, London, under the aegis of Bernal. In 1962, the virus group moved to the Medical Research Council Laboratory for Molecular Biology in Cambridge, where a great deal of the work described in the following paper was carried out. It is noteworthy that, through all the present-day virus crystallography, one can trace the visionary influence of Bernal and Frankuchen.

Before starting work on TMV, Rosalind Franklin had played a major role in the elucidation of the structure of DNA. One of her motivations in turning to TMV was an interest in the companion nucleic acid RNA. Of the systems containing RNA in an organized form TMV seemed to be ripest for attack. Franklin started out with the ambitious resolve to determine the structure of TMV by means of the then newly discovered method of isomorphous replacement. It was quickly established that the virus had 49 subunits in three turns [8], that the virus was hollow, and that the nucleic acid was located at a radius of 4.0 nm [6]. These results were obtained by analysis of the low-resolution part of the zero layer-line (<1.0 nm^{-1}), where it can be shown that the data are effectively the diffraction pattern from a centrosymmetric structure (ie, the phase problem reduces to a sign problem).

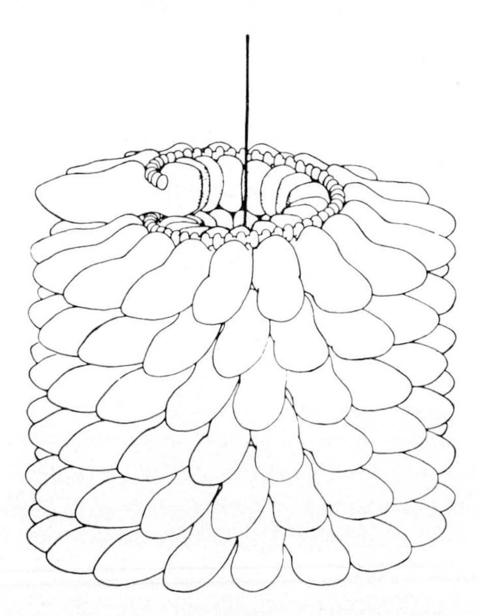

Fig. 1. Diagram of tobacco mosaic virus drawn by D. L. D. Caspar. The six-turn segment shown corresponds to 1/20 the length of the virus. The RNA chain, with three bases per protein subunit, is coiled between the turns of the 2.3-nm pitch helix of protein subunits. The diameter of the RNA is 8.0 nm. There are 16 1/3 subunits per turn. The inner diameter of the particle is 4.0 nm and the outer diameter is 18.0 nm. From Caspar [2], with permission.

The signs of the peaks on the inner part of the zero layer-line were determined independently by Caspar [7] and by Franklin and Holmes [8]. Together with Klug, Franklin showed that the virus surface is marked by a deep helical groove [9]. Using two heavy-atom derivatives, methyl-mercury [10] and osmic acid, she proceeded to analyze the third layer-line in order to arrive at a low-resolution helical projection of the virus. With the death of Rosalind Franklin in 1958, the project lost some impetus. Moreover, in 1958 the necessary methodology, particularly in computer development, was not available. As it turned out, determining the structure of TMV entailed developing new methods in theoretical and practical crystallography and in biochemistry so that nearly 20 years were to elapse before detailed structural information about the RNA and its environment could be obtained. Klug and Holmes continued the structural analysis of the orientated gels by means of the method of isomorphous replacement. In particular, in an unpublished study Klug was able to show that the RNA was single-stranded at a time when it was widely supposed, on the basis of end-group analyses, that the RNA was in many segments. With the establishment of Klug's group in Cambridge the necessary technical milieu became available to extend Franklin's early work.

Microcrystals of protein disks were first reported by Macleod et al [11]. Reproducible crystals were produced by Finch et al [12] in the MRC laboratory, Cambridge. Furthermore, they were able to demonstrate that the symmetry of the disk was 17-fold. This work opened the way to obtaining an atomic model of TMV by means of X-ray crystallography. The disk consists of two rings of protein subunits, each with the same polarity and each containing 17 subunits. The disks crystallize in the orthorhombic space group $P22_12_1$. The molecular weight of the asymmetric unit is $17 \times 2 \times 17{,}400 = 592{,}280$ daltons. The solution of a structure of this size presents formidable problems. To solve the structure of the disk to 0.28 nm resolution it was necessary for Bloomer et al [13] to measure the intensities of 2×10^6 reflections. The analysis at 0.5 nm [14] showed the entire polypeptide chain except for a section within 4.0-nm radius which was not visible. This observation reflects the fact that in the disk (but not in the virus) there is a flexible segment of a length of about 25 residues. The high-resolution structure [13] allowed all the residues to be fitted into the map except for the flexible segments 89–113 and the C-terminal residues 155–158.

The preparation of heavy-atom derivatives for the solution of the phase problem presented major difficulties so that much of the analysis was carried out with a single mercury derivative [13, 14]. Using the formulation of Crowther, Jack [15] was able to demonstrate the power of the noncrystallographic 17-fold symmetry as a supplement to the heavy-atom method for the solution of the phase problem. By setting up the equivalent formulation in real space Bricogne [16] made it possible to use this method for the high-resolution analysis of the structure of the disk [13].

Parallel to this work the analysis of the X-ray fiber diffraction progressed. Here it soon became clear that the limiting factor would be the availability of well-defined single heavy-atom derivatives of the virus. Both genetic and chemical modifications to the virus were used [17–19]. Initially, single heavy-atom derivatives were favored because the location of the heavy atoms in three dimensions from fiber diffraction data proved difficult and only the simplest cases were soluble. In 1968 Holmes moved to Heidelberg and work on the intact virus was transferred to Heidelberg. The determination of the structure of TMV from the X-ray fiber diagrams by means of the method of isomorphous replacement is a unique problem. The data consist essentially of the cylindrical average of the square of the Fourier transform of a single particle [20]. Thus the data at each point in reciprocal space consist of the sum of squares of a number of Bessel function terms, the actual number of Bessel functions which contribute being determined by the helical symmetry of the particle and the reciprocal space radius of the point in question. At low radii (small scattering angles) only one Bessel function contributes to the intensity. Theoretical studies by Klug et al [21] had shown that in principle it is possible to regenerate the three-dimension structure if the problem of "separating the Bessel functions" can be solved. Franklin and Klug realized that, to a resolution of about 1.0 nm on account of the high symmetry of TMV (49-fold), only one Bessel function could contribute to the intensity. As a consequence it can be shown that to this resolution the intensities are cylindrically symmetrical so that no information is lost by cylindrical averaging. Using this idea Barrett et al [22] were able to calculate a three-dimensional electron density map of the virus with a nominal resolution of 1.0 nm. This map showed the general appearance of the nucleic acid running circumferentially at 4.0 nm radius, and the overall appearance of the subunit including strong radial features which were later shown to be alpha-helices. The structure of the virus extends proximally to a radius of 2.2 nm, in sharp contrast to the situation in the disk. Holmes [23] had shown that it is possible to separate Bessel functions by means of heavy-atom derivatives. The problem of separating Bessel functions by means of isomorphous replacement is a multidimensional analog of the phase problem and can formally be solved by a multidimensional Harker construction. Locating the best intersection of the Harker hyperspheres by a multidimensional search for every point in reciprocal space is a very time-con-

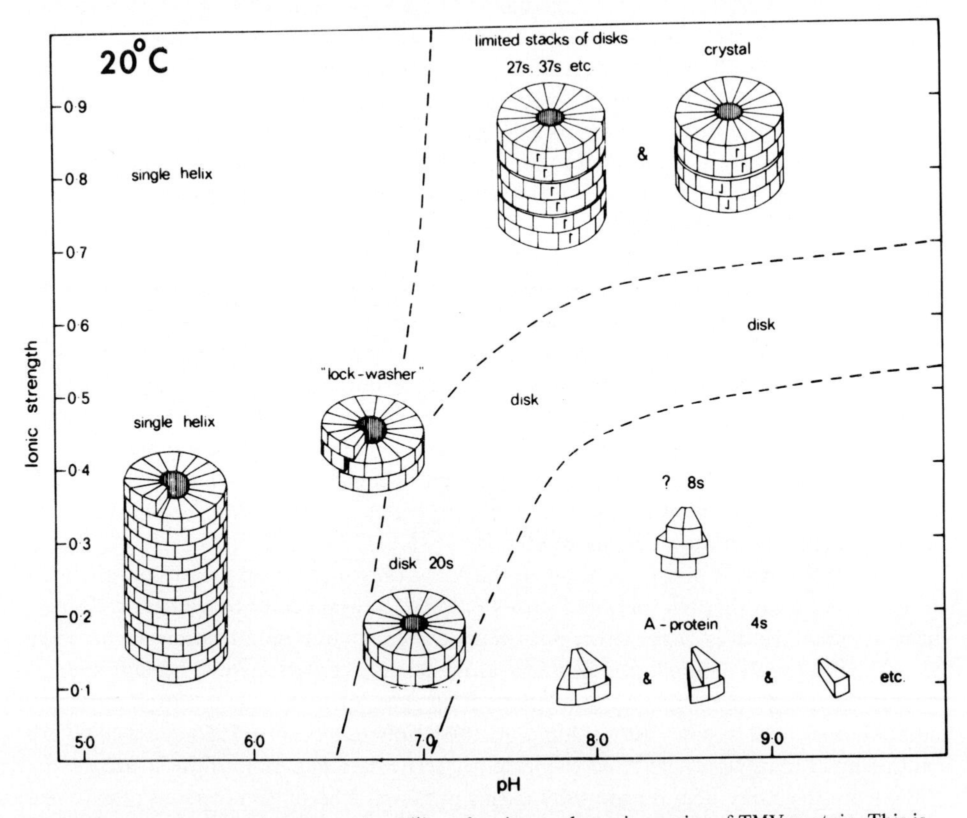

Fig. 2. Diagram showing the ranges of stability of various polymeric species of TMV protein. This is not a conventional phase diagram, since the polymorphic forms shown are in equilibrium. The boundaries indicate regions in which a particular form predominates. The boundaries are approximately correct for a protein concentration of 5 mg/ml at 20°C. From Klug and Durham [29], with permission.

suming computing problem. An algebraic solution was developed by Stubbs and Diamond [24]. Using this method Holmes et al [25] analyzed the data out to a resolution of 0.67 nm, which is the two-Bessel-function limit. The resulting map showed some of the polypeptide chain, in particular the two radial alpha-helices which are such a prominent feature of the structure of the TMV subunit. It allowed a preliminary chain tracing (the 0.5 nm resolution map of the protein disk showed subsequently that the alpha-helices had been correctly identified but not correctly linked together). Further work admitting three Bessel functions, by Stubbs, Warren, and Holmes [26], yielded an electron density map at 0.4 nm resolution by the use of six heavy-atom derivatives, which taken together with the detailed map of the subunit in the disk produced by Bloomer et al [13] yields a considerable amount of stereochemical information about the nature of the protein-RNA interaction in TMV assembly.

POLYMORPHISM OF TMV COAT PROTEIN

The isolated nucleoprotein particle can readily be taken apart [27] and reassembled [28]. In addition, a number of polymorphs of the coat protein have been described [29] (Fig. 2). Moreover, Butler and Klug [30] were able to show that the disk is the main precursor for the initiation and growth of the virus.

The major polymorphic forms of the protein *without* nucleic acid are:

the A-protein (predominantly a trimer);
the disk (containing two rings of 17 subunits);
a rod-like structure made from stacked disks;
a helical virus-like structure (16 1/3 subunits per turn);
a helical virus-like structure (17 1/3 subunits per turn [31]).

At neutral pH and low ionic strength the disk is the majority species. By raising the salt concentration the formation of the stacked disk is encouraged. The stacked disk may apparently be trapped by mild proteolytic cleavage of the flexible segment in a form which can no longer be taken apart [32]. This demonstrates the important role of the flexible segment in the aggregation and disaggregation of the disk. On lowering of the pH, the disks build a virus-like helix even in the absence of nucleic acid. This is thought to come about through the formation of an intermediate "lockwasher" [29]. The lockwasher may form by dislocating along either of two intersubunit boundaries to form a 16 1/3 or 17 1/3 helix [31]. In fact both forms are found, which lends credence to the idea that the helix can be built from dislocated disks. In the presence of RNA only the 16 1/3 form is found.

A remarkable feature of all these polymorphs is the conservative nature of the side-to-side bonding between the subunits, which scarcely alters. In contrast, the up-and-down bonding between subunits is markedly different between the disk forms and the helical forms [14] (Fig. 3). The subunits are 0.3 nm further apart in the disk than in the helix and are displaced 1.0 nm circumferentially.

ASSEMBLY OF THE VIRUS

During the assembly of the virus the coat protein has to specifically recognize the viral nucleic acid and at the same time has to be able to tolerate the variation of sequence found when encapsulating the RNA. Butler and Klug [30] showed by kinetic analysis that

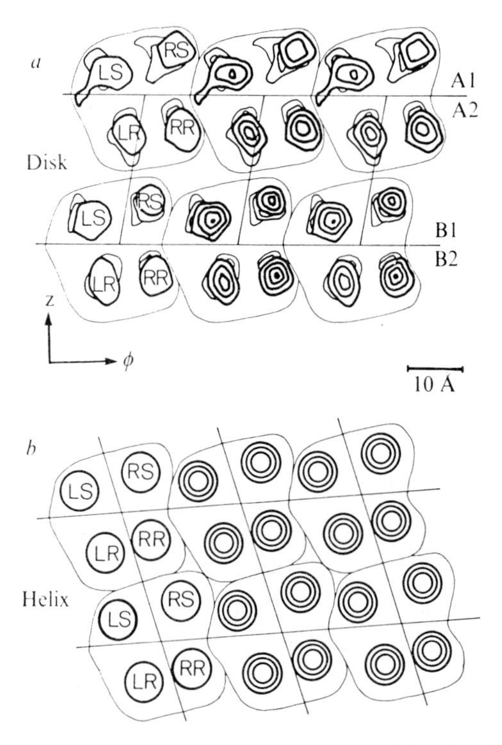

Fig. 3. Packing of subunits and alpha-helical rods in the protein disk and in the virus. a, Superposed cylindrical sections of the electron density map of the disk at radii 5.7, 5.85, and 6.0 nm. The outline of the subunit is shown. b, Schematic diagram of the arrangement of subunits in the helix at the same radius (from Champness et al [14], with permission). Note the change in the surface lattice in going from the disk to the helix, which involves a movement of about 1.0 nm circumferentially at this radius. The side-to-side contacts remain the same.

the disk is the major species responsible for coating the RNA during in vitro assembly. Assembly starts by the insertion of a specific RNA loop (the initiation loop) into the disk [33]. A specific sequence of the RNA (the assembly origin) binds to the disk [34–36], thereby mediating the formation of the lockwasher and initiating the growth of the helix. The assembly origin of the RNA is about 100 nucleotides long and the initial binding is thought to occur between the turns of the disk. Close to the center of the assembly origin is the sequence AGAAGAAGUUGUUGAUGA. In this and the surrounding sequences there is a strong tendency for every third residue to be G. The assembly origin has a very low C content and occurs about 15% of the length away from the 3′ end. Growth takes place fastest in the 5′ direction [37, 38]. Growth in the 3′ direction goes on at the same time but rather more slowly and possibly by a different mechanism [38]. Rather unexpectedly, as growth proceeds towards the 5′-hydroxyl end the 5′ end is dragged through the central hole of the nascent virus (Fig. 4) so that both the 3′ and 5′ ends are initially at the same end of the virus [39, 40].

STRUCTURE OF THE DISK

The general appearance of the subunit, as deduced from the high-resolution studies of Bloomer et al [13], is shown in Figure 5. Both the C and N termini are located distally.

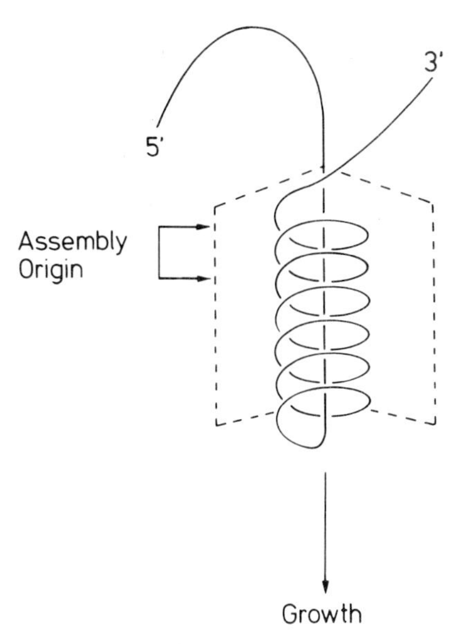

Fig. 4. Diagram of the encapsulation of RNA during the growth of TMV particles. The dominant direction of growth (indicated by an arrow) is downwards from the assembly origin (towards the 5′ end). The RNA is pulled through the hollow core of the nascent virus. From Butler et al [39], with permission.

Starting from the N terminus the chain builds a short helical structure between residues 8 and 15. Then follows a short section of extended chain (16–18) which forms part of the distal beta-pleated sheet. The chain continues into the LS (left-slew) helix (20–32) which runs proximally to a radius of 5.0 nm. After a tight turn (33–38) the chain proceeds distally along the RS (right-slew) helix (38–48), which is parallel to the LS helix. The chain now forms part of a beta-sheet. At 74 the RR (right-radial) helix starts. This helix, which is tilted about 20° to the horizontal, runs proximally and up to a radius of about 4.5 nm, where the density fades out. Then follows a flexible segment containing 24 residues, which has been shown by NMR investigations [41] to be highly mobile. After the flexible segment (89–113) the LR (left-radial) helix runs distally roughly horizontally from a radius of 4.0 nm to 7.0 nm (114–134). Then the chain runs through a short alpha-helical segment to the C terminus. The last four residues (155–158) are not visible and are in a random coil. The four main alpha-helices include about 60 residues of the total of 158. The helices are packed as in a segment of a left-handed, four-stranded supercoiled bundle with hydrophobic contacts along the center. The distal ends of the four helices are connected transversely by a strip of beta-sheet shown schematically in Figure 6. Residues 53–54, 16, 18, 68–71, 137–139, and 135–136 are involved. In the central region of each subunit on the distal side of the beta-sheet is a cluster of aromatic residues, which gives rise to a continuous belt of hydrophobic interactions encircling each ring of the disk.

STRUCTURE OF THE VIRUS

The structure of the subunit in the virus is substantially the same as the disk except for the flexible loop and the vertical intersubunit contacts. Proximally from 4.0 nm radius, where the density in the disk fades out, one sees the LR and RR helices continuing to a

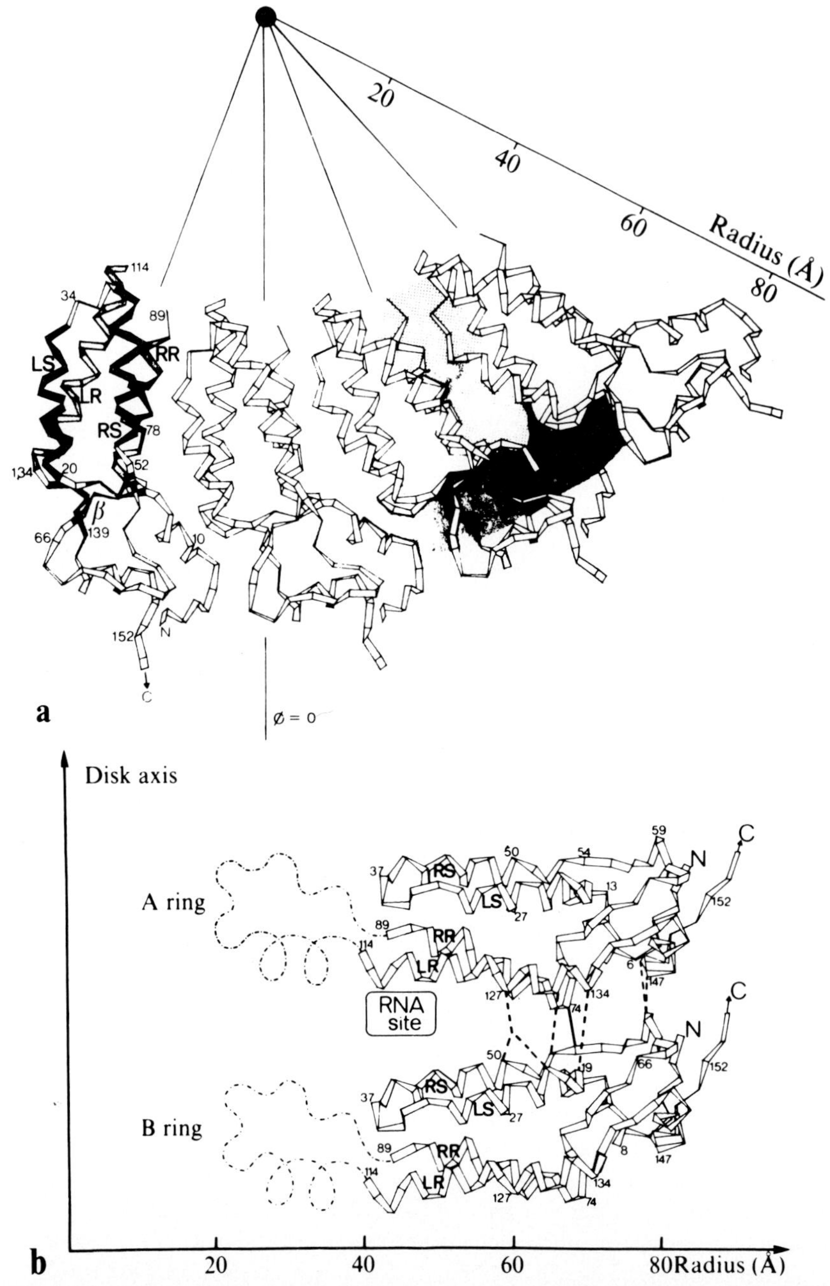

Fig. 5. Plan (a) and elevation (b) of the TMV disk. a, Four adjacent subunits of the A ring are viewed from above. The N and C termini of the chain are marked on the extreme left subunit together with the four main helices, left and right slewed (LS and RS) in the upper half of the molecule with right and left radial (RR and LR) below, and the beta-sheet, which connects all four helices at their distal ends. The subunit interface contains alternating patches of polar residues (shown by a stippling) and hydrophobic residues (shown with solid shading). Distally from the beta-sheet is the hydrophobic girdle, which extends circumferentially across the whole width of the subunit. From Bloomer et al [13], with permission. b, Side view through a sector of the disk showing the disposition of the subunits in the A and B rings and the axial contacts between them. There are three regions of contact, indicated by a solid line for the hydrophobic contact of pro 54 with ala 74 and val 75, by dashed lines for the hydrogen bonds between thr 59 and ser 147 and 148, and further dashed lines for the extended salt-bridge system. All these bonds are broken in the transition to the helical structure of the virus. The proximal region of the subunit, which has no structure in the disk form, is shown schematically by a broken line in approximately the configuration it takes in the virus. From Bloomer et al, with permission.

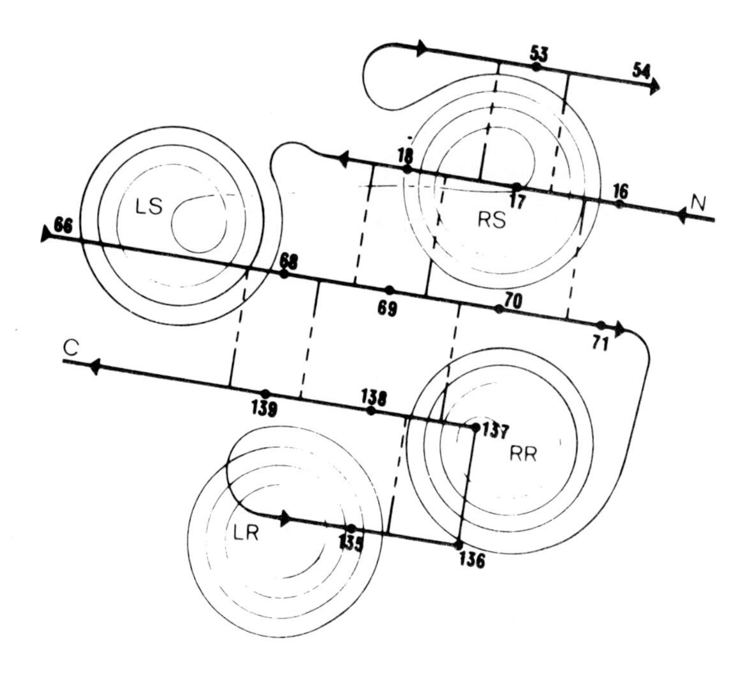

Fig. 6. Schematic view of the antiparallel beta-sheet. Shown is the connectivity of the four strands and of the four principle alpha-helices. The direction of view is radially towards the center. The beta-sheet is a narrow, twisted strip which extends across the subunit, bracing together the four alpha-helices. At the bottom of the beta-sheet is a type II beta-bend involving the invariant gly 137. From Bloomer et al [13], with permission.

low radius (2.5 nm) (Fig. 7), where they are joined together by a strong vertical column of density (the V-column) which was identified by Stubbs et al [26] as an alpha-helix, although more recent studies show this identification to be somewhat uncertain. In particular, the number of residues involved is less than was proposed by Stubbs et al. In addition to this density, which can unequivocally be ascribed to the protein, there is a ribbon of density running roughly circumferentially at a radius of 4.0 nm *underneath* the LR helix which can be ascribed to the nucleic acid. Recent studies by Mandelkow et al [42] of the structure of the helical protein aggregate *without* RNA substantiate the assignment of this density to the RNA. The studies of Mandelkow et al have led to some small revisions of the RNA conformation from the form reported by Stubbs et al, but the essential features remain unchanged. A diagram of the RNA conformation is shown in Figure 8. Each protein subunit binds three bases which are numbered 1, 2, and 3 from the 5′ end. Two of the bases appear to be in the *anti* conformation and one in the *syn,* but this is not very well determined. The sugar puckers are all C3′ endo. The direction of the nucleic acid agrees with that independently determined by Wilson et al [43]. The bases form a claw-like structure around the LR helix (Fig. 9). Two bases are at a radius of about 3.8 nm and one base is at a larger radius (4.5 nm). The phosphates are clustered around a radius of 4.0 nm and appear to bond to the adjoining RS and RR helices from the neighboring subunit in the next turn down. Thus the RNA binding site lies between the rings of a disk, in agreement with the model for assembly proposed by Butler et al [33].

RELATIONSHIP BETWEEN THE DISK AND VIRUS MAPS

Gilbert [44] estimated the shift involved in changing from the disk to the helix by assuming that the protein subunit packing at a radius of 8.0 nm stayed the same in the azimuthal direction and that no rearrangement of the subunit structure took place. One ob-

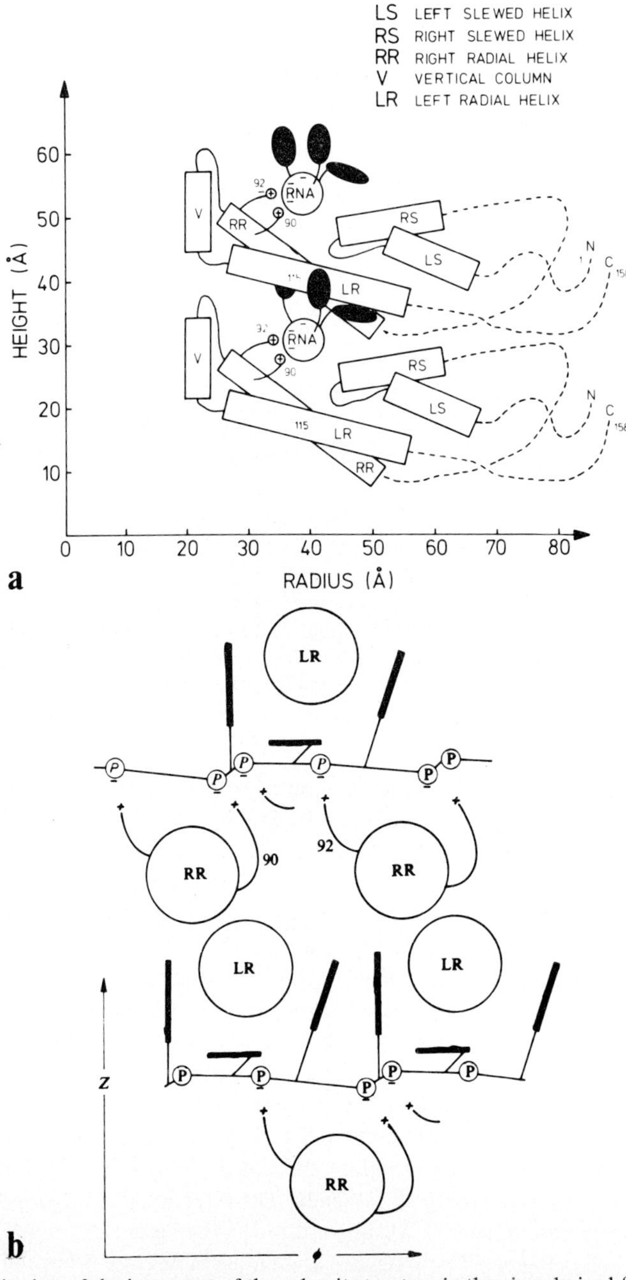

Fig. 7. Schematic view of the inner part of the subunit structure in the virus derived from the electron density maps at 0.4-nm resolution calculated from X-ray fiber diffraction patterns of the virus. a, A sector of the virus viewed circumferentially showing the extended LR and RR helices and the V-column joining them together. This part of the structure is not seen in the disk, although it is represented by high density in the helical virus. Comparison with the disk structure shows that the LR helix should be displaced distally by one turn so that the bottom of the V-column will probably occur at about residue 108 rather than 112. An important component of the RNA binding site is the interaction with the LR helix between residues 114 and 123. This part of the LR helix already exists in the disk. The detailed structure of the V-column is at present unclear. b, Diagram of the RNA binding site at a radius of 4.0 nm. The direction of view is radially towards the center. Note how the bases (thick lines) group around the LR helix in the shape of a claw. The base lying horizontally is at a larger radius (circa 4.8 nm). The phosphate groups form salt bridges with the subunit underneath so that the RNA binding site lies between turns of the virus helix. a and b from Stubbs et al [26], with permission.

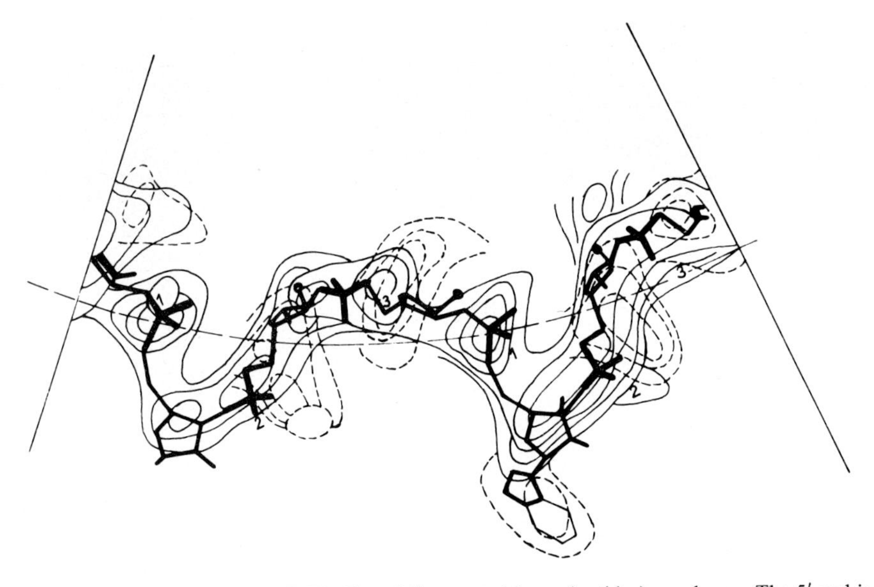

Fig. 8. The RNA configuration in TMV. Two full repeats (six nucleotides) are shown. The 5′ end is to the left. The direction of view is from above with the center of the virus at the top of the diagram. The mean radius of the RNA (4.0 nm) is indicated with a broken-line circle. The nucleotides have been numbered 1, 2, and 3. The phosphate groups are shown with a heavy line. Bases 1 and 3 rise up vertically out of the plane of the diagram and have been omitted for clarity. The electron density associated with the bases is shown by broken-line contours. From Mandelkow et al [42].

tains the value 0.32 nm for the radial displacement necessitated by changing the symmetry from 17-fold to 16 1/3-fold. Comparing heavy-atom positions in the virus and in the disk supports the assumption that the change from disk to helix is approximately a rigid body transformation [45] so that Gilbert's assumptions seem to be justified.

Bloomer [13] has analyzed the relationship between the A-disk and the B-disk and the virus and has shown that, in addition to the radial movement, the subunits must be rotated by 10° (A-disk) or 20° (B-disk) about an axis perpendicular to the common disk or helix axis in order to bring them into the same orientation as the virus. Recently Bricogne and Bloomer, in an unpublished study, have determined the transformation necessary to bring the A- and B-disk density into coincidence with the virus density by a least-squares procedure.

The tilt of the subunits in the virus helix gives rise to the characteristic "Christmas tree" appearance of the virus particle [43], which may be used to define the polarity of the virus. The point of the Christmas tree is taken as "up" in structural studies and "down" in assembly and elongation studies!

THE RNA BINDING SITE

The following account of the binding site should be taken as plausible but not proven.

If we apply Bricogne and Bloomer's transformation to the LR helix we find that it amounts to a translation of about 0.32 nm in the radial direction in the neighborhood of the nucleic acid. Applying this transformation to the LR helix in the A-ring of the disk results in the relationship between the LR helix and RNA shown in Figure 10. Comparing the resulting structure with the virus electron density indicates that only minor revisions will be necessary to take account of local distortions. Comparing the model so produced with the model fitted to the helix electron density by Stubbs et al (Fig. 7), we

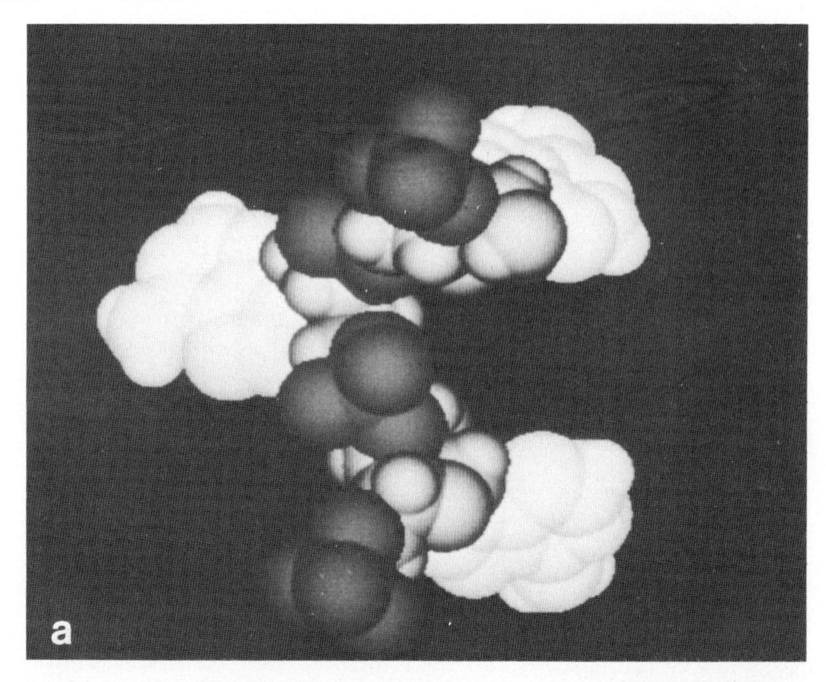

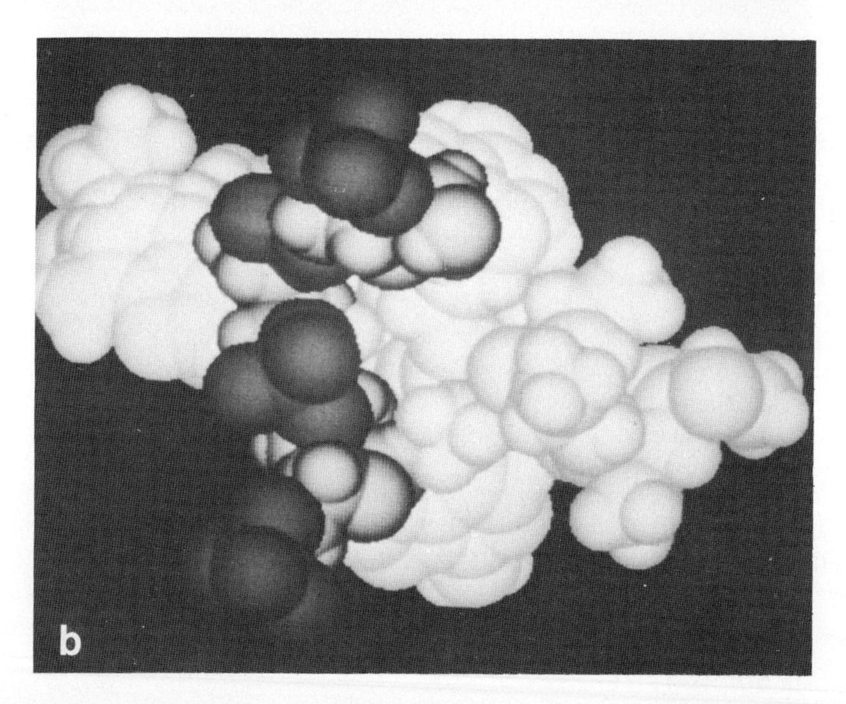

Fig. 9. A computer drawing (Dr. R. J. Feldmann, NIH from coordinates provided by Dr. G. Stubbs, Brandeis) of the relationship between the LR helix and the RNA claw. a, The three-nucleotide repeat in the claw configuration. The bases are white, the sugars light grey, and the phosphate groups dark grey. The direction of viewing is from underneath with the center of the virus on the right. b, The RNA claw gripping the LR helix.

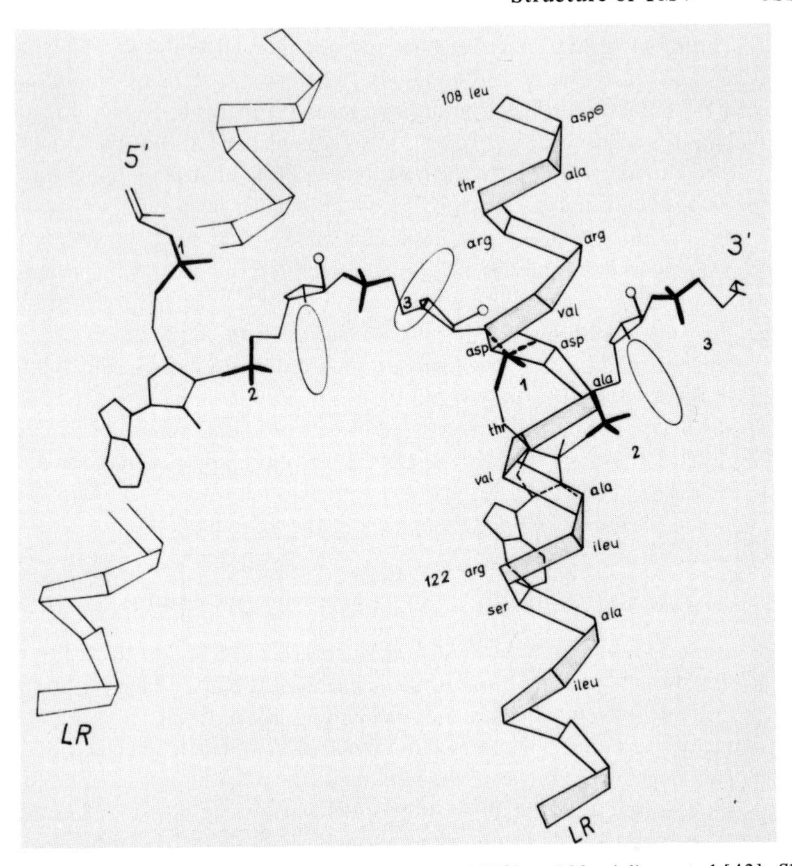

Fig. 10. The RNA binding site. Data are from Bloomer et al [13] and Mandelkow et al [42]. Shown are the RNA and the LR helix. The direction of view is from above with the center of the virus at the top of the diagram (as in Fig. 8). Two repeats are shown. The LR helix runs above the RNA. In the left repeat, part of the LR helix has been omitted for clarity. The bases which rise up out of the plane of the diagram are shown by ellipses. The 2′ OH groups are shown with small circles. Note how the 2′ OH groups of bases 1 and 3 come close to Asp 115 and 116 of the LR helix.

find that the transformed disk structure is one turn further out (3.6 residues) than the Stubbs, Warren, and Holmes model. A similar shift has already been proposed by Bloomer et al [13]. Given the high quality of the electron density map of the disk, this comparison strongly suggests that the virus model needs revision, namely, that the LR helix in the virus model should be translated one turn distally.

With the virus model adjusted in this way, some interesting features of the RNA binding site reveal themselves: An important component of the binding site seems to be provided by the invariant pair of aspartate residues (115, 116). These probably hydrogen-bond to the RNA ribose 2′ OH on residues 1 and 3. Such a ribose binding site is typical of alcohol dehydrogenase, lactate dehydrogenase, and malate dehydrogenase [46–48]; here this binding motif may be used twice per protein subunit. The accompanying bases find themselves in hydrophobic pockets close up against the alpha-helix, as was suggested by Stubbs et al [26]. For base 1 the methylene chain of Arg 113 appears to be an important component of the hydrophobic site. The other component of the site is the β-carbon of Asp 115. The guanidinium group of Arg 113 might also be involved in a salt bridge to a phosphate group. Base 3 appears to bind against Val 114 and Ala 117. Base 2 is at a larger radius and may form a hydrogen bond to Ser 123. The sugar of residue 2 seems to lie face down against a hydrophobic surface formed by residues 119 (Val) and 120 (Ala). If the 2′ OH is able to form a hydrogen bond, it must be to the neighboring slewed hairpin joining the LS or RS helices from the subunit underneath, possibly from Asn 33.

The effect of the tight binding of residues 1 and 3 to the LR helix is to pucker the RNA so that the phosphate residues are grouped together, thereby making them a target for the formation of salt bridges with arginines. Stubbs et al [26] have suggested that the salt bridges are formed with arginine residues 90 and 92 from the RR helix of the subunit directly underneath, and possibly with residue 41 from the RS helix of this subunit. The assignment of the invariant residues 90 and 92 to the phosphate binding site is plausible, but higher resolution studies of the virus structure are needed before such an assignment can be proven. Unfortunately, this part of the RR helix cannot be seen in the disk structure [13]. The disk structure indicates that Arg 41 is not such a strong candidate for the phosphate binding site since it lies on the wrong face of the RS helix.

A HYPOTHETICAL TRIGGER MECHANISM FOR THE DISK-HELIX TRANSITION

The surprising result from the studies summarized above is that the major component of the RNA binding site in TMV is one alpha-helix, the LR helix, which binds three bases. In the disk structure, which is the template for binding RNA, this helix continues as a finger proximally to a radius of about 4.0 nm (residue 114) before the electron density peters out. This finger could nucleate the binding of the RNA by means of three kinds of interaction: 1) a stereospecific interaction of the aspartate groups 115, 116 with two of the ribose groups; 2) a hydrophobic interaction with the three bases forcing them into the shape of a claw around the alpha-helix (Fig. 9); 3) specific hydrogen bonding which is still to be determined. Present studies suggest a possible hydrogen bond between ser 123 and base 2 which would favor a hydrogen bond acceptor in this position.

The binding is accompanied by the formation of salt bridges between Arg 90 and 92 from the neighboring subunit one turn deeper in the virus helix and the phosphate groups of the nucleic acid. Initially 90 and 92 are part of the random-coil segment adjoining the RR helix, but apparently through the binding to the phosphate groups this structure is stabilized and as a result the V-column and the rest of the LR and RR helices, which are not seen in the disk, can build. However, the V-columns from adjoining subunits are in close contact and it seems probable that they can only fit together in one way, namely, as in the virus helix. The change in relative heights in passing from the disk to the helix is 0.14 nm per subunit, and it is conceivable that the packing of the V-columns could not tolerate this much distortion. As a result, therefore, of the packing of the V-columns the disk structure may be strained and may attempt to transform into the helix. The contact between the A- and B-rings of the disk is mediated by a complex system of hydrogen bonds (Fig. 6) which must break when the helix is formed. The disk-helix transition is therefore reminiscent of the R-T transformation in hemoglobin [49]. The disk is the low-affinity form for RNA because one cannot make the salt bridges to the phosphates without allowing the V-column to build, thereby destabilizing the disk. The helix is the tight-binding form. However, to reach the helix it is necessary to break the network of hydrogen bonds between the A and B rings of the disk. The balance between these effects determines the binding of the RNA and makes the RNA binding highly cooperative. Perhaps it is relevant in this context that

Lomonossoff and Butler [50] report that the incorporation of RNA during in vitro reassembly is quantized, the quantum of binding being close to 50 bases or 100 bases. This is essentially the number of bases in one or two turns, which would seem to show that binding proceeds through the addition of single or double rings, thereby supporting the idea that the disk is the major precursor of virus growth. Through such cooperativity one could also explain the specificity of the binding. Two properties of the assembly origin seem noteworthy: the requirement that every third residue should be G; and the requirement that the sequence should be long. If we postulate a binding mechanism whereby the selectivity for any one site is small (eg, AGU may have twice the binding constant of UUU) but where the *simultaneous* binding to 16 sites produces the disk-helix transition, then one sees at once why the assembly origin can be so specifically recognized – it has the selectivity 2^{16}.

Stubbs et al [26] have pointed out that the inner structure of the virus (the V-column and the adjoining pieces of LR and RR helix which build on binding RNA) is destabilized by a high concentration of carboxyl groups in this region, which would probably explain the anomalous pKs appearing during the disk-helix transition [2, 49]. The important function of the energetically unfavorable carboxylate-carboxylate interactions may be to help keep part of the RNA binding site in a random-coil configuration until the RNA binds.

ACKNOWLEDGMENTS

I gratefully acknowledge the help I have received from Dr. Gerald Stubbs and Dr. Anne Bloomer, in particular their generously making unpublished results available to me. I am grateful to Drs. Stubbs, Bloomer, and Klug for reading the manuscript and for making many helpful suggestions.

REFERENCES

1. Anderer FA: Adv Protein Chem 18:1, 1963.
2. Caspar DLD: Adv Protein Chem 18:37, 1963.
3. Bernal JD, Fankuchen I: J Gen Physiol 25:111, 1941.
4. Cochran W, Crick FHC, Vand V: Acta Crystallogr 5:581, 1952.
5. Watson JD: Biochim Biophys Acta 13:10, 1954.
6. Franklin REF: Nature (Lond) 177:929, 1956.
7. Caspar DLD: Nature (Lond) 177:928, 1956.
8. Franklin REF, Holmes KC: Acta Crystallogr 11:213, 1958.
9. Franklin REF, Klug A: Biochim Biophys Acta 19:403, 1956.
10. Fraenkel-Conrat H: In Benesch R, Benesch RE (eds): "Sulphur in Proteins." New York: Academic, 1959, p 339.
11. Macleod R, Hills GJ, Markham R: Nature (Lond) 200:932, 1963.
12. Finch JT, Leberman R, Chang Y-S, Klug A: Nature (Lond) 212:349, 1966.
13. Bloomer AC, Champness JN, Bricogne G, Staden R, Klug A: Nature (Lond) 276:362, 1978.
14. Champness JN, Bloomer AC, Bricogne G, Butler PJG, Klug A: Nature (Lond) 259:20, 1976.
15. Jack A: Acta Crystallogr A29:545, 1973.
16. Bricogne G: Acta Crystallogr A32:832, 1976.
17. Wittmann HG: Z Vererbungslehre 95:333, 1964.
18. Perham RN, Thomas JO: J Mol Biol 62:415, 1971.
19. Gallwitz U, King L, Perham RN: J Mol Biol 87:257, 1974.
20. Franklin RE, Klug A: Acta Crystallogr 8:777, 1955.
21. Klug A, Crick FHC, Wyckoff HW: Acta Crystallogr 11:199, 1958.
22. Barrett AN, Barrington Leigh J, Holmes KC, Leberman R, Mandelkow E, von Sengbusch P: Cold Spring Harbor Symp Quant Biol 36:433, 1971.

23. Holmes KC: PhD Thesis, University of London 1959.
24. Stubbs GJ, Diamond R: Acta Crystallogr A31:709, 1975.
25. Holmes KC, Stubbs GJ, Mandelkow E, Gallwitz U: Nature (Lond) 254:192, 1975.
26. Stubbs GJ, Warren S, Holmes KC: Nature (Lond) 267:216, 1977.
27. Schramm G: Z Naturforsch 2b:112,249, 1947.
28. Fraenkel-Conrat H, Williams RC: Proc Natl Acad Sci USA 41:690, 1955.
29. Klug A, Durham ACH: Cold Spring Harbor Symp Quant Biol 36:449, 1971.
30. Butler BJG, Klug A: Nature New Biol 229:47, 1971.
31. Mandelkow E, Holmes KC, Gallwitz U: J Mol Biol 102:265, 1976.
32. Durham ACH FEBS Lett 25:147, 1972.
33. Butler PJG, Bloomer AC, Briconge G, Champness JN, Graham J, Guilley H, Klug A, Zimmern D: In Markham R, Horne R (eds): "Structure-Function Relationships of Proteins." 3rd John Innes Symposium. Amsterdam: North Holland-Elsevier, 1976.
34. Zimmern D, Butler PJG: Cell 11:455, 1977.
35. Zimmern D: Cell 11:463, 1977.
36. Jonard G, Richards KE, Guilley H, Hirth L: Cell 11:483, 1977.
37. Zimmern D, Wilson TMA: FEBS Lett 71:294, 1976.
38. Lomonossoff P, Butler PJG: Eur J Biochem 93:157, 1979.
39. Butler PJG, Finch JT, Zimmern D: Nature (Lond) 265:217, 1977.
40. Lebeurier G, Nicolaieff A, Richards KE: Proc Natl Acad Sci USA 74:149, 1977.
41. Jardetzsky O, Akasaka K, Vogel D, Morris S, Holmes KC: Nature (Lond) 273:564, 1978.
42. Mandelkow E, Stubbs GJ, Warren S: J Mol Biol (Submitted for publication).
43. Wilson TMA, Perham RN, Finch JT, Butler PJG: FEBS Lett 64:285, 1976.
44. Gilbert PFC: PhD thesis, University of Cambridge, 1970.
45. Graham J, Butler PJG: Eur J Biochem 83:528, 1978.
46. Branden CI, Eklund H, Nordstrom B, Boiwe T, Soderlund G, Zeppezauer E, Ohlsson F, Akeson A: Proc Natl Acad Sci USA 70:2439, 1973.
47. Smiley IE, Koekack R, Adams MJ, Rossmann MG: J Mol Biol 55:467, 1971.
48. Hill E, Tsernoglou D, Webb L, Banaszak LJ: J Mol Biol 72:577, 1972.
49. Durham ACH, Klug A: Nature New Biol 229:42, 1971.
50. Lomonossoff GP, Butler PJG: J Mol Biol 126:877, 1978.

Journal of Supramolecular Structure 12:321–334 (1979)
Biological Recognition and Assembly 171–184

α-Bungarotoxin Binding Sites (Acetylcholine Receptors) in Denervated Mammalian Sarcolemma

Ulka R. Tipnis and Sudarshan K. Malhotra

Biological Sciences Electron Microscopy Laboratory, University of Alberta, Edmonton, Canada T6G 2E9

The nonsynaptic sarcolemma of denervated skeletal muscle of rat shows an abundance of ~15 nm intramembranous particles on the P face. These particles are either singly distributed or are in clusters, and they are essentially lacking from the comparable freeze-fractures of the innervated sarcolemma. Autoradiographic studies using ^{125}I-α-bungarotoxin (BGT) on 1 μ-thick sections, and freeze-etch studies using ferritin-α-BGT conjugates on membrane fractions, show that the distribution of the label corresponds to the distribution of the 15-nm particles in the nonsynaptic sarcolemma. On the basis of these results and existing physiologic and biochemical data, it is suggested that the 15-nm intramembranous particles are components of the α-BGT binding sites, ie, acetylcholine (Ach) receptors, in the nonsynaptic sarcolemma of denervated muscle and that the two types of distributions represent two spatial manifestations of Ach receptor molecules. The significance of these findings in relation to synapse formation in denervated muscle is discussed.

Key words: **denervated sarcolemma, nonsynaptic acetylcholine receptors, ^{125}I-α-bungarotoxin, ferritin-α-bungarotoxin, electron microscopy, freeze-fracture, freeze-etching, autoradiography**

Denervation of the adult mammalian skeletal muscle is being extensively applied to understanding the nerve-muscle interaction [1–6]. Physiologic and biochemical changes occurring in the sarcolemma of denervated muscle include a fall in the resting membrane potential and an increase in membrane resistance [7, 8] and a decrease in acetylcholinesterase activity (AchE), mainly the 16S form of AchE [6, 9]. Among the three known molecular forms of AchE, namely 4S, 10S, and 16S in mammalian muscle, the induction of the 16S form is under neural control [10]. Denervation also results in an increased sensitivity to acetylcholine (Ach) in nonsynaptic sarcolemma with a concomitant increase in the number of Ach receptors [11]. Earlier, we reported changes in the sarcolemma of denervated rat muscle examined by freeze-fracturing technique [4] and described the

Received May 24, 1979; accepted October 15, 1979.

appearance of intramembranous particles (~15 nm) on the cytoplasmic fracture of the membrane (P face) in the nonsynaptic sarcolemma. These intramembranous particles were either singly distributed or in clusters. It was hypothesized that these ~15 nm particles which appear in the nonsynaptic region following denervation are the Ach receptors. The present paper describes the work primarily directed towards testing the above hypothesis by use of marker, α-bungarotoxin (α-BGT), which binds to the Ach receptors in a specific and irreversible manner [12]. α-bungarotoxin, when conjugated to ferritin, can be visualized by electron microscopy or, when conjugated to ^{125}I, can be visualized by autoradiography at light or electron microscope level. The results of such an investigation are included in this paper.

It should be pointed out that in mammalian muscle the size of intramembranous particles thought to correspond to the Ach receptor complex is 11–15 nm [6, 37]. In Xenopus embryonic muscle and cultured myotubes of chick, such particles are 10–19 nm [38, 39]. These sizes differ from the 7 nm given for the Ach receptor complex in the electroplaques [40]. Though the significance of these differences remains to be investigated, the intramembranous particles in the segments of cultured myotubes of chick containing identifiable regions of high acetylcholine sensitivity measure 10–19 nm [39].

MATERIALS AND METHODS

Denervation Procedure

The lumbricals and extensor digitorum longus (EDL) of rats of the Sprague-Dawley strain weighing 100–120 g were used in this study. The animals were anesthetized and denervated by transection of the sciatic nerve in the upper thigh region and sacrificed by cervical decapitation 2 weeks after denervation. The rationale for using such a period of denervation was based upon our previous morphologic and histochemical studies, which showed marked changes in the nonsynaptic sarcolemma after 2 weeks of denervation [4, 6]. Also, the incorporation of Ach receptors in the nonsynaptic sarcolemma is optimum 2 weeks after denervation [11]. The muscles of normal innervated rats were used as controls.

Preparation of Muscle Membrane Fraction

The muscles were homogenized and crude membrane fractions were isolated from control, as well as denervated animals, according to the procedure of Boegman et al [13].

Preparation of α-Bungarotoxin Conjugates

α-BGT supplied by Miami Serpentarium, Florida, was used for the following conjugates.

Iodination of α-BGT. α-BGT was iodinated with 1 mCi ^{125}I using the method of Greenwood and Hunter [14]. The iodination was carried out by Radiopharmacy Centre, University of Alberta. The specific activity and protein concentration of iodinated protein were 2.192×10^5 Ci/mole and 8.4 μg/ml, respectively.

Preparation of ferritin-α-bungarotoxin (Ft-α-BGT). Ferritin (6X and cadmium-free) obtained from Polysciences was conjugated to α-BGT by using glutaraldehyde, according to the method of Hourani et al [15].

Incubation of Muscles With ^{125}I-α-BGT

The EDL muscles from denervated and innervated rats were tied at both ends to a wooden stick and immediately transferred to an oxygenated Kreb's ringer containing ^{125}I-α-BGT (2×10^{-7} M). The muscles were incubated for 2 h. For determining the specific binding of α-BGT, another set of EDL from each of the innervated and denervated rats was incubated initially in Kreb's ringer containing d-tubocurarine (10^{-4} M) for 1 h, followed by incubation in ^{125}I-α-BGT for 2 h. During incubation, the oxygen was continuously bubbled through the medium maintained at 37°C in water bath. After incubation, the muscles were thoroughly washed by repeated changes of buffer for 1 h and fixed in 2% glutaraldehyde buffered with 0.1 M phosphate buffer (pH 7.2). The tissue was then given several changes of buffer and gently dried on filter paper and weighed. The radioactive counting of these tissues was done in a gamma counter (Baird-Atomic). The tissue was then postfixed in OsO_4 and routinely processed for preparation of 1 μ-thick Araldite sections for autoradiography (see below).

Incubation of Membranes in ^{125}I-α-BGT

The membranes were assayed for their Ach receptor activity by the filter assay procedure suggested by Klett et al [16]: A 1 ml portion of Kreb's ringer contained 50 μg of membrane protein, ^{125}I-α-BGT (2×10^{-9} M), and 1% (w/v) Tween 80. Control samples were incubated with α-BGT (10^{-4} M) for 1 h prior to incubation in ^{125}I-α-BGT. The mixture was incubated at room temperature (20–22°C) for varying periods and then filtered through DE81 cellulose anion exchange filter disk (Whatman). The dried filter disks were counted for the radioactivity in a gamma counter.

Preparative Procedures for Microscopy

Light microscope autoradiography. Araldite sections 1 μ thick were cut from blocks of innervated and denervated muscles which had previously been incubated in ^{125}I-α-BGT as described above. The sections were coated with 1:1 diluted Ilford L_4 emulsion on gelatinized slides. After exposure for one week, the slides were developed and fixed in 25% sodium thiosulphate. The sections were stained with 2% phenylenediamine and examined under phase contrast microscope.

Incubation of muscle membranes with Ft-α-BGT. The muscle homogenate (1000 g) and the crude membrane fractions were incubated with Ft-α-BGT (0.6 μg/ml) for 1 h. For determination of specific binding, the fractions were first incubated in α-BGT (10^{-4} M) or d-tubocurarine (10^{-4} M), followed by incubation with the conjugate. The suspension of crude membranes was washed by filtration through millipore as described by Karlin et al [17]. Suspension of the homogenate was centrifuged at 1,000 g for 10 min. The supernatant was discarded and the pellet was washed several times with buffer. The samples were then fixed and processed for electron microscopy as described above. Thin sections were examined, unstained or stained with uranyl acetate and lead citrate, in Phillips EM 300 electron microscope.

Freeze-fracture and freeze-etch preparations. A portion of the homogenate incubated in Ft-α-BGT was used for freeze-etching. The pelleted samples were fixed in 2% glutaraldehyde buffered with 0.1 M phosphate (pH 7.2). The material was rinsed with buffer and finally with distilled water. Freeze-fracturing and etching were done in Balzers BA 360M and the etching period lasted up to 2 min. The replicas were washed in 40% chromic acid and rinsed in several changes of water and examined in Phillips EM 300 electron microscope.

RESULTS

Freeze-Fracture Studies of Innervated and Denervated Sarcolemma

A comprehensive account of the alterations in the structure of sarcolemma visualized in the freeze-fractured replicas of denervated muscle is given elsewhere [4–6]; therefore, only those features which are pertinent to the present investigation are mentioned below.

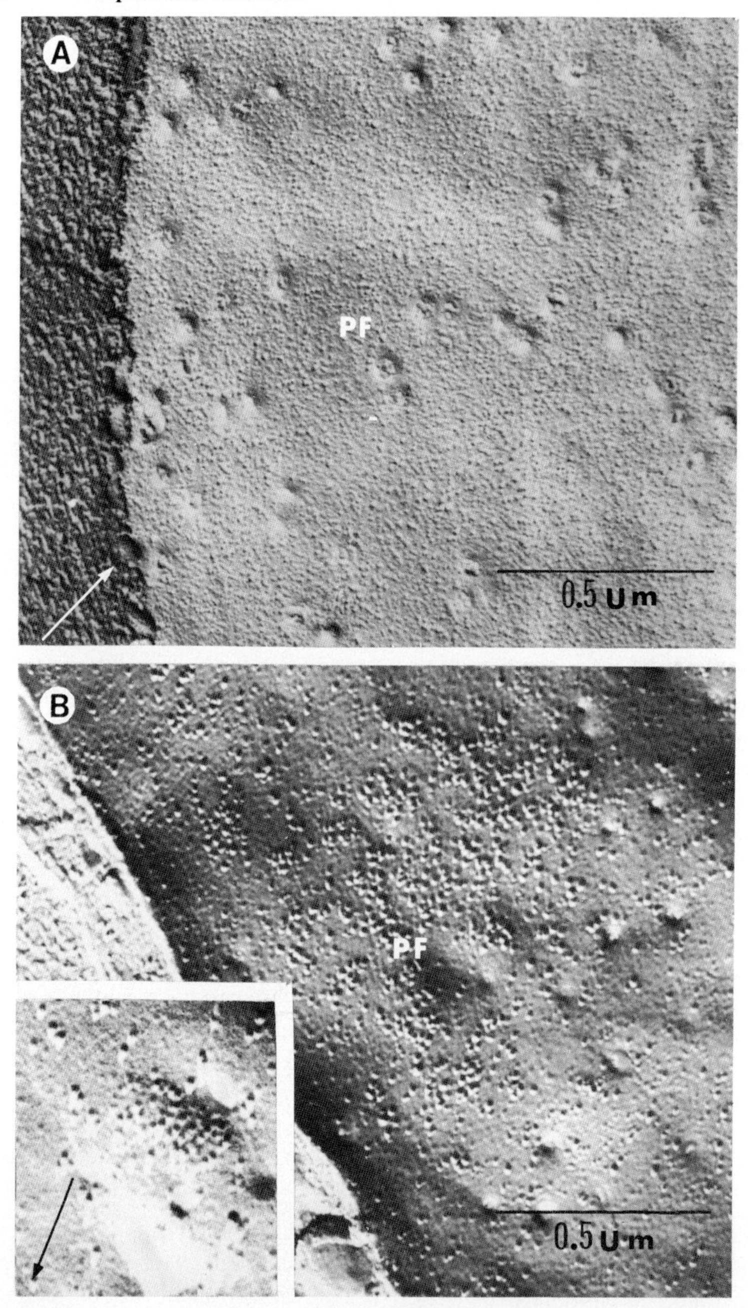

Fig. 1. A. Freeze-fractured preparation of a normal lumbrical muscle, showing the convex fractured face (PF) of nonjunctional sarcolemma. The intramembranous particles (~8 nm) are distributed uniformly over the entire fractured face. B. Fractured face (PF) of the sarcolemma from a lumbrical muscle of rat denervated for 2 weeks. A large number of particles are apparent in the denervated muscle which are not discerned in the normal (innervated) muscle. These particles are bigger (~15 nm) than those seen on this face in the normal muscle and it is likely that they represent extrajunctional acetylcholine receptors in denervated muscle. The inset is a PF of the denervated EDL muscle showing an aggregate with a number of 15 nm particles. Arrow in lower left corner indicates direction of shadowing. From Malhotra [22], with permission of Plenum Publishing Corporation, New York.

In innervated sarcolemma, the P face (PF) shows randomly dispersed intramembranous particles which are approximately 8 nm in diameter ($\sim 2{,}000/\mu^2$) (Fig. 1A). In contrast, the corresponding face of the denervated sarcolemma shows an abundance of ~15 nm (15–18 nm) intramembranous particles. These particles are dispersed singly or in aggregates. These aggregates may be small, with as few as 4–10 particles, or large, with approximately 25–100 particles (Fig. 1B). The average density of the particles on the PF (convex fracture) is ~400 particles/μ^2 and is predemoninantly made up by the 15-nm particles. A precise correlation between the intramembranous particles and the Ach receptors remains to be determined. It is of interest, however, that there are ~1150 α-BGT binding sites/μ^2 in the denervated extrajunctional sarcolemma. This estimate is based upon the assumption that each receptor binds one ^{125}I-α-BGT molecule (Tipnis and Malhotra, unpublished data), yet there may be more than one binding site per receptor molecule [43]. On the basis of existing physiological and biochemical evidence [11, 18], it has been hypothesized that these particles are components of the α-BGT binding sites (Ach receptors) [4, 5] and the two types of distributions of particles noted in these freeze-fractured replicas represent two distinct topographic distributions of receptors in the nonsynaptic sarcolemma of denervated muscle.

The following results refer to the experiments designed to test the above hypothesis by localization of Ach receptors through the use of ^{125}I-α-BGT and Ft-α-BGT conjugates.

Incorporation of ^{125}I-α-BGT Into Muscle

Incubation of innervated and denervated muscles in media containing ^{125}I-α-BGT shows a marked increase in the binding of toxin by denervated muscle over the innervated muscle (Table I). Preincubation of the muscles with d-tubocurarine leads to a marked sup-

TABLE I. Specific Binding of EDL Muscles of Innervated and Denervated Rats (^{125}I-α-BGT counts/min/mg muscle)

Experiment	^{125}I-α-BGT	d-tubocurarine and ^{125}I-α-BGT	Specific labeling
Innervated	604	292	312
Denervated	2394	1186	1208

The EDL muscles from innervated and denervated rats were incubated in oxygenated Kreb's ringer with 2×10^{-7} moles of ^{125}I-α-BGT. For determining the specificity of binding, another set of muscles was incubated with d-tubocurarine (10^{-4} moles) for 1 h and subsequently incubated with ^{125}I-α-BGT.

TABLE II. Distribution of Autoradiographic Grains in Nonsynaptic Region over 1-μ-Thick Sections of Innervated and Denervated Muscles Labeled With ^{125}I-α-BGT (number of silver grains /μ^2)

Experiment	^{125}I-α-BGT	d-Tubocurarine and ^{125}I-α-BGT
Innervated	0.6	0.5
Denervated	5	1.4

Control in both innervated and denervated muscles represents the preincubation of muscles in d-tubocurarine followed by incubation in ^{125}I-α-BGT. The muscles were incubated under the same conditions as mentioned in Table I.

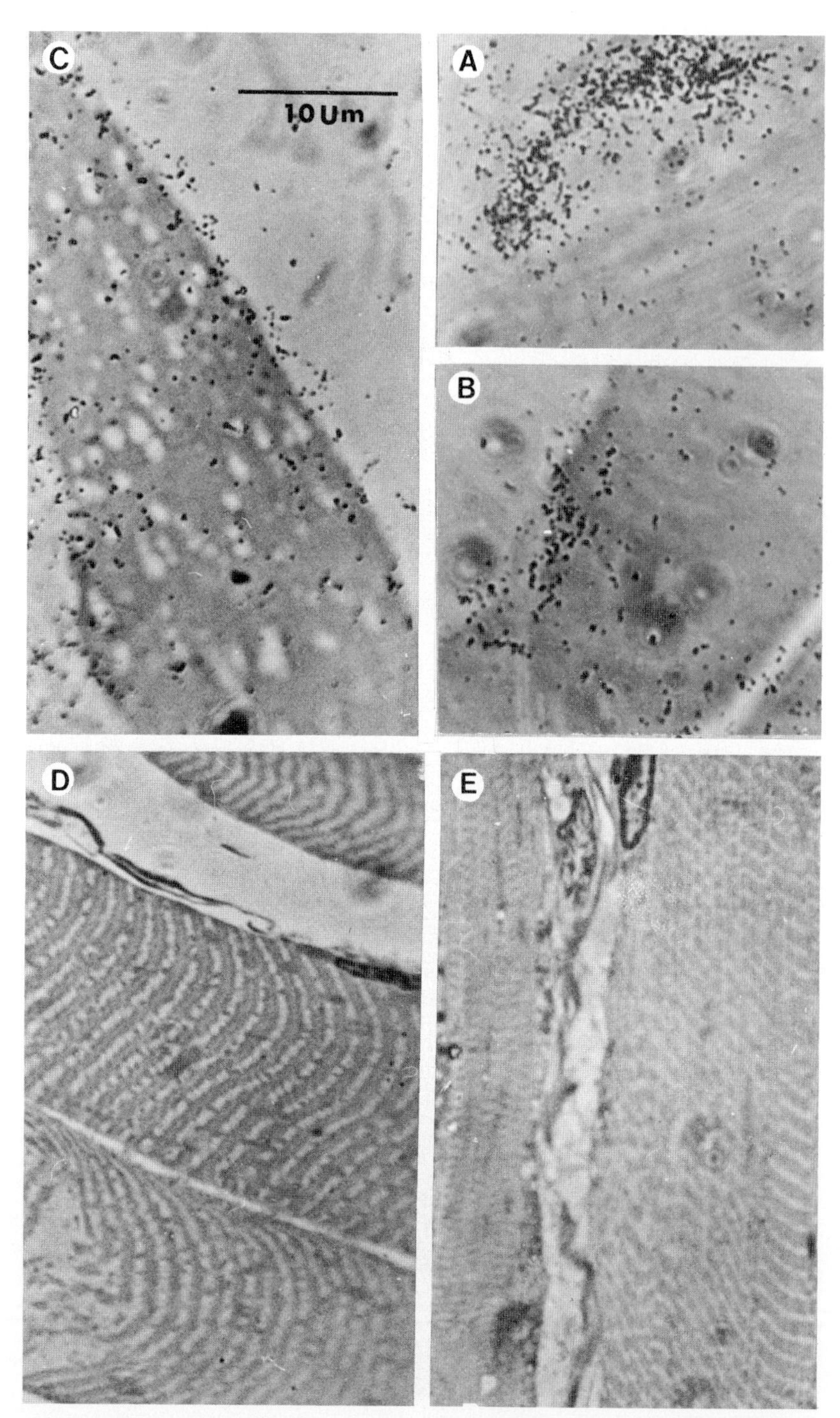

Fig. 2. Comparison of the specific incorporation of the labeled toxin in denervated vs innervated muscle. Sections of denervated muscle (1 μ thick) showing synaptic region with high density of grains (A) and nonsynaptic regions with uniformly dispersed grains (C) and clusters of grains (B). Sections of innervated muscle (D) or d-tubocurarine-treated denervated muscle (E) do not show grains. Calibration line in C applies to all illustrations in this figure.

pression in the binding of the toxin (Table I). Since d-tubocurarine is known to be a specific ligand for Ach receptors [19] and binds to the same site as the α-BGT, the data on the specific incorporation of ^{125}I-α-BGT indicate the labeling of the Ach receptors. As seen in Table I, the specific incorporation of ^{125}I-α-BGT in denervated muscle is approximately four times more than that in the innervated muscle. In both innervated and denervated muscle, d-tubocurarine inhibits toxin labeling by about 50%, which is in agreement with previous findings [41, 42].

Light Microscope Autoradiography

Table II shows the extent of silver grains seen in the nonsynaptic sarcolemma of 1 μ-thick sections. Silver grain counts given are after subtraction of the background grains. In each experiment, the background grains have been counted in areas located approximately 5 μ̂ away from the tissue. The background counts in sections of ^{125}I-α-BGT labeled muscle are comparable to the background grains encountered in sections prepared from muscles incubated in cold α-BGT and processed for autoradiography.

The distribution of silver grains in 1-μ-thick sections from innervated and denervated muscle is displayed in Figure 2. The silver grains in the nonsynaptic sarcolemma of denervated muscle appear either dispersed singly (Fig. 2C) or clustered (Fig. 2B). The number of silver grains per square micrometer is approximately 5–10 times higher in the clusters than outside such regions. Such areas of higher density do not represent the synaptic regions where the density of silver grains is far more in innervated as well as denervated muscle (Fig. 2A).

In contrast, the nonsynaptic regions of the innervated muscle have very few grains and their number is close to the background density (Fig 2D). Also, sections of denervated muscle treated with d-tubocurarine prior to incubation in ^{125}I-α-BGT are generally lacking in silver grains (Fig. 2E). It is concluded from the above data that the silver grains in the nonsynaptic sarcolemma of denervated muscles are far more numerous and represent much more binding of ^{125}I-α-BGT to specific sites on denervated muscle than in the corresponding regions of the innervated muscle.

Incorporation of ^{125}I-α-BGT Into Crude Membrane Fractions

The presence of sarcolemma in the crude membrane fraction was ascertained by assaying for Ach receptor activity. (Though the Ach receptors have been reported to reside in the Golgi apparatus during synthesis [20], it is assumed that in intact cells the receptors are exposed at the surface only in the plasma membrane.) In the membrane fractions from both innervated and denervated muscle, the binding of the toxin is linear during the first 10 min, after which saturation occurs (Fig. 3A). The specific activity in denervated membranes is 2.99×10^{-2} pmoles/μg as compared to 9.2×10^{-4} pmoles/μg in innervated preparation. Figure 3B shows that the binding of ^{125}I-α-BGT to denervated membranes is specific, as it is greatly inhibited by preincubation with cold α-BGT.

Transmission Electron Microscopy of Membranes Incubated With Ft-α-BGT

Crude membrane fractions show membrane vesicles ranging from 0.2 to 2 μ in diameter. Many of these vesicles from preparations incubated in Ft-α-BGT conjugate and when filtered through millipore get trapped in the filter along with nonspecifically bound ferritin as reported by Karlin et al [17]. In the present study, however, vesicles trapped in the filter were not considered and only vesicles lying above the filter were examined. Denervated crude membranes incubated in ferritin conjugate show ferritin associated with the membrane of the vesicle. Ferritin particles may be situated slightly removed

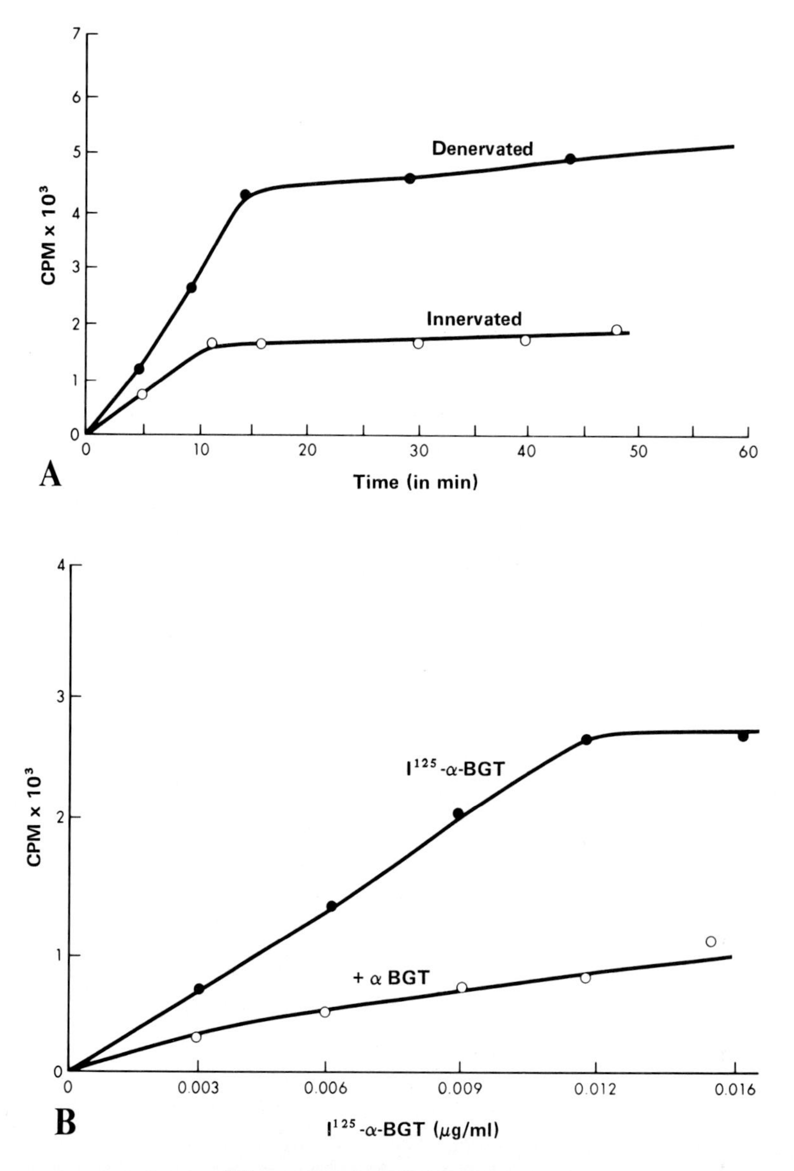

Fig. 3. A. Time-course binding of ^{125}I-α-BGT to crude sarcolemmal preparations from innervated and denervated muscles. The membranes from innervated and denervated muscles were incubated in ^{125}I-α-BGT. The incubation medium contained oxygenated Kreb's ringer, 50 μg/ml membrane protein, and ^{125}I-α-BGT (2×10^{-9} moles). The membranes were filtered through DE81 cellulose according to the method of Klett et al [16]. The filter paper was dried and the radioactivity was counted in a gamma counter. The specific activity of ^{125}I-α-BGT was 2.192×10^5 Ci/mole. B. Specific binding of ^{125}I-α-BGT to denervated crude sarcolemmal preparation. The membrane preparation (50 μg/ml of protein) was incubated in oxygenated Kreb's ringer containing ^{125}I-α-BGT (2×10^{-9} moles) for 1 hour at room temperature. The membranes were filtered through DE81 cellulose according to the method of Klett et al [16]. The filter was dried and the radioactivity counted in a gamma counter. For nonspecific binding, the aliquots were incubated in medium containing α-BGT (0.1 mg/ml) for 1 h. The specific activity of ^{125}I-α-BGT was 2.192×10^5 Ci/mole.

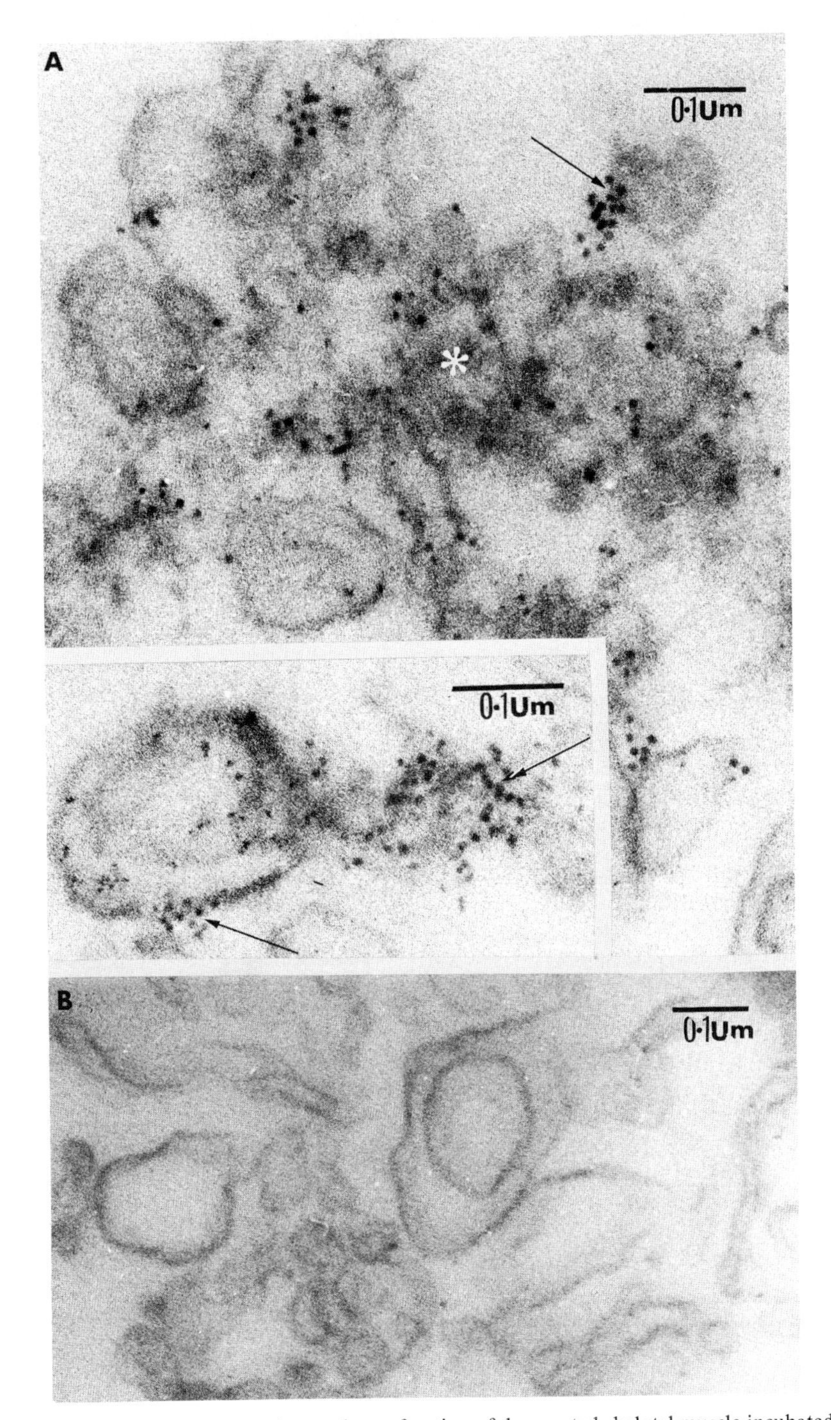

Fig. 4. A. Thin sections from a crude membrane fraction of denervated skeletal muscle incubated in Kreb's ringer containing 0.6 μg/ml of Ft-α-BGT conjugate showing ferritin binding. Several membrane vesicles are seen labeled. Arrows indicate clusters of ferritin particles. *Some unidentifiable material that could be cell debris. Such regions have not been considered in the estimation of Ft-α-BGT binding sites. B. Control from d-tubocurarine-incubated material showing paucity of ferritin binding.

from the surface of the membrane (~5–7 nm). The particles are seen as single molecules bound to the membrane or in small clusters (Fig. 4A). Several vesicles without the associated ferritin molecules are also encountered in electron micrographs. Estimation of the vesicles from four experiments indicates that 60% of the vesicles are labeled (Table III) whereas, in d-tubocurarine-treated controls, the number of vesicles showing associated ferritin is reduced to 11% of the total vesicles (Table IV). Small, dense particles are sometimes encountered inside the vesicles, both in the experimental and the d-tubocurarine-treated material. These are generally smaller than the ferritin molecules and their nature is not known. It appears that these particles do not result from the preparative fixation procedure, as membrane vesicles which have not been incubated in medium containing Ft-α-BGT conjugate do not show these particles. It is therefore assumed for the present that they represent degraded ferritin molecules, and the membranes are leaky to these particles.

TABLE III. Counts of Membrane Vesicles (denervated) Labeled with Ft-α-BGT

Experiment No.	Total vesicles counted	Labeled vesicles	Unlabeled vesicles
1	31	25	6
2	146	82	64
3	115	70	45
4	99	52	47
Total	391	229	162

Crude membrane fractions (~50 μg protein) were incubated in Kreb's ringer containing Ft-α-BGT conjugate (0.6 μg/ml). The membranes were filtered through millipore and, subsequent to washing, were processed for electron microscopy and embedded in Araldite. The vesicles lying above the filter were randomly counted in thin sections. It is apparent that nearly 60% of the vesicles were labeled. The number of ferritin molecules on the labeled vesicles may vary from 2 or more. The presence of ferritin molecules in the background is rare as the millipore filters were soaked in 2% albumin to minimize nonspecific binding [17].

TABLE IV. Counts of Membrane Vesicles (denervated) Incubated With d-Tubocurarine Followed by Incubation in Ft-α-BGT

Experiment No.	Total vesicles counted	Labeled vesicles	Unlabeled vesicles
1	95	15	80
2	153	18	135
3	252	11	241
4	116	23	93
Total	616	67	549

The procedure in these experiments was similar to the one outlined in Table III except in these experiments, the membrane fractions were incubated in d-tubocurarine (10^{-4} moles prior to incubation in Ft-α-BGT conjugate. The counting of more than 600 vesicles indicate that only 11% of these were labeled.

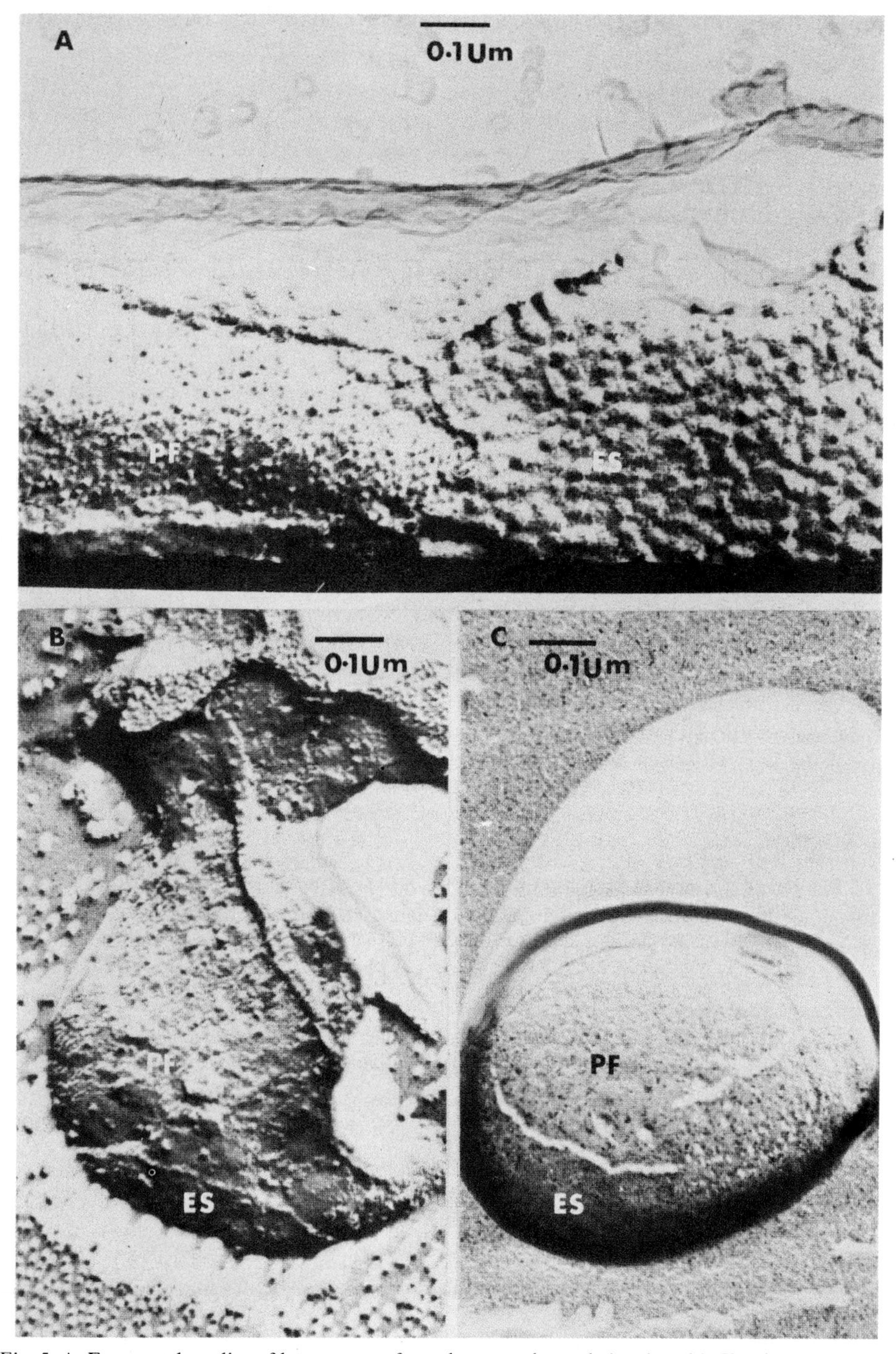

Fig. 5. A. Freeze-etch replica of homogenate from denervated muscle incubated in Kreb's ringer containing 0.6 μg/ml of Ft-α-BGT conjugate showing bumpy appearance of the etched face (ES), presumably due to binding to the conjugate. B, C. Controls without incubation in conjugate showing relatively smooth etched face (ES). Arrow in lower left corner indicates direction of shadowing.

Freeze-Etching of Ft-α-BGT-Labelled and Unlabelled Homogenate of Denervated Muscle

Initially, experiments were conducted on Ft-α-BGT-labeled crude membrane preparation, but despite several (40–50) attempts, satisfactory fractures of such membranes were not obtained. Therefore, muscle homogenate was investigated as a source of Ach receptor containing membranes. The results reported in this section are based on studies on Ft-α-BGT-labeled homogenate.

Replicas of freeze-etched membranes in muscle homogenate incubated with Ft-α-BGT show vesicles with a fractured face (PF) and an etched face (ES). The etched face often displays densely packed bumps (Fig. 5A) comparable to the size of ferritin molecules discernible in metallic replicas [21]. Some of these bumps are larger than individual ferritin molecules and may represent clusters of ferritin. It is emphasized that only some of the vesicles show bumpy etched face, whereas others have relatively smooth etched face. Also, comparable membrane faces in replicas of the homogenate without incubation in the Ft-α-BGT show relatively smooth etched faces (Figs. 5B, 5C). It is therefore concluded that the bumps on the etched face of the membranes displayed in Figure 5A represent ferritin particles presumably bound to the Ach receptors. A comparison of the number of labeled and etched vesicles with the number of labeled vesicles in thin sections would be valuable, but the etched vesicles are not of as frequent occurrence as the labeled vesicles in thin sections. This discrepancy is presumably due to difficulties in getting large areas of etched vesicles in replicas.

It should be remarked that membrane vesicles with scanty intramembranous particles on the fractured face are seen in the replicas. It is conceivable that such fractured faces represent regions of the membrane that are deficient in particles. Alternatively, they may represent inside-out vesicles and the two fractured faces show an asymmetric distribution of intramembranous particles, there being a few on one half [22].

DISCUSSION

The results from autoradiography of ^{125}I-α-BGT and labeling with Ft-α-BGT lead to the conclusion that the Ach receptors are either distributed singly or clustered in the nonsynaptic sarcolemma of denervated muscle.

The experiments on ^{125}I-α-BGT binding were primarily undertaken to ascertain the specificity of incorporated radioactivity in both muscle and crude membrane fractions (Figs. 3A, 3B; Tables I and II). The filter assay of Klett et al [16], employed in the present study, is based on the use of anion exchange cellulose filters for filtering the Ach receptor preparation. The filter binds anionic molecules while ensuring the elimination of cationic substances like unbound α-BGT. In both muscle and the crude membrane fraction from denervated animals, there is a marked increase in incorporated radioactivity. This binding, which is inhibited both by d-tubocurarine and cold α-BGT, is considered specific. The qualitative data based on light microscope autoradiography demonstrates the increased labeling in extrajunctional regions and supports the quantitative data reported in Table I. These findings, therefore, are in agreement with several physiologic studies that have demonstrated the extrajunctional sensitivity to acetylcholine in denervated muscle [18, 11, 23]. This extrajunctional sensitivity results from the incorporation of newly synthesized receptors [24–26].

The findings on localization of Ach receptors, obtained by using ^{125}I-α-BGT and Ft-α-BGT at the level of light and electron microscopy, respectively, indicate that there are two distinct populations of Ach receptors, viz., singly dispersed and clustered. The labeling of receptors with Ft-α-BGT is never as much as that observed in Ach receptor-rich preparations from *Torpedo* electroplaque [17]. This difference, however, is likely, as there is an extremely high concentration of Ach receptors in *Torpedo* electroplaque (40,000–50,000/μ^2) [27] compared to the relatively low concentration (1,695 receptors/μ^2) in the nonsynaptic region of denervated sarcolemma [11]. The membrane vesicles from synaptic sarcolemma, which would be expected to show dense labeling of Ft-α-BGT, may not have been encountered and possible explanations for the difficulty in finding such vesicles are 1) that membrane vesicles representing the synaptic sarcolemma are few and are lost during isolation; and 2) that the membrane preparation is filtered through millipore which traps unbound ferritin as well as vesicles.

The ferritin molecules may appear to be slightly removed from the surface of the membrane in thin sections (Fig. 4A), and this is consistent with the recent studies on Ach receptor-rich membranes from *Torpedo* in which it has been reported that the receptor molecules traverse the membrane and project approximately 5.5 nm outside the membrane bilayer [28].

The presence of Ach receptors in the nonsynaptic sarcolemma is a property of adult denervated muscle, as well as that of developing muscle cell [29, 30]. The nonuniformity of Ach receptor distribution reported in the present study has also been demonstrated in uninnervated developing muscles from chick and *Xenopus laevis* [31–33]. The presence of Ach receptor clusters in denervated muscle suggests that in relation to sarcolemma and Ach receptors, the muscle cell reverts to its embryonic state. Ach receptor clusters were thought to be the site of synapse formation [31]. It is pertinent to mention in this regard that the formation of synapse beside the original endplates is known to occur either by muscle injury or by removal of a portion of the muscle with the original nerve [34, 35]. However, recently Anderson and Cohen [33] have followed the distribution of Ach receptors (labeled with α-BGT) during synapse formation formation on myocytes cultured from *Xenopus laevis.* Their studies indicate that the clusters of Ach receptors are not the site of synapse formation.

The significance of the clustered Ach receptors is not yet clear. Axelrod et al [36] found that the receptors in the clusters are not free to diffuse into the surrounding areas. It is not as yet clear whether cluster formation occurs by movement and aggregation of individual particles or if there are distinct sites on the membrane where bulk incorporation of Ach receptor molecules takes place. Studies are in progress to further elucidate the role of the clustered Ach receptors in mammalian skeletal muscle.

ACKNOWLEDGMENTS

This investigation has been supported through grants awarded by the Natural Science and Engineering Research Council of Canada (A 5021). This work was part of a PhD dissertation submitted by U.R. Tipnis in the Department of Zoology.

Mr. C. Saunders has provided invaluable technical assistance in the preparation of the material for electron microscopy. We are grateful to Dr. G.W. Stemke for his help in the initial stages in iodination and for the use of his gamma counter.

REFERENCES

1. Miledi R, Slater CR: Proc Roy Soc B 169:289, 1968.
2. Miledi R, Slater CR: J Cell Sci 3:49, 1968.

3. Pollack MS, Bird JWC: Am J Physiol 215:716, 1968.
4. Tipnis U, Malhotra SK: FEBS Lett 69:141, 1976.
5. Tipnis U, Malhotra SK: Cytobios 19:181, 1977.
6. Malhotra SK, Tipnis U: Proc R Soc Lond B 203:59, 1978.
7. Albuquerque EX, Thesleff S: Acta Physiol Scand 73:471, 1969.
8. Albuquerque EX, McIssac RI: Exp Neurol 26:183, 1970.
9. Hall ZW: J Neurobiol 4:343, 1973.
10. Koenig J, Vigny M: Nature 271:75, 1978.
11. Hartzell HC, Fambrough DM: J Gen Physiol 60:248, 1972.
12. Changeux JP, Kasai M, Lee CY: Proc Natl Acad Sci USA 67:1241, 1970.
13. Boegman RJ, Manery JF, Pinteric L: Biochim Biophys Acta 203:506, 1970.
14. Greenwood FC, Hunter WM: Biochem J 89:114, 1963.
15. Hourani BT, Torain BF, Henkart MP, Carter RL, Marchesi VT, Fischbach GD: J Cell Sci 16:473, 1974.
16. Klett RP, Fulpius BW, Cooper D, Smith M, Reich E, Possani LD: Biol Chem 248:6841, 1973.
17. Karlin A, Holtzman E, Valderrama R, Damle V, Hsu K, Reyes F: J Cell Biol 76:577, 1978.
18. Axelsson J, Thesleff S: J Physiol 147:178, 1959.
19. Prives JM, Reiter MJ, Cowburn DA, Karlin A: Mol Pharmacol 8:786, 1972.
20. Fambrough DM, Devreotes PN: J Cell Biol 76:237, 1978.
21. Shotton D, Thompson K, Wofsky L, Branton D: J Cell Biol 76:512, 1978.
22. Malhotra SK: In Roodyn DB (ed): "Structure of Biological Membranes: Functional Characterization." New York: Plenum, 1978, p 221.
23. Uchitel O, Robbins N: Brain Res 153:539, 1978.
24. Grampp W, Harris JB, Thesleff S: J Physiol 221:743, 1972.
25. Devreotes PN, Fambrough DM: Cold Spring Harbor Symp Quant Biol 40:237, 1975.
26. Devreotes PN, Fambrough DM: Proc Natl Acad Sci USA 73:161, 1976.
27. Bourgeois JP, Popot JL, Ryter A, Changeux JP: J Cell Biol 79:200, 1979.
28. Klymkowsky MW, Stroud RM: J Mol Biol 128:319, 1979.
29. Land BR, Podleski TR, Salpeter EE, Salpeter MM: J Physiol 269:155, 1977.
30. Smilowitz H, Fischbach GD: Dev Biol 66:539, 1978.
31. Sytkowski AJ, Vogel Z, Nirenberg MW: Proc Natl Acad Sci USA 70:270, 1973.
32. Vogel Z, Daniels MP: J Cell Biol 69:501, 1976.
33. Anderson MJ, Cohen MW: J Physiol 268:757, 1977.
34. Miledi R: Nature 199:1191, 1963.
35. Gwyn DG, Aitken JT: J Anat 100:111, 1966.
36. Axelrod D, Ravdin P, Koppel DE, Schlessinger J, Webb WW, Elson EL, Podleski TR: Proc Natl Acad Sci USA 73:4594, 1976.
37. Rash JE, Ellisman MH: J Cell Biol 63:567, 1974.
38. Bridgmen PC, Greenberg AS: Abstracts, Society for Neuroscience, Ninth Annual Meeting, Atlanta, Georgia, 1979, p 476.
39. Yee AG, Fischback GD, Karnovsky MJ: Proc Natl Acad Sci USA 75:3004, 1978.
40. Cartaud J, Benedetti EL, Sobel A, Changeux J-P: J Cell Sci 29:313, 1978.
41. Miledi R, Potter LT: Nature (Lond) 233:599, 1971.
42. Berg DK, Kelly RB, Sargent PB, Williamson P, Hall ZW: Proc Natl Acad Sci USA 69:147, 1972.
43. Karlin A, Damle V, Valderrama R, Hamilton S, Wise D, McLaughlin M: Fed Proc 37:121, 1978.

Journal of Supramolecular Structure 12:481–489 (1979)
Biological Recognition and Assembly 185–193

Analysis of Cell Surface Interactions by Measurements of Lateral Mobility

Elliot L. Elson and Jeffry A. Reidler

Department of Chemistry, Cornell University, Ithaca, New York 14853

Interactions of cell surface components with one another and with structures inside and outside the cell may have important physiological functions in the transmission of signals and the assembly of specialized structures. These interactions may be detected and analyzed through their effects on the lateral mobility of cell surface molecules. Measurements by a fluorescence photobleaching method have shown that in general lipid-like molecules diffuse rapidly and freely through the plasma membrane, whereas proteins move much more slowly or appear to be immobile. This dichotomy has been supposed to result from forces beyond the viscosity of the lipid bilayer, which specifically retard the diffusion of membrane proteins. This general picture should be qualified, however, by noting that the lateral mobility of lipid-like molecules can be influenced in detail by changes in the state of the plasma membrane such as result from mitosis or fertilization. The interactions of cell surface proteins that limit their lateral mobility are unknown. The effects of binding concanavalin A to localized regions of cell surface show that these interactions can vary in subtle and complex ways. It may soon be useful to interpret mobility experiments in terms of simple reaction models that attempt to describe surface interactions in physicochemical terms. More experimental data are needed to carry out this program and to relate interactions that affect mobility to the structural connections between cell surface components and the cytoskeleton, which have been detected by biochemical methods and electron and immunofluorescence microscopy.

Key words: **fluorescence photobleaching, cell surface, cytoskeleton, lateral mobility, membrane interactions**

INTRODUCTION

Important physiological functions that occur at the surface of an animal cell depend on dynamic interactions of molecules embedded in the plasma membrane with one another and with molecules and structures inside and outside the cell. For example, it has recently been recognized that activation of specific processes in various kinds of cells by polypeptide hormones such as insulin and epidermal growth factor (EGF) [1, 2] and on mast cells and

Dr. Elson is now at Department of Biological Chemistry, Division of Biology and Biomedical Sciences, Washington University School of Medicine, St. Louis, MO 63110.

Dr. Reidler is now at Department of Structural Biology, Stanford University School of Medicine, Stanford University Medical Center, Stanford, CA 94305.

Received November 26, 1979; accepted November 28, 1979.

0091-7419/79/1204-0481$02.00

basophils by immunoglobulin E [3] may involve the rapid association in the plasma membrane of ligand receptor complexes. Apparently a limited aggregation of these complexes is needed to transmit into the cells the signals required to trigger their specific responses. It also now appears that formation of (perhaps somewhat larger) aggregates is required for clearing ligand–receptor complexes from the plasma membrane by a specific endocytic pathway that involves coated vesicles [4].

Similarly, the formation of specialized structures at the cell surface must require interactions among plasma membrane molecules or with cytoplasmic components. These specialized structures are involved in various cellular functions and include, for example, subsynaptic concentrations of acetylcholine receptor, an intrinsic membrane protein that mediates neuromuscular signal transmission [5]; fibers of the extrinsic glycoprotein, fibronectin, which contributes to the formation of a pericellular connective matrix [6]; and coated regions of the plasma membrane, which may contain receptors of various kinds as well as the protein clathrin [7] and which mediate the specific endocytosis of a wide range of ligand–receptor complexes, including those of insulin, EGF, low-density lipoprotein, and α_2-macroglobulin [8–11].

Characterization of the specific interactions of cell surface components and of the pathways by which specialized surface structures are assembled present important problems for cell physiology. Recently, methods have been developed that allow a physicochemical approach to these problems. This paper discusses the use of measurements of macroscopic lateral mobility to gain mechanistic information about cell surface interactions and assembly processes. This approach is based simply on the idea that the lateral mobility of a molecule in the plasma membrane will be retarded to the extent that it interacts with other surface or cytoplasmic molecules or supramolecular structures. As we shall see, the nature and extent of this retardation depends both on the mobility of the interacting structures and on the strength and kinetics of the interactions. Therefore, in favorable circumstances it may be possible to obtain information about these several properties from mobility measurements.

APPROACH AND METHOD

The lateral and rotational mobility of cell surface molecules can be measured in different ways using methods based on fluorescence [12], magnetic resonance [13], or other properties [14]. We use a fluorescence photobleaching recovery (FPR) method that is relatively simple, quantitative, and readily applicable to individual cells living in culture without apparent serious perturbation of cellular properties [15]. A cell surface component is labeled with a fluorescent tag. The fluorescence from a small open region of the cell surface is measured using a microscope equipped with a sensitive photomultiplier tube [16]. The measured fluroescence (corrected for background) is proportional to the number of labeled molecules in the surface observation region. A brief intense pulse of light is used to "destroy" by irreversible photolysis a fraction of the labeled molecules in the region. Thus the fluorescence in the region is momentarily decreased below its initial level by an amount that depends on the intensity and duration of the bleaching pulse as well as the photochemical properties of the fluorophore. In the absence of further bleaching the measured fluorescence will recover due to transport of unbleached fluorophores from adjacent regions of the surface into the observed region. Hence, from this measured rate of fluorescence recovery, the rate of transport may be deduced [17]. Recovery may occur after diffusion or systematic flow of labeled molecules. Up to now almost all measurements that have been made are accounted for simply in terms of diffusion.

As a simple model for an interaction involving a mobile surface component, we may consider the reaction

$$A + B \underset{k_b}{\overset{k_f}{\rightleftharpoons}} C$$

We shall suppose that A is a nonfluorescent molecule or structure, which because of its size or its connection with other cellular components, is either immobile or slowly moving with diffusion coefficient D_A. The membrane component B is fluorescent and and, when free, moves with a diffusion coefficient, D_B. The effective mobility of B is, however, retarded by its interaction with A to form the complex C. For simplicity we shall suppose that interaction with B does not change the low mobility of A. Hence $D_A = D_C << D_B$. The slower the rate of diffusion of C and the stronger the interaction of A and B, the greater will be the retardation of B. Hence, the retardation depends on D_C and on the equilibrium constant $K = k_f/k_b = \bar{C}_C/\bar{C}_A\bar{C}_B$, where $\bar{C}_X$ is the equilibrium concentration of X.

The kinetics of the interaction may be reflected in the experimental measurements in two different ways. If there is a substantial change in the fluorescence excitation or emission of B due to its interaction with A, then the chemical relaxation can be observed directly. This type of effect has been demonstrated in a study of the interaction of ethidium bromide with DNA by a closely related fluorescence fluctuation method [18]. The kinetics of reaction (1) may also be revealed by the nature of the retarding effect on the mobility of B. This will depend on the relative magnitudes of the characteristic times for fluorescence recovery by simple diffusion of B and by chemical relaxation. In an FPR experiment the characteristic time for diffusion of B, τ_B, is given as

$$\tau_B = w^2/4D_B$$

In this equation w is the radius of the observation region on the cell surface [17]. The characteristic time for chemical relaxation is given, assuming linearized chemical kinetics, by the well-known formula:

$$\tau_{chem} = R^{-1} = [k_f(\bar{C}_A + \bar{C}_B) + k_b]^{-1}$$

If the chemical reaction is slow compared to diffusional recovery of B – ie, $\tau_{chem} >> \tau_B$, – a given B molecule will be observed, for the most part, to be either free or complexed with A, but only rarely will it experience both states during the measurement interval. Therefore, if $D_B >> D_A$, the B molecules will appear to reside in two distinct mobility classes: fast, due to diffusion of free B with diffusion coefficient D_B, and slow, due to diffusion of C with diffusion coefficient D_C. For D_C sufficiently small, there may be negligible recovery of C during the measurement period, so that B molecules involved in the complex will appear to be effectively immobile. If, however, the chemical reaction is fast compared to the diffusional recovery of B – ie, $\tau_{chem} << \tau_B$, – each B molecule will react many times with A during the observed fluorescence recovery time. Then all B molecules will be retarded to a comparable extent, so that the FPR experiment will reveal a single mobility class with effective diffusion coefficient D_e [19]:

$$D_e = D_C f_C + D_B f_B$$

where

$$f_C = \frac{\bar{C}_C}{\bar{C}_B+\bar{C}_C} = \frac{K\,\bar{C}_A}{1+K\,\bar{C}_A} \quad \text{and} \quad f_B = 1 - f_C = \frac{\bar{C}_B}{\bar{C}_B+\bar{C}_B} = \frac{1}{1+K\,\bar{C}_A}$$

Hence, under these conditions, K may be determined by measurements of the effective diffusion coefficient of B at various values of $\bar{C}_A$. This principle has been applied by Borejdo to study the interaction of actin and myosin using a closely related fluorescence fluctuation approach [20]. A detailed theoretical analysis of the measurement of the kinetics of reaction [1] by this fluorescence fluctuation approach has been presented [21]. The analysis of the measurement of this system by FPR is fundamentally similar, although important differences in detail must be taken into account [19].

Since the characteristic time for diffusional recovery, τ_B (but not τ_{chem}), depends on the dimension of the observation region, the time scale of the experiment may be changed by varying w. This raises the possibility of probing the kinetics of a cell surface reaction by increasing w to pass from a condition in which two mobility classes are observed ($\tau_{chem} \gg \tau_B$) to a condition in which B seems to move at a single "effective" rate ($\tau_{chem} \ll \tau_B$). Up to now this principle does not appear to have been experimentally applied. Nevertheless, experimental results are now available that indicate the presence of some interactions that are fast and others that are slow under conditions typically used for FPR measurements of surface mobility on animal cells (time scale ~ 0.1 sec to ~ 100 sec).

It is important also to ask to what extent the mobility measurements might be influenced by technical problems with the FPR method. Recently, there have appeared reports of photo-induced cross-linking of cell surface proteins by irradiation of fluorescein bound to surface molecules [22, 23]. In principle, this type of cross-linking could reduce the mobility of, and possibly even immobilize, cell surface proteins. Unfortunately, the conditions under which these experiments must be carried out in order to obtain sufficient material for biochemical analysis (relatively low intensities of excitation and long periods of exposure) are not comparable to those of FPR measurements (higher intensities, shorter exposure). Moreover, it appears that increasing intensity and decreasing the duration of exposure at constant total excitation energy decreases the amount of cross-linking [23]. Yet mobility measurements carried out using different excitation intensities and lengths of exposure show no effect of these variables on the measured mobilities [24]. This and other evidence [24, 25] leads us to conclude, at least provisionally, that photo-induced cross-linking of surface components does not strongly perturb the dynamic properties measured by FPR.

Cell Surface Lipids and Proteins

Different kinds of interactions seem to govern the rates of diffusion of cell surface lipids and proteins. Lipid-like molecules are typically observed to diffuse rapidly, and apparently homogeneously, in the plasma membranes of cultured cells. The diffusion coefficient of the lipid probe 3,3′-dioctadecylindocarbocyanine ("diI") is approximately 10^{-8} cm^2/sec in a number of different kinds of cells [26–28]. Usually the diI fluorescence recovers after a photo-bleaching pulse to nearly its initial prebleach level, indicating that most of the observed fluorescence is in a freely mobile form. In contrast, cell surface proteins appear to exist in both mobile and immobile forms, with the mobile molecules moving substantially more slowly than diI. Both nonspecifically labeled cell surface "proteins" [26] and several different surface receptors, including those that bind immunoglobulin E on mast cells [27], α-bungarotoxin on myotubes (acetylcholine receptor) [29], and insulin and epidermal growth factor on fibroblasts [30], have diffusion coefficients less than 10^{-9} cm^2/sec. Moreover, incomplete recovery of the fluorescence to its prebleach level indicates that some portion of each of these membrane components is immobile on the time scale of the measurement.

These results and other work with which they are consistent [14, 31] indicate that the mobility of the lipid probel diI is determined by the viscosity of the lipid bilayer matrix in which it is embedded. It would be simplest to suppose that the mobility of cell surface proteins is also limited only by the viscosity of this matrix. In fact, this does appear to be a reasonable view of the behavior of rhodopsin in amphibian rod outer segment disk membranes [32, 33]. The experimental results just described, however, argue on two counts against applying this viewpoint to typical animal cell surface membranes. First, a fraction of the proteins in these membranes is constrained, even to the extent that the proteins appear immobile in our measurements. Second, even the mobile proteins move far more slowly than the lipid probe and more slowly than would be expected simply from the greater viscious resistance due to the greater size of the protein molecules [33]. It has therefore been argued that interactions and forces in addition to the viscous resistance of the lipid bilayer must be restraining the mobility of membrane proteins [34]. The structural basis of these restraining interactions is entirely unknown. Recently, biochemical, electron microscopic, and immunofluorescence evidence of links between the cell surface and the cytoskeleton has been presented [35–40]. The effect of these links on mobility has yet to be demonstrated, however, Attempts to use agents such as colchicine and cytochalasin B, which disrupt (respectively) microtubules and microfilaments, have not led to definitive conclusions about the role of these cytoskeletal elements in determining the mobilities of cell surface components [26, 34]. Similarly, little can yet be said about the dynamic characteristics of these interactions. From the viewpoint of the simple reaction model described above, cell surface proteins would seem to experience both slow and fast interactions to account, respectively, for the apparent immobilization of a fraction and the slow diffusion of the balance of these molecules.

A detailed analysis of the dynamics of well-characterized classes of cell surface proteins has yet to be performed. This should include measurements of the diffusion of membrane proteins reconstituted into model membrane systems to establish a baseline condition for the mobility of those molecules in the absence of restraints by cytostructural components [41–45]. Experiments of this kind must contend with questions of the fidelity of the reconstituted system to the natural disposition of the membrane protein. It could also be valuable to attempt a more detailed analysis on cell surfaces of the range of mobilities experienced by specific proteins and of the degree of variation of mobility at different positions on the cell. This kind of analysis is complicated by systematic and random errors, which are difficult to eliminate from measurements on living cells.

Modulation of Lipid and Protein Mobilities

The contrast between cell surface lipids, which more freely and rapidly in the plasma membrane, and proteins, which are constrained by unknown forces to much slower rates of diffusion, is consistent in measurements up to now [12, 34]. Nevertheless, interesting examples of modulation of lipid mobility by changes in the physiological condition of a cell or by interaction with a specific membrane component have recently been discovered. Edidin and Johnson [46] have shown that both diI and surface antigens on mouse ova become less mobile upon fertilization. The mechanism of this effect is still unknown. Edidin and Johnson speculate, however, that the immobilization of cell surface components may be needed for several processes involved in embryonic development [46]. Studies on neuroblastoma cells have shown that the mobilities both of lipid probes and of surface antigens change in a systematic way over a two- to threefold range during the cell cycle [47]. The diffusion coefficients of both kinds of molecules are at a minimum during mitosis and increase during G_1. Then the diffusion coefficients of the lipid probes remain constant during S and G_2, before diminishing again during M. In contrast, the diffusion coefficients of surface antigens decrease gradually during S and G_2, with a small further reduction occurring during M. The mobilities of two structurally distinct lipid probes change similarly through the cell cycle. Therefore it seems likely that this results from a general change in membrane fluidity rather than from specific interactions with cellular structures. Moreover, the differences in behavior of the lipid probes and the membrane antigens indicate that factors in addition to lipid fluidity must influence the dynamic properties of these molecules (presumed to be mainly cell surface proteins).

Examples have recently have been found in which the mobilities of lipid probes are diminished by apparently specific interactions with identified membrane components. These were discovered in measurements of the diffusion of fluorescein-labeled ganglioside and ceramide analogs in the plasma membranes of cells infected by vesicular stomatitis and sindbis viruses [48]. Viral glycoprotein at the surface of infected cells may spontaneously form patches or can be induced to do so with bivalent antibody directed against the viral glycoproteins. Double-label fluorescence microscopic observations revealed that the fluorescein-labeled lipid probes were selectively concentrated into regions rich in viral glycoprotein. Measurements of the mobilities of the analogs in regions of high and low viral glycoprotein concentration indicated a complex pattern of behavior with decreases both in diffusion coefficients and in the fraction of mobile molecules. The mobility of diI, however, was the same in regions of high and low viral glycoprotein concentration, suggesting that the effects on the mobility of the analogs resulted from specific interactions rather than a more general effect on local membrane fluidity.

These results show that, although lipid-like molecules typically do move more freely than do proteins through the plasma membrane, nevertheless, even lipid probes may experience various kinds of interactions that influence their mobilities in detail.

The forces in addition to membrane viscosity that retard the motion of typical cell surface proteins may result from interactions with other cell surface components or with structures inside or outside the cell. In the plane of the membrane short-lived, possibly nonspecific, interactions can uniformly reduce the mobility of cell surface proteins without immobilizing them [44]. Long-lived specific interactions due, for example, to a cross-linking antibody [27] or to the binding of insulin or ECF to their receptors [30], can induce the formation of large immobile aggregates. Whether the immobility of these aggregates results from increased viscous resistance due to their larger size or from the enhancement of interactions with slow-moving or stationary structures is still uncertain [34].

Little is known yet about the effects on mobility of contact between the plasma membrane and the extracellular connective matrix. It has been shown that fibronectin fibers do not impede the diffusion of lipid probes and cell surface antigens [28]. Concanavalin A, however, is immobilized by binding directly to immobile fibronectin fibers [28].

Structural and biochemical evidence implicate interactions between cell surface components and the cytoskeleton both in immobilization [39] and in dynamic redistribution – eg, cap formation [37, 49, 50]. It is typically supposed that stable linkages with the cytoskeleton are established in response to cross-linking surface proteins [39, 36, 49]. Even in the absence of an externally applied cross-linking agent, however, there may be interactions between the cytoskeleton and membrane proteins that are stable enough to survive gentle extraction by Triton X-100 [51]. It is reasonable to suppose that the linkages revealed in these studies must also influence cell surface mobility and that measurements of the lateral diffusion of membrane proteins should help to characterize the kinds of interactions involved. For example, it might be supposed that the immobilized fraction of labeled cell surface proteins indicated by incomplete recovery of fluorescence after photo-bleaching is due to the formation of stable cytoskeletal links. It is perhaps surprising, therefore, that the effects of anti-cytoskeletal agents such as colchicine and cytocholasin B have been so unrevealing [34].

An example of the potential complexity and subtlety of interactions affecting surface mobility has been provided by a study of the influence of concanavalin A (conA) on the lateral diffusion of plasma membrane antigens [52]. The conA was confined to localized regions of 3T3 cell surfaces by having first been reacted with platelets. Then the conA-platelets were bound in clumps to circumscribed regions of the cell surface. FPR measurements at various distances from the platelet clumps showed no effect on lateral diffusion until some 4% of the dorsal cell surface was covered by platelets. Beyond that point the diffusion coefficient of the antigens was reduced sixfold and remained at this reduced value at higher fractions of surface coverage by the conA-platelets. There was no effect observed on the fraction of *immobile* antigens as measured by the extent of fluorescence recovered, near 50% on average under all conditions tested. The extent of inhibition of mobility was not correlated with distance from the platelet clumps. When the experiment was repeated in the presence of anti-microtubule agents, a similar pattern of behavior was observed, except that antigen mobility was reduced three- rather than sixfold.

These measurements indicate that localized perturbations of the state of conA-receptors can inhibit mobility over the entire cell surface. The threshold and plateau of the inhibition suggest that the responsible interactions are initiated in a highly cooperative process. This process could be a polymerization or assembly of, for example, a cytoskeletal component, or it could be the activation of a diffusable enzyme; either could propagate the effect over the entire cell surface [52, 34].

It is interesting to consider the effects of local conA binding from the viewpoint of the simple reaction model, equation (1). It is simplest to suppose that the lack of effect on the fraction of recovered fluorescence, and therefore of immobile antigens, is due to failure of conA to influence long-lived interactions that constrain these molecules. (This analysis is complicated by the heterogeneous specificities of the anti-cell surface antibodies used to detect surface mobility in these experiments [52]. It is also possible that some sets of antigens were immobilized by short-lived interactions for which the binding equilibrium lay far to the right; $f_C \sim 1$, $f_B \sim 0$. Detected in isolation these molecules would appear to be entirely immobile rather than divided into mobile and immobile fractions. It is therefore important to determine the effect of local conA binding on the mobility of well-defined, homogeneous surface components.) The model further suggests that the sixfold reduction in the diffusion coefficients of the mobile antigens (B) could result from enhancement of interactions with anchorage components (A), which are rapid compared to the typical fluorescence recovery time (10–500 sec in these experiments). This enhancement could result from increasing the concentration of A available for reaction with B or by increasing the equilibrium constant K.

The threefold inhibition of mobility with characteristic threshold and plateau induced in the presence of anti-microtubule agents suggests that microtubules are not directly involved in the interactions of cell surface antigens that respond to localized conA binding. These interactions seem to be enhanced similarly in the presence and absence of the anti-microtubule agents. The twofold decreases in the extent of immobilization in the presence of these agents indicates that microtubules do have some role – perhaps in further stabilizing the surface-modulating assembly [49].

CONCLUSIONS

Measurements of lateral mobility indicate that cell surface components experience both transient and long-lived interactions. These interactions may modulate the kinetics of dynamic physiological processes such as the aggregation of hormone- and immunoglobulin-receptor complexes [1–4]. They may also be involved in the assembly of specialized cell surface structures [29]. It should eventually be possible to extend the characterization of these interactions by interpreting experimental measurements in terms of reaction models, of which equation (1) embodies an especially simple example. That model, for example, could be tested directly if the concentrations of components A and B could be systematically varied on the cell surface. Interpretation of interactions that limit mobility in structural and biochemical terms is a further challenge – one that will require the combined application of many experimental techniques. It is particularly interesting now to relate the interactions detected through their effects on mobility with those that have already been revealed by biochemical analysis and electron and immunofluorescence microscopy [35–40, 50, 51].

ACKNOWLEDGMENTS

The authors would like to thank their co-workers who have participated in these studies, especially J. Schlessinger and W.W. Webb, who have contributed to most of the work involving FPR; D. Koppel and D. Axelrod, with whom the early studies were carried out; and S. deLaat, H. Wiegandt, J. Lenard, P. Keller, M. Schlesinger, and D. Johnson, who collaborated with us in the later work involving ganglioside analogs, cell cycle variations, and studies of virus infected cells. Work in the authors' laboratory described in this paper was supported by NIH grants CA 14454, GM 21661. J.R. was supported by an NIH traineeship.

REFERENCES

1. Kahn CR, Baird KL, Jarrett DB, Flier JS: Proc Natl Acad Sci USA 75:4209–4213, 1978.
2. Shechter Y, Hernaez L, Schlessinger J, Cuatrecasas P: Nature 278:835–838, 1979.
3. Segal D, Taurog J, Metzger H: Proc Natl Acad Sci USA 75:2993–2997, 1977.
4. Schlessinger J, Schechter Y, Willingham MC, Pastan I: Proc Natl Acad Sci USA 75:2659–2663, 1978.
5. Fertuck HC, Salpeter MM: Proc Natl Acad Sci USA 71:1376–1378, 1974.
6. Yamada KM, Olden K: Nature 275:179–184, 1978.
7. Pearse BMF: J Mol Biol 97:93–98, 1975.
8. Anderson RGW, Goldstein JL, Brown MS: Proc Natl Acad Sci USA 73:2434–2438, 1976.
9. Gordon P, Carpentier J-L, Cohen S, Orci L, Proc Natl Acad Sci USA 75:5025–5029, 1978.
10. Maxfield FR, Schlessinger J, Shechter Y, Pastan I, Willingham MC: Cell 14:805–810, 1978.
11. Brown MS, Goldstein JL: Proc Natl Acad Sci USA 76:3330–3337, 1979.

12. Schlessinger J, Elson EL: In Ehrenstein G, Lecar H (eds): "Biophysical Methods." New York: Academic Press (in press).
13. Chan SI, Bocian DF, Petersen NO: In Grell E (ed): "Membrane Spectroscopy." New York: Springer, Verlag (in press).
14. Edidin M: Ann Rev Biophys Bioeng 3:179–201, 1974.
15. Elson EL, Schlessinger J, Koppel DE, Axelrod D, Webb WW: In Marchesi VT (ed): "Membranes and Neoplasia: New Approaches and Strategies." New York: Alan R. Liss, 1976, pp 137–147.
16. Koppel DE, Axelrod D, Schlessinger J, Elson EL, Webb WW: Biophys J 16:1315–1329, 1976.
17. Axelrod D, Koppel DE, Schlessinger J, Elson E, Webb WW: Biophys J 16:1055–1069, 1976.
18. Magde D, Elson EL, Webb WW: Biopolymers 13:29–61, 1974.
19. Elson EL: in preparation.
20. Borejdo J: Biopolymers 18:2807–2820, 1979.
21. Elson EL, Magde D: Biopolymers 13:1–27, 1974.
22. Lepcock JR, Thompson JE, Kruuv J, Wallach DFH: Biochem Biophys Res Commun 85:344–350, 1978.
23. Sheetz MP, Koppel DE: Proc Natl Acad Sci USA 76:3314–3317, 1979.
24. Jacobson K, Hou Y, Wojcieszyn J: Exp Cell Res 116:179–189, 1978.
25. Wolf D, Dragsten P: personal communication.
26. Schlessinger J, Axelrod D, Koppel DE, Webb WW, Elson EL: Science 195:307–309, 1977.
27. Schlessinger J, Webb WW, Elson EL, Metzger H: Nature 264:550–552, 1976.
28. Schlessinger J, Barak LS, Hammes GG, Yamada KM, Pastan I, Webb WW, Elson EL: Proc Natl Acad Sci USA 74:2909–2913, 1977.
29. Axelrod D, Ravdin P, Koppel DE, Schlessinger J, Webb WW, Elson EL, Podleski TR: Proc Natl Acad Sci USA 73:4594–4598, 1976.
30. Schlessinger J, Shechter, Y, Cautrecasas P, Willingham MC, Pastan I: Proc Natl Acad Sci USA 75:5353–5357, 1978.
31. Fahey PF, Webb WW: Biochemistry 17:3046–3053, 1978.
32. Poo MM, Cone RA: Nature 247:438–441, 1974.
33. Saffman PG, Delbruck M: Proc Natl Acad Sci USA 72:3111–3113, 1975.
34. Elson EL, Schlessinger J: In Schmitt FO and Worden FG (eds): "The Neurosciences, Fourth Study Program." Cambridge, Massachusetts: MIT Press, 1979, pp 691–701.
35. Koch GLE, Smith MJ: Nature 273:274–278, 1978.
36. Flanagan J, Koch GLE: Nature 273:278–281, 1978.
37. Albertini DF, Berlin RD, Oliver JM: J Cell Sci 26:57–75, 1977.
38. Bretscher A, Weber K: Proc Natl Acad Sci USA 76:2321–2325, 1979.
39. Ash JF, Louvard D, Singer SJ: Proc Natl Acad Sci USA 74:5584–5588, 1977.
40. Geiger B: Cell 18:193–205, 1979.
41. Wu ES, Jacobson K, Szoka F, Portis A: Biochemistry 17:5543–5550, 1978.
42. Darszon A, Vandenberg CA, Ellisman MH, Montal M: J Cell Biol 81:446–452, 1979.
43. Racker E: J Supramol Struct 6:215–228, 1977.
44. Wolf DE, Schlessinger J, Elson EL, Webb WW, Blumenthal R, Henkart P: Biochemistry 16:3476–3483, 1977.
45. Smith LM, Parce JW, Smith BA, McConnel HM: Proc Natl Acad Sci USA 76:4177–4179, 1979.
46. Johnson M, Edidin M: Nature 272:448–450, 1978.
47. deLaat SW, Van der Saag PT, Elson EL, Schlessinger J: Proc Natl Acad Sci USA (in press).
48. Reidler J, Wiegandt H, Schlessinger M, Elson EL: in preparation.
49. Edelman GM: Science 192:218–226, 1976.
50. Condeelis J: J Cell Biol 80:751–758, 1979.
51. Ben-Ze'ev A, Duerr A, Solomon F, Penman S: Cell 17:859–865, 1979.
52. Schlessinger J, Elson EL, Webb WW, Yahara I, Rutishauser U, Edelman GM: Proc Natl Acad Sci USA 74:1110–1114, 1977.

Journal of Supramolecular Structure 12:491–504 (1979)
Biological Recognition and Assembly 195–208

Selective Protein Transport: Characterization and Solubilization of the Phosvitin Receptor From Chicken Oocytes

John W. Woods and Thomas F. Roth

Department of Biological Sciences, University of Maryland Baltimore County, Catonsville, Maryland 21228

Phosvitin (PV), a subunit of a female-specific protein, vitellogenin, binds to oocyte membranes with a KD of 10^{-6} M. Binding reaches equilibrium within 30 min after incubation at 25°C. Bound ^{125}I-PV dissociates from the membrane with a $t_{1/2}$ of 13 h when incubated in buffer. However, when ^{125}I-PV-labeled membranes are incubated in buffer containing 10^{-5} M unlabeled PV, 50% of the initially bound ^{125}I-PV dissociates from the membrane within 10 min. These results support the conclusion that PV binds to a membrane-associated receptor.

Solubilization studies show that Triton X-100 solubilizes up to 45% of the total membrane-bound ^{125}I-PV. Gel-exclusion chromatography of the solubilized material yields a 500,000 dalton ^{125}I-PV-containing complex separated from free ^{125}I-PV. The 500,000 dalton complex completely dissociates to yield free ^{125}I-PV when incubated with excess unlabeled PV. However, when incubated with 1) no addition, 2) IgG, or 3) serum albumin, the extent of dissociation is significantly reduced and is consistent with that which would be predicted on the basis of the observed dissociation rate in the absence of unlabeled PV. These results suggest that bound ^{125}I-PV can only be displaced by unlabeled PV.

These results also indicate that the 500,000 dalton species is a solubilized PV-receptor complex and that it is possible to solubilize the PV-receptor in an active form.

Key words: oocyte protein transport, receptor solubilization, phosvitin receptor

Selective protein transport mediated by coated pits and coated vesicles appears to be a ubiquitous biologic process and is particularly manifest in oocytes sequestering vitellogenin. Vitellogenin is a maternal serum protein which is stored in the oocyte until needed in embryogenesis, at which time it is degraded to provide nutrients for the developing embryo. Detailed studies suggesting that coated pits and coated vesicles mediate the transport of vitellogenin have been carried out in the Mosquito [1], in Xenopus laevis [2, 3], in Saturnid moth [4], in Cecropia moths [5], and in the domestic chicken [6]. In addition to vitellogenin, immunoglobulins are also apparently transported from the mother to her offspring by coated pits and coated vesides [7]. This maternofetal

Received May 7, 1979; accepted November 19, 1979.

0091-7419/79/1204-0491$02.60

transport of immunoglobulins has been postulated by Brambell [8] to provide the newborn with passive immunity to its septic environment until it is immunocompetent. Specific examples include IgG transport across the chicken oocyte membrane [9], IgG transport across the chicken oocyte membrane [9], IgG transport into rabbit fetuses via the yolk sac [10], and IgG transport across rat illial cells [11]. Coated pits and coated vesicles have also been shown to mediate protein transport in systems unrelated to oogenesis or reproduction. Low-density lipoprotein, a serum cholesterol carrier, appears to enter fibroblasts exclusively via coated vesicles [12]. Epidermal growth factor also appears to enter cells via a coated-vesicle-mediated mechanism [13].

Proteins appear to be selected for specific transport by receptors which recognize and bind specific proteins from the extracellular millieu. Receptors have been characterized which mediate binding in a variety of protein transport systems. Examples include the vitellogenin receptor on chicken oocyte membranes [14], the vitellogenin receptor on Xenopus oocytes [15], IgG receptors on chick yolk sac [16], IgG receptors on rabbit yolk sac [17, 18], and IgG receptors on rat illial cells [19]. In addition, receptors have been characterized for low-density lipoprotein [20] and epidermal growth factor [21] on fibroblasts. The techniques used to study the receptors involved in protein transport were taken from prior studies of receptors for insulin [22], IgE [23], acetylcholine [24], and other molecules. For a detailed review on the development and use of these methods, see Cuatrecasas and Hollenberg [25].

The objective of this study is to elucidate further the mechanism of vitellogenin transport into developing chicken oocytes. We present evidence that phosvitin, a subunit of vitellogenin, is selectively bound to receptors on developing oocytes. We also present evidence suggesting that Triton X-100 solubilizes the phosvitin receptor in an active form.

MATERIALS

Agarose 0.5 M, Agarose 5 M, and Bio Gel P-10 were obtained from Bio-Rad Laboratories. Carrier-free $Na^{125}I$ was obtained from the Amersham Radiochemical Centre. Bovine serum albumin (BSA), chicken serum albumin, and phosvitin (PV) were from Sigma Co. Siliclad was obtained from Clay Adams. All other chemicals were of reagent grade and were obtained from commercial sources. Live White Leghorn laying hens and oocytes were purchased from a local slaughter house.

Chicken IgG, kindly provided by Dr. Carol Linden, was purified from egg yolks using a modification of the method of Bernardi and Cook [16, 26].

All experimental procedures were carried out in an incubation buffer (IB) consisting of 0.01 M Mes (2-N-morpholinoethane sulfonic acid), pH 6.0, 0.14 M NaCl, 5 mM KCl, 0.83 mM $MgSo_4$, 0.13 mM $CaCl_2$, plus 0.02% sodium azide. Where indicated, bovine serum albumin was added to a final concentration of 5 mg/ml (IB-BSA).

METHODS

Iodination of Phosvitin

Phosvitin was iodinated by the procedure of Pressman and Eisen [27]. In a typical iodination, 5 mCi of Na ^{125}I were first oxidized to $^{125}ICl^-$ with $NaNO_2$ in 1 M HCl. The pH of the solution was then adjusted to neutrality and 10 mg of PV was added in 0.1 M borate buffer, pH 8.0. After the reaction was allowed to proceed for 4 min at 0°, the reaction was quenched with sodium meta-bisulfite, and ^{125}I-PV was separated from other reactants by chromatography on a Bio Gel P-10 column that had been prerun with unlabeled PV. These procedures routinely yielded ^{125}I-PB preparation with specific activities of approximately 10^8 cpm/mg.

Determination of Phosvitin Concentration

Phosvitin concentration was determined by absorbance at 280 nm using a value of $E_{1cm}^{1\%} = 4.4$, obtained by measuring the OD_{280} of various dilutions of 10 mg/ml solutions of PV. The 10 mg/ml solutions were prepared by weighing lyophilized PV. Molar concentrations were based on an average molecular weight of 34,000 daltons for the two classes of phosvitin [30].

Isolation of Oocyte Membranes

The ovaries were removed from freshly killed White Leghorn laying hens and placed in IB-BSA at 0°C. Oocytes approximately 1.5–2 cm in diameter were slit, drained of yolk and returned to ice-cold IB-BSA. Adherent yolk was removed by gentle shaking in the IB-BSA solution. The membrane complex consisting of the oocyte plasma membrane, a fibrous viteline layer, a monolayer of follicular epithelium cells, and an acellular basement lamella was dissected free of the overlaying connective tissue and placed in fresh IB-BSA. Dissected membranes were homogenized with 2–3 passes of a loose-fitting pestle in a glass homogenizer. Tissue fragments were harvested by centrifugation for 5,000 g-min at 4° and resuspended in a minimal volume of IB-BSA. In a typical experiment, 36 oocytes were dissected and the homogenized membranes resuspended in a final volume of 3.6 ml such that 100 μl of homogenate was approximately equivalent to the tissue obtained from one oocyte.

Assay of PV Binding to Oocyte Membranes

Siliclad-coated glass tubes (12 × 75 mm) were used for all incubations. In a typical binding experiment, 100 μl of homogenized oocyte membranes, an amount equivalent to the membranes from one oocyte, were placed in a tube. ^{125}I-PV and other proteins, when appropriate, were added to yield a final volume of 250 μl. Except where noted in the figure legends, incubations were carried out for 60 min in a shaking water bath at 25°C. Details of protein concentration are given in the figure legends. Incubations were routinely terminated by the addition of 3 ml of IB-BSA at 4°, followed by centrifugation for 5,000 g-min at 4° to harvest the membranes. The membrane pellet was resuspended in successive 3 ml volumes of IB-BSA and washed as described above a total of five times. ^{125}I radioactivity in the wash supernatants and final pellets was determined in a well-type gamma counter.

In our experiments, specific ^{125}I-PV binding was defined as that which was displaced by 50-fold to 100-fold molar excess of unlabeled PV, an amount sufficient to displace 98–99% of the specifically bound ^{125}I-PV. Each experiment was conducted using parallel sets of incubations, the first containing only ^{125}I-PV plus homogenized oocyte membranes and the second containing ^{125}I-PV, membranes, and unlabeled PV. The bound ^{125}I-PV resulting from the first set of tubes represents total ^{125}I-PV binding; the bound ^{125}I-PV in the second set of tubes represents nonspecific ^{125}I-PV binding. By subtracting the nonspecific from the total radioactivity bound, we arrive at values for ^{125}I-PV specifically bound. In these experiments nonspecific ^{125}I-PV binding typically represents 15–20% of the total ^{125}I-PV binding.

Solubilization Procedure

Except as noted in the figure legends, solubilization procedures were carried out as follows. Homogenized oocyte membranes obtained from a single oocyte were suspended in 333 μl of 1% (w/v) Triton X-100 in IB-BSA and incubated for 30 min at 25°. Following incubation, three samples were pooled to yield 1 ml, and insoluble material was removed by centrifugation at 100,000 g for 30 min at 4°.

Column Chromatography and Solubilized Receptors

Siliclad-treated glass columns were filled with Agarose 0.5 M (0.8 × 32 cm column) or Agarose 5 M (0.9 × 53 cm column) and equilibrated with IB containing 0.1% (w/v) Triton X-100 and 0.1 mg/ml bovine serum albumin at 4°. Elution with buffer of the same composition was carried out at a flow rate of 4 ml/h for the Agarose 0.5 M columns and 2 ml/h for the Agarose 5 M columns. Columns were routinely calibrated with ferritin (MW 480,000), bovine IgG (MW 160,000), and bovine serum albumin (MW 64,000).

RESULTS

Characterization of Phosvitin Binding to Oocyte Membranes

Association kinetics. An initial series of experiments was designed to determine the kinetic properties of phosvitin (PV) binding to homogenized oocyte membranes. To determine the rate at which PV binding reached equilibrium, identical aliquots of homogenized oocyte membranes were incubated with a low concentration (3.8×10^{-7} M) of ^{125}I-VP at 25° for various times (Fig. 1). Specific binding increased rapidly for the first 20 min of incubation; after 30 min there was no further increase. Nonspecific binding

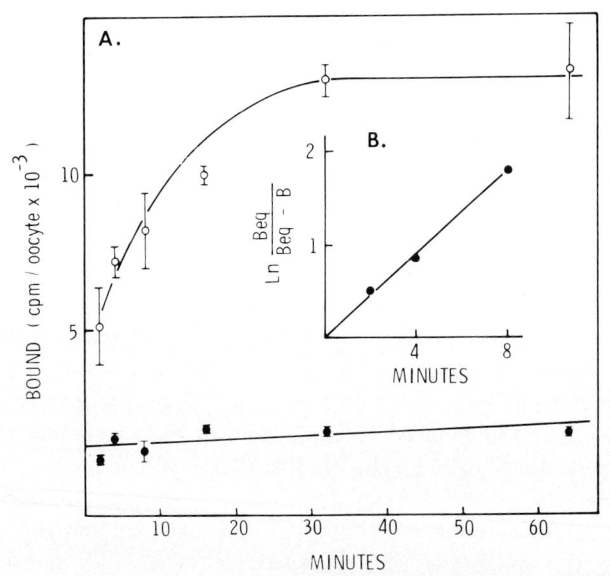

Fig. 1. Time course of ^{125}I-PV binding to oocyte membranes. Homogenized membranes were incubated with 3.8×10^{-7} M ^{125}I-PV (10^6 cpm) for the indicated times at 25°. A: Specific (○) and nonspecific (●) binding was quantitated as described in Materials and Methods. B: Replot of early time points (see text). Error bars indicate standard deviation of triplicate determinations.

increased linearly and was less than 20% of specific binding at all times. All subsequent experiments were carried out at receptor concentrations approximately equal to that in the time course experiment and with the ligand concentration equal to or greater than that used in the time course experiment. Hence, since subsequent experiments were incubated for times greater than 60 min at 25°, the binding reactions will have reached equilibrium.

The association rate constant, k_1, was determined by the method of Kitabgi et al [28]. Originally developed to analyze the interaction between neurotension and synaptic membranes, this method assumes that the free-ligand concentration equals total ligand concentration. In our experiments, this assumption is only valid for short incubation times. Thus, according to Kitabgi [28], for short incubation times the interaction of ^{125}I-PV and oocyte membranes can be described as a pseudo-first-order reaction by the equation:

$$\ln \frac{[RL_{eg}]}{[RL_{eg}] - [RL]} = ([L] k_1 + k_{-1}) \cdot t$$

$[RL_{eg}]$ is the concentration of bound PV at equilibrium, [RL] is the concentration of bound PV at time t, [L] is the toal concentration of PV, k_1 is the association rate constant and k_{-1} is the dissociation rate constant. When $\ln \frac{RL_{eg}]}{[RL_{eg}]-[RL]}$ was plotted as a function of time (Fig. 1B), a straight line was obtained with a slope of

$$3.8 \times 10^{-7}\ M (k_1) + k_{-1} = 3.7 \times 10^{-3}\ sec^{-1}$$

Dissociation kinetics. The dissociation rate of ^{125}I-PV from oocyte membranes was determined at 25° and 4° by measuring the dissociation of bound ^{125}I-PV in the presence of excess unlabeled PV. Under these conditions, the association reaction of ^{125}I-PV with the receptor is blocked and the dissociation reaction can be described by the equation

$$\frac{d[RT]}{dt} = -k_{-1}\ [RL]$$

thus,

$$k_{-1} = \frac{1}{t} \ln \frac{[RL]0}{[RL]}$$

and

$$k_{-1} = \frac{\ln 2}{t_{1/2}},$$

where $t_{1/2}$ = half-time of dissociation.

The results (Fig. 2A) show that dissociation at 4° is linear when plotted as a semilogarithmic plot of percentage bound versus time, and the $t_{1/2}$ (4°) is approximately 13 h. The dissociation rate at 25° appeared to be biphasic. When corrected for the contribution of the slow component ($t_{1/2}$ = 13 h), the data for the rapid component yielded a straight line on a semilogarithmic plot with a $t_{1/2}$ of approximately 10 min. Thus, the dissociation rate constant, (k_{-1}), equals 1.1×10^{-3} sec. Using the association kinetic data and the experimentally determined value of k_{-1}, the association rate constant was calculated as $6.8 \times 10^3\ M sec^{-1}$. From the values of the rate constants, the dissociation constant, K_D, was calculated as

$$K_D = \frac{k_{-1}}{k_{+1}} = 1.6 \times 10^{-7}$$

The presence of a slowly dissociating component at 25°, which represents about 20% of the bound ^{125}I-PV, is at present unexplained. However, since the data for the dissociation experiment was not corrected for nonspecific binding, and nonspecific binding represents 15–20% of the total ^{125}I-PV bound at the ligand concentrations used in this experiment (data not shown), the slowly dissociating component may represent dissociation of nonspecifically bound ^{125}I-PV. Alternatively, the slowly dissociating component may represent the receptor-ligand complex undergoing a transition to form a more tightly coupled receptor-ligand complex, as was proposed for neurotoxin binding to the acetylcholine receptor [24] $[R + L \underset{k_{-1}}{\overset{k_1}{\rightleftharpoons}} RL \underset{k_{-2}}{\overset{k_2}{\rightleftharpoons}} RL]$. If this is true, we have calculated the K_D only for the first half of the overall reaction: $K_D = \frac{k_{-1}}{k_1}$. Further experiments will be required to determine the significance of the slowly dissociating component.

In a similar experiment, dissociation of bound ^{125}I-PV was determined in the absence of excess unlabeled PV (Fig. 2A). In this case, reassociation of ligand and receptor is possible. This experiment was done to determine the extent of dissociation that would occur during the washing procedure used to separate free and bound ligand. At 4° in the absence of unlabeled PV, the half-time of dissociation was about 13 h. Thus, less than 5% of the bound ^{125}I-PV would be expected to dissociate from the membrane during the 30-min washing procedure at 4°C. These results also suggest that a significant amount of ^{125}I-PV receptor complexes may be expected to remain after 20 h, which is approximately the maximum time required for the column chromatography experiments in the following sections.

Displacement assay. In order to assay the ability of unlabeled PV to competitively displace bound ^{125}I-PV from oocyte membranes, an additional displacement experiment was carried out. The results (Fig. 2B) show that at the lowest concentration tested (1.1×10^{-6} M), 84% of the initially bound ^{125}I-PV remained bound after a 60-min incubation at 25°. However, at the highest concentration of unlabeled PV tested (3.9×10^{-4} M), only 19% of the initially bound ^{125}I-PV remained associated. This experiment suggests that at least 80% of the initially bound ^{125}I-PV was bound in a freely reversible manner.

Equilibrium studies. The apparent dissociation constant, K_D, of PV binding to homogenized oocyte membrane was determined under steady-state conditions. In this experiment, the concentration dependence of PV binding was assayed by adding increasing amounts of unlabeled PV to a constant amount of ^{125}I-PV and incubating with oocyte membranes for 2 h at 25°. A Scatchard plot of the data (Fig. 3) gives an apparent K_D of 3.3×10^{-6} M. The minimum number of binding sites per oocyte calculated from these data is approximately 6.5×10^{13}. This number is most likely a low estimate in view of the loss of tissue which undoubtedly occurs during the isolation and assay procedures.

Control for wash procedure. In these experiments the separation of free and bound ligand was routinely accomplished by washing the oocyte tissue with five 3-ml changes of cold (4°) IB-BSA. That this procedure removes most free ^{125}I-PV from the membranes was shown when oocyte membranes were first incubated with ^{125}I-PV and then subjected to repeated washing. The results (Fig. 4) show that successive washes remove progressively smaller proportions of the remaining activity, suggesting that free ^{125}I-PV comprises a smaller proportion of the total ^{125}I-PV remaining after each wash. After five repetitions, each subsequent was removed approximately 5% of the activity associated with the preceding pellet, suggesting that no free ^{125}I-PV remains associated with the pellet. In view of the very slow dissociation rate observed at 4° and of the fact that each wash requires only 5 min to complete, the 5% loss of activity in washes 6–10 may be due to a failure to pellet 100% of the membranes.

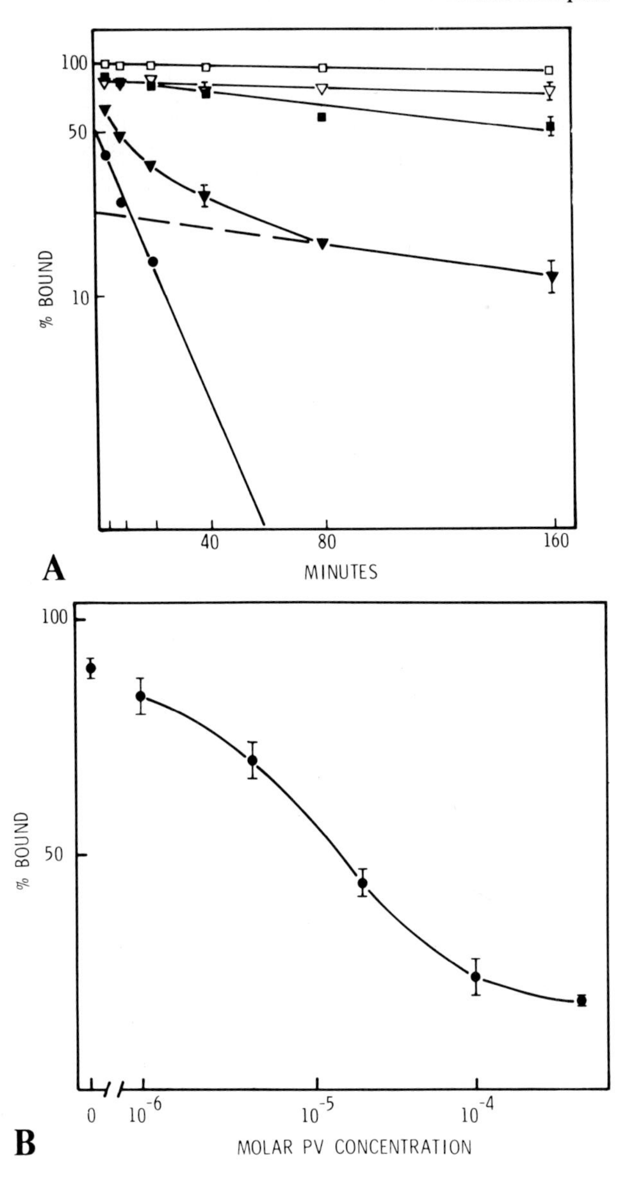

Fig. 2. Dissociation of bound ^{125}I-PV from oocyte membranes. Homogenized membranes were incubated with 3.5×10^{-6} M ^{125}I-PV (96.8×10^5 cpm) for 60 min at 25°. Unbound PV was removed by washing as described in Materials and Methods. A: Time course of dissociation of bound ^{125}I-PV. Washed membranes, containing 3×10^4 cpm of bound ^{125}I-PV were resuspended in 1 ml of IB-BSA in the presence (▼, ■) or absence (▽, □) of 3×10^{-4} M unlabeled PV and incubated for the indicated times at 25° (▼, ▽) or 4° (■, □). The fast component (●) of the 25° plus unlabeled PV data was calculated by subtraction of the contribution of the slowly dissociating component (dashed line) from the early time points. B: Displacement of bound ^{125}I-PV by unlabeled PV. Washed membranes, containing 4×10^4 cpm of bound ^{125}I-PV were resuspended in 1 ml of IB-BSA containing 0 to 3×10^4 M unlabeled PV. After incubation for 60 min at 25°, the reaction was terminated by centrifugation at 4° for 5,000 g-min. Percentage bound was calculated as: $\frac{(^{125}\text{I in pellet})}{(\text{total } ^{125}\text{I})} \times 100$. Error bars indicate standard deviations of triplicate determinations.

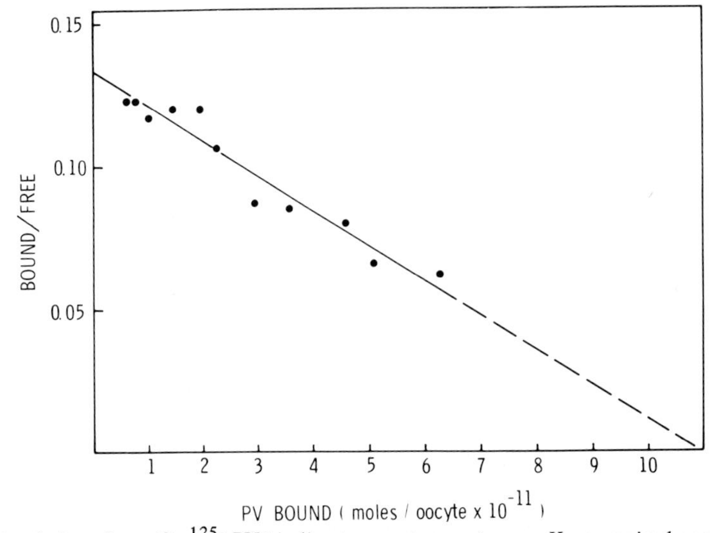

Fig. 3. Scatchard plot of specific ^{125}I-PV binding to oocyte membranes. Homogenized oocyte membranes were incubated for 2 h at 25° with a constant amount of ^{125}I-PV (3.8×10^{-7} M, 3.5×10^{5} cpm) to which increasing concentrations of unlabeled PV had been added to yield final PV concentrations of 3.8×10^{-7} M to 4.4×10^{-6} M. Bound and free ^{125}I-PV were separated by washing as described in Materials and Methods. Bound PV was that remaining with the pellet after the final wash, and free PV was the total PV recovered in the wash supernatants. Nonspecific binding, quantitated in the presence of 50 X molar excess unlabeled PV was subtracted from total bound PV. Each datum point is the mean of triplicate determinations. The best fit of the data to a straight line was calculated using linear regression analyses (r = 0.96) $K_D = 3.3 \times 10^{-6}$ M, number of receptors/oocyte = 6.5×10^{13}.

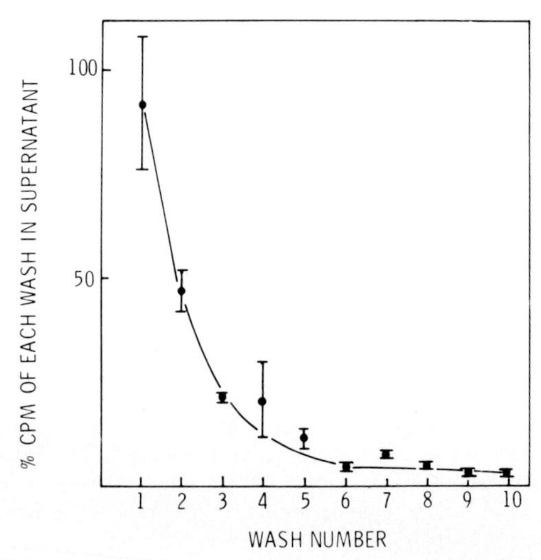

Fig. 4. Elution of free ^{125}I-PV from oocyte membranes. Homogenized oocyte membranes were incubated with 2.9×10^{-6} M ^{125}I-PV (4×10^{6} cpm) for 60 min at 25°. Following incubation, the membranes were diluted by addition of 3 ml of IB-BSA at 0°. The membranes were then harvested by centrifugation at 5,000 g-min, and the pellet was immediately resuspended in 3 ml of fresh IB-BSA for each of 10 washes. Following the final wash, 1.4×10^{5} cpm of ^{125}I-PV remained bound to the membrane pellet. Percentage cpm in each wash supernatant was calculated as the percentage of the total cpm initially present in each wash before centrifugation. Error bars indicate standard deviations of triplicate determinations.

Solubilization of Membrane-Bound ^{125}I-PV With Triton X-100

Low-speed centrifugation assay. In order to determine the degree to which Triton X-100 solubilized bound ^{125}I-PV, washed membranes containing bound ^{125}I-PV were incubated with various concentrations of Triton X-100. These experiments showed (Table 1) that Triton concentrations of 0.05–2% solubilized approximately 40–50% of the initially bound ^{125}I-PV. Triton concentrations higher than 2% appeared to solubilize only 25–35% of the initially abound material, even though more of the oocyte tissue appeared to be solubilized, since the final centrifugation yielded a much smaller pellet. The significance of this observation is not known; however, it is possible that high concentrations of Triton form large protein-mycelle aggregates which are segmentable under the conditions used in these experiments. In control experiments where membranes were incubated in the absence of Triton, only 8% of the initially bound ^{125}I-PV was released.

Gel exclusion chromatography assay. If Triton X-100 solubilizes bound ^{125}I-PV and the receptor as a complex, this PV-receptor complex should elute in a position different from free PV on an appropriate gel exclusion column. The elution profile on an Agarose 0.5 M column of material solubilized by 1% Triton X-100 from membranes incubated with ^{125}I-PV is shown in Figure 5A. The ^{125}I-PV activity eluted from the column as two peaks, one containing approximately 40% of the total activity eluted in the void volume and the other eluted in the position of free ^{125}I-PV. Because the ^{125}I-PV-containing material eluting the void volume has a higher molecular weight than free ^{125}I-PV, we shall tentatively refer to this high-molecular-weight material as solubilized PV-receptor complex.

In order to demonstrate the reversible nature of phosvitin binding to the solubilized PV-receptor complexes, these complexes were incubated in the presence or absence of excess unlabeled PV (2.9×10^{-5} M) prior to chromatography on Agarose 5 M columns.

TABLE I. Solubilization of Bound ^{125}I-PV by Triton X-100

Triton X-100 (%w/v)	% Solubilized
0.0	9.4 ± 0.7
0.01	23.4 ± 1.6
0.05	40.5 ± 0.9
0.1	45.0 ± 0.1
0.5	39.3 ± 2.5
1.0	45.0 ± 4.5
2.0	47.2 ± 0.8
3.5	33.4 ± 0.3
5.0	27.5 ± 6.4

Homogenized oocyte membranes were incubated with 5.8×10^{-6} M ^{125}I-PV ^{125}I-PV (2.4×10^{6} cpm) for 60 min at 25°. Unbound PV was removed as described in Materials and Methods. These washed membranes, containing 4×10^{4} cpm, were resuspended in 1 ml of IB-BSA with 0–5% w/v Triton X-100 and incubated for 30 min at 24°. Following incubation, the membranes were harvested by centrifugation at 4° for 5,000g-min. The percentage solubilized was calculated as: $\frac{(^{125}\text{I in supernatant})}{(\text{total } ^{125}\text{I})} \times 100$. The means of three determinations ± standard deviation are given.

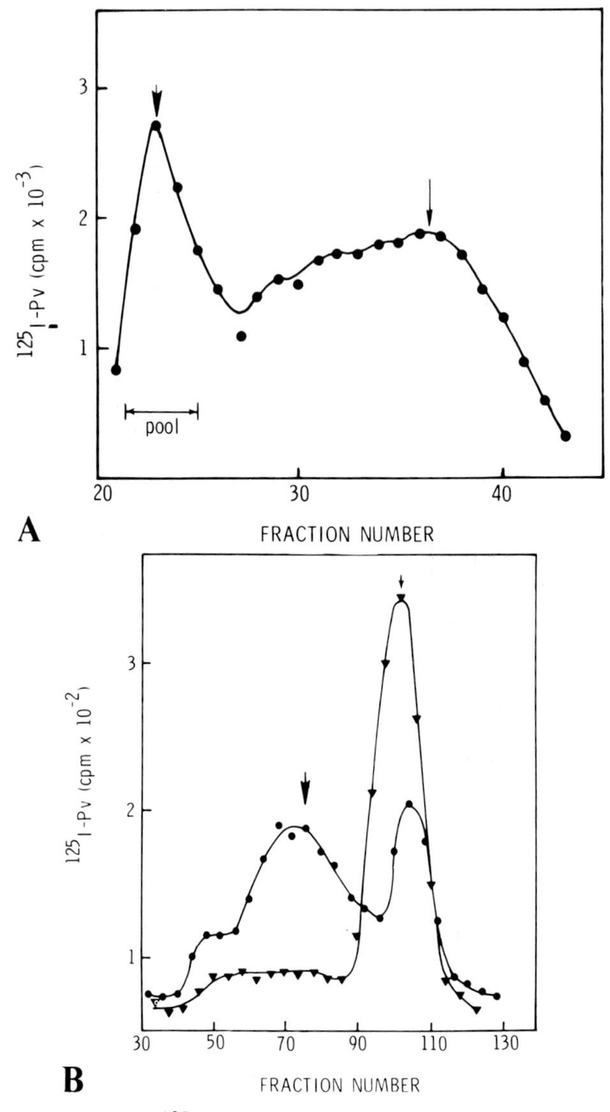

Fig. 5. Elution profiles of solubilized ^{125}I-PV receptor complexes. A: Three tubes containing 2.9 × 10^{-6} M ^{125}I-PV (94 × 10^6 cpm) were incubated for 90 min at 25° with homogenized oocyte membranes. Unbound ^{125}I-PV was removed by washing as described in Materials and Methods. The washed pellets, each of which contained 4 × 10^4 cpm, were pooled, suspended in 1 ml of 1% w/v Triton X-100 in IB-BSA, and incubated a further 30 min at 25°. Insoluble material was removed by centrifugation at 100,000 g for 30 min at 4°, and the soluble supernatant (5 × 10^4 cpm) was applied to an Agarose 0.5 M column. The larger arrow indicates void volume and the smaller arrow indicates the elution position of unbound ^{125}I-PV. B: Elution of solubilized ^{125}I-PV receptor complexes on an Agarose 5 M column. Triton-solubilized ^{125}I-PV receptor complexes (1 × 10^4 cpm) from fraction 22–25 in panel (A) above were divided into two identical aliquots and incubated with (▾) or without (•) 2.9 × 10^{-5} M ^{125}I-PV for 30 min at 25° immediately prior to application to the Agarose 5 M column. The large arrow indicates the elution position of ferritin and the smaller arrow indicates the elution position of unbound ^{125}I-PV. Incubations of ^{125}I-PV receptor complexes with either 10^{-5} M IgG or 10^{-5} serum albumen for 30 min at 25° prior to chromatography on the Agarose 5 M column resulted in elution profiles that were similar to those obtained when the ^{125}I-PV receptor complexes were incubated in the absence of unlabeled PV.

The elution profiles of ^{125}I-PV radioactivity (Fig. 5B) show that the sample incubated in the absence of unalbeled PV eluted as two peaks, both of which are well within the included volume of these columns. The major peak eluted at a position slightly larger than ferritin (480,000 daltons). The minor peak eluted at the same position as free ^{125}I-PV. In contrast, the sample which was incubated with excess unlabeled PV prior to chromatography yielded a single peak in the position of free ^{125}I-PV, indicating that unlabeled PV displaced bound ^{125}I-PV from the solubilized receptor.

The preceding experiment demonstrated that the material eluting in the void volume of an Agarose 0.5 M column contains a PV-binding component. This component must be heterogeneous, since it eluted as a broad symmetrical peak on an Agarose 5 M column. The average apparent molecular weight of this component, based on its elution position compared to ferritin and IgG, is approximately 500,000 daltons. However, since it is not known how much Triton is associated with the complex or the conformation of the complex, the true molecular weight is not known.

Additional experiments which demonstrate the specificity of the PV–receptor interaction were also carried out. Solubilized ^{125}I-PV receptor complexes were incubated with chicken serum albumin (1×10^{-5} M) or chicken IgG (1×10^{-5} M) for 30 min at 25°C and then chromatographed on Agarose 5 M columns. The results show no displacement of bound ^{125}I-PV and were similar to those obtained when ^{125}I-PV receptor complexes incubated in buffer alone were chromatographed on Agarose 5 M columns (Fig. 5B).

Demonstration of Soluble PV Receptor-Binding Activity

To determine whether Triton X-100 can solublize the PV receptor in a form capable of binding free ligand, homogenized membranes were first extracted with 1% Triton X-100 The solubilized material was assayed for PV-binding activity by incubation with ^{125}I-PV in the presence or absence of excess unlabeled PV for 60 min at 25°C, followed by chromatography on two identical Agarose 0.5 M columns (Fig. 6A). Samples incubated in the absence of excess unlabeled PV yielded a high-molecular-weight component which eluted in the void volume; however, more than 90% of eluted ^{125}I-PV was located in the same position as free ^{125}I-PV. The sample that was incubated in the presence of excess unlabeled PV eluted as a single peak in the same position as free ^{125}I-PV, indicating that no ^{125}I-PV receptor complexes had been formed.

^{125}I-PV binding to the solubilized receptor was shown to be reversible by incubation of the ^{125}I-PV receptor complexes in the presence or absence of unlabeled PV (2.9×10^{-5} M) for 30 min at 25°. Chromatography on two identical Agarose 5 M columns (Fig. 6B) indicates that the ^{125}I-PV receptor complexes incubated in the absence of unlabeled PV eluted as two peaks, one having a molecular weight slightly greater than ferritin and one eluting in the same position as free ^{125}I-PV. In contrast, the ^{125}I-PV receptor complexes, which were incubated in the presence of excess unlabeled PV prior to chromatography, eluted primarily as a single peak in the same position as free ^{125}I-PV.

The specificity of ^{125}I-PV binding to soluble receptors was demonstrated in additional experiments in which ^{125}I-PV soluble receptor complexes were incubated with excess chicken IgG (10^{-5} M) or excess chicken serum albumin (10^{-5} M) for 30 min at 25° prior to chromatography on Agarose 5 M columns. The results, which show no displacement of bound ^{125}I-PV by either IgG or albumin, are similar to that shown in Figure 6B, in which the soluble ^{125}I-PV receptor complexes were incubated with buffer alone prior to chromatography. These results suggest that only PV and not IgG or serum albumin can displace ^{125}I-PV from the solubilized receptor.

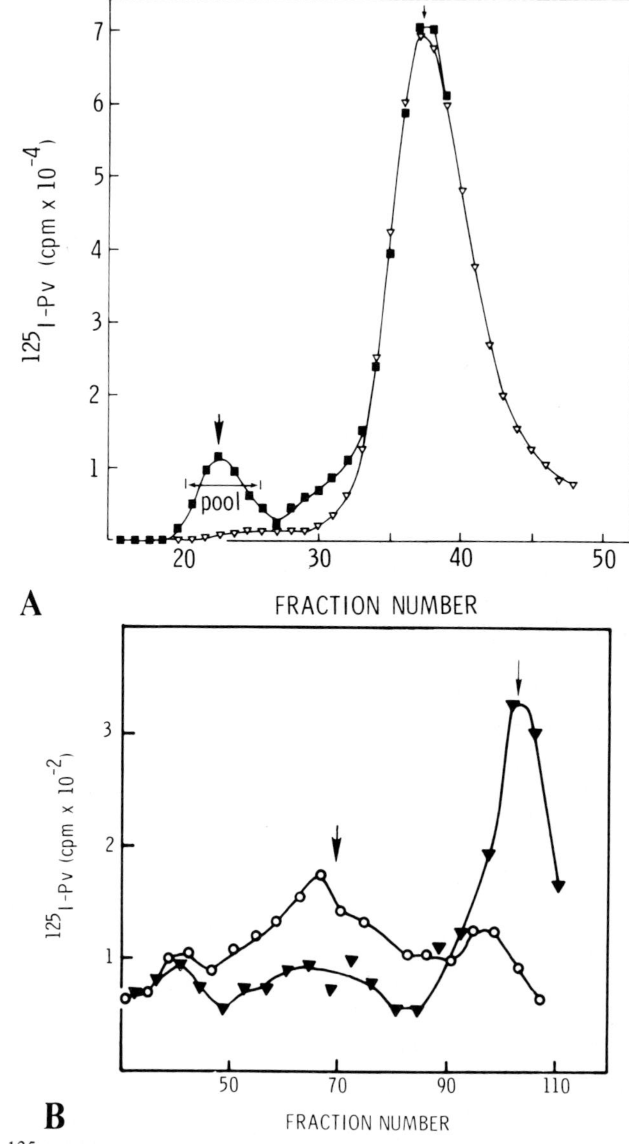

Fig. 6. Assay of ^{125}I-PV binding to previously solubilized receptors. A: Homogenized oocyte membranes from three oocytes were solubilized by incubation in 1 ml of 1% w/v Triton X-100 in IB-BSA for 30 min at 25°. Insoluble material was removed by centrifugation at 100,000 g for 30 min at 4°. The soluble supernatant fraction was divided into two equal aliquots and incubated with 5.9×10^{-7} M ^{125}I-PV (6.8×10^5 cpm) in the presence (▿) or absence (▪) of 2.9×10^{-5} M unlabeled PV for 60 min at 25°. Following incubation, the samples were immediately applied to two identical Agarose 0.5 M columns. The large arrow indicates the void volume and the small arrow indicates the elution position of unbound ^{125}I-PV. B: Assay for specific displacement of ^{125}I-PV bound to soluble receptors by unlabeled PV. Soluble receptor ^{125}I-PV complexes were obtained by pooling fractions 21–25 from panel (A). This pooled material (4×10^3 cpm) was divided into two equal aliquots and incubated in the presence (▾) or absence (○) of 2.9×10^{-5} M unlabeled PV for 30 min at 25° immediately prior to chromatography on the Agarose 5 M columns. The large arrow indicates the elution position of ferritin and the small arrow indicates the elution position of unbound ^{125}I-PV. In similar experiments, incubation of soluble receptor ^{125}I-PV complexes with either 10^{-5} M, IgG, or 10^{-5} M serum albumen for 30 min at 25° prior to chromatography on Agarose 5 M columns gave elution profiles which were similar to those obtained when soluble receptor ^{125}I-PV complexes were incubated in the absence of unlabeled PV (○).

DISCUSSION

A previous report from our laboratory showed that vitellogenin binds specifically to oocyte tissue [14]. On the basis of the present data, we postulate that there are specific receptors for phosvitin on chicken oocyte membranes. PV is a subunit of a maternal serum protein, vitellogenin. In addition, we have evidence that both vitellogenin and PV bind to oocyte membranes with a K_D of about 5×10^{-7} M, and that both molecules apparently bind to the same receptor [29]. Taken together, these results suggest that the PV component of vitellogenin is recognized by the receptor and hence mediates the transport of vitellogenin into the developing oocyte.

In the present report the dissociation constant, K_D, of PV binding to oocyte membranes was determined by two separate and independent methods. One method, based on the equilibrium binding of PV and membrane-associated receptors yielded a $K_D = 3.3 \times 10^{-6}$ M. The second method, based on the independent determinations of the association rate constant k_1 and the dissociation rate constant k_{-1} yield a $K_D = 1.6 \times 10^{-7}$ M, where $K_D = \frac{k_{-1}}{k_{+1}}$. These two values are probably not significantly different, given the errors intrinsic in the assumptions made in these calculations. The K_D determined from the equilibrium experiment is probably more accurate, because fewer assumptions were made in calculating the K_D from equilibrium data.

A major finding of the present study is that phosvitin receptor-binding sites in oocyte tissue can be solubilized and assayed in the soluble state. Our assay for soluble PV-binding activity depends on the apparent increase in molecular weight as determined by gel exclusion chromatography of ^{125}I-PV when it is bound to the solubilized receptor. Free ^{125}I-PV eluted as a 80,000 dalton globular protein, whereas bound ^{125}I-PV elutes in a broad peak centered at 500,000 daltons. However, the true molecular weight of the ^{125}I-PV receptor complex is not known since the amount of Triton bound to the complex and the conformation of the complex are not known.

Gel exclusion chromatography has been used in a similar manner to assay a number of detergent-solubilized membrane-associated receptors. Specific examples include the insulin receptor [22], the IgE receptor [23], the β-adrenergic receptor [31], the parathyroid hormone receptor [32], the vasopression receptor [33], and the follitropic receptor [34].

Soluble PV-binding activity was shown to be specific. When ^{125}I-PV was incubated with solubilized oocyte membranes in the presence or absence of excess unlabeled PV, ^{125}I-PV receptor complex could be detected only in the absence of excess unlabeled PV. Specificity of binding was also shown by displacement experiments. We observed that only unlabeled PV could displace bound ^{125}I-PV (see Figs. 5B and 6B). Incubation of soluble ^{125}I-PV receptor complexes with buffer alone, serum albumin, or IgG resulted in dissociation similar to that predicted on the basis of the experiment shown in Figure 2A. The results shown in Figure 2A suggest that under similar conditions (30 min incubation, 25°, no unlabeled PV) up to 30% of the bound ^{125}I-PV would dissociate from the membrane-bound receptor. IgG and serum albumin were used in our displacement experiments because both are present in the serum and hence in vivo they would be exposed to the receptor at the same time as PV in the form of vitellogenin. In addition, IgG is specifically transported into the developing oocyte at a parallel rate to PV but at an additional developmental stage [9]. Serum albumin is apparently not transported into the oocyte. Therefore, the PV receptor must be able to discriminate between PV, IgG, and serum albumin in vivo. Our results demonstrate that the PV receptor can discriminate between IgG, albumin, and PV in vitro.

In conclusion, we have shown that a specific receptor for PV is associated with oocyte membranes. This receptor appears to mediate the specific transport of vitellogenin into the developing oocyte [28]. In addition, we have shown the PV receptor can be solubilized with the non-ionic detergent Triton X-100, and the solublized receptor retains the ability to specifically recognize PV.

ACKNOWLEDGMENTS

We are grateful to Dr. Carol Linden for critical reading of the manuscript. This investigation was supported in part by NIH grants HD 09549 and HD 11519 from the National Institute of Child Health and Human Development.

REFERENCES

1. Roth TF, Porter KR: J Cell Biol 20:313, 1964.
2. Wallace RA, Dumont JN: J Cell Physiol (Suppl)27:73, 1968.
3. Jared DW, Dumont JH, Wallace RA: Dev Biol 35:19, 1973.
4. Telfer WH: J Biophys Biochem Cytochem 9:747, 1961.
5. Melius ME, Telfer WH: J Morphol 129:1, 1969.
6. Roth TF, Cutting JA, Atlas SD: J Supramol Struct 4:527, 1976.
7. Wild AE: Phil Trans R Soc B 271:395, 1975.
8. Brambell FWR: Lancet 2:1087, 1966.
9. Cutting JA, Roth FR: Biochim Biophys Acta 289:951, 1973.
10. Slade B: IRCS Med Sci 3:235, 1975.
11. Rodewald R: J Cell Biol 58:189, 1973.
12. Anderson RGW, Goldstein JL, Brown MS: Proc Natl Acad Sci USA 73:2434, 1976.
13. Gorden P, Carpenter J, Cohen S, Orci L: Proc Natl Acad Sci USA 75:5025, 1978.
14. Yusko SC, Roth TF: J Supramol Struct 4:89, 1976.
15. Wallace RA, Jared Dw: J Cell Biol 69:345, 1976.
16. Linden CD, Roth TF: J Cell Sci 33:317, 1978.
17. Sonoda S, Schlamowitz M: J Immunol 108:1345, 1972.
18. Tsaj DD, Schlamowtz M: J Immunol 115:939, 1975.
19. Rodewald R: J Cell Biol 71:666, 1976.
20. Goldstein JL, Brown MS: Curr Topics Cell Reg 11:147, 1976.
21. Carpenter G, Lemback K, Morrison M, Cohen S: J Biol Chem 250:4297, 1975.
22. Cautrecasas P: J Biol Chem 247:1980, 1972.
23. Rossi G, Newman SA, Metzger H: J Biol Chem 252:704, 1977.
24. Klett RP, Fulpius BW, Cooper D, Smith M, Riech E, Possani LD: J Biol Chem 248:6841, 1973.
25. Cautrecasas P, Hollenberg MD: Adv Protein Chem 30:251, 1976.
26. Bernardi G, Cook WH: Biochim Biophys Acta 44:86, 1960.
27. Pressman D, Eisen HN: J Immunol 64:273, 1950.
28. Kitabgi P, Carraway R, Van Rietschoten J, Granier C, Morgat JL, Menez A, Leeman S, Freychet P: PNAS 74:1846, 1977.
29. Yusko SC, Roth TF, Smith T: J Supramol Struct (Submitted).
30. Clarke RC: Biochem J 118:537, 1970.
31. Caron MG, Lefkowitz RJ: J Biol Chem 251:2374, 1976.
32. Malbon CC, Zull JE: Biochem Biophys Res Commun 66:179, 1975.
33. Roy C, Raerison R, Bockert J, Jard S: J Biol Chem 250:7885, 1975.
34. Abbou-Issa H, Reichert LE: J Biol Chem 252:4166, 1977.

Biological Recognition and Assembly 209–231 (1980)

Protein Switches in Muscle Contraction

Carolyn Cohen, Peter J. Vibert, Roger W. Craig, and George N. Phillips, Jr.

Rosenstiel Basic Medical Sciences Research Center, Brandeis University, Waltham, Massachusetts 02254

Two types of Ca^{2+}-sensitive protein complexes control the contraction of muscle: Troponin (TN) and tropomyosin (TM) are associated with the thin actin filaments, and a specific light chain is a regulatory subunit of myosin itself. Most muscles have both types of regulation. X-ray diffraction diagrams from whole muscle have shown changes in the position of tropomyosin and changes in the pattern of myosin crossbridge attachment associated with different states of the regulatory switches. The full interpretation of these diagrams is often ambiguous, however, and structural studies of the purified proteins provide essential information. Recent crystallographic results reveal that the TM molecule has unusual local domains of marginal stability, leading to extensive motions of the tropomyosin filaments. Electron microscopy of negatively stained thin filaments decorated with subfragments of scallop myosin yields unusually detailed images that show marked conformational changes in myosin crossbridges that are dependent on the presence or absence of the regulatory light chain. These observations suggest that both the special dynamic design of tropomyosin and the striking structural changes in the myosin crossbridges are significant clues for detailed models for the regulatory mechanism.

Key words: **muscle regulation, tropomyosin, myosin light chains**

MUSCLE AS A MACHINE

One of the most natural problems for the use of models or analogs in scientific theory is the analysis of how muscle works. The "animal as machine" metaphor has scarcely been questioned in the successive developments of our understanding of muscle contraction. Yet this view comes from a tradition several centuries old. The machine analogy began to be used once the laws of mechanics were formulated, and the systems in which it was used most successfully were those, in fact, in which motion was involved. Harvey's work is often cited in this connection. He conceived of the heart as a pump, and the blood vessels as a hydraulic system, and this analogy worked *precisely* because the heart does act as a pump and the blood vessels as such a system [1].

Received May 22, 1979; accepted May 22, 1979.

Let us illustrate these ideas. The question is sometimes asked whether Nature is an artist or an engineer. The first vision is shown by a detail of Michelangelo's famous "Creation of Adam" in the Sistine Chapel (Fig. 1). Here we see that God's right hand "touches life into Adam's left" [2]. This mystical (or Platonic) symbolization of the hand can be contrasted with the opposing philosophy of Nature (Fig. 2). Here we see a crude drawing by Ambroise Paré, field surgeon and physician to François I. Now the hand is seen as a machine with gears and springs operated by muscles, which would also be conceived of as machines. This mechanistic or reductionist philosophy of biology has been a major trend in scientific thinking since the Renaissance, and it is held by most molecular and structural biologists today.

Fig. 1. God's right hand touching life into Adam's left in Michelangelo's "Creation of Adam." From Weyl [2].

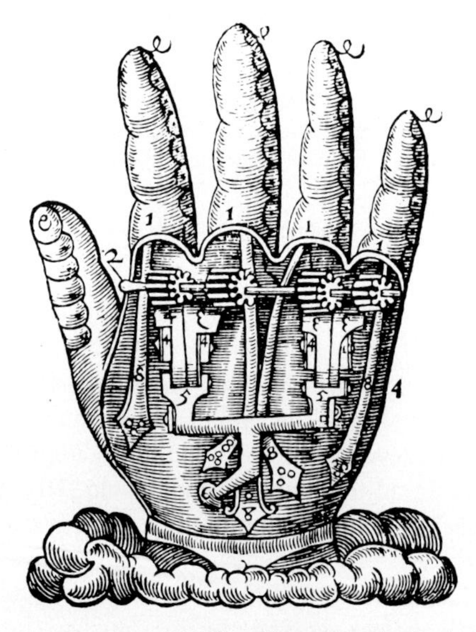

Fig. 2. "The forme of a hand made artificially of irone" by Ambroise Paré. From Paré [42].

Muscle as Machine is in many aspects a happy metaphor. The work of Huxley and Hanson [3] and A. F. Huxley and Niedergerke [4] has shown that muscle shortens when two sets of filaments move past one another. The protein myosin forms the so-called thick filaments in the A band of cross-striated muscle (Fig. 3). The thin actin-containing filaments in the I bands are driven into the fixed array of myosin by crossbridges that form part of the myosin molecules. There is strong evidence that directed movement and tension generation are due to a specific interaction between the myosin molecules and actin, that is, the crossbridges and the actin filaments (Fig. 4) [5]. A particular conformational cycle of the crossbridges has been envisaged, in the course of which energy for the contractile process is generated: Each time a bridge goes through a cycle, a single molecule of adenosine triphosphate (ATP) would be split. The precise motions of the crossbridges during this cycle have not yet been defined, but the process has been compared to a kind of *rowing* mechanism, in which the thin filaments are pulled into the array of thick filaments by the repetitive action of the crossbridges (Fig. 5). One of the major areas of interest in muscle research today is the relationship between the physical changes in the crossbridge and the energetics of contraction. We wish to answer the questions: What is the precise nature of the conformational changes, and how is the chemical energy of the ATP transduced into mechanical work? But there is another aspect of the muscle machine, and that is how muscle is switched on and off. Here again we can fruitfully bring the machine analogy to the molecular level, yet point out some basic features that are not yet understood.

THE THIN-FILAMENT SWITCH

In all muscles the level of Ca^{2+} controls contraction. Muscle maintains its resting state because certain regulatory proteins are calcium-sensitive switches. The discovery of the first of these switches was made by Ebashi in 1963 [6]. Ebashi performed a very simple and dramatic experiment. He compared the calcium sensitivity of purified rabbit actin and myosin to natural actomyosin prepared by gentle extraction from muscle. It turned out that the pure system split ATP regardless of the level of calcium; there was no calcium control. But natural actomyosin behaves differently. It behaves, in fact, as do myofibrils or intact muscle: It is under calcium control. Above a critical level of calcium (about 10^{-6}M) "superprecipitation" (the test tube analog of contraction) of the proteins takes place – or shortening in the case of myofibrils or muscle. Below this level of calcium the proteins do not interact in the presence of ATP. The muscle is switched off and is at rest. Ebashi identified this regulatory system as consisting of two kinds of protein: One, a calcium-binding complex of three globular units, he discovered and called troponin; the other is the fibrous protein tropomyosin. All muscles had been known to contain tropomyosin since its discovery in the 1940s by Bailey, but its function was not understood [7]. It was shown in 1966 that when tropomyosin is precipitated with divalent cations paracrystals with a repeat of 395 Å were formed [8] (Fig. 6). By 1969 Ebashi and his colleagues had formulated a working model in which tropomyosin was located in the two grooves of actin and positioned the troponin complex every 385 Å on the thin filaments [9] (Fig. 7). And he also formulated a general mechanism for the calcium-controlled switch. He postulated that some conformational change in troponin was triggered by calcium. This in turn caused the release of ATPase inhibition. That is, the myosin heads could then attach to actin, and muscle was switched "on."

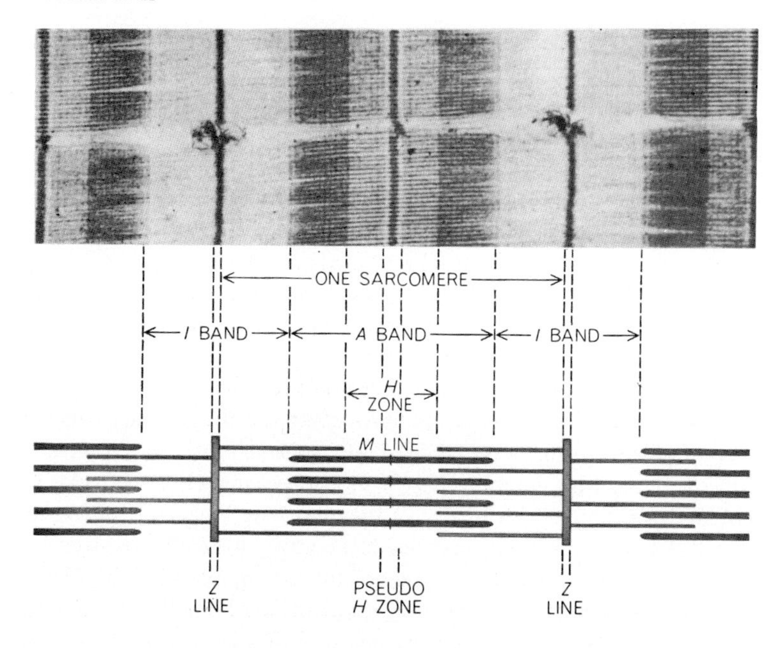

Fig. 3. Electron micrograph of myofibril, showing thick filament array in A band, and thin filaments in I band. Figures 3 and 4 from Huxley [43], with permission.

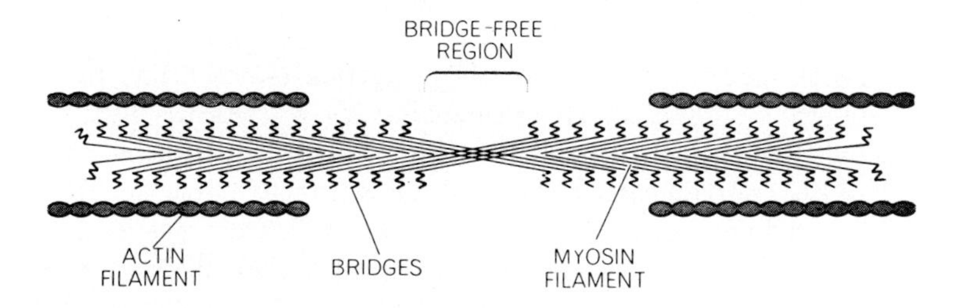

Fig. 4. Myosin crossbridges in thick filaments are arranged in a bipolar manner so that they can interact with the polar actin filaments to produce directed motion.

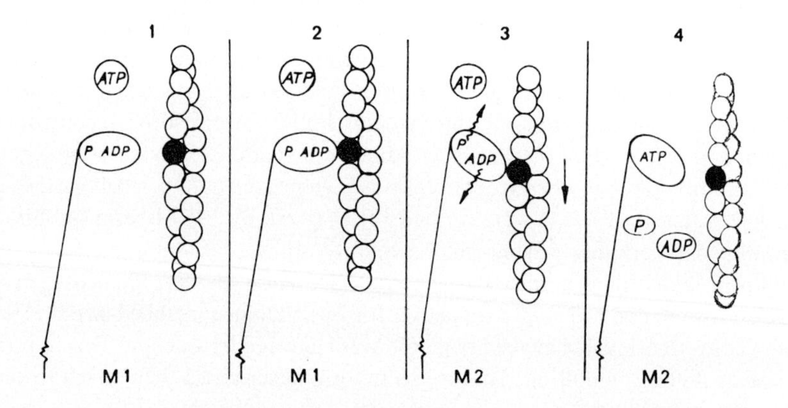

Fig. 5. Diagram of possible crossbridge cycle in contraction showing "rowing" action of myosin heads and splitting of ATP. From Mannherz et al [44], with permission.

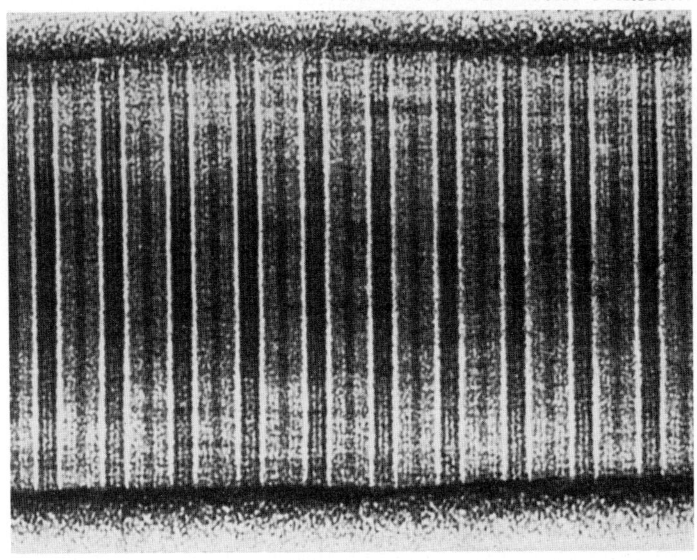

Fig. 6. Electron micrograph of Mg paracrystal of tropomyosin showing axial repeat of 395 Å. Negatively stained with uranyl acetate. Figures 6, 9, 10, 12, 13, and 17 from Cohen [45], with permission.

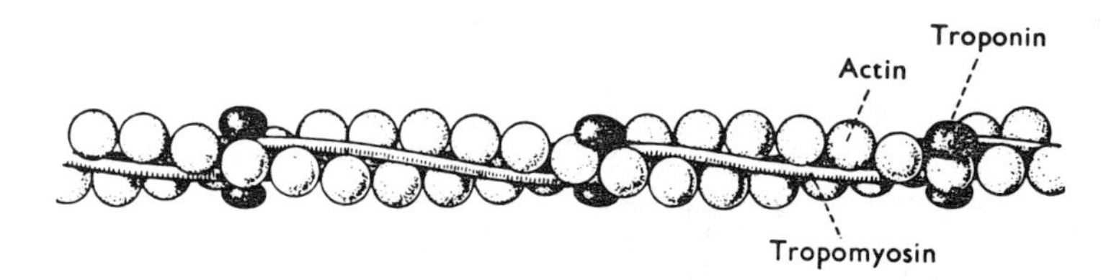

Fig. 7. Model for arrangement of proteins in the thin filament. Tropomyosin filaments wind in the grooves of the actin helix and provide binding sites for troponin at 385 Å intervals. From Ebashi et al [9], with permission.

This picture was further defined by structural studies. X-ray diffraction diagrams from various types of intact muscle revealed the position of tropomyosin when the muscle was at rest, when it was in the active state, and when crossbridges were attached in rigor (that is, fixed in the absence of ATP) [10–12]. The marked differences in intensities of the X-ray diagram indicated a movement of tropomyosin in different states of activity of muscle. It appeared that upon activation, or when crossbridges were attached in rigor, tropomyosin moved from a position nearer the edge of the actin filament to one deeper into the groove. An additional structural technique was crucial in the formulation of a model for this mechanism. Three-dimensional reconstruction of electron micrographs of actin "decorated" with myosin heads (S1 subunits) showed the approximate position and shape of the crossbridge portions of myosin when these are attached in rigor [13].

On the basis of this finding, the so-called steric blocking model was advanced [10, 14] (Fig. 8). At rest, tropomyosin would block the site on actin where myosin cross-bridges attach. At the critical calcium level troponin changes its conformation and tropomyosin moves deeper into the groove of the actin helix and exposes the sites, which then interact with myosin. The muscle is switched "on": Force is developed, and ATP splitting takes place. When the calcium level is reduced, the troponin complex undergoes its shape change, and tropomyosin rolls out to its "blocking" position. The muscle is switched "off" and is at rest. This model accounts also for the cooperativity discovered by Bremel and Weber [15]. The binding of calcium ions to troponin switches on about seven actin monomers. Tropomyosin has an effective length of about 400 Å in muscle, and so can span seven actin subunits. One troponin would then effect the blocking or unblocking of seven actin monomers because the tropomyosin molecule is linked to this number of subunits.

Further structural details of this mechanism are emerging. There are three troponin components: TnT binds tropomyosin; TnI binds to actin and can inhibit contraction; and the component TnC binds calcium [16]. The interaction of these subunits with one another and with actin is such that one may picture the model in more detail (Fig. 9). Binding of calcium to TnC tightens the linkages among the components of the complex while weakening the linkage of TnI to actin [17]. Tropomyosin then moves deeper into the actin groove. Lowering the calcium level reverses this effect: The conformational change in TnC is such that the linkages to TnI and TnT are weakened, and TnI binds more tightly to actin. Tropomyosin is then fixed to the actin monomers in its blocking position, and the muscle is switched "off."

The amino acid sequence of all the proteins in the thin filament switch of rabbit muscle is now known [18–22]. TnC has in fact been crystallized [23], and aspects of its three-dimensional structure have been inferred from certain sequence homologies to the calcium-binding protein parvalbumin [24]. Neither of the other two components nor the whole complex have been crystallized as yet, but a number of structural inferences are being made on the basis of the sequences and various spectroscopic, hydrodynamic, and physical chemical studies of the components. It is fair to say, however, that at this time we are far from visualizing this critical complex.

Tropomyosin is at present the most fully understood protein of the regulatory switch on the thin filaments, and its structure reveals a number of surprises. The molecule has a deceptively simple design. It is a two-chain α-helical coiled-coil (Fig. 10). The complete sequence of the alpha chain of rabbit tropomyosin, comprising 284 residues, has been determined [18, 19]. (Tropomyosin is present in all muscles, and most species have two populations with slightly different sequences termed alpha and beta.) The sequence itself has accounted for certain features of the structure. There is a short-range periodicity of non-polar residues as envisaged in Crick's model for the coiled-coil [25], so that a regular pattern of interactions stabilizes the molecules. The two identical chains in the molecule run in the same direction and have little or no relative stagger. A most interesting feature of the longer-range periodicity of tropomyosin is the absence of an exact sevenfold repeat in the structure. Analyses of the sequence as well as the band pattern in paracrystals of tropomyosin reveal, however, a 14-fold repeat of acidic and nonpolar residues [26]. This feature of the structure has been related to the switching mechanism in a very specific way. A recent model postulated by McLachlan and Stewart [27] interprets the fourteen quasi-equivalent zones on tropomyosin as two sets of alternative binding sites for actin (Fig. 11). When tropomyosin is bound to the actin filament it would make seven half-twists in one molecular length, so that quasi-equivalent regions of the coiled-coil would link to seven identical actin monomers in the "off" position. When muscle is switched "on," the tropomyosin would make a quarter turn, and the seven quasi-equivalent sites of the molecule would interact with a second set of seven sites on actin. These necessarily speculative models assume a uniformly α-helical tropomyosin molecule that makes highly regular symmetrical interactions.

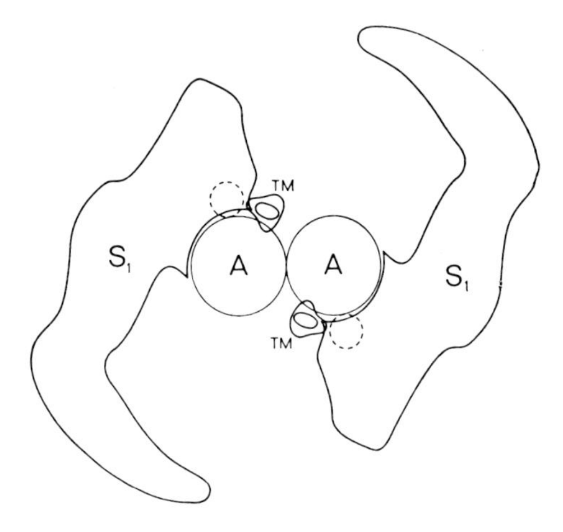

Fig. 8. The steric blocking model of muscle regulation. In the "off" state tropomyosin prevents the myosin heads from attaching to actin. In the "on" state, tropomyosin moves deeper into the grooves exposing these binding sites. From Huxley [10], with permission.

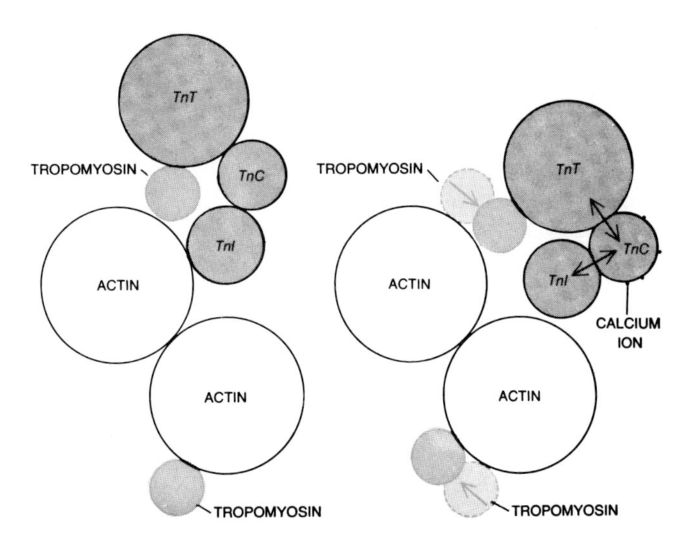

Fig. 9. Model for the interaction of troponin subunits in regulation. In the "off" state (left) at low levels of Ca^{2+}, the troponin subunits are relatively weakly linked to one another and TnI is tightly linked to actin. In the "on" state (right) binding of Ca^{2+} to TnC causes a "tightening" of the complex and weakens the linkage between TnI and actin. Tropomyosin then moves deeper into the actin groove, exposing the site on actin where myosin binds.

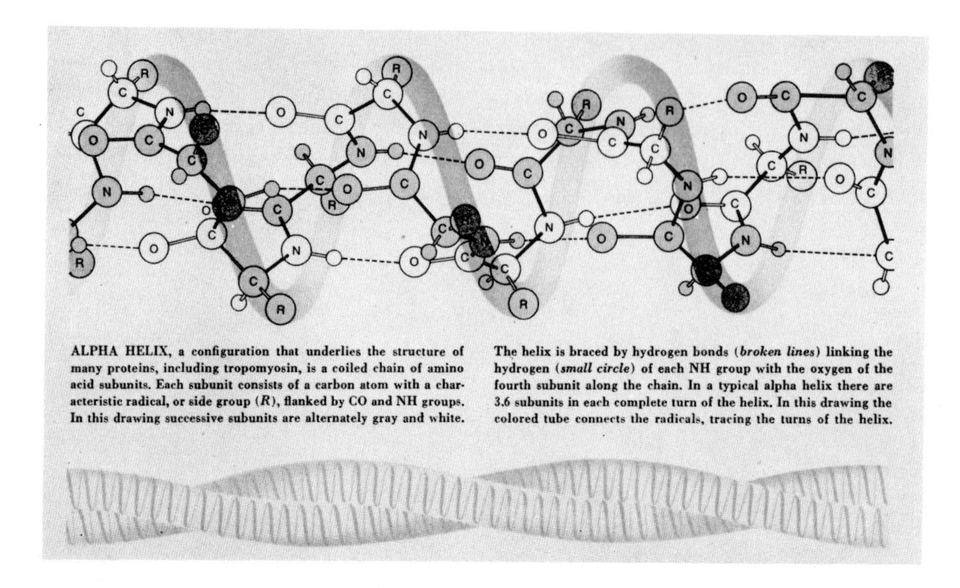

Fig. 10. α-Helical coiled-coil. The structure is stabilized by regular interactions between the two polypeptide chains. Optimal packing of amino acid residues at the line of contact is achieved by a left-handed coiled-coil.

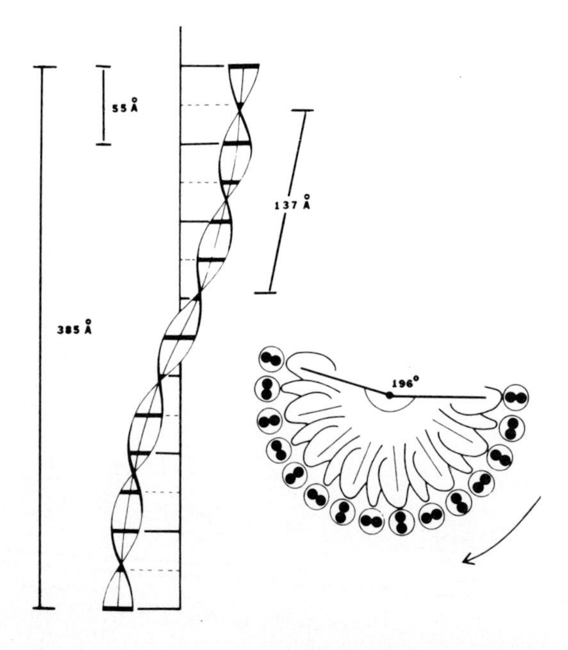

Fig. 11. Model for regular interactions of tropomyosin and actin. Each tropomyosin molecule would have seven (or 14) quasi-equivalent bonding sites that link to the seven actin subunits. From McLachlan and Stewart [27], with permission.

For many years one of the projects in our laboratory has been the X-ray crystallographic analysis of the structure of tropomyosin. Crystals of tropomyosin are unique in a number of ways (Fig. 12). They have the highest water content yet observed in any protein crystal: 95% of the volume in the lattice is water. The crystals thus tend to disorder readily and diffract very weakly beyond 15 Å. Thus, we can determine at present only a low-resolution structure. Nevertheless – in contrast to almost all other protein crystals – interactions in the crystal are not adventitious. Rather, the molecules bond end-to-end to form filaments, and this linkage is the same as that in the tropomyosin filaments of muscle. That is to say, the tropomyosin crystal is made up of filaments very like those that run in the grooves of the actin helix (Fig. 13). The determination of the structure of this crystal, therefore, can give insight into the role of tropomyosin in muscle.

In early studies one projection of the crystal was solved, and it was shown that the molecules are bonded head-to-tail to form filaments that are bent into gentle sinusoids, implying a supercoiling of the coiled-coil structure (Fig. 14) [28]. Thus tropomyosin filaments in the crystal are coiled-coil coils. The structure of tropomyosin has now been determined by single-crystal diffraction analysis to Bragg spacings of 20 Å [29]. The approximate path of the molecule in three dimensions has been established; the tropomyosin filaments wind along a line parallel to the body diagonal of the unit cell. Since the approximate path can now be traced, the effective molecular length has been determined as 410.3 ± 1.4 Å. This information can be used to deduce the number of residues involved in the end-to-end bonding region; this is found to be about 8.6 residues. The three-dimensional winding of the filament is not in fact helical, but has the form of an ellipse when projected down the axis of the molecule.

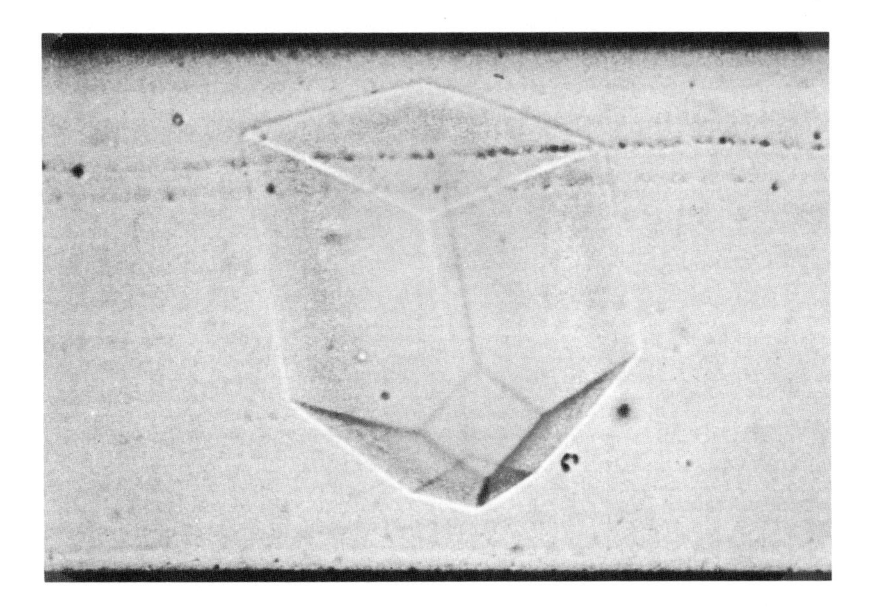

Fig. 12. Tropomyosin crystal photographed with polarized light. The crystals are remarkable in that 95% of their volume is water.

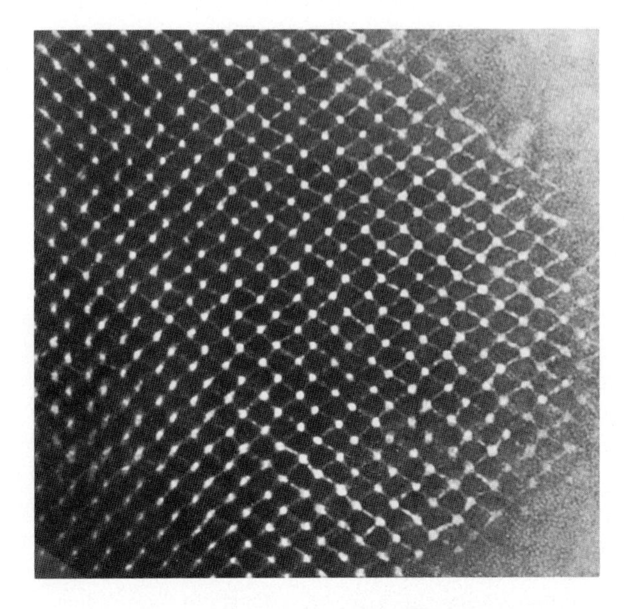

Fig. 13. Electron micrograph of a tropomyosin crystal. The projected structure is seen as a net of wavy cross-connected strands having long and short "arms." The period along the strands is about 400 Å.

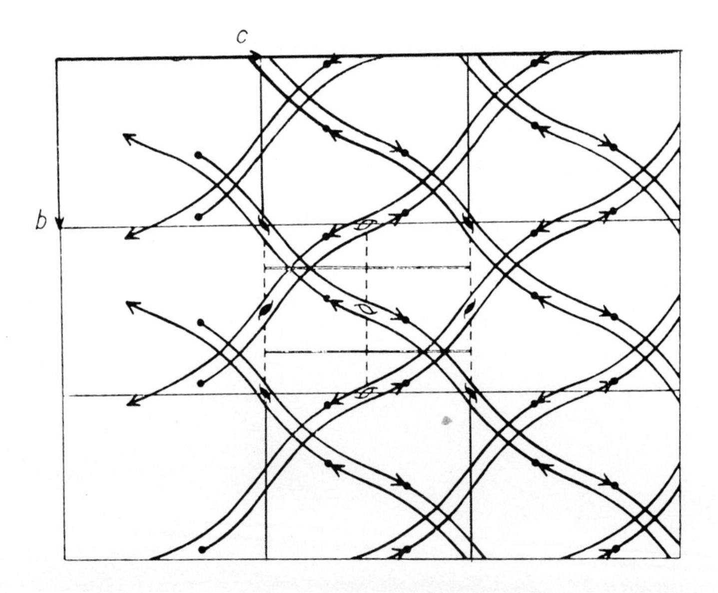

Fig. 14. Molecular configuration and symmetry elements in one projection of tropomyosin crystal lattice. From Caspar et al [28], with permission. (The position of the ends of the molecule was not known at the time of publication.)

The single –SH group on each chain of the molecule has been located within the unit cell by the use of mercury derivatives. The mercury serves as a marker for the location of the cysteine 190 residue along the filament but leaves an ambiguity in the direction of the polypeptide chain in the filament. Interpretation of the electron density maps shows that the best fit between the sequence and the electron density is found when the chain direction is taken as illustrated (Fig. 15). This choice means that the intermolecular overlap would occur in the short arm of the filament near a crossover region (see Fig. 13). Mercury labeling of beta-tropomyosin (which contains an additional pair of –SH groups) should allow confirmation of this solution of the chain direction.

The run of the filaments seems quite continuous, including the dense region identified as molecular ends, where the head-to-tail overlap of about nine residues must take place. Here the filament shows only a slight swelling in all directions. If the molecular ends were simple alpha-helical coiled-coils linked by overlapping of the "broad" faces [27], one would expect to see some discontinuity in this region of the map. An alternative explanation is the molecule might not maintain the alpha-helical conformation at the ends, so that the chains would be joined by *intermeshing* to form a small globular domain. This kind of link is consistent with both the electron density map and chemical characteristics of the molecule [30]. Tropomyosin has been seen as the archetype of the alpha-helical coiled-coil, but its functional design to form filaments by end-to-end bonding may have produced a specialized structure for the molecular ends.

It does appear that there are no major deviations from the coiled-coil structure at 4°C. However, this observation does not mean that there are no local regions of "flexibility." The filament seems to be slightly kinked in several places on the map. These deviations from a uniform structure may be more pronounced at physiologic temperatures. X-ray photographs taken of crystals at 30°C rather than 4°C essentially eliminate reflections arising from the long arms but not from the short arms. This result confirms that there is a difference in stability of the two arms of the filament and shows that the magnitude of this effect is considerable at physiologic temperatures.

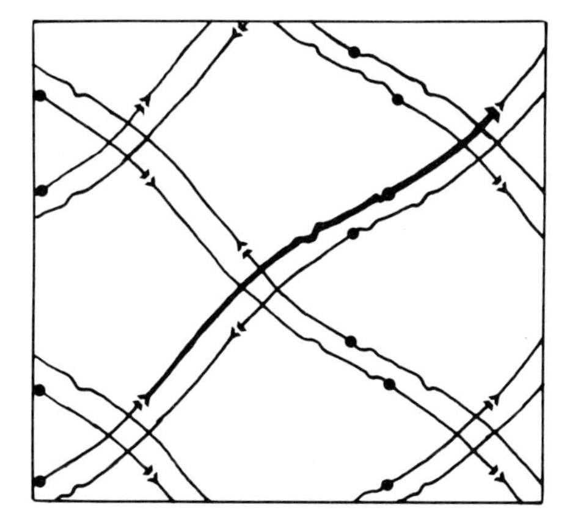

Fig. 15. Projection of tropomyosin crystal showing position and orientation of molecules in the filaments. C terminal is arrowhead; N terminal is tail. Kinked regions show possible deviations from regular coiled-coil. Circles mark the location of cysteine 190.

The pitch of the coiled-coil in the molecules has recently been established by comparing diffraction patterns calculated from model structures with the observed X-ray patterns (Phillips, G.N., unpublished). The coiled-coil makes three complete turns in one molecular length, which leads to a pitch of 137 Å (Fig. 16). The *hand* of the filament supercoiling (coiled-coil coil) is also determined by these results, since the α-helical coiled-coil is known to be left-handed [25]. The sense of the supercoiling is seen to be right-handed: that is, of handedness opposite to the coiled-coil. This is, in fact, the sense required for the tropomyosin filament wound in the grooves of the right-handed actin helix. The pitch of 137 Å determined for the molecular coiled-coil in the crystal is also precisely that predicted for tropomyosin [27, 30], so that regular interactions with actin may be achieved (see Fig. 11).

These structural results – even at low resolution – can provide more detailed insight into the role of tropomyosin in regulation. The motions of tropomyosin depend on the precise structure(s) of the molecule and its specific interactions with actin (Fig. 17). The slight irregularities of the filament seen in the electron density maps and the motions of the filament deduced from the diffuse scattering analysis imply local flexible regions of the molecule, and also indicate that there may be regions of the molecule other than the ends that are not in stable alpha-helical conformations. The picture that is beginning to emerge from our crystallographic analysis and other studies is that of a less regular and rigid structure than hitherto envisaged. In its regulatory role in muscle tropomyosin moves on the actin helix. The molecular motions displayed by tropomyosin supercoils in the crystal lattice may reflect to some extent the real nature of the dynamic elements in the structure. The effective length of 410 Å would place the tropomyosin at a large radius on the actin helix (say 40 Å) if it were fully extended in the blocking position, since the axial repeat of the filaments in muscle is 385 Å [14]. In the "on" state, when tropomyosin moves deeper into the actin groove, the radius must be less, and the molecule if bound to actin must therefore bend in some way. It is likely that at some stage of the contractile cycle the tropomyosin molecule may be rather loosely wound about the actin helix, perhaps held strongly at only a few sites, and locally kinked at other regions. In view of the unusual characteristics of the molecule that have been deduced thus far we suggest, therefore, that the motions of tropomyosin in muscle depend on two aspects of its design: the regular coiled-coil structure with its quasi-equivalent regions that interact with actin; and the aperiodic features. The latter, uncommon domains may play a critical part in the function of this highly specialized molecule.

The crystallography is now being extended to higher resolution, and cocrystals of tropomyosin and the troponin complex are under study (Fig. 18). The thin filament regulatory switch is thus accessible to detailed structural analysis.

THE MYOSIN SWITCH

Following the discovery of thin-filament regulation in vertebrate muscle by Ebashi, it was assumed and in some cases demonstrated that various other types of muscle also contained troponin, conceived of as the essential switch for muscle contraction. This generalization failed when A. G. Szent-Györgyi and colleagues discovered that molluscan muscle lacked troponin, yet was switched "on" and "off" by calcium in the same range as vertebrate muscles [31]. This remarkable finding came about when our two laboratories were working on the study of paramyosin, an alpha-helical protein found in large amounts in molluscs. In his efforts to prepare "native" thin filaments from these animals, Szent-Györgyi found that the thin filaments from clam muscle complexed with rabbit myosin,

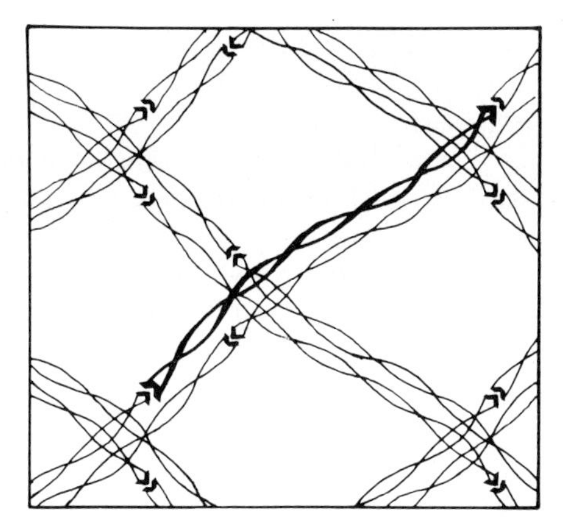

Fig. 16. Coiled-coil of tropomyosin molecule in crystal lattice. The pitch is 137 Å so that there are six half-turns in one molecular length. From G. N. Phillips (unpublished).

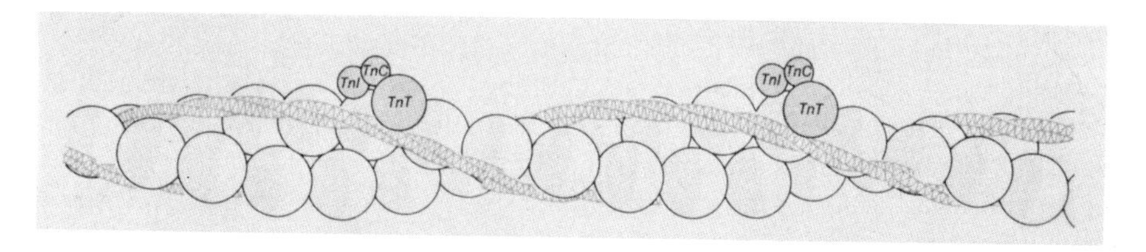

Fig. 17. Diagram of thin filament showing tropomyosin (note coiled-coil has *wrong* hand here) wound in grooves of actin helix, positioning troponin complex.

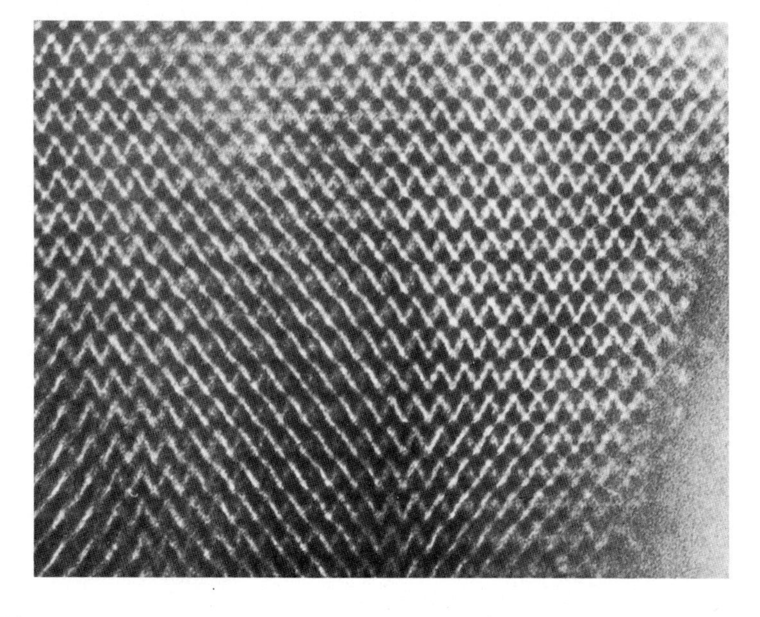

Fig. 18. Electron micrograph of tropomyosin crystal showing troponin bound to long arms of net. From Cohen et al [46], with permission.

but that the ATPase activity of the hybrid actomyosin was *not* sensitive to calcium [31]. There were two possibilities: either that the thin filaments had been changed or denatured in some way, or – most remarkable – that molluscan myosin itself controlled regulation. This latter was shown to be the case [32]. Thus, in molluscan muscles troponin is in fact absent; there is no thin-filament regulation, and the special design of the myosin molecule accounts for calcium control of contraction.

Comparative studies carried out by Szent-Györgyi and colleagues have shown that most animals display both regulatory systems [33]. The exceptions appear to be vertebrate striated muscle, regulated only by thin filaments, and molluscs and a few other small groups, regulated only by myosin. Vertebrate smooth muscle, surprisingly, is also regulated only by myosin [34].

In order to comprehend regulation by myosin we must know something of the structure of the molecule. The myosin molecule has a most unusual form – related to its dual function both as an enzyme and as a structural protein. This large molecule (about 500,000 daltons) is basically dimeric in structure. It consists of two globular domains, or subfragment 1 (S1) regions, attached to a tail about 1,500 Å long (Fig. 19). The conformation of the tail is very like tropomyosin: a two-chain alpha-helical coiled-coil. The globular head regions contain the actin-binding and ATP-splitting sites. In the thick-filament assembly the tail portions of the molecule, which confer the solubility properties to myosin, are built into the "backbone" of the filament; and the heads are at the surface, where they can interact with actin. (The head regions and any portions of the tail that reach out are often referred to as "crossbridges.") During the past decade the significance of small, noncovalently bonded light chains has been recognized. All myosins contain two "essential" light chains (necessary for ATPase activity) and in addition two other light chains. These are called "regulatory" in the case of molluscan myosins. Thus, myosin comprises two heavy chains and four light chains of two types. The light chains are located on each of the two heads of the molecule.

During the past decade, a series of basic discoveries have been made which led to inferences as to the mechanism of regulation.

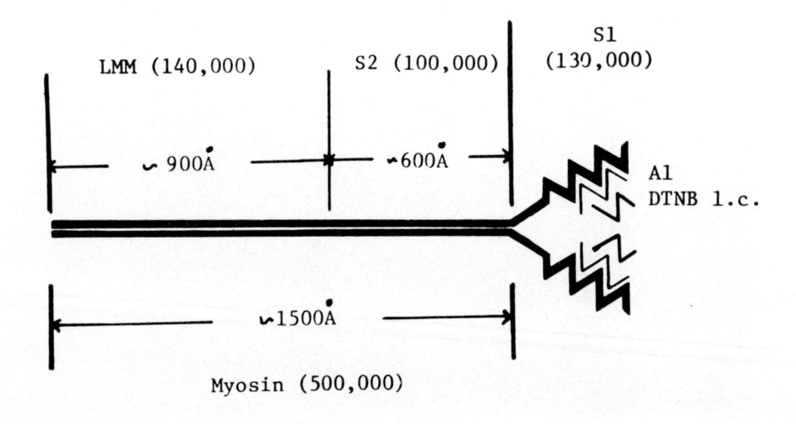

Fig. 19. Schematic diagram of myosin molecule, showing two pairs of noncovalently bonded light chains. From Lowey [47], with permission.

1. The crucial biochemical finding was that removal of one regulatory light chain per molluscan myosin molecule (by ethylenediaminetetraacetic acid) completely abolishes calcium sensitivity. (It has turned out that the only calcium-regulated myosin from which a regulatory light chain can be reversibly detached is that from scallop. Consequently, scallop myosin has been the focus of the analysis of this mode of regulation [35, 36]). The fact that complete loss of regulation is achieved when only one of the two identical light chains is removed led to the idea that there might be certain cooperative interactions between the two heads of the molecule.

2. This notion is strengthened by the fact that S1 from scallop myosin – that is, the heads alone – are not regulated by calcium, although the light chain can be shown to be attached. In contrast, the proteolytic subfragment HMM, which contains both heads and part of the tail, is calcium-sensitive.

3. Calcium binding is coupled to calcium sensitivity: Myosin with both regulatory light chains binds two moles of calcium. The isolated light chain essential for calcium sensitivity does not by itself specifically bind calcium. This finding implies that the calcium-binding sites require the interaction of the light chain and the heavy chain(s).

4. The light chain appears to have a relatively elongated structure. The size (and shape) of the head region of myosin are not yet clearly established, but results from shadowing and three-dimensional reconstruction indicate a length of about 150 Å. The light chain itself appears to be of the order of 80–100 Å in length [37]. This regulatory light chain thus has the potential for interacting over a considerable length of myosin head.

5. Recently a single-headed myosin has been produced, and it has been shown to be calcium-sensitive (Stafford, Szentkiralyi, and Szent-Györgyi, in preparation). Thus, two heads are not needed for calcium sensitivity. Rather, what may be needed is a myosin head with a single regulatory light chain that can interact with part of the tail region (subfragment 2 of the myosin molecule).

The general scheme that emerges is that the interaction of the regulatory light chain with the heavy chain forming the globular head of myosin somehow blocks interaction with actin; and that this blocking is removed by the binding of calcium. Since the presence of a portion of myosin far removed (by about 150 Å) from the end of the head appears essential for this mechanism, it would appear that the conformation of the head may be affected by the critical light chain and may also be influenced by the presence of the second head of myosin. But what information do we have of structural changes dependent upon the light chain?

A recent spectroscopic study of this problem has revealed that the light chain does not appear to move relative to its heavy chain, depending on the level of calcium [38]. Moreover, neither the light- nor the heavy-chain conformations change very much with the calcium level. (This is in marked contrast, for example, with the dramatic effect of calcium on the conformation of TnC [39].) Removal of the light chain does not alter the conformation of the head region of myosin – at least with respect to alpha-helix content. Thus, the spectroscopic studies show little large-scale change in myosin conformation. In collaboration with Szent-Györgyi and his group we have begun to carry out more direct studies of the structural events that may be associated with myosin-linked regulation.

The scallop muscle is well suited to X-ray diffraction analysis and gives a detailed (and somewhat unusual) X-ray diffraction diagram. Thus far we have examined only one stable state of the crossbridge cycle (apart from relaxed), namely, rigor. Here, some or all of the bridges are "frozen" into interactions with actin. The small-angle X-ray pattern from rigor scallop muscle shows strong diffraction that arises from the myosin crossbridges attached to actin (Fig. 20). When the light chain was removed from such samples by treatment with EDTA, a striking increase in intensity was observed on certain layer lines of the pattern. This effect was reversible when the light chain was diffused back into the sample. Thus we can say that there is a structural change effected by removal of the regulatory light chain, and that this corresponds to some change in the number, distribution, or conformations of myosin crossbridges associated with actin. Unfortunately, as is often the case with such X-ray diffraction studies, the patterns do not tell us directly what this structural change may be. It appears from stiffness measurements (Simmons and Szent-Györgyi, unpublished) that removal of the light chain does not greatly affect the number of myosin heads bound to actin in rigor. Therefore, the change we are observing by X-ray diffraction must be related to some *conformational* change in the crossbridges themselves that had not been detected by spectroscopic measurements and/or to a change in *distribution* of crossbridges among the available actin sites. We should emphasize that the X-ray studies were made on muscle in rigor, that is, without ATP and hence fixed to actin; and that the X-ray changes we are observing may therefore be related to such structural changes as angle of attachment, which might not be in question in some of the spectroscopic studies cited.

In order to visualize the possible conformational changes more directly we have undertaken electron microscopy of this system. The most striking findings have been obtained by negative staining of thin filaments "decorated" with scallop myosin. In fact, for technical reasons scallop heavy meromyosin and S1 were used, with and without their regulatory light chains present. The familiar arrowhead pattern originally obtained by Moore et al [13] with rabbit S1 – or a pattern closely similar to it – has been observed

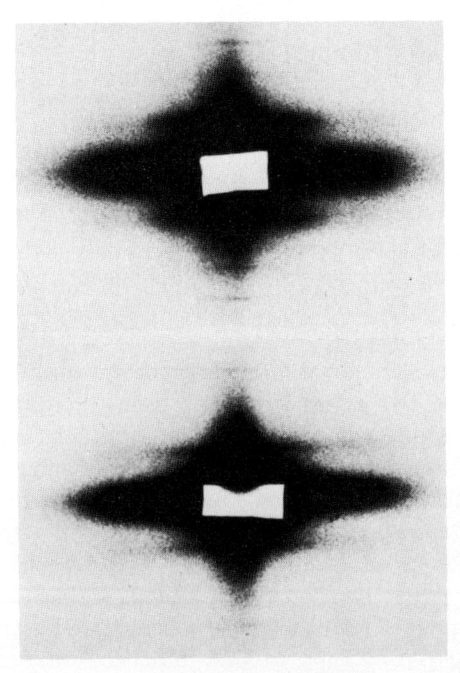

Fig. 20. X-ray diffraction patterns from scallop striated muscle. Above: In normal rigor (regulatory light chain present). Below: In rigor after "desensitization" (one regulatory light chain per myosin removed). The inner part of the 385-Å layer line changes intensity reversibly when the light chain is removed or replaced, indicating that the pattern of myosin-actin attachment is modified by the interactions between the light chains and the myosin head. From Vibert et al [48], with permission.

by means of myosin fragments from which the regulatory light chain has been removed (Figs. 21 and 22). We call this conformation of myosin "blunted." (It corresponds to the tilted and skewed model of attachment of the crossbridges.) When the regulatory light chain is present, the arrowheads look distinctly different, their trailing edges being clearly more "barbed" (Figs. 23 and 24). Thus, we have demonstrated a different conformation of the actomyosin complex that depends directly on the presence of the regulatory light chain (Craig, Szent-Györgyi, Beese, Flicker, Vibert, and Cohen, in preparation). Moreover, using lower concentrations of the myosin subfragments, one can observe the attachment of individual heads or pairs of heads without the confusion of superposition that arises in a more fully decorated filament (Fig. 25). It appears that both heads of HMM attach to the same actin strand of a single thin filament – and usually to adjacent actin monomers. The heads appear to be long and narrow, and distinctly curved. (The latter finding contrasts with the observations of Elliott and Offer [40] on isolated shadowed myosin molecules and of Takahashi [41] on negatively stained myosin, where the head is seen to be curved only occasionally.) This suggests that binding of the heads to actin in rigor could induce a large-scale shape change. The differing curvature of the two heads – that is, the apparent elongation and marked bending of the leading head of the pair – suggests also that they may have considerable flexibility. These findings are certainly of general significance for the structure of myosin, but for our studies of the mechanism of myosin regulation, they show that there are two forms of the myosin head when attached to actin – "barbed" and "blunted" – that are associated with the presence or absence of the regulatory light chain. The barbed form may be required for the "off" position of the myosin switch. As with the X-ray diffraction experiments, however, the results are suggestive, but we require additional information for a full interpretation of these images. In the case of a fully decorated filament, each "arrowhead" represents the superposition of different views of many myosin heads because of the helical symmetry of actin [13]; so the interpretation is not at all straightforward. The fact that consistent differences are observed, however, is crucial. Three-dimensional reconstruction of the barbed and blunted forms is under way and should reveal more precisely the structural differences giving rise to these two different pictures. Certainly both the X-ray diffraction and electron microscope results suggest that the mechanism of myosin-linked regulation is accessible to structural analysis.

THE TWO SWITCHES

Although we do not yet have a satisfactory model for myosin-linked regulation, we can compare the mechanisms, as far as possible, in the two switches (Fig. 26). While the detailed structures, and the nature of the transitions, remain to be clarified, on a gross level the steric blocking model gives a plausible picture of how thin-filament regulation might work. The essential aspect of the scheme is that a protein complex whose structure and interactions depend on the level of Ca^{2+} in the medium is bound to the actin helix by a cable-like fibrous protein, whose movement controls the availability of the sites on seven actin monomers. Some of the structural features of tropomyosin show regularity, possibly related to regular linkages with actin. The more dynamic features of the molecules – such as regions of marginal stability – may be associated with the movements known to occur during activation. We should emphasize that it would be a mistake to expect that such a model can account for various biochemical findings that are associated with function at a different level of structure.

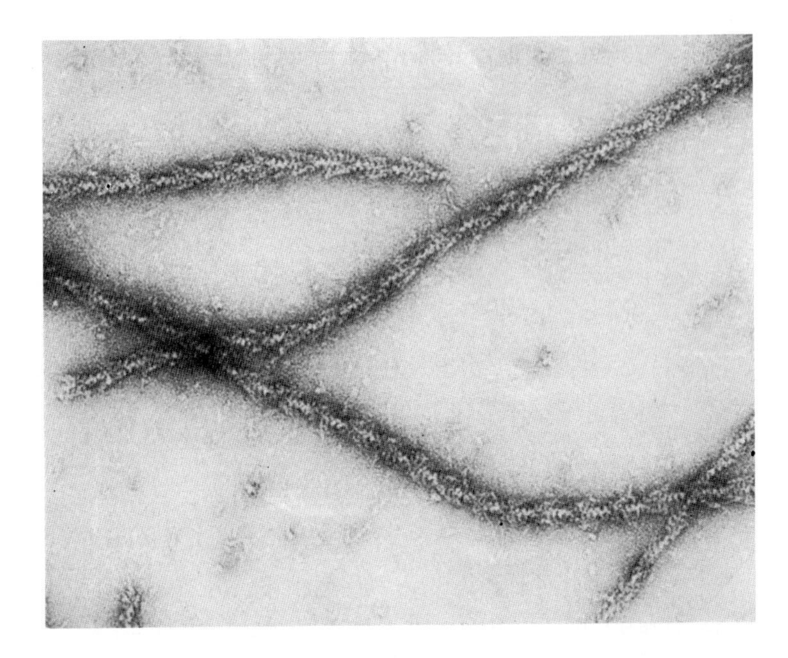

Fig. 21. Decoration using scallop S1 from which the light chain has been removed. The arrowheads are "blunt," and their edges somewhat convex. The arrowhead repeat is that of the actin helix – about 385 Å. Figures 21–25 from R. Craig et al (in preparation; see text).

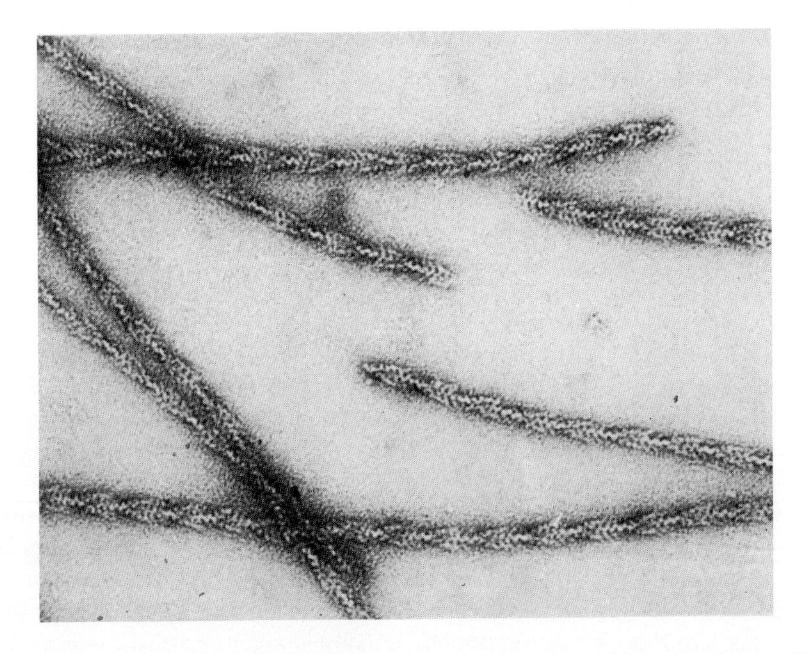

Fig. 22. Decoration with scallop HMM ("desensitized") containing no regulatory light chains. "Blunted" arrowheads.

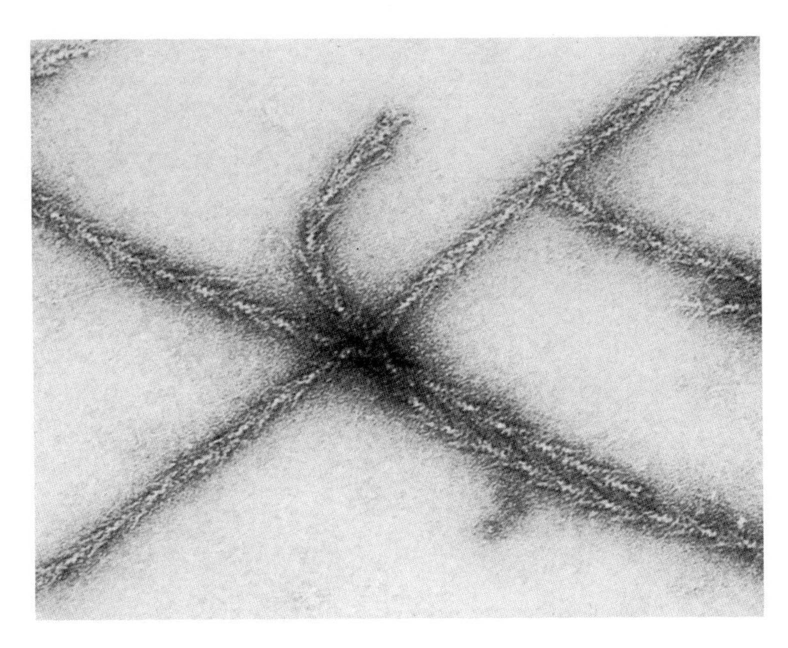

Fig. 23. Decoration of scallop thin filaments using scallop myosin S1 with regulatory light chain present. The arrowheads appear flared, with sharply "barbed" trailing edges.

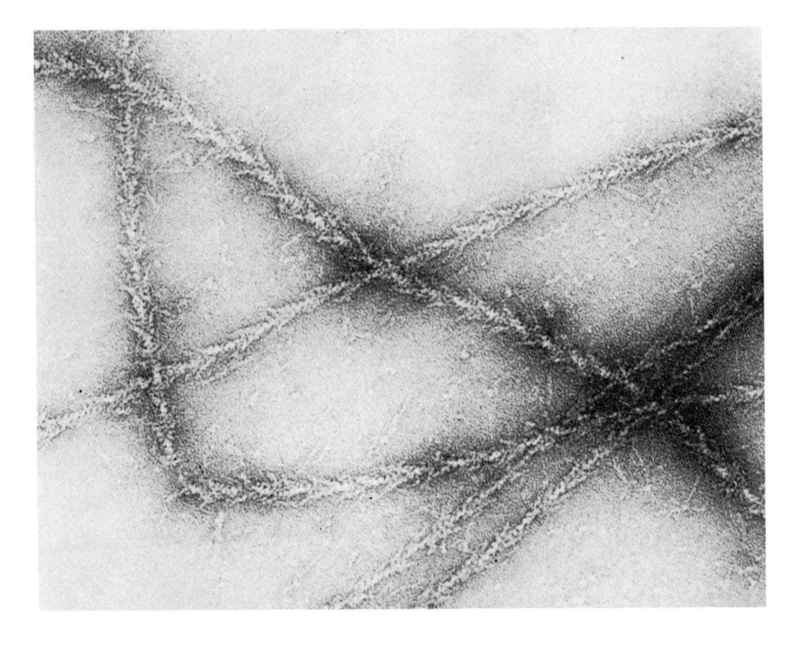

Fig. 24. Decoration with scallop HMM ("sensitive") containing two regulatory light chains. "Barbed" arrowheads.

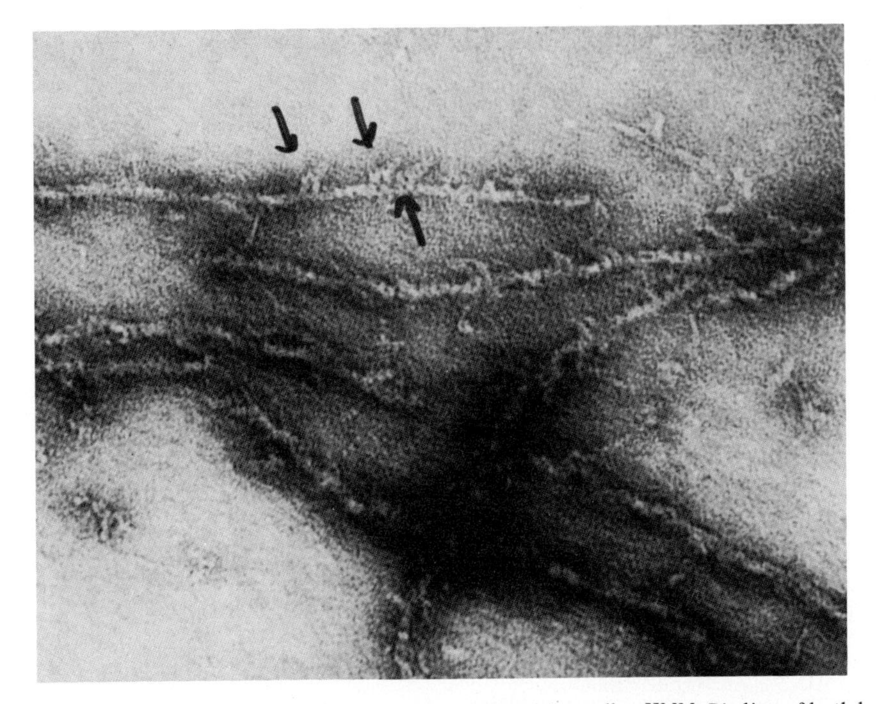

Fig. 25. Partial decoration of scallop thin filaments using sensitive scallop HMM. Binding of both heads of an HMM subfragment to adjacent actin monomers on one long-pitch strand of a thin filament is seen.

The role of tropomyosin in myosin-linked regulation is not at all clear. Tropomyosin is present in all muscles, and it moves also when molluscan myosin is activated, although no troponin is present. This motion has been pictured as due to "pushing" by the cross-bridges. Indeed, if in the molluscan thin filament tropomyosin were bound more loosely (and its thermal stability is strikingly less than that of tropomyosin from vertebrates), it could easily allow some heads to attach. Present findings do not indicate that tropomyosin is part of the myosin switch. So tropomyosin may have a dual role in all muscles – one essential to the thin-filament switch, and the other a structural role in affecting actin-actin interactions – or even in limiting the length of thin filaments.

We now have to put together the information necessary for a satisfactory model for the myosin switch. Here also, it is worth pointing out that the interactions of proteins that change conformation upon the binding of Ca^{2+} ions are a central part of the mechanism. The Ca^{2+}-binding site is not present on an isolated regulatory light chain, as it is on isolated TnC, but it is generated by the interaction of this light chain with the myosin head. We do not yet know where this critical region is: It appears that another nonspecific divalent cation-binding site already located on the sequence may have the function of linking the light chain to the myosin head. We would picture that specific Ca^{2+} binding by the regulatory region affects the conformation at another portion of the myosin head – nearer the end of the molecule – so that the *blocking* of actin binding is released. In this sense, the light chain interacting with the myosin head is somewhat like TnC interacting with the actin: The information is transmitted by protein-protein interactions over some distance. Now, in the case of the thin-filament switch, the rolling mechanism for tropomyosin accounts for actin's accessibility: Here myosin is always turned on. And it is actin that is eventually turned on and off in response to the Ca^{2+} level. With the myosin switch, actin is always on

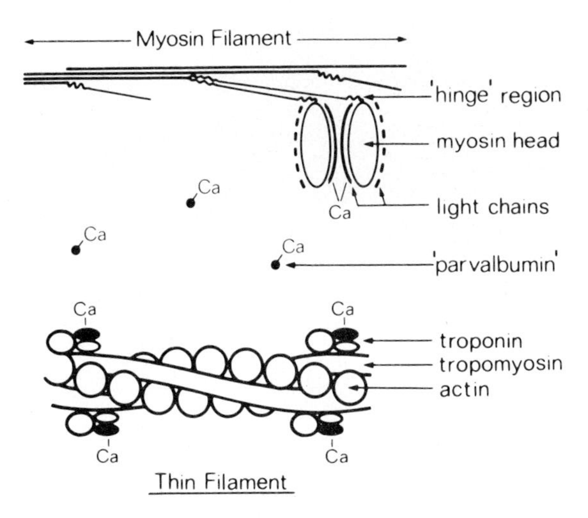

Fig. 26. Protein components of the thin and thick filament regulatory switches. In thin-filament regulation, Ca^{2+} binds to the TnC component of the troponin complexes, which are located every 385 Å along each of the two long-pitch helical strands of actin monomers. Altered interactions between the troponin components, tropomyosin and actin, lead to movement of the rod-like tropomyosin molecules into a new position in the "groove" between the actin strands. As a result, binding of myosin to the seven actin monomers that are regulated by one troponin/tropomyosin complex is no longer inhibited. In thick-filament regulation, the Ca^{2+}-binding site is located on the myosin molecule, and probably requires interactions between the regulatory light chain and the myosin head (heavy chain). Ca^{2+} binding releases the inhibition that prevents myosin from binding to actin in relaxed muscle. The precise locations of the regulatory light chain and the Ca^{2+}-binding site on the myosin head are not yet established, but the light chain may be long enough to span most of the length of head. It is possible that the light chain functions in a manner analagous to the troponin/tropomyosin complex; the Ca^{2+}-binding region may control the position of a segment of the chain some distance away that can sterically block the actin-binding site on the myosin head. From Kendrick-Jones and Jakes [49], with permission.

(tropomyosin not being a critical factor), and it is myosin that is turned on and off in response to the Ca^{2+} level. If we push the analogy further, one could say that the site on myosin that binds to actin may be blocked by the regulatory light chain when Ca^{2+} is not bound; and that Ca^{2+} binding (occurring some distance away) causes the conformational changes of the head, so that the light chain moves to another position. With our present limited knowledge, it might even be pictured as rolling away from the myosin-actin linkage site. So here again we could, by analogy, say that there is a steric blocking model in myosin-linked regulation. But unlike the thin-filament switch, we have not yet defined the parts and the movements sufficiently to build the model. We expect that an explanation – *at this level* – will soon be forthcoming.

Both protein switches will have to be seen in far greater detail for their function to be understood. But it would be a mistake to assume that their design is comparable to that of the machinery of everyday life. Machina carnis (the "animal machine") is an assembly of proteins whose design has features – not yet appreciated – that are undoubtedly more intricate than the human engineer's constructions. The last illustration (Fig. 27) summarizes this point of view. This cartoon shows the regulatory switches in the muscles of both vertebrates and molluscs. And there in the corner is a little Venus, who reminds us, as, we hope, do the structures we have tried to describe, that the vision of Nature as an artist should not be dismissed entirely.

Fig. 27. Nature as artist. The two regulatory switches are activated by similar levels of Ca^{2+}. The thin-filament switch is found in many invertebrate muscles, but has been most extensively characterized in the skeletal muscle of vertebrates. The myosin-linked system was first identified in the adductor muscles of bivalve molluscs such as the clam (formerly known as Venus mercenaria) and the scallop. Cartoon by A. B. Tulp; from Kendrick-Jones and Jakes [49], with permission.

ACKNOWLEDGMENTS

This study was supported by grants from the National Institutes of Health (AM 17346), the National Science Foundation (PCM76-10558), and the Muscular Dystrophy Association. P.J.V. is an Established Investigator of the American Heart Association; G.N.P. is the recipient of a National Research Service Award (GM 06148); R.C. received a Postdoctoral Fellowship from the Muscular Dystrophy Association.

REFERENCES

1. Jacob F: "The Logic of Life" (Spillander B, transl). New York, 1973.
2. Weyl H: "Symmetry." Princeton: Princeton University Press, 1952, p 23.
3. Huxley HE, Hanson J: Nature 173:973, 1954.
4. Huxley AF, Niedergerke R: Nature 173:971, 1954.
5. Huxley HE: Science 164:1256, 1969.
6. Ebashi S: Nature 200:1010, 1963.
7. Bailey K: Biochem J 43:271, 1948.
8. Cohen C, Longley W: Science 152:794, 1966.
9. Ebashi S, Endo M, Ohtsuki I: Q Rev Biophys 2:351, 1969.
10. Huxley HE: Cold Spring Harbor Symp Quant Biol 37:361, 1972.

11. Haselgrove JC: Cold Spring Harbor Symp Quant Biol 37:341, 1973.
12. Vibert PJ, Haselgrove JC, Lowy J, Poulsen FR: J Mol Biol 71:754, 1972.
13. Moore PB, Huxley HE, DeRosier DJ: J Mol Biol 50:279, 1970.
14. Parry DAD, Squire JM: J Mol Biol 75:33, 1973.
15. Bremel RD, Weber A: Nature 238:97, 1972.
16. Greaser ML, Gergely J: J Biol Chem 246:4226, 1971.
17. Hitchcock SE, Huxley HE, Szent-Györgyi AG: J Mol Biol 80:825, 1973.
18. Sodek J, Hodges RS, Smillie LB: J Biol Chem 253:1129, 1978.
19. Stone D, Smillie LB: J Biol Chem 253:1137, 1978.
20. Wilkinson JM, Grand RJA: Biochem J 149:493, 1975.
21. Pearlstone JR, Johnson P, Carpenter MR, Smillie LB: J Biol Chem 252:983, 1977.
22. Collins JH, Greaser ML, Potter JD, Horn MJ: J Biol Chem 252:6356, 1977.
23. Mercola D, Bullard B, Priest J: Nature 254:634, 1975.
24. Weeds AG, McLachlan AD: Nature 252:646, 1974.
25. Crick FHC: Acta Cryst 6:685, 1953.
26. Parry DAD: Biochem Biophys Res Commun 57:216, 1974.
27. McLachlan AD, Stewart M: J Mol Biol 103:271, 1976.
28. Caspar DLD, Cohen C, Longley W: J Mol Biol 41:87, 1969.
29. Phillips GN Jr, Lattman EE, Cummins P, Lee KY, Cohen C: Nature 278:413, 1979.
30. Parry DAD: J Mol Biol 98:519, 1975.
31. Szent-Györgyi AG, Cohen C, Kendrick-Jones J: J Mol Biol 56:239, 1971.
32. Kendrick-Jones J, Lehman W, Szent-Györgyi AG: J Mol Biol 54:313, 1970.
33. Lehman W, Szent-Györgyi AG: J Gen Physiol 66:1, 1975.
34. Bremel RD: Nature 252:405, 1974.
35. Szent-Györgyi AG, Szentkiralyi EM, Kendrick-Jones J: J Mol Biol 74:179, 1973.
36. Kendrick-Jones J, Szentkiralyi EM, Szent-Györgyi AG: J Mol Biol 104:747, 1976.
37. Stafford WF III, Szent-Györgyi AG: Biochemistry 17:607, 1978.
38. Chantler PD, Szent-Györgyi AG: Biochemistry 17:5440, 1978.
39. Potter JD, Seidel JC, Leavis PL, Lehrer SS, Gergely J: J Biol Chem 251:7551, 1976.
40. Elliott A, Offer G: J Mol Biol 123:505, 1978.
41. Takahashi KJ: Biochemistry (Tokyo) 83:905, 1978.
42. Paré A: "Collected Works." Pound Ridge, New York: Milford House, 1968.
43. Huxley H: Sci Am 213:18, 1965.
44. Mannherz HG, Barrington Leigh J, Holmes KC, Rosenbaum G: Nature New Biol 241:226, 1973.
45. Cohen C: Sci Am 233:36, 1975.
46. Cohen C, Caspar DLD, Parry DAD, Lucas RM: Cold Spring Harbor Symp Quant Biol 36:205, 1971.
47. Lowey S: In "Fibrous Proteins: Scientific, Industrial and Medical Aspects." Meeting of February 12–16, 1979, at Massey University, Palmerston North New Zealand. Creamer LK, Parry DAD, Editors. New York: Academic (In press).
48. Vibert P, Szent-Györgyi AG, Craig R, Wray J, Cohen C: Nature 273:64, 1978.
49. Kendrick-Jones J, Jakes R: Trends Biochem Sci 1:281, 1976.

Biological Recognition and Assembly 233–258 (1980)

Structural Studies of the Assembly of Simple Viruses

Lee Makowski

Rosenstiel Basic Medical Sciences Research Center, Brandeis University, Waltham, Massachusetts 02254

The principles of structural design and the bonding properties of structural proteins form a basis for the study of virus assembly. Virus coat proteins are designed specifically to interact with one another and with the viral nucleic acid to form a stable virus particle. The process of assembly is controlled by the switching of protein subunit conformation, which can alter the binding properties of the subunits. The self-assembly processes of several simple viruses in vitro have significantly different rates of assembly and specificities for their viral nucleic acid. It is possible that many viruses have multiple pathways for assembly, each pathway exhibiting somewhat different characteristics but all resulting in identical infectious virus particles.

Key words: virus assembly, virus structure, tobacco mosaic virus, tomato bushy stunt virus

CONTENTS

Received May 18, 1979; accepted May 21, 1979.

I. INTRODUCTION

Self-assembly of simple viruses from mixtures of structural proteins and nucleic acid in vitro has demonstrated the dynamic character of coat proteins that are designed to interact with one another and with the viral nucleic acid. The same proteins that make up a virus particle stable enough to protect the viral nucleic acid in a multitude of environmental conditions encountered outside the host cell must also act, perhaps in concert with other viral components, to accomplish both the assembly of the virus particle and the infection of the host cell.

The assembly of simple viruses requires the controlled association of only a few components; the viral nucleic acid, the major coat protein, and in some cases a few copies of a second structural protein. In contrast to the complex bacteriophages in which dozens of structural components come together in a controlled fashion throughout a complex assembly pathway [1, 2], there are perhaps only a few steps in the assembly pathway of simple viruses. The simplicity of these structures offers the possibility of studying the steps in virion assembly in detail.

The structural principles that govern the design of simple viruses were first set down by Caspar and Klug [3]. The expression of these principles at the molecular level has been probed by X-ray diffraction studies of tobacco mosaic virus [4], tobacco mosaic virus protein disk [5], tomato bushy stunt virus [6], and southern bean mosaic virus [7]. Examination of the molecular structures of virus particles and intermediates in the assembly process may help elucidate the mechanisms involved in virus assembly. However, as of yet there is no well-defined theoretical framework comparable to our ideas about virus structure that can be used to evaluate studies of virus assembly.

It is worth asking the question "What constitutes a description of virus assembly?" From a structural point of view this might involve a description of the conformation and aggregation state of the structural proteins, nucleic acid, and other contributing components at each stage in the assembly. This type of description can be made at any spatial and temporal resolution. As we increase the resolution of this description – both in time and space – we are likely to gain further insight into the forces driving the assembly process. Ultimately, atomic resolution may be required to understand the control of assembly which, at least in some cases, is mediated through small changes in the subunit conformation.

Assembly of even the simplest viruses cannot proceed by a simple adding of protein building blocks as was assumed early in the study of tobacco mosaic virus (TMV) [8]. This view of assembly denies the need for control of the assembly process, which must be an intrinsic part of all purposeful biologic activities [9]. The discovery of the role of the TMV protein disk in the assembly of TMV particles [10] led to the realization that the assembly of this simple helical virus was far more complex than had been anticipated. Virus coat proteins are designed to bind to one another, but uncontrolled formation of protein aggregates cannot productively contribute to virus assembly. An efficient control mechanism for the assembly of many viruses may involve the catalysis of a given assembly step by the product of the previous step [11, 12].

In this paper the presumed steps in virus assembly are examined in light of the high-resolution images of virus particles and virus protein assemblies that have been produced by X-ray diffraction studies. The steps of virus assembly are considered more or less chronologically, beginning with the association of virus proteins with one another and the control of the state of aggregation. All subsequent steps in assembly are considered in the section on assembly of coat protein and nucleic acid. Disassembly must involve structural transitions relevant to the assembly and stability of the virus particle and will be discussed in a separate section. The remainder of the virus life cycle is not considered except in passing.

Virus assembly is a dynamic process, the understanding of which requires an understanding of virus structure. As a background for the discussion of assembly, Section II contains a review of the structural design principles relevant to virus structure and descriptions of several of the recently produced high-resolution images of virus particles. The molecular forces involved in the stabilization of virus structure are considered in Section III.

Discussion is limited to viruses that contain a single major coat protein. Viruses with a few copies of a second structural protein such as the icosahedral RNA phages or the filamentous bacteriophages are included. Viruses such as polio or ϕX174 are built on equally simple structural principles but contain several protein species in the coat. The nucleocapsids of enveloped viruses are contained within a lipid or lipid-protein layer. These and other more complex viruses are not considered explicitly, although some of the discussion may be relevant to their pathways of assembly. Table I lists the viruses that are considered in this review according to their basic morphology, along with the abbreviations of their names that are used here. The emphasis throughout the paper is on those viruses that are best characterized structurally.

TABLE I. Viruses Discussed in This Paper

Rigid-rod plant viruses		Filamentous bacteriophages	
TMV	Tobacco mosaic virus		Pf1
TRV	Tobacco rattle virus		Xf
BSMV	Barley stripe mosaic virus		fd
Icosahedral plant viruses		Icosahedral RNA phages	
TBSV	Tomato bushy stunt virus		R17
SBMV	Southern bean mosaic virus		MS2
CCMV	Cowpea chlorotic mottle virus		M12
BMV	Brome mosaic virus		f2
BBMV	Broad bean mottle virus		fr
TYMV	Turnip yellow mosaic virus		Qβ
CMV	Cucumber mosaic virus		
Rigid-rod plant virus with rounded ends		Icosahedral animal viruses	
AMV	Alfalfa mosaic virus		Polyoma
			SV-40
			Papilloma

II. PRINCIPLES OF STRUCTURAL DESIGN

A. Quasi-equivalence

The simplest viruses are built from many copies of chemically identical protein subunits. This has the advantage of requiring a minimum amount of information to be encoded in the viral nucleic acid. A virus particle is formed through the interaction of these protein subunits with one another and with the nucleic acid. The equilibrium association product of the viral components will occupy a minimum on the free-energy surface for the system. In forming a stable structure, each subunit makes as many bonds as possible with other subunits. Since all the subunits are chemically identical, it seems likely that this requirement will result in a structure in which all the subunits are in equivalent positions relative to other subunits in the particle. However, as Caspar and Klug [3] pointed out, the subunits need not be in *exactly* equivalent environments to minimize the free energy of the system.

There is a limited number of ways in which a closed shell may be constructed about a nucleic acid through the symmetrical placement of protein subunits. Crick and Watson [13], observing that all simple viruses are either rod-shaped or spherical, suggested that on the basis of crystallographic data (for example, see Watson [8] and Caspar [14]) all rod-shaped viruses are constructed from helical arrangements of identical subunits and all spherical viruses from cubic arrangements of subunits. Caspar and Klug [3] recognized the advantages of icosahedral symmetry over the other cubic symmetries.

Although it is possible to arrange any number of identical subunits in identical positions on a helix, no more than 60 subunits can be arranged in equivalent positions to construct a closed spherical or polyhedral shell. However, chemical data show that many isometric virus particles are composed of far more than 60 copies of a single coat protein subunit. Reasoning that identical subunits that cannot interact with one another in exactly equivalent ways are likely to interact in almost equivalent or quasi-equivalent ways, Caspar and Klug [3] enumerated all the ways in which protein subunits can be arranged in quasi-equivalent positions on an icosahedral surface lattice. There is no upper limit to the number of subunits that can be arranged on an icosahedral surface lattice in this way.

Protein subunits in one position on a surface lattice will assume somewhat different conformations and interact with other subunits in somewhat different ways from those protein subunits in other quasi-equivalent but not exactly equivalent positions on the lattice. Proteins that are capable of assuming several conformations and interacting with one another in a variety of ways might be expected to interact with one another to form a variety of different structures. Polymorphic forms of virus coat protein aggregates are very common. Indeed, the ability of coat proteins to aggregate into more than one structure may be a requirement for the control of virus assembly. The creation of each polymorphic form is driven by the same force – the formation of bonds between subunits. The modes of interaction in the different aggregates must be different. However, the stronger a given bond is, the more likely it is to be conserved in the formation of different aggregates.

B. Helical Viruses

There are at least three classes of simple helical viruses; the rigid rod viruses of which TMV is the most well known member [15]; the flexuous plant viruses such as potato virus X [16]; and the filamentous bacteriophages [17]. Simple helical animal viruses are either very rare or nonexistent, perhaps because of the susceptibility of long rod-shaped particles to breakage by shear. TMV is undoubtedly the most extensively studied virus from the point of view of structure and assembly. Filamentous bacteriophage Pf1 is the only other helical virus on which there have been detailed structural studies. The structures of both TMV and Pf1 will be described below.

The nucleic acid in a helical virus extends the entire length of the virus particle and generally determines the virus particle length. For instance, tobacco rattle virus (TRV) has a split genome – two RNA molecules of different lengths give rise to two virus particles, one longer than the other, both of which are found in infectious preparations of TRV [16, 18]. Usually the nucleic acid molecule conforms to the same symmetry as the protein subunits. In TMV there are exactly three RNA bases per coat protein. However, filamentous bacteriophage fd has a nonintegral ratio of 2.35 ± 0.1 DNA bases per coat protein [19], indicating that the DNA cannot conform to the same symmetry as the coat protein. In contrast, filamentous bacteriophages Pf1 and Xf apparently have integer ratios of nucleotides to coat proteins [20, 19].

With the exception of the subunits near the ends of the particle, all coat proteins in a regular helical virus are arranged in equivalent environments. Minor structural variations among subunits may result from the interaction of the subunits with the nucleic acid, since the sequence of bases in the nucleic acid must necessarily vary along its length. In some cases, further deviation from exact equivalence of subunits may result in a lower energy configuration. In the *Dahlemense* strain of TMV the subunits take up 98 quasi-equivalent positions, forming a periodically perturbed helical array [21].

1. Tobacco mosaic virus. X-ray diffraction studies of intact TMV particles [4] and the disk aggregate of TMV protein [5] have produced high-resolution images of these two structures. Diagrams of these structures can be found in the paper by Holmes (this volume). In the virus the coat protein is arranged on a 23-Å pitch helix with 16 1/3 subunits per turn [22]. Three nucleotides are associated with each coat protein. There is a hole 40 Å in diameter through the center of the particle, and the coat protein extends from about 20 Å radius out to about 90 Å radius [23]. The single-stranded RNA molecule is completely enclosed by protein and winds between adjacent layers of the protein helix at a radius of about 40 Å. The RNA is separated from the axial hole of the virus by a short length of peptide that is probably two or three turns of a rather irregular α-helix [4]. A solid core of four α-helices extends from about 30 Å to 60 Å radius. From the similarity of the coat protein in the disk and virus helix, it is likely that the α-helical core is enclosed in a "hydrophobic girdle" made up largely of aromatic residues and forming a solid layer of hydrophobic residues encircling the particle [5].

The disk aggregate of TMV protein is an essential intermediate in virus assembly [10] and consists of two rings of 17 coat proteins each. The transformation from disk to helix conserves the side-to-side interactions between subunits but completely changes the interactions in the axial direction between subunits on adjacent turns. At radii of more than 40 Å from the particle axis the conformation of the coat protein is very similar in the disk and helix – one structure being derived from the other by a simple rigid-body transformation [24] and perhaps minor conformational changes. At distances less than 40 Å from the particle axis the coat protein is disordered in the disk [5, 25] and exhibits unusual mobility [26]. The inside portions of the upper and lower layers of protein tilt away from one another like a pair of jaws awaiting interaction with the RNA [25].

2. Filamentous bacteriophages. The architecture of the filamentous bacteriophages is very different from that of the helical plant viruses. These virus particles are, depending on the strain, 1–2 μ long and only about 60 Å in diameter [17]. The major coat protein is very small, 45–50 residues, and encloses a circular, single-stranded DNA molecule. A few copies of a second structural protein are present in the particles, probably located at one end. The structures of these viruses have been studied by X-ray diffraction from oriented fibers [27–29]. Based on a comparison of X-ray patterns from different strains, all filamentous phages appear to have similar structures. Not all the phages have the same helical symmetry. In Pf1, Xf, and probably several other phages the subunits are arranged on a 15-Å pitch helix with 5.4 subunits per turn ([30]; however, see also Marvin et al [27]). The symmetry of fd is not certain but probably includes a five-fold rotation axis as one of its symmetry elements [30].

The molecular structure of Pf1 has been determined to 7-Å resolution [29, 31, 32]. The hydrated nucleic acid occupies most of the volume in the center of the particle out to about 10 Å radius. The coat protein is arranged into two α-helical segments each about 30 Å long. One segment, which is centered at about 15 Å radius, is almost parallel to the particle axis. The other, at about 25 Å radius, is tilted 25° to the particle axis. The connection between the two helices is not clear in the electron density map, and they may be arranged in either a parallel or antiparallel configuration. The middle of the coat protein sequence is exceedingly hydrophobic [33]. The confluence of hydrophobic side chains in the region between the two α-helical layers of about 20 Å radius produces a solid layer of hydrophobic groups about the DNA [29].

C. Icosahedral Viruses

A large number of plant, bacterial, and animal viruses of widely varying complexity are built with icosahedral symmetry. The simple icosahedral viruses include the small plant viruses such as tomato bushy stunt virus (TBSV) and southern bean mosaic virus (SBMV); the single-stranded RNA phages that contain one copy of a second structural protein [34]; and the papovaviruses, animal viruses that contain cellular histones [35]. The small plant viruses are the best characterized structurally. Both TBSV [6] and SBMV [7] are being studied by X-ray crystallography. Although the assembly of the RNA phages has been studied in some detail [34, 36], the structure of these particles is relatively poorly characterized. The papovaviruses include the oncogenic viruses polyoma, SV-40, and the papilloma viruses. Three-dimensional reconstructions from electron micrographs of negatively stained virus have been produced for polyoma [37] and human papilloma virus [38]. The structures of tubular variants of human papilloma virus have also been analyzed [39, 40].

Caspar and Klug [3] showed that symmetrical, closed shells with quasi-equivalent subunits can be built from 60T subunits where T, the triangulation number, is limited to being one of a set of integers, the smallest ones being T = 1, 3, 4, 7, 9, 12, 13 ... The mirror images of some surface lattices define another lattice distinct from the original. For these skew classes of surface lattice, the handedness of the lattice must also be specified. For instance, the coat proteins of rabbit papilloma virus are arranged on a surface lattice with T = 7*l* (levo), while in human papilloma virus they are arranged with T = 7d (dextro) [41].

Clustering of subunits on the surface of the virus particle can produce a variety of appearances from the same surface lattice. If each subunit projects from the virus surface, a total of 60T bumps or morphologic units would be seen on the surface. However, the subunits can also cluster into dimers, producing 30T morphologic units; or trimers, producing 20T morphologic units. If the subunits cluster into hexamers and pentamers, 10(T–1) hexamer + 12 pentamer morphologic units are formed. For instance, dimer clustering on the surface of TBSV (T = 3) produces 90 morphologic units; hexamer-pentamer clusters on the surface of Polyoma and the papilloma viruses (all T = 7) produces 72 morphologic units.

The packing of nucleic acid in icosahedral viruses poses an interesting problem. It is topologically impossible for a single strand of nucleic acid to be arranged with icosahedral symmetry. Although none of the single-stranded RNA conforms to icosahedral symmetry in TBSV [6], fully one-third of the RNA may be ordered with icosahedral symmetry in SBMV [7]. The secondary structure of cowpea chlorotic mottle virus RNA does not change upon extraction from virus particles [42], suggesting that the presence of an icosahedral shell around it does not significantly alter its configuration. In bacteriophages T4, λ, and P22 the double-stranded DNA is wound much like a skein of yarn, the strands being locally parallel and hexagonally packed, completely filling the phage heads [43–45].

Only recently, as a result of high-resolution virus crystallography has it been possible to observe the different conformations and modes of interaction that coat proteins adopt in the distinct, quasi-equivalent surface lattice positions. The structures of TBSV and SBMV provide considerable insight into the way in which coat proteins are designed explicitly to conform to the various quasi-equivalent positions.

1. Tomato bushy stunt virus. Harrison and co-workers [6] have solved the structure of TBSV to 2.9-Å resolution by X-ray crystallography. The subunits of TBSV are constructed in two domains connected by a short hinge section of peptide. One of these domains packs into a compact shell about 30 Å thick that entirely encloses the RNA. The other domain projects out from the shell. These projecting domains are combined in pairs making up the 90 morphologic units on the strict and local two-fold axes of the particle. Both domains are made up largely of β-sheet structure. The conformations of the two domains appear to be essentially identical in the three quasi-equivalent positions on the surface lattice. In two of these positions, the angle between the domains is approximately the same. In the third position, the hinge conformation is slightly different and the angle between the two domains is about 20° different from the other two positions. Furthermore, about 30 residues on each protein in this third position spiral around the three-fold axes in a novel β-annulus configuration.

A substantial portion of each protein subunit does not conform to the icosahedral symmetry of the surface lattice and appears to be disordered. The disordered protein, which is located entirely within the particle interior, amounts to about 50–60 residues in the subunits involved in the β-annulus structure and 80–90 residues in the other subunits. Small-angle neutron scattering [46] suggests that much of this protein is at small radius, separated from the protein coat by the bulk of the RNA, and connected to the major part of the protein by relatively extended segments of peptide. The RNA in the virus particle is also disordered – none of it is apparent in the electron density map [6]. It seems likely that the disordered protein is involved in RNA binding. It is also possible that the concentration of protein at small radius may be involved in the assembly process.

The TBSV coat protein is specifically adapted to taking on significantly different configurations, requiring only small conformational changes. These conformational changes are limited to two small regions of the protein – the hinge and the β-annulus arm.

2. Southern bean mosaic virus. SBMV has also been studied by X-ray crystallography. At present, the structure has been solved to 5-Å resolution [7], but work at higher resolution is in progress. The coat protein appears to have three domains: a large domain containing five α-helices, a small one containing two α-helices, and a second small one that appears to contain some β-structure. In contrast to TBSV, approximately 36% of the RNA in the virus interior appears to be ordered with the icosahedral symmetry of the surface lattice. The RNA is separated from the protein by a small gap with only a few direct RNA-protein contacts apparent in the electron density map.

D. Polymorphism

Polymorphism – the capacity of the same molecules to form different structures – provides the basis for the functioning of many protein assemblies [47]. Polymorphism may occur within a single structure as in the icosahedral viruses with triangulation numbers greater than one, but usually polymorphism refers to the different dispositions of identical subunits in different structures. In the assembly of a virus any intermediates in the assembly pathway are likely to be structures in which the conformation of a given protein and its interactions with its neighbors are somewhat different than those in the complete virus particle. Study of the structures of polymorphic assemblies, even those not directly involved in the assembly process, may provide relevant information about the flexibility of the protein subunits and the modes of interaction of which they are capable.

In the study of virus structure there are two distinct classes of polymorphic structures – those which contain nucleic acid and those which do not. Virus coat proteins are designed to interact with one another and, depending on their environment, they can be induced to aggregate into a wide variety of structures. For the most part these will be considered in Section IVA. Complete viruses also undergo structural transitions, depending on solvent conditions, and these can provide information about the types of interactions important to virus particle stability.

Specific protein-protein interactions such as those which occur between coat proteins generally occur between interfaces complementary to one another on a molecular level. They are often as tightly packed as the interiors of globular proteins [48]. Polymorphic structures may be formed either by the utilization of several distinct sets of complementary surfaces or by conformational changes in the subunits that alter the interaction on a given surface, thereby causing a change in the mode of aggregation.

1. Tobacco mosiac virus. Three independent structures of TMV protein are visualized in two different assemblies: the virus helix, where all the subunits are equivalent, and the protein disk, in which the proteins in the two layers are not in equivalent positions. Surprisingly, the subunits in the two layers of the disk are virtually identical, the molecular model of one layer fitting into the electron density map of the second layer with adjustments of not more than about 1 Å [5]. The major difference between subunits in the disk and in the virus particle is the disorder in the 25 residues at smallest radius in the disk [4, 5].

The side-to-side interaction between TMV proteins is the strongest interaction in the various aggregates that the protein forms [3]. Whereas this interaction is conserved in the transition from TMV protein disk to TMV helix (with or without RNA), the axial bonds between subunits on adjacent turns are entirely reorganized.

2. Filamentous bacteriophage Pf1. The intact filamentous bacteriophage Pf1 appears capable of switching between states with slightly different helical symmetries [49, 50]. The subunits of Pf1 are usually observed arranged on a helix with a pitch of 15 Å and 5.4 subunits per turn [30]. If the fibers used for diffraction experiments are prepared at 5°C rather than 20°C, there is a slight tightening of the subunit packing to 5.5 subunits per turn. Other variants with 5.38 or 5.46 units have also been observed ([50]; note the discrepancy in the symmetries of Pf1 derived by various investigators [19, 20, 30, 50]). This transition must involve a change in the interactions among the α-helical segments of the coat protein. The major interaction between α-helices occurs between α-helical segments in the inner and outer layers, which cross at an angle of about 20° along most of their length [29]. The transition from 5.4 to 5.5 units per turn will involve a rotation of both α-helical segments by about 6° around an axis perpendicular to the particle axis so that the inner helix becomes almost exactly parallel to the virus particle axis. This will occur with a decrease in the crossing angle between the α-helices in the inner and outer layers of not more than 2–3°. The small change in preparative conditions required to cause this transition indicates that the energy needed to bring it about is very small.

3. Swelling of icosahedral plant viruses. Many icosahedral plant viruses are observed to swell with an increase in radius of as much as 10% when divalent cations are removed [51]. This swelling usually renders the virus particles susceptible to nuclease digestion and degradation by proteases and lowers the infectivity. The radius of SBMV increases by about 9% when Ca^{++} and Mg^{++} are removed (Caspar, personal communication). This swelling may play a role in virus particle disassociation during infection [51, 52] and will be discussed in detail in the section on disassembly.

4. Alfalfa mosaic virus. An exceptional case of polymorphism is found in the particles of Alfalfa Mosaic Virus. This multicomponent virus consists of several cylindrical particles of different lengths, each with a diameter of 150–180 Å. The ends of the cylindrical particles are rounded [53–55]. Only one coat protein is found in the particles, and this must serve to form both the rounded ends and the cylinders. The smallest particles observed are icosahedral and contain 60 coat proteins (T = 1). An icosahedron has 12 five-fold axes and 20 three-fold axes. By systematically adding hexamers, cylinders can be constructed continuous with the hemispherical rounded ends, which will each contain six pentamers. Cylinders constructed from pentamers of subunits are also possible [40]. The lengths of the particles formed are determined, at least to some extent, by the size of the RNA molecule they contain. Similar structures have been observed in the tubular variants of papilloma virus coat proteins [39, 40].

III. STABILIZING FORCES

A. Protein-Protein Interactions

In his classic paper on protein denaturation Kauzman [56] described the various forces important for protein stability. Many of these forces are also important in interactions between proteins. Generally, the forces are of two types: polar and nonpolar or hydrophobic. The polar interactions, which include salt linkages and hydrogen bonds, are directional, electrostatic interactions. Hydrophobic interactions are relatively nondirectional, driven mainly by the large entropy decrease involved in the contact of a hydrophobic group with water. Chothia and Janin [48] have suggested that the stability of protein-protein interactions is due mainly to hydrophobic interactions, while the specificity is dictated by polar interactions.

1. Tobacco mosaic virus. The most complete description of the contacts between coat proteins is that of the TMV protein disk [5]. The disk is a relatively open structure compared to the TMV helix, with only four regions of interaction between the subunits in the top and bottom layers. Three of these involve salt bridges, the other is a single hydrophobic interaction. One of the salt bridge contacts is an extended system involving nine residues and including two water molecules that are trapped within the interface between layers. The side-to-side interactions between subunits involve four patches which are alternately hydrophobic and hydrophilic. These interactions are conserved during the disk-to-helix transformation.

2. Tomato bushy stunt virus. Sixty of the protein subunits of TBSV are involved in a remarkable β-annulus interaction. At each of the 20 three-fold axes, arms from three of the proteins interdigitate, winding around the axis in a configuration rather like the overlapping flaps of a cardboard carton [6]. The chemical nature of the intersubunit bonds is not yet clear, since the amino acid sequence has not been completed. The conserved interaction between the projecting domains involves apposed β-sheets. In nonconserved interfaces, side chains that appear to make important interactions in one state are rather disordered in the other states.

The conserved local three-fold interfaces have one particularly interesting feature – an ion, perhaps calcium, which is coordinated by side chains from both subunits and probably linked to an EDTA-induced swelling observed at pH 7.5 [6].

B. Protein-Nucleic Acid Interactions

Electrostastic interactions between the negatively charged phosphates of the nucleic acids and the basic residues arginine and lysine are expected to be important in the interactions between proteins and nucleic acids. Specific interactions between coat proteins and nucleic acid involve binding of an essentially random sequence of nucleotides to the protein. Although specific recognition sequences of nucleic acids may be important for the initiation of the assembly of some viruses, the major part of the nucleotide sequence must be defined by the coding requirements of virus proteins rather than the structural requirements for virus assembly. Disorder of protein chains in the region of nucleic acid binding sites has been observed frequently and is likely to play a crucial role in protein-nucleic acid interactions.

1. Tobacco mosaic virus. Fiber diffraction studies of intact TMV particles have provided a clear image of the interaction of TMV coat protein and RNA [4]. The RNA lies between 30 and 50 Å from the particle axis, enclosed between subunits on two adjacent turns of the TMV helix. The three phosphates form ion pairs with three arginines – two on the right radial α-helix of one subunit, one on the left radial α-helix of a subunit on an adjacent turn of the TMV helix. The RNA bases lie flat against this left radial α-helix. The interactions between the flat rings of the RNA bases and the protein side chains of the α-helix are largely hydrophobic and thus relatively nonspecific, allowing any base to fit into the binding sites. Little or no disorder is apparent in the binding site, and the protein and RNA conformations must be very similar for all nucelotide sequences. There is, in fact, little difference between the protein conformation in the TMV particle (with RNA) and the TMV protein helix (without RNA) [57].

In contrast, about 25 residues of the coat protein near the RNA binding site are disordered in the TMV protein disk [5, 25, 26]. This disordered loop contains all the arginines that form ion pairs with the RNA phosphates. Whether the disorder in this region is required for RNA recognition or for binding or for accommodating any sequence of RNA into the binding site is not clear.

2. Tomato bushy stunt virus. The interaction of RNA and protein in TBSV is almost entirely confined to the region of protein that is disordered in the virus particle [6]. The amount of disordered protein is not certain, but from molecular weight estimates [58, 59] and analysis of the electron density map [6], it may amount to a mass almost equal to that of the RNA. Chauvin et al [46] have shown that much of the disordered protein is located 50–80 Å from the center of the particle; the bulk of the RNA occupies the region between 80 and 110 Å radius; the icosahedrally ordered protein coat extends outward from about 110 Å radius. The inward-facing part of the protein shell presents a relatively smooth surface to the interior with no obvious clefts that might act as RNA-binding sites [6]. Small-angle X-ray [60] and neutron scattering [46] show that the RNA extends to radii immediately internal to the coat protein surface. Apparently the RNA does not interact with this internal surface in a unique fashion, rendering it invisible in the high-resolution electron density map.

3. Southern bean mosaic virus. About one-third of the RNA in SBMV appears to be ordered with the icosahedral symmetry of the coat protein [7]. It is not possible to trace a single chain in the electron density map and in some places a number of chains seem to meet at a single point. Suck et al [7] have interpreted this as the superposition of several distinct positionings of the RNA in the virus interior that have been superimposed by crystallographic and noncrystallographic averaging. Although some regions of contact between the protein and RNA exist, the RNA is mostly separated from the protein shell. There is no evidence to indicate whether or not a disordered region of coat protein exists in the interior of SBMV as it does in TBSV.

IV. VIRUS ASSEMBLY

A. Association of Protein Subunits

Virus coat proteins are designed to interact with one another. Monomers of few, if any, of the coat proteins of simple viruses are found to be the dominant species in solution, except under strongly dissociating conditions. The coat protein of TMV has a surface which, although not particularly hydrophobic, has more hydrophobic groups than are normally found on the surface of a globular protein [5]. It would be surprising if this were not true of of many virus coat proteins.

The aggregates that coat proteins form under physiologic conditions are of interest, since these are the structures that must interact with the nucleic acid during assembly. A study of the states of aggregation under a broad range of conditions allows an assessment of the types of interactions important in virus particle structure and assembly. In particular, the aggregation state that the coat proteins assume is intimately related to the conformation of the individual proteins. The ability of coat proteins to switch conformational state and bonding properties in response to environmental variables and thus control the mode of aggregation is of primary importance in the control of virus assembly [15].

1. Tobacco mosaic virus. The aggregates that the TMV proteins form under a broad range of conditions have been extensively studied (for reviews see Caspar [15] and Butler and Durham [61]). The different types of aggregates that the coat proteins form exist in equilibrium with one another. At high pH and low ionic strength, the dominant species are small 4S and 8S aggregates usually referred to as A-protein. Lowering pH or increasing ionic strength leads to the formation of larger aggregates. Increasing the ionic strength at high pH leads to the formation of disks and then crystals of disks and larger aggregates such as the stacked disk. At low pH, helical aggregates with structures very similar to the protein in the intact virus [57] are formed. Two types of helical aggregates have been observed [62]. One of these has 16-1/3 subunits per turn as in the intact virus. The other has 17-1/3 subunits per turn.

The conformations and modes of interaction of the proteins appear to be different in each of the TMV protein aggregates. This has been shown directly in the case of the helical aggregates and the disk by comparison of the structures as seen by electron microscopy [63] and X-ray diffraction [4, 5, 57]. A similar structural comparison cannot be made with the smaller protein aggregates. Although the A-protein may be limited to forming two-layered structures by a pairing distortion similar to that observed in the disk [64, 65], at least some conformational changes occur during the formation of disk from A-protein. This transformation also involves the uptake of one proton per protein monomer [66]. A second proton per monomer is bound during the formation of the protein helix [67].

The state of aggregation of the TMV protein is apparently controlled by the state of ionization of two abnormally titrating carboxylic acid groups [15]. These groups, which are probably carboxyl-carboxylate, hydrogen-bonded pairs, titrate with pK's near neutrality. Ionization of these groups by the removal of protons at increased pH will result in an electrostatic repulsion between the groups presumably causing changes in the monomer conformation. The conformational changes determine the binding properties of the subunits and the modes of aggregation that they assume.

In the protein helix there are two abnormally titrating groups and two bound protons per subunit. These bound protons are required for the stability of the protein helix. Increasing the pH causes the removal of these protons and the breakdown of the helix [15]. Similarly, removal of a single proton per subunit from the disk will cause it to dissociate into smaller aggregates [66]. The intact virus apparently has three abnormally titrating groups [67, 68], which although important for virus stability are not essential, since they titrate reversibly without the breakdown of the virus particle [15].

The positions of the two abnormally titrating groups in TMV protein are not certain. One is probably formed by Asp 115 and Asp 116 [69] near the end of the left-radial helix at a radius of about 30 Å [4], close to the RNA-binding site and adjacent to the region of protein which is disordered in the disk (residues 90–113) [5]. These residues are part of a concentration of negatively charged groups in the virus called the "carboxyl cage." The second abnormally titrating group may also be associated with this region [4, 67].

The kinetics of polymerization of the various TMV protein aggregates are very sensitive to experimental conditions. Near pH 6.7, the half-life of helical rods goes from days to minutes within a few tenths of a pH unit [70]. Similarly, A-protein formed under different conditions appears capable of forming disks at very different rates [71–73]. Scheele and Schuster [74] observed that the turbidity of TMV protein solutions forming a helix could "overshoot" its equilibrium value. This is caused by the formation of very long metastable helices with relatively long lifetimes under conditions where growth of helices is much more rapid than nucleation. Metastable aggregates can also be formed when A-protein polymerizes to disks and stacks of disks [75, 76].

2. Other helical viruses. TRV is another rigid-rod, helical plant virus. In contrast to TMV, no conditions have been found under which TRV coat protein forms helical aggregates [18]. Only small aggregates (and perhaps monomers) and protein disks have been observed. There is some uncertainty as to the structure of the disk aggregates. They may have two layers as in the case of TMV [18], but recent data suggest they are larger, made up of three or four layers of protein [77]. In contrast to TMV, disks of TRV protein are found at higher pH than the dissociated protein [78, 79]. Although the TRV protein aggregates do not appear to have any abnormally titrating groups, there are one or two groups with unexpected pK values in the intact virions [80].

Coat protein from the structurally similar barley stripe mosaic virus (BSMV) aggregates into a variety of stable forms including both disk and helical aggregates [81]. Monomers with altered conformations making them incapable of forming ordered aggregates also may exist in solution.

The common occurrence of disk aggregates of coat proteins from helical viruses [82] and the importance of the disk in TMV assembly suggest that the assembly pathways of many helical plant viruses may be similar to that of TMV.

3. Bromoviruses. The bromoviruses, which include cowpea chlorotic mottle virus (CCMV) and brome mosaic virus (BMV), are small icosahedral plant viruses (for reviews see Bancroft [51] and Lane [83]). The coat proteins of these viruses are arranged on a T = 3 icosahedral surface lattice, the subunits exhibiting hexamer-pentamer clustering. The physical and chemical properties of all the bromoviruses are quite similar.

Under various conditions of pH and ionic strength, the aggregation behavior of CCMV protein is found to be analogous to that of TMV protein in that the aggregation state is essentially independent of pH except for one region, in this case at about pH 5.5, where pH is the dominant factor [84]. At alkaline pH, 3S proteins, mainly dimers [85], are the dominant species. These aggregate to form tubes or sheets at low ionic strengths. At low pH, empty capsids closely resembling intact virions are found. Empty BMV capsids have radii larger than the intact BMV particles but smaller than the swollen form of BMV [86]. Again, at low pH, low ionic strength leads to formation of larger, more complex aggregates. No intermediates between the small 3S aggregates and the 50S empty capsids were found [84]. At conditions under which the CCMV virus particle has been reassembled in vitro empty capsids are not generally found [87, 88].

The effect of protein concentration on the equilibrium between the 3S and 50S aggregates of CCMV protein shows that the capsid formation behaves much like a crystallization process [84]. Beyond a certain total protein concentration, the concentration of the small 3S aggregates remains at a "critical" concentration and all further protein goes into 50S capsids. This "critical" concentration is strongly dependent on pH between pH 5 and 6.

Both intact BMV particles [52, 89] and CCMV particles [90] have abnormally titrating groups with pH values near neutrality. The loss of two protons per protein subunit is associated with a reversible swelling of the virus particles at about pH 6.5 for BMV [51, 52] and pH 6.75 for CCMV [90, 91]. The titration behavior of the protein aggregates of CCMV depends on the state of aggregation of the protein [90]. On the basis of titration data, Jacrot (90) suggested that there were two carboxyl groups per subunit with raised pK values in the 50S capsid that were not observed in the 3S aggregates. These data may also be interpreted as indicating that several carboxyl pK values are raised slightly by capsid formation [52]. In either case, the effect of proton binding on the electrostatic interactions between subunits is likely to be critical in determining the aggregation state.

4. Other icosahedral plant viruses. The icosahedral plant viruses can be broken down into two classes on the basis of whether or not the coat proteins are capable of forming complete capsids in the absence of nucleic acid. Kaper [92, 93] has used this as an indication of the relative importance of protein-protein and protein-nucleic acid interactions in the stability of the virus particle. Turnip yellow mosaic virus (TYMV) is the paradigm of those viruses in which protein-protein interactions dominate. It is one of the physically stablest viruses and empty shells are usually associated with natural preparations of TYMV. Cucumber mosaic virus (CMV) is typical of viruses with strong RNA-protein interactions and coat proteins that do not form empty capsids.

Usually, only small aggregates (dimers and/or trimers) and complete shells are observed in studies of plant virus coat proteins under mild conditions. Other, intermediate-sized aggregates can be induced to form by variations in solvent. For instance, in addition to capsids and small aggregates, eggplant mosaic virus coat protein has been observed, at pH 4, to form 12–14S threads that appear to be linear aggregates of protein in negative stain [94]. CMV coat protein precipitates at low ionic strength but no organized aggregates have been identified [51]. Kaper's suggestion [92] that protein-protein interactions are important to TYMV stability and that protein-RNA interactions are important to the stability of CMV particles is supported by small-angle neutron diffraction [95], which indicates that there is little or no interpenetration of RNA and protein in TYMV (in contrast to previous results [96]), but that an interpenetration of about 20 Å occurs in CMV particles.

5. Icosahedral RNA phages. The RNA phages, including R17, MS2, M12, f2, fr, and Qβ, have particles in which 180 copies of the principal coat protein are arranged on a T = 3 icosahedral surface lattice (for reviews, see Hohn and Hohn [34] and Knolle and Hohn [36]) exhibiting hexamer-pentamer clustering [97]. The first four of the phages – R17, MS2, M12, and f2 – have coat proteins with amino acid sequences differing by at most one amino acid [98, 99] and will be considered here as a group. The virus particles contain one copy of a second structural protein [100, 101] that functions both in assembly [102, 103] and phage attachment to host [103, 104].

Empty shells of coat protein can be formed from the dissociated protein in vitro [105], and at neutral pH these exist in equilibrium with low-molecular-weight aggregates [106]. At low ionic strengths larger, more complex aggregates such as multishells are formed. Aggregates with molecular weights intermediate between the low-molecular-weight aggregates (monomers?) and empty shells are observed in buffers containing guanidine hydrochloride [107]. The empty shells formed by reaggregation of dissociated protein sediment faster than capsids formed by alkaline degradation of intact phage [108] and probably contain 10–20 extra copies of coat protein that are trapped in the interior of the particle during reassociation [109]. The stability of the empty shells is comparable to that of the intact virions [106], although RNA will induce assembly under conditions where empty shells will not form. During infection, the RNA is transferred into the host cell, leaving the empty protein shell loosely attached to the sides of the F-pilus [110]. Disassembly of the capsid does not occur during infection. Thus, the requirements for capsid stability are likely to be different than in the case of plant viruses.

B. Assembly of Coat Protein With Nucleic Acid

In this section, the steps involved in the assembly of intact virus particles through the interaction of viral coat proteins and nucleic acids are described for several specific viruses. Even for the few simple viruses considered here, it should be clear that the assembly pathways taken by different viruses are very different indeed. These differences are dictated both by the virus particle structure and its interactions with the host cell.

1. Specificity of protein-nucleic acid interactions. Viral nucleic acids promote the formation of intact, infectious virus particles under conditions where the coat proteins do not form empty virus-like particles. Virus-like particles containing foreign nucleic acids can also be formed under many conditions. Any combination of coat protein and nucleic acid from CCMV, BMV, or BBMV was found to assemble into virus-like particles that were shown to be infective, the host range being specified by the identity of the RNA [88]. Furthermore, RNA from TMV, f2, or ribosomes could also be used to nucleate particle formation. Fragments of RNA as small as 23 residues were found to nucleate the formation of spherical particles, although the smallest polynucleotides resulted in the formation of small, T = 1 particles. Hohn [108] found similar results for bacteriophage fr, for which he showed that particles could be formed using TYMV RNA, poly U, or even polyvinyl sulfate.

Fraenkel-Conrat and Singer [111] showed that helical nucleoprotein particles resulted from mixing TMV, 4S A-protein aggregates with either poly A or poly I. Reassembly of TMV proceeds much faster using disk aggregates [10] which interact with TMV RNA to form nucleoprotein rods at least two orders of magnitude faster than with any synthetic polynucleotide. When mixed with fragments of TMV RNA, the TMV disk is able to discriminate between the fragments and bind to only a few [112]. When disks are mixed with fragments of TMV RNA longer than 800 nucleotides, the interaction is suggestive of a unique nucleation site [113]. This nucleation site on TMV RNA has been identified and sequenced [114–116].

The existence of a unique protein-binding site on viral nucleic acid which could initiate assembly would seem to be advantageous; it could provide additional control of assembly, perhaps by specifying the order in which the nucleic acid is packaged; and it could minimize indiscriminant coating of host cell nucleic acid by the coat protein. It is clear that unique nucleic acid sites for the initiation of assembly exist on some viral genomes. However, the balance of evidence at present is that most coat proteins bind to a wide variety of nucleic acids.

2. Tobacco mosaic virus. The initial interaction between TMV protein disks and TMV RNA probably involves two disks [117] and a unique initiation site on the RNA [114]. When small amounts of TMV disk are mixed with TMV RNA, a unique region of the RNA becomes protected from nuclease digestion [114]. The protected RNA fragments are 50–500 nucleotides long and combine with coat protein into nucleoprotein rods with the same protein-RNA ratio as intact virus. Smaller TMV protein aggregates (A-protein) are ineffective in forming nuclease-resistant complexes. When one disk equivalent of protein for each RNA molecule is used in these experiments, virtually all the protein is used to produce rods of a length greater than three disk lengths, while only 25% of the RNA molecules are coated at all. This suggests that the initial addition of protein is a highly cooperative process. After the initial interaction of RNA and protein, the addition of several more disks to form stable nucleoprotein rods could be very rapid [114].

A sequence of about 100 RNA bases is common to all the nucleotide fragments protected through the limited interaction of TMV disks and TMV RNA [115]. These 100 bases are located about 800 residues from the 3′ end of the RNA. Additional bases in the longer fragments extended about 30 bases toward the 3′ end of 400 bases toward the 5′ end, establishing the primary direction of elongation of the nucleoprotein rods as toward the 5′ end. The nucleotide sequence of the first 100 nucleotides coated has been determined [115, 116]. This fragment contains an extended region with an approximate repeat every three bases:

GAA GAA GUU GUU GAU GAG UUC AUG GAA GAU

which correlates with the ratio of three nucleotides per coat protein in the virus. Perhaps equally important is the secondary structure of the RNA. The periodic sequence of bases given above is likely to form a small hairpin loop of the proper size to fit into the center of the TMV disk.

The structural transformation that the interaction of RNA causes within the disk is remarkable. The two layers of protein in the disk are tilted away from each other, forming a cylindrical cleft facing the disk axis [25]. Also within the central cavity are the 25 amino acid residues per protein that are disordered [5]. The interaction of the RNA with the TMV protein disk (or disks) ultimately results in its transformation to a helical form. During this transformation, the side-to-side interactions between two pairs of protein subunits are broken and another one formed; the tilt of the subunits relative to the axis is shifted by 10° and 20°, respectively, in the two layers [5]; the interactions between subunits in different layers is reorganized, breaking up the extensive salt-bridge systems observed in the disk and causing the formation of another set of specific interactions in the helix.

The formation of short, presumably helical nucleoprotein rods by the interaction of TMV protein disk with TMV RNA acts as a catalyst for further elongation which, after the initial binding, is a relatively slow process requiring perhaps 6 sec per disk [10]. Electron microscopy of intermediates in the elongation process [118, 119] has shown that during elongation both ends of the RNA extend from the same end of the growing nucleoprotein particle. As elongation proceeds, the 5′ end of the RNA is pulled up and through the central channel of the growing rod. Coating of the 3′ end is much slower and may not occur until the 5′ end is entirely coated. It may not be irrelevant that the last section of RNA that is coated codes for the TMV coat protein. It is possible that coat proteins are continually synthesized in vivo during the encapsulation of the remainder of the RNA.

3. Icosahedral plant viruses. Study of the assembly of spherical plant viruses has proven more difficult than the study of TMV assembly. No obvious candidates for intermediates in the assembly process have been identified, and clearly nonphysiologic conditions have usually been required for reassembly in vitro (for reviews, see Bancroft [51] and Kaper [93]).

Recently, CCMV has been reassembled under mild conditions [85]. The assembly process involved the mixing of dimers (3S) of CCMV coat protein with CCMV RNA. The pH of the solution containing the 3S protein was dropped from about pH 7.5 to 6.0 during the mixing. Potentiometric titration experiments [90] have demonstrated hysteresis through the pH region where 3S aggregates form empty capsids. By rapid reduction of the pH, a metastable form of the 3S aggregate is obtained that combines with CCMV RNA to give a high yield of infectious particles. The equilibrium form of 3S aggregate does not combine with the RNA under mild conditions.

No obvious specificity was revealed in these experiments [85]. A number of other plant virus RNAs and synthetic polynucleotides were coated with efficiency similar to CCMV RNA.

The assembly of CCMV contrasts strikingly with that of TMV. In CCMV, the protein dimer is apparently the active unit in assembly. The requirement of a rapid pH drop during the assembly leaves open the question of how the metastable protein aggregates differ from the equilibrium 3S aggregates. The lack of specificity in the assembly process is also in contrast to that found in TMV. Local factors in the plant cell may ensure that the CCMV protein subunits encapsulate mainly the proper RNA molecule [85].

4. Icosahedral RNA phages. The assembly of icosahedral RNA phages has some similarities to that of the icosahedral plant viruses. Both processes are relatively insensitive to the origin of the RNA involved, and no well-defined protein intermediates have been identified. Unlike the complex DNA phages, where DNA is packaged into a preformed precursor particle [2], the assembly of RNA phages is apparently nucleated by the RNA [36]. The presence of RNA induces the formation of spherical, virus-like particles under conditions in which the coat protein normally does not form empty shells. Infectious particles also contain one copy of a second structural protein, the A-protein. Not to be confused with the TMV A-protein, which is the common name given to the equilibrium mixture of small aggregates of coat proteins sedimenting at 4S, this protein is chemically distinct from the major coat protein.

Reassembly of infectious particles of R17 in vitro has required the RNA coat protein and A-protein to be mixed under denaturing conditions and renatured from 5.7 M urea [120]. Particles of Qβ have been reassembled by denaturation from 8 M guanidine hydrochloride [121]. These conditions of reassembly are clearly nonphysiologic, leaving uncertain the relationship of these assembly processes to assembly in vivo.

Initiation of assembly may take place by the formation of an intermediate complex consisting of the RNA molecule and a few copies of coat protein [108]. The A-protein probably also acts early in assembly [102]. No infectious particles were formed by adding A-protein to preformed particles that lacked A-protein.

5. Filamentous bacteriophages. Assembly of filamentous bacteriophages may vary depending on the virus strain, the assembly of fd being the most extensively studied (for reviews, see Marvin and Hohn [17] and Marvin and Wachtel [31]). Prior to assembly, the fd coat proteins are inserted into the cell membrane with their amino terminal end extending outside the membrane and the carboxyl end of the cytoplasmic side [122]. After insertion into the membrane, a leader sequence of 23 amino acids is cleaved from the amino terminal end. The coat proteins may be present as dimers in the cell membrane [123]. While still in the cytoplasm, the viral DNA is coated by gene V protein to form a long, rod-shaped nucleoprotein complex. Formation of the complete virus particle occurs by the extrusion of the DNA through the cell membrane during which the gene V protein is displaced and the coat protein binds to the nucleic acid. The extrusion of DNA through the cell membrane occurs without the lysis of the cell. During this process, there may be significant reorganization of the secondary structure of the coat protein [122]. The basic groups near the carboxyl end of the coat protein are closely associated with the DNA in the virus particle [31, 32]. These groups are in the region of the coat protein, which extends into the cell cytoplasm when the protein is in the cell membrane and are probably involved in the binding of protein to DNA.

C. Stability and Disassembly of Virus Particles

Virus particles are designed to be stable under a wide variety of conditions outside the host cell. However, on encountering the host cell, they must be able to deliver their nucleic acid intact into the host cell.

Bacteriophage capsids do not actually enter the host cell. Most strains of both the isometric RNA phages [110, 124] and the filamentous phages [17] bind specifically to the bacterial sex pili. The isometric phages bind to the sides of the pili; the filamentous phages to the ends. After binding, the viral nucleic acid enters the sex pilus and travels down the pilus into the cytoplasm, leaving the empty capsid loosely bound to the cell exterior.

Many nonenveloped plant and animal viruses apparently enter the host cell intact, some in pinocytotic vesicles. The mechanism by which the usually very stable virus particles are disassembled within the host cell is not clear.

1. Tobacco mosaic virus. The release of RNA from TMV particles in vitro requires nonphysiologic conditions such as the use of detergents, dimethyl sulfoxide, urea, alkaline pH, or high temperatures [125, 126]. One possible pathway for virus particle disruption in vivo would involve the interaction of the particle with the cell membrane or other lipid-containing structure which might act to disrupt bonds between subunits in a way similar to nonpolar solvents or detergents [15]. Another possible factor is the very low concentration of free Ca^{++} in plant cell cytoplasm [68].

Intact TMV particles have three groups per protein subunit that titrate with pKs near neutrality [67, 68]. Each of these groups has significant metal-ion binding. The transition from extracellular to intracellular Ca^{++} concentrations would cause a significant change in the charge on the TMV particle [68] which may in part explain the observed stabilizing effect of Ca^{++} on the TMV particle [125]. The low level of Ca^{++} within the host cell may contribute to the disassembly of the virus particle.

2. Swelling of spherical viruses. The stability of spherical viruses may be altered by changes in pH or divalent cation concentration. Several spherical plant virus particles undergo a structural transition when divalent cations are removed and pH is raised above about pH 7 [51]. This swelling usually renders the virus susceptible to attack by proteases and nucleases and lowers its infectivity. In most cases, the transition is fully reversible.

Near pH 6.75, the sedimentation constant of particles of CCMV decreases from 83S to 73S as pH is increased [42, 127, 128]. This swelling occurs with the release of about two protons per protein subunit [90]. Over the pH range of the transition, no significant changes in RNA secondary structure are evident [91]. Raising the temperature results in a gradual but substantial reduction in virus particle sedimentation constant [42]. Magnesium ions reduce the swelling and completely inhibit the temperature-dependent part [90]. A well-defined hysteresis in the titration curve is observed over the range of the structural transition [90]. This hysteresis is removed by the presence of Mg^{++} – perhaps, as suggested by Durham [52], because the Mg^{++} may be binding to the same sites as the protons involved in the structural transition, which increases the stability of the compact form of the virion.

A similar, but not entirely reversible swelling occurs with BMV [51]. BMV particles swollen in the presence of EDTA by an increase in pH from 5.8 to 7.2 do not fully recontract when the pH is returned to 5.8 [129]. In the presence of $MgCl_2$, fresh virus has a slightly larger radius than with EDTA. Small-angle neutron diffraction indicates that at pH 5.5, although the overall size of BMV particles remains constant, the radial extent of the RNA changes under the compacting action of high ionic strength or spermine [86]. At neutral pH, both the overall size of the BMV particle and the extension of the RNA vary and are strongly dependent on environmental conditions. Swelling is maximal in the absence of divalent cations. Reversible swelling occurs only when they are present. The thickness of the BMV protein coat does not change on swelling. However, lack of reversibility extends to both RNA and overall size of the particle and is accompanied by a small increase in the capsid thickness.

At pH 7.5, removal of divalent cations from SBMV results in a swelling [130] which amounts to an increase in radius of about 9% (Caspar, personal communication). This transition causes a substantial disordering of the protein subunits of the icosahedral shell, and preliminary neutron magnetic resonance (NMR) experiments indicate that the flexibility of at least part of the coat protein is increased by swelling.

Treatment of polyoma virus particles with EGTA and dithiothreitol at pH 7.5 – 10.5 resulted in the dissociation of the virus particles [131]. Removal of bound Ca^{++} by the EGTA was required for this dissociation. After breakdown of the virus particle, most of the coat protein was found present as capsomeres that sedimented at 12S.

Possible interpretations of the response of spherical viruses to pH and divalent cations include 1) the effect of these variables on protein-protein interactions, perhaps through the presence of carboxyl-carboxylate pairs which are ionized near neutrality and repel each other at high pH values [51, 127]; and 2) effects on the interaction of protein with nucleic acid [67, 86, 90].

TBSV undergoes an EDTA-induced swelling at pH 7.5 [6]. Examination of the high-resolution electron density map of TBSV indicates that on one of the conserved interfaces an ion, perhaps calcium, is coordinated by side chains from two adjacent protein subunits. This is the most direct evidence available implicating variations in the interactions between protein subunits in the swelling of icosahedral viruses.

V. DISCUSSION

Virus assembly is a dynamic process in which components that are designed to react with one another come together to form a single structure that is far more stable physically and chemically than any of the components on its own. The driving force in assembly is the formation of bonds – both protein-protein and protein-nucleic acid bonds. In general, coat proteins are capable of several modes of interaction. The protein subunit bonding properties are controlled by their conformations, which can be altered by small changes in their surroundings – for instance, the presence of nucleic acid or a change in pH – to transform the subunit bonding potential and trigger the quick assembly of a virus particle or capsid. The complete virus particle is relatively insensitive to changes in its environment until it encounters a host cell, an action which necessarily results at some point in the spontaneous release of the viral nucleic acid.

The assembly of a virus can be thought of as a sequence of transitions between protein and nucleoprotein aggregates of ever increasing size. Experiments on virus assembly can be roughly divided into two groups: those that study structures, both of intermediates and of the complete virion; and those that study the transitions between stable states of the virus components. Both types of experiment are essential in developing a description of the in vivo pathways for virus assembly.

A. Pathways

The in vitro assembly of viral components to form infectious virus particles, although an important accomplishment, does not necessarily follow the same pathway as virus assembly in vivo. Infectious TMV particles can be assembled from RNA and either A-protein or protein disk [10]. Although both pathways result in the same product, they have significantly different characteristics: assembly of A-protein and RNA is relatively nonspecific for the source of RNA; assembly of protein disk with RNA includes a highly specific RNA recognition step and progresses much faster than assembly with A-protein. The characteristics of TMV assembly from protein disk and RNA seem particularly advantageous. This does not, however, insure that assembly in vivo follows this pathway. A direct demonstration that this is in fact the case may be quite difficult.

A specific nucleic acid recognition step has not been demonstrated for any assembly pathway of a simple icosahedral virus. CCMV can be assembled from a metastable form of coat protein and RNA under mild conditions [85], but this assembly process is not specific for CCMV RNA. It is possible that there are other pathways that are highly specific for viral RNA. At present, however, there is no experimental evidence suggesting conditions under which this might occur.

If specific recognition of viral nucleic acids occurs in the assembly of icosahedral viruses, it may involve a relatively large, perhaps metastable, aggregate of the coat protein. For a region of nucleic acid to act as a unique nucleation site in must be large enough to ensure that there are no similar sites in the remainder of the virus nucleic acid and few, if any, in the host cell nucleic acids. The TMV disk interacts with a minimum of 50 nucleotides so that a relatively long sequence can be used to produce the necessary specific binding region. In the case of icosahedral viruses, dimers or trimers often appear to be the units of action during assembly in vitro. Since these small aggregates are unlikely to interact specifically with more than perhaps a dozen nucleotides, the degree of specificity of action must be much smaller than for TMV disks. If pathways exist in which CCMV proteins react specifically with CCMV RNA, they might be found at lower ionic strengths, where larger aggregates of CCMV protein have been observed [84].

These examples show that during interaction of coat protein with nucleic acid, both the aggregation state and conformation of the coat protein may be very important in determining the properties of the assembly process. The conformation of the nucleic acid may be equally important. Even though several pathways may result in infectious virus particles, there may be characteristics of the in vivo assembly pathway as crucial for the survival of the virus as, for instance, the protease resistance of the intact virus particle. These characteristics may not be common to all assembly pathways. The speed of assembly, specific control, and nucleic acid recognition may all depend on the exact pathway followed.

B. Structures

High-resolution structural information can be obtained only for stable, isolatable states of the viral components. These include assembly intermediates, coat protein aggregates not directly involved in the assembly process, empty capsids, and intact virus particles. Since virus particles are intrinsically stable under a wide variety of conditions, most structural studies have concentrated on these. The results obtained from studies on the TMV disk suggest that the study of assembly intermediates will also be very informative.

One pathway for TMV assembly can be represented as follows: monomer → A-protein → disk + RNA → virus particle. Both the disk and the virus particle are stable under well-defined conditions and high-resolution X-ray diffraction studies have produced a great deal of information about their structures and the differences in the protein subunit conformations in the two structures. Structural work has not been possible on the low-molecular-weight aggregates, and differences in protein conformations in these structures can only be inferred from the larger assemblies. Metastable structures, for instance, the lockwasher aggregate [10], may also occur in the pathway, but the configuration of protein and nucleic acid in these forms, and indeed even their existence, is difficult to assess.

The conformation of virus coat proteins appears to change during most, if not all, the transitions between assembly intermediates. Although some of the intermediates may prove exceedingly difficult to study structurally, an indication of the flexibility and possible modes of interaction of a coat protein may be obtained by studying more easily accessible structures, even though they might not be directly involved in the assembly process. The most common structure of this type is the empty protein capsid. Protein conformation in the capsid may be different from that in the intact virus particle as demonstrated by the fact that the BMV capsid has a radius intermediate in size between that of the compact and swollen forms of the virus particle [86].

Many of the subtleties of the action and interactions of virus coat proteins can be visualized only at atomic resolution. For instance, it is possible to infer the presence of a cation-binding site at a protein-protein interface from the reaction of icosahedral viruses to changes in pH and Ca^{++} concentrations [51]. However, a high-resolution electron density map is required to actually observe the ion-binding site [6]. Similarly, the disorder in the RNA-binding site of the TMV protein disk became apparent only in the 5-Å resolution electron density map [25]. It is reasonable to assume that many of the effects that express themselves during virus assembly involve small changes in subunit conformation: The stability of protein molecules can be strongly affected by single amino acid substitutions that have virtually no effect on the protein secondary structure [132, 133]; the ionization of carboxyl-carboxylate pairs has a marked effect on the stability of virus particles and/or the conformation of protein subunits.

Although there are considerable problems involved in determining the sequence of structural changes occurring in proteins during virus assembly, we know far more about this process than about the role of the viral nucleic acid during assembly. There is some indication that the secondary structure of TMV RNA is important during the initial interaction between protein and RNA [114, 115], and a high-resolution image of the RNA within the virus particle has been produced [4]. The existence of some icosahedrally ordered RNA in SBMV [7] may provide information about the configuration of RNA in that virus. However, on the whole little is known about the role of the nucleic acid during assembly. Any influence which the structure of the nucleic acid might have on the control or specificity of assembly has not been explored.

C. Structural Transitions

During a structural transition between two intermediate states in the assembly process there may be movement of the viral components relative to one another and, perhaps, changes in the conformation of one or more of the components involved. A transition is characterized by descriptions of the initial and final states, by the equilibrium distribution between the initial and final states, and the reaction kinetics and dynamics. The equilibrium is determined by the changes in free energy and entropy during the transition. Both the kinetics and the equilibrium distribution are strongly dependent on the solution conditions, which can affect the conformation of the protein subunits and alter the energy requirements for the transition.

One of the best-characterized structural transitions between two aggregation states of a virus protein is the TMV protein disk-to-helix transition. The equilibrium between protein disk and protein helix has been extensively studied. However, neither the equilibrium behavior nor the kinetics of the transition are yet fully characterized, in part because of the existence of metastable forms and the difficulties involved in obtaining true equilibrium mixtures. The structural work on these two aggregates will soon provide a detailed accounting of which saltbridges are broken and which are formed, which hydrophobic contacts are altered and which are conserved during the transformation. Even so, the dynamics of the process may remain obscure. It will be a considerable challenge to account completely on a molecular level for the details of this transformation. A computer model-building molecular dynamics approach similar to that used in studies of protein folding [134, 135] would be very informative. A similar approach might also be able to account for and characterize the disorder in the RNA-binding site of the protein disk. The computational requirements for such a calculation on a large molecular aggregate are, at least presently, prohibitive.

The switching of protein subunits between different conformational states is an important method of controlling the assembly process [11, 12, 15]. The requirements for control are expressed in the nature of the structural transitions that occur during assembly. These transitions can be quite complex, involving changes in subunit conformation, changes in nucleic acid conformation, interactions between subunits, and binding of nucleic acid to the protein subunits. A given step in the assembly process requires that the subunits be in a specific conformation. In the assembly of TMV, a negative switch [15] prevents the proteins from forming helix until they interact with the RNA. Similarly, the coat proteins of most simple isometric viruses do not by themselves form icosahedral shells under physiologic conditions. The interaction between coat proteins and nucleic acid triggers structural changes in the subunits and in their bonding properties, effectively catalyzing the formation of the virus particle.

Once fully assembled, the virus particle is stable under a broad range of conditions. In many cases, however, the conditions under which the nucleic acid is released from the virus particle into the host cell cytoplasm during infection cannot be grossly different from the conditions under which the particle was assembled. The release of nucleic acid remains a poorly characterized phenomenon, particularly in the case of the plant viruses, which are the viruses best understood structurally.

D. Viewpoints

Our understanding of the structure of viruses and of macromolecular assemblies in general is based on three sets of concepts: geometry, thermodynamics, and the efficiency of design dictated by evolutionary selection [3]. The use of many copies of a single protein to build a virus particle minimizes the coding requirements for the viral nucleic acid. A stable structure is one in which the components interact to form as many intersubunit bonds as possible and occupy a minimum in the free-energy surface for the system. This requirement generally forces the system to take on a symmetric form in which all the subunits are in equivalent or quasi-equivalent positions relative to one another. The geometry of symmetric structures allows a delineation of all possible arrangements of subunits, and considerations of design efficiency suggest which of these arrangements are most advantageous.

Similar considerations must be made in studying the assembly process. Each stable intermediate in the process will be subject to the same geometric and thermodynamic considerations as complete structures. The evolutionary advantage of a given intermediate cannot be evaluated without consideration of the entire assembly pathway. For instance, the TMV protein disk might be thought to be an unlikely intermediate in the formation of the virus helix until the additional speed and specificity it provides are taken into account. A specific pathway may be advantageous because it provides delicate control mechanisms for the assembly, or because of the speed with which it occurs, or because it provides additional protection from host cell defense mechanisms or specific recognition of viral nucleic acid. Some of these are difficult to evaluate in vitro. Others may not always be advantageous. For instance, it may be that in some cases nonspecific coating of host cell nucleic acids provides necessary disruptions of host cell functions.

The same considerations that are the basis of our understanding of virus structure are crucial for an understanding of virus assembly. The structural studies of TBSV [6] and SBMV [7] have shown that the protein subunits in these viruses are designed explicitly for fitting into the quasi-equivalent positions of the virus surface lattice. It is reasonable to expect that details of the subunit structures are also designed specifically to fulfill the requirements for assembly.

One additional consideration involved in the study of assembly pathways is the existence of more than one pathway for the assembly of many viruses. There may be only one arrangement of viral protein and nucleic acid that is recognizably a virus particle, but there are likely to be several ways by which this arrangement can be formed. Just as coat proteins can be assembled into more than one sort of polymorphic form it is quite possible that a single structure can be constructed by several different pathways. Multiple pathways provide an adaptable way for biologic systems at all levels of organization to adapt.

VI. ACKNOWLEDGMENTS

I would like to thank D. L. D. Caspar, G. Stubbs, and T. Wagenknecht for comments on this paper, and acknowledge financial support from National Institutes of Health Young Investigators Award No. CA24407 and an Alfred P. Sloan Foundation fellowship.

VII. REFERENCES

1. Eiserling FA, Dickson RC: Ann Rev Biochem 41:467, 1972.
2. Casjens S, King J: Ann Rev Biochem 44:555, 1975.
3. Caspar DLD, Klug A: Cold Spring Harbor Symp Quant Biol 27:1, 1962.
4. Stubbs G, Warren S, Holmes KC: Nature 267:216, 1977.
5. Bloomer AC, Champness JN, Bricogne G, Staden R, Klug A: Nature 276:362, 1968.
6. Harrison SC, Olson AJ, Schutt CE, Winkler FK, Bricogne G: Nature 276:368, 1978.
7. Suck D, Rayment I, Johnson JE, Rossmann MG: Virology 85:187, 1978.
8. Watson JD: Biochim Biophys Acta 13:10, 1954.
9. Atkinson DE: "Cellular Energy Metabolism and Its Regulation." New York: Academic, 1977, Chap 1.
10. Butler PJG, Klug A: Nature, New Biol 229:47, 1971.
11. Caspar DLD: "Proceedings of the Third John Innes Symposium," 1976, p 85.
12. Kellenberger E: Phil Trans Roy Soc Lond B276:3, 1976.
13. Crick FHC, Watson JD: Nature 177:473, 1956.
14. Caspar DLD: Nature 177:475, 1956.
15. Caspar DLD: Adv Protein Chem 18:37, 1963.
16. Matthews REF: "Plant Virology." New York: Academic, 1970.
17. Marvin DA, Hohn B: Bacteriol Rev 33:172, 1969.
18. Abou Haidar M: Phil Trans Roy Soc Lond B276:165, 1976.
19. Day LA, Wiseman RL: In Denhardt DT, Dressler D, Ray DS (eds): "The Single-Stranded DNA Phages." Cold Spring Harbor, New York: Cold Spring Harbor Laboratories, 1978, p 605.
20. Wiseman RL, Day LA: J Mol Biol 102:549, 1976.
21. Caspar DLD, Holmes KC: J Mol Biol 46:99, 1969.
22. Franklin RE, Holmes KC: Acta Crystallogr 11:213, 1958.
23. Caspar DLD: Nature 177:928, 1956.

24. Graham J, Butler PJG: Eur J Biochem 83:523, 1978.
25. Champness JN, Bloomer AC, Bricogne G, Butler PJG, Klug A: Nature 259: 20, 1976.
26. Jardetzky O, Kasaka KA, Vogel D, Morris S, Holmes KC: Nature 273:564, 1978.
27. Marvin DA, Wiseman RL, Wachtel EJ: J Mol Biol 82:121, 1974.
28. Marvin DA, Pigram WJ, Wiseman RL, Wachtel EJ, Marvin FJ: J Mol Biol 88:581, 1974.
29. Makowski L, Caspar DLD, Marvin D: J Mol Biol (In press).
30. Makowski L, Caspar DLD: In Denhardt DT, Dressler D, Ray DS (eds): "The Single-Stranded DNA Phages." Cold Spring Harbor, New York: Cold Spring Harbor Laboratories, 1978, p 627.
31. Marvin DA, Wachtel EJ: Nature 253:19, 1975.
32. Marvin DA, Wachtel EJ: Phil Trans Roy Soc Lond B276:81, 1976.
33. Nakashima Y, Wiseman RL, Konigsberg W, Marvin DA: Nature 253:68, 1975.
34. Hohn T, Hohn B: Adv Virus Res 16:43, 1970.
35. Tooze J: "The Molecular Biology of Tumor Viruses." Cold Spring Harbor, New York: Cold Spring Harbor Laboratories, 1973.
36. Knolle P, Hohn T: In Zinder ND (ed): "RNA Phages." Cold Spring Harbor, New York: Cold Spring Harbor Laboratories, 1975, p 147.
37. Finch JT: J Gen Virol 24:359, 1974.
38. Crowther RA, Amos LA: Cold Spring Harbor Symp Quant Biol 36:489, 1971.
39. Kiselev NA, Klug A: J Mol Biol 40:155, 1969.
40. Klug A: In Engstrom A, Strandberg B (eds): "Symmetry and Function of Biological Systems at the Macromolecular Level." Nobel Symposium, No. 11. New York: Wiley Interscience, 1969, p 313.
41. Finch JT, Crawford LV: In Fraenkel-Conrat H, Wagner RR (eds): "Comprehensive Virology." New York: Plenum, vol 5, p 119.
42. Adolph KW: J Gen Virol 28:137, 1975.
43. Earnshaw W, Casjens S, Harrison SC: J Mol Biol 104:387, 1976.
44. Earnshaw W, King J, Harrison SC: Cell 14:559, 1978.
45. Earnshaw W, Harrison SC: Nature 268:598, 1977.
46. Chauvin C, Witz J, Jacrot B: J Mol Biol 124:641, 1978.
47. Caspar DLD, Cohen C: In Engstrom A, Strandberg B (eds): "Symmetry and Function of Biological Systems at the Macromolecular Level." Nobel Symposium, No. 11. New York: Wiley Interscience, 1969, p 393.
48. Chothia C, Janin J: Nature 256:705, 1975.
49. Wachtel EJ, Marvin FJ, Marvin DA: J Mol Biol 107:379, 1976.
50. Marvin DA: In Denhardt DT, Dressler D, Ray DS (eds): "The Single-Stranded DNA Phages." Cold Spring Harbor, New York: Cold Spring Harbor Laboratories, 1978, p 583.
51. Bancroft JB: Adv Virus Res 16:99, 1970.
52. Durham ACH, Hendry DA, von Wechmar MB: Virology 77:524, 1977.
53. Hull R: Adv Virus Res 15:365, 1969.
54. Mellema JE, van den Berg HJN: J Supramol Structure 2:17, 1974.
55. Mellema JE: J Mol Biol 94:643, 1975.
56. Kauzman W: Adv Protein Chem 14:1, 1959.
57. Mandelkow E, Stubbs G, Warren S: Submitted for publication.
58. Weber K, Rosenbusch J, Harrison SC: Virology 41:763, 1970.
59. Ziegler A, Harrison SC, Leberman R: Virology 59:509, 1972.
60. Harrison SC: J Mol Biol 42:457, 1969.
61. Butler PJG, Durham ACH: Adv Protein Res 31:187, 1977.
62. Mandelkow E, Holmes KC, Gallwitz U: J Mol Biol 102:265, 1976.
63. Durham ACH, Finch JT, Klug A: Nature, New Biol 229:37, 1971.
64. Durham ACH, Klug A: Nature, New Biol 229:42, 1971.
65. Durham ACH, Klug A: J Mol Biol 67:315, 1972.
66. Vogel D, Jaenicke R: Eur J Biochem 41:607, 1974.
67. Durham ACH, Vogel D, DeMarcillac GD: Eur J Biochem 79:151, 1977.
68. Durham ACH, Hendry DA: Virology 77:510, 1977.
69. Butler PJG, Durham ACH: J Mol Biol 72:19, 1972.
70. Schuster TM, Scheele RB, Khairallah LH: J Mol Biol 127:461, 1979.
71. Richards KE, Williams RC: Proc Natl Acad Sci USA 69:1121, 1972.
72. Butler PJG Klug A: Proc Natl Acad Sci USA 69:2950, 1972.

73. Vogel D: Biochem Biophys Res Commun 52:335, 1973.
74. Scheele RB, Schuster TM: Biopolymers 13:275, 1974.
75. Adiarte AL, Vogel D, Jaenicke R: Biochem Biophys Res Commun 63:432, 1975.
76. Vogel D, Durham ACH, DeMarcillac GD: Eur J Biochem 79:161, 1977.
77. Mayo MA, DeMarcillac GD: Virology 76:560, 1970.
78. Fritsch C, Witz J, Abou Haidar M: FEBS Lett 29:211, 1973.
79. Morris TJ, Semancik JS: Virology 53:215, 1973.
80. Durham ACH, Abou Haidar M: Virology 77:520, 1977.
81. Atabekov JG, Norikov Vk, Kiselev NA, Kaftanova AS, Egorov AM: Virology 36:620, 1968.
82. Hirth L: In Fraenkel-Conrat H, Wagner RR (eds): "Comprehensive Virology." New York: Plenum, 1975, vol 6, p 39.
83. Lane LC: Adv Virus Res 19:152, 1974.
84. Adolph KW, Butler PJG: J Mol Biol 88:327, 1974.
85. Adolph KW, Butler PJG: J Mol Biol 109:345, 1977.
86. Chauvin C, Pfeiffer P, Witz J, Jacrot B: Virology 88:138, 1978.
87. Bancroft JB, Hiebert E: Virology 32:354, 1967.
88. Hiebert E, Bancroft JB, Bracker CE: Virology 34:492, 1968.
89. Incardona NL, Keasberg P: Biophys J 4:11, 1964.
90. Jacrot B: J Mol Biol 95:433, 1975.
91. Adolph KW: J Gen Virol 28:147, 1975.
92. Kaper JM: Virology 55:299, 1973.
93. Kaper JM: "The Chemical Basis of Virus Dissociation and Reassembly." New York: Elsevier, 1975.
94. Briand JP, Bouley JP, Witz J: Virology 76:664, 1977.
95. Jacrot B, Chauvin C, Witz J: Nature 266:417, 1977.
96. Klug A, Longley W, Leberman R: J Mol Biol 15:315, 1966.
97. Vasquez C. Granboulan N, Franklin RM: J Bacteriol 92:1779, 1966.
98. Wittmann-Liebold B, Wittman HG: Mol Gen Genet 100:358, 1967.
99. Konigsberg W, Maita T, Katze J, Weber K: Nature 227:271, 1970.
100. Steitz JA: J Mol Biol 33:923, 1968.
101. Steitz JA: J Mol Biol 33:937, 1968.
102. Kaerner HC: J Mol Biol 53:515, 1970.
103. Verbraeken E, Fiers W: FEBS Lett 28:89, 1972.
104. Kozak M, Nathans D: Nature New Biol 234:209, 1971.
105. Herrmann R, Schubert D, Rudolph U: Biochem Biophys Res Commun 30:576, 1968.
106. Matthews KS, Cole RD: J Mol Biol 65:1, 1972.
107. O'Callaghan R, Bradley R, Paranchych W: Virology 54:476, 1973.
108. Hohn T: Eur J Biochem 8:552, 1969.
109. Zipper P, Schubert D, Vogt J: Eur J Biochem 36:301, 1973.
110. Crawford EM, Gesteland RF: Virology 22:165, 1964.
111. Fraenkel-Conrat H, Singer B: Virology 23:354, 1964.
112. Guilley H, Jonard G, Richards KE, Hirth L: Phil Trans Roy Soc Lond B276:181, 1976.
113. Tyulkina IG, Nazarova GN, Kaftanova AS, Ledneva RK, Bogdanov AA, Atabekov JG: Virology 63:15, 1975.
114. Zimmern D, Butler PJG: Cell 11:455, 1977.
115. Zimmern D: Cell 11:463, 1977.
116. Jonard G, Richards KE, Guilley H, Hirth L: Cell 11:483, 1977.
117. Butler PJG: J Mol Biol 72:25, 1972.
118. Lebeurier G, Nicolaieff A, Richards KE: Proc Natl Acad Sci USA 74:149, 1977.
119. Butler PJG, Finch JT, Zimmern D: Nature 265:217, 1977.
120. Roberts JW, Steitz JA: Proc Natl Acad Sci USA 58:1416, 1967.
121. Hung PP, Overby LR: Biochemistry 8:820, 1969.
122. Webster RE, Cashman JS: In Denhardt DT, Dressler D, Ray DS (eds): "The Single-Stranded DNA Phages." Cold Spring Harbor, New York; Cold Spring Harbor Laboratories, 1978, p 557.
123. Tanford C, Reynolds JA: Biochim Biophys Acta 457:133, 1976.
124. Edgell MH, Ginoza W: Virology 27:23, 1965.
125. Powell CA: Virology 64:75, 1975.
126. Nicolaieff A, Lebeurier G, Morel M-C, Hirth L: J Gen Virol 26:295, 1975.

127. Bancroft JB, Hills GJ, Markham R: Virology 31:354, 1967.
128. Bancroft JB, Hiebert E, Rees MW, Markham R: Virology 34:224, 1968.
129. Zulauf M: J Mol Biol 114:259, 1977.
130. Hsu CH, Sehgal OP, Pickett EE: Virology 69:587, 1976.
131. Brady JN, Winston VD, Consigli RA: J Virol 23:717, 1977.
132. Perutz MF, Raidt H: Nature 255:256, 1975.
133. Grütter MG, Hawkes RB, Matthews BW: Nature 277:667, 1979.
134. Levitt M, Warshel A: Nature 253:694, 1975.
135. McCammon JA, Gellin BR, Karplus M: Nature 267:585, 1977.

Biological Recognition and Assembly 259–273 (1980)

Synthesis and Assembly of Viral Membrane Proteins

Harvey F. Lodish

Department of Biology, Massachusetts Institute of Technology, Cambridge, Massachusetts

We and others have been utilizing the RNA-containing enveloped viruses vesicular stomatitis virus (VSV) and Sindbis virus to probe several fundamental problems in biogenesis of membranes and, in particular, in the biosynthesis of glycoproteins [1–10]. First, all membrane proteins are asymmetrically oriented with respect to a phospholipid bilayer. How is this asymmetry achieved? Specifically, in the case of glycoproteins which span the phospholipid bilayer, how are they synthesized on and glycosylated by the rough endoplasmic reticulum membrane? How do they achieve a transmembrane disposition? Second, how do membrane proteins move from their site of synthesis to their ultimate location in the cell? In the specific case of glycoproteins found on the cell surface, how and when do they move from their site of synthesis in the rough endoplasmic reticulum? How do plasma membrane proteins achieve their final distribution in the surface membrane?

There are a number of advantages of using viruses to investigate these and other problems of membrane biogenesis. Viruses contain few structural proteins, including usually one (VSV: G protein) or two (Sindbis: E_1 and E_2) integral membrane surface glycoproteins and in some cases an internal peripheral membrane protein (VSV: M protein). In contrast, an average mammalian cell contains several hundred membrane proteins. Further, these viruses do not contain enough genetic information to encode all of the enzymes necessary for their biogenesis and therefore must rely on the host cell for many functions. In general, host cell protein synthesis is inhibited during viral infection, making the study of specific proteins much easier.

It is assumed that the mode of biogenesis of these viral proteins will be similar to the manner in which normal cells produced those surface glycoproteins that have a transmembrane disposition similar to that of the viral glycoproteins, such as the red cell protein glycophorin and the major histocompatability antigen [11, 12].

In this paper we shall summarize our progress in these areas; emphasis will be placed on the use of viral temperature-sensitive mutants to elucidate both processes.

TRANSMEMBRANE BIOSYNTHESIS OF THE VSV G PROTEIN

VSV encodes five polypeptides, all of which are structural components of the virion, and directs the synthesis of five corresponding mRNA species. Messenger RNAs for four of

Received April 19, 1979; accepted April 19, 1979.

them (N, NS, M, and L) are translated on free polyribosomes, and the newly made polypeptides are soluble in the cell cytoplasm. G mRNA, by contrast, is exclusively bound to the endoplasmic reticulum (ER) and at all stages of its maturation G itself is bound to membranes [2]. Immediately after its synthesis G spans the ER, and is thus transmembranous. About 30 amino acids at the very COOH terminus remain exposed to the cytoplasm [3, 13, 14]. The balance of the polypeptide, including the NH_2 terminus and the two asparagine-linked carbohydrate chains, face the lumen of the ER and are protected from extravesicular protease digestion by the permeability barrier of the ER membrane [14, 15].

The G mRNA appears to be bound to the membrane via the nascent G chain, since treatment with puromycin, which causes premature termination of the growing polypeptide, causes release of the G mRNA from the ER. Unlike the case of mRNAs encoding secretory proteins, there apparently is not an ionic linkage between the ribosomes and membranes, since dislodging of the mRNA does not require solutions of high salt [4].

Several recent experiments have established that the growing G chain is extruded across the ER membrane into the lumen, similar to the way in which a nascent secretory protein is processed.

1. G protein synthesized in vitro in the absence of membranes (G_0) contains 16 amino acids at the NH_2 terminus that are absent from the form of G made either in the presence of ER membranes or by the cell (G_1) (Fig. 1) [14]. Most of these residues are highly hydrophobic and resemble in structure and function "signal" peptides found at the NH_2 terminus of presecretory proteins [14, 16].

2. During synchronized in vitro protein synthesis in wheat germ extracts ER membranes must be added to the protein synthesis reaction before the nascent chain is about 80 amino acids in length in order for the G molecule to be subsequently inserted into the ER bilayer and to be glycosylated. Since about 30–40 of these 80 NH_2-terminal residues would, at this key time, still be imbedded in the large ribosome subunit, this result establishes that it is the 40–50 most NH_2-terminal residues, which contain the 16 cleaved amino acids, that are crucial in directing proper interaction of the nascent chain with the membrane [5]. Presumably, if the nascent chain is of greater length when the membranes are added, the NH_2 terminus is folded in such a fashion that it cannot interact with the postulated ER receptors.

BIOGENESIS OF TWO SINDBIS VIRUS GLYCOPROTEINS

Sindbis provides a rather different and very important system with which to study biogenesis of transmembrane glycoproteins. Sindbis, like VSV, is a lipid-enveloped virus, but it contains only three structural proteins. Two are envelope glycoproteins which are integral membrane proteins (E_1 and E_2); one (E_2), like VSV G, spans the lipid membrane with a few amino acids, at the COOH terminus, exposed to the cytoplasmic surface. E_1 is also imbedded in the lipid membrane near the COOH terminus, but may not be transmembrane [6, 17]. The third protein, core (C), is internal to the membrane and, like VSV N, is complexed to the viral RNA genome. In marked contrast to the case of VSV-infected cells, all three proteins are synthesized from one polyadenylated mRNA, 26S, which contains the nucleotide sequences found at the 3′ end of the virion 42S RNA [18, 19]. Both in cell-free systems and in infected cells, a single initiation site is used for the synthesis of all three proteins encoded by 26S RNA [20, 21]. The core protein is synthesized first, followed by the two envelope glycoproteins (E_1 and PE_2, a precursor to E_2). C, E_1, and PE_2 are derived by proteolytic cleavage of the nascent chain. The gene order in Semliki Forest virus, a close

Ribosome binding site, mRNA

met	lys	cys	leu	leu	tyr	leu
1	2	3	4	5	6	7

G_0:

met	lys	cys	leu	leu	tyr	leu	ala	phe	leu	phe	ileu	his	val	asn	cys	lys	phe	___	ileu	val	phe	pro
1	2	3	4	5	6	7	8	9	10	11	12	13	14	15	16	17	18	19	20	21	22	23

G_1:

lys	phe	___	ileu	val	phe	pro
1	2	3	4	5	6	7

G_1':

lys	phe	___	___	___	phe	___
1	2	3	4	5	6	7

G_2:

lys	phe	trp	ileu	val	phe	pro
1	2	3	4	5	6	7

Fig. 1. Amino-terminal sequences of different forms of the VSV glycoprotein (from Ref. 14). VSV G_0 is the form synthesized in a wheat germ cell-free system in the absence of membranes; its amino-terminal sequence is identical to that determined from the sequence of the ribosome binding site of G mRNA [33]. G_1 is the form synthesized in the wheat germ cell-free system in the presence of endoplasmic reticulum vesicles; G_1' is the product of digestion of G_1, contained in ER vesicles, with extravesicular protease [3, 13]. G_2 is the virion form of G with the finished carbohydrate chains.

relative of Sindbis, is core–PE_2–E_1 [22]. Thus one RNA encodes two very different types of proteins: a soluble cytoplasmic protein (C) and two integral membrane proteins.

During infection the 26S RNA is found mainly in membrane-bound polysomes which synthesize all three virion proteins [6]. Attachment of these polysomes to membranes, like those synthesizing VSV G, is mediated by forces similar to those which bind polysomes synthesizing secretory proteins to membranes, since treatment with puromycin will dislodge the polysomes from the membrane vesicles. Vesicles containing Sindbis 26S RNA in polysomes will direct cell-free synthesis of all three Sindbis structural proteins, C_1, PE_2, and E_1. All of the newly made C protein is on the outside (cytoplasmic side) of the vesicles, while E_1 and PE_2 are sequestered in the vesicles. About 30 amino acids of E_2 remain exposed to the cytoplasm, as is the case with synthesis of VSV G [6, 10].

A model [6] explaining these and other results on the translation of 26S RNA is shown in Figure 2. A ribosome begins translating the core protein at the 5′ end of a free 26S mRNA. As soon as the core protein is finished, it is removed by a protease, thus exposing the amino terminus of the nascent envelope protein, PE_2. The amino terminus of the PE_2 protein initiates an interaction with the membrane, and is subsequently transferred into and across the membrane, presumably through a protein channel. As in the case of VSV G protein, this interaction leads to the binding of the polysome to the endoplasmic reticulum. As the ribosome continues to transverse the mRNA, completed PE_2 is cleaved and remains imbedded in or sequestered by the membrane. E_1 is then translated and inserted into the membrane, again attaching the polysome to the membrane (Fig. 2).

Since the same ribosome synthesizes both soluble (C) and membrane (E_1 and PE_2) proteins, it is clear that the specificity of binding of the 26S RNA to membranes in such a way as to transfer only the glycoprotein into the membranes cannot reside in the ribosomes alone. The 60S ribosome subunit may interact directly with membrane proteins, but this

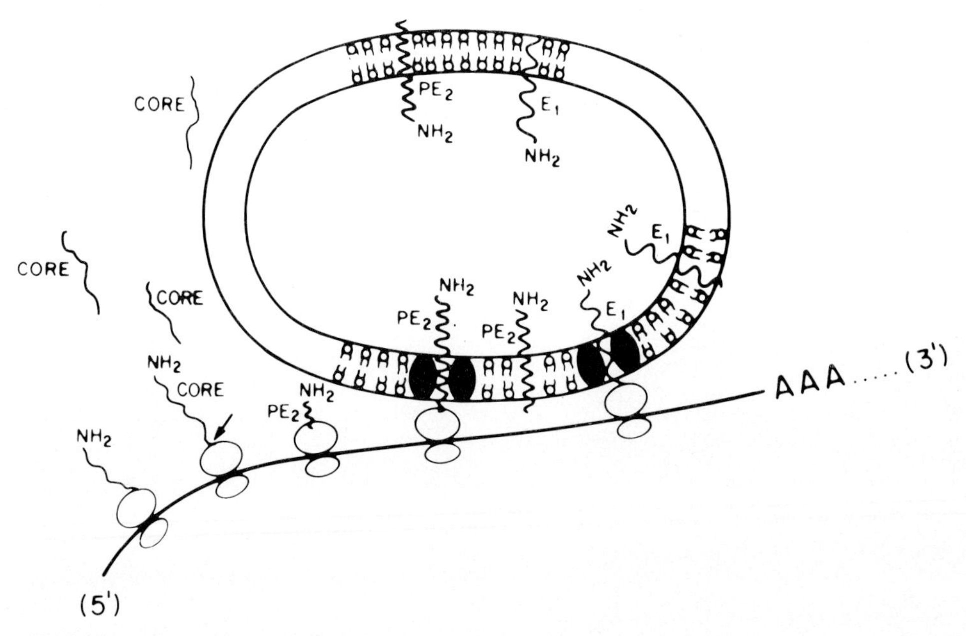

Fig. 2. Model for synthesis of Sindbis virus proteins, subsequent nascent cleavage, and sequestration of the envelope proteins PE_2 and E_1 by the endoplasmic reticulum. From Wirth et al [6], with permission.

is neither the primary interaction nor is it sufficient to result in specific insertion of the membrane proteins. Similarly, binding of the mRNA directly to the membrane cannot alone account for the specificity of this interaction. We conclude that the nascent chain of the two glycoproteins – the only other possibility – determines the specific insertion of the membrane proteins and has a major role in the binding of polysomal 26S RNA to membranes.

INTERACTION OF NASCENT SINDBIS GLYCOPROTEIN AND THE ENDOPLASMIC RETICULUM

Strong support for this model has come recently from two sources. First, Garoff et al [10] have done in vitro translation and ER membrane addition experiments on the 26S RNA similar to those done with VSV G mRNA [5]. In synchronized in vitro translation of 26S mRNA, ER membranes can be added as late as the time when the soluble C protein has just been completed and still allow subsequent normal membrane insertion of the E_1 and E_2 proteins. If membrane addition is delayed beyond this point, however, the E_1 and E_2 protein subsequently made are not inserted into the ER phospholipid bilayer, nor is E_1 cleaved from PE_2. This is consistent with the notion that it is the NH_2 terminus of nascent PE_2 that is crucial in directing interaction of the complex of ribosomes, mRNA, and growing polypeptide to ER membranes. If the NH_2 segment of nascent PE_2 is too long, it cannot interact productively with the ER receptors.

Second, we have investigated the properties of a temperature-sensitive mutant of Sindbis virus, ts2 [23]. This mutant fails to cleave the structural proteins at the nonpermissive temperature, resulting in the production of a polyprotein of 130,000 molecular weight (referred to as the ts2 protein). The order of the proteins in this polypeptide is presumed to reflect the gene order which is NH_2–core–PE_2–E_1–COOH [21, 22, 24]. Therefore, in the ts2 protein the amino terminus sequence of each glycoprotein, E_1 and PE_2, is internal.

Figure 3 shows that the majority of virus-specific RNA in cells infected with ts2 at 41°C, the nonpermissive (Fig. 3A) temperature, or at 30°C, the permissive (Fig. 3B) temperature, was membrane-bound. Here cells were labeled with [^{3}H]uridine, lysed, and fractionated on a linear sucrose gradient, with a dense sucrose cushion. In this system, the membranes migrate to the 55%/40% sucrose interface and the soluble material, including the polysomes, is distributed in the linear gradient. Further studies established that indeed the 26S RNA, representing the majority of the virus-specific RNA, is also bound to membranes [23]. Moreover, as shown in Figure 4, about 85% of the [^{35}S]methionine radioactivity in the ts2 protein is also associated with the membrane fraction. In this study, infected cells were labeled with [^{35}S]methionine and fractionated as before; the labeled polypeptides from membrane and soluble fractions were analyzed by SDS gel electrophoresis (Fig. 4, inset).

These results were surprising to us. As the NH_2 terminus of PE_2 is buried within the ts2 protein, the 26S RNA might not be expected to bind to membranes. Thus, the crucial question was to ascertain the topologic relationship between the ts2 protein and the isolated membrane vesicles. To determine this, membrane vesicles were isolated as described above from cells infected with ts2 at 41°C and pulse-labeled with [^{35}S]methionine for 10 min prior to cell lysis. We have previously shown that in vesicles isolated from the endoplasmic reticulum, the polysomes are found on the outside of the vesicles while the lumen is inside [6]. These isolated vesicles were digested with low levels of chymotrypsin. If the ts2 protein is entirely on the cytoplasmic side of the endoplasmic reticulum, it should be totally

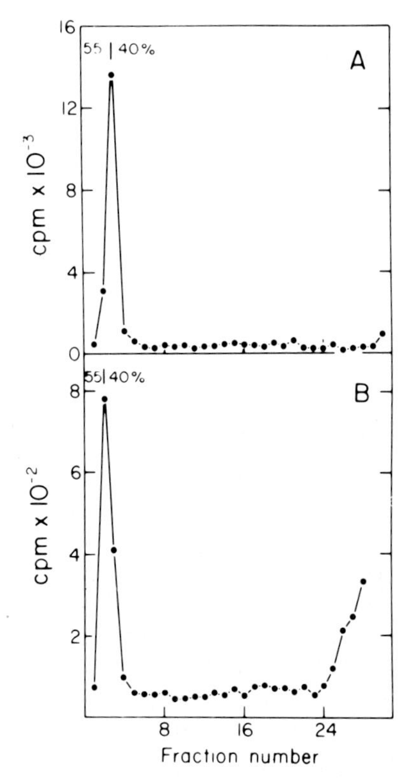

Fig. 3. Subcellular fractionation of Sindbis-specific RNA. Two roller bottles of chick embryo fibroblasts (CEF) (8×10^7 cells per bottle) were infected with Sindbis ts2 virus at a MOI of 10–50 in medium containing 5% calf serum and 1.5 μg/ml of actinomycin D. After adsorption (30 min) and an incubation at 30°C, one of the infected cultures (A) was shifted to 41°C; the other remained at 30°C (B). [^{3}H]uridine (5μ Ci/ml) was added to each culture. The cells were harvested 4.5 h after infection and lysed by Dounce homogenization, and the postnuclear supernatant was layered onto a 15–46% linear sucrose gradient with a 55% sucrose cushion in SS buffer [6] (15–40%/55%). After 4.5 h centrifugation at 26,000 rpm in an SW27 rotor the gradients were fractionated; 1.5-ml fractions were collected and the trichloroacetic acid-insoluble radioactivity was determined for 100-μl aliquots of each fraction. Sedimentation is from right to left.

removed by proteolytic digestion. If the entire protein is inserted through the bilayer into the lumen, then it should be resistant to proteolysis, unless the membrane vesicles are destroyed by pretreatment with detergent. If the protein is partially inserted, then it should be cleaved to a smaller size by proteolysis.

Figure 5 shows that the ts2 protein is absent after proteolytic digestion of the membrane vesicles from pulse-labeled infected cells. No smaller fragments are found after proteolysis that could account for the radioactivity associated with the ts2 protein before proteolysis.

As an internal control for the proteolysis experiment, cells were coinfected with wild-type Sindbis and the ts2 mutant at the nonpermissive temperature. As can be seen in Figure 5, in coinfected cells, the ts2 protein, PE_2, E_1, and core protein are synthesized and localized to microsomes. If membrane vesicles from [^{35}S]methionine pulse-labeled cells are digested with chymotrypsin, some 90% of the radioactivity associated with the ts2 and core proteins is destroyed. By contrast, over 80% of the radioactivity associated with PE_2 and E_1 is protected from proteolysis. PE_2 has an increased electrophoretic mobility, as described previously [6], due to loss of the COOH-terminal segment [17]. If the vesicles are permeabilized with detergent before proteolysis, all of the proteins are digested. On the basis of these results, we conclude that all of each molecule of ts2 protein is located on the cytoplasmic side of the endoplasmic reticulum membrane. In this respect, the ts2 protein resembles the core protein and not PE_2 or E_1 from cells infected with wild-type virus. Although the envelope protein sequences are present, they are not inserted into the membrane and are not protected from proteolysis. This result is consistent with the notion that the proteolytic cleavage between core and PE_2 exposes a sequence, presumably at the

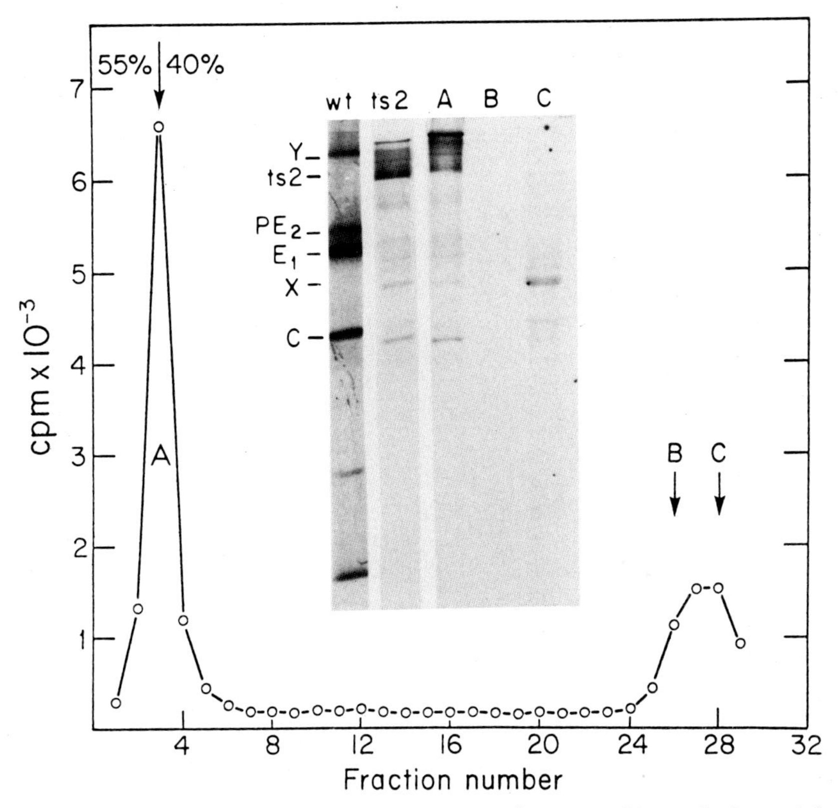

Fig. 4. The ts2 protein is membrane-associated. CEF (5 × 10^6 cells per 100-mm plate) were infected with ts2 virus at a MOI of 10, in minimum essential medium (MEM) containing 5% calf serum and 1 μg/ml actinomycin D. After absorption (30 min) and 1 h incubation at 30°C, the infected monolayers were transferred to 41°C. At 4.5 h after infection, the culture medium was removed and the monolayer washed twice with MEM containing 2% dialyzed calf serum but without methionine. The cultures were incubated for 10 min at 41°C in the above medium containing 20 μCi/ml of [^{35}S]methionine. The cells were harvested and lysed by Dounce homogenization, and the postnuclear supernatant was layered onto a 15–40%/55% sucrose gradient in SS buffer. After centrifugation for 4.5 h at 26,500 rpm in a SW27 rotor at 4°C, the gradient was fractionated; 1.5-ml fractions were collected and the TCA-insoluble radioactivity from 100-μl aliquots was determined. The direction of sedimentation is from right to left. Inset: The proteins in the fractions indicated were concentrated by acetone precipitation and resolved by 10% polyacrylamide gel electrophoresis (PAGE). Shown is an autoradiograph of the dried gel. Exposure time: 7 days. X and Y indicate the positions of host cell proteins. wt, Proteins made in wild-type infected cells after a 10-min pulse of [^{35}S]methionine; ts 2, cytoplasmic extract of cells infected with ts2 and labeled with [^{35}S]methionine for 10 min at 41°C; A–C, proteins contained in the membrane and soluble fractions of the cytoplasmic extract of cells infected with ts2 as described above.

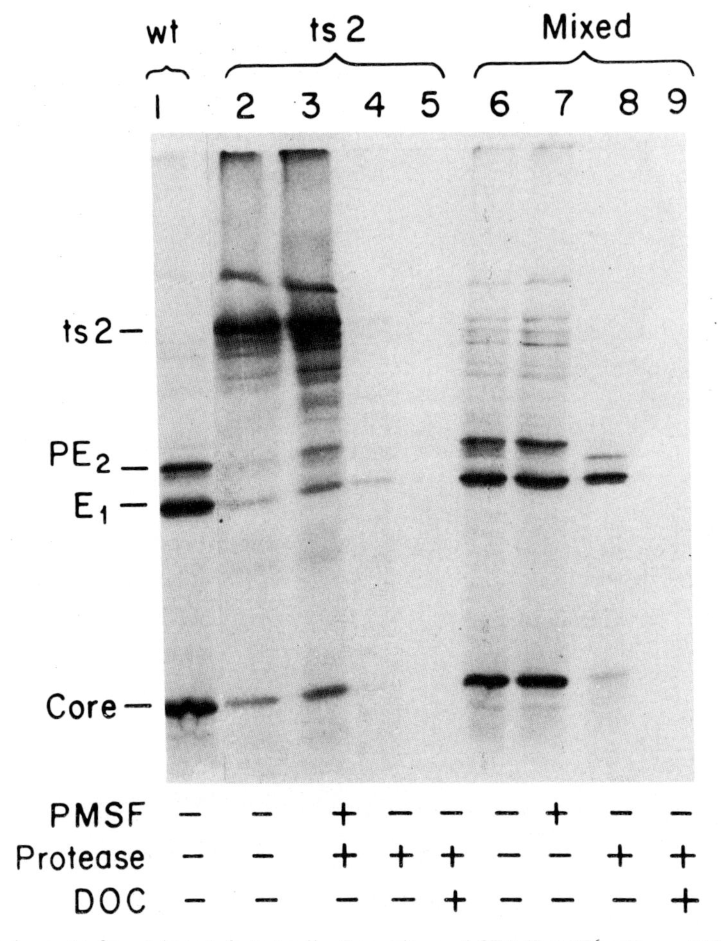

Fig. 5. Topology of ts2 protein in infected cells. One culture of CEF (5×10^6 cells per 100-mm dish) was infected with ts2, as described in Figure 4. A second culture was infected with both ts2 and wild-type at a MOI of 10 for each virus. After absorption (30 min) and 1 h incubation at 30°C the culture was shifted to 41°C. At 4.5 h after infection the culture was pulse-labeled with [^{35}S]methionine (10 min, 50 μCi/ml) and fractionated, as described in the legend to Figure 4. The membrane fraction sedimenting at the 40%/55% sucrose interface was isolated, diluted with phosphate-buffered saline (PBS) buffer, and pelleted for 30 min at 4°C (26,500 rpm in an SW27 rotor). The membrane fractions were resuspended in PBS, and aliquots were incubated as described below. All incubations were for 30 min at 30°C. Chymotrypsin (Worthington, final concentration 10 μg/ml), PMSF (Sigma, final concentration 20 μg/ml) and DOC (Sigma, final concentration 1% w/v) were added to the reaction as indicated by "+". At the end of the incubation period, the reactions were heated for 5 min at 100°C. The proteins were concentrated by acetone precipitation and resolved by electrophoresis in a 10% SDS-polyacrylamide gel. Shown is an autoradiograph of the dried gel.

amino terminus of PE_2, which is essential for proper insertion of PE_2 into the endoplasmic reticulum membrane. At least one other proteolytic cleavage does not occur in ts2-infected cells: that between PE_2 and E_1. Presumably the principal defect of the ts2 mutation is the inhibition of the C-PE_2 cleavage; the cleavage between E_1 and PE_2 probably does not occur, since PE_2 is not properly inserted into the membrane.

The majority of the 26S mRNA is bound to membranes in cells infected with ts2 at the nonpermissive temperature. Presumably this is due to an indirect binding of the mRNA via the nascent ts2 protein, which itself binds to membranes (due to its content of hydrophobic sequences). Thus the binding of mRNA to membranes may in fact be only indirectly related to function. In the case of the ts2 26S mRNA there is no detectable insertion of the protein into membranes, yet the mRNA exhibits characteristics similar to the wild-type 26S RNA, where insertion of the envelope protein does occur. Thus, the presence of mRNA in membrane polysomes may accompany the insertion of a nascent polypeptide into the endoplasmic reticulum, but binding alone is not sufficient to direct this insertion.

MOVEMENT OF GLYCOPROTEINS FROM THE ROUGH ER TO THE PLASMA MEMBRANE

This remains one of the most mysterious, least understood problems in membrane biogenesis. These proteins do not appear on the cell surface until 30–45 min after their synthesis [25]. The polypeptides move first to an intracellular smooth membrane, presumably the Golgi, where glycosylation is completed [2]. This "terminal" glycosylation is, in fact, an extremely complex process, involving first the removal of certain sugars (mannose and glucose) from the oligosaccharide, followed by stepwise addition of the peripheral sugars N-acetylglucosomine, galactose, and sialic acid [9, 26–28]. In what manner and by what force these integral membrane proteins are channeled, first to the Golgi and then to the surface, is obscure.

One approach to elucidating this process is by studying ts viral mutants in the complementation group corresponding to the viral glycoprotein. A number of ts VSV mutants in this complementation group (V) have been isolated. In the ones that have been studied in detail, it is clear that the viral G protein is synthesized in normal amounts at the nonpermissive temperature. The protein receives the normal "core" sugars (N-acetylglucosamine, and mannose), is inserted normally into the rough microsomes, and is metabolically stable. However, it does not move to the cell surface and is not found in budding virus particles [29–32, 34]. Cell fractionation studies on cells infected by one such mutant, ts045, showed that the G polypeptide remained localized in the rough endoplasmic reticulum [30]. Hopefully, sequence analysis of the wild-type and mutant G proteins, and of the G protein from temperature-resistant revertants, will shed light on the cause of the inability of the mutated G to move from the rough ER.

MECHANISM AND SPECIFICITY OF BUDDING OF VSV PARTICLES FROM THE CELL SURFACE

A VSV particle, like those of most lipid-containing animal viruses, is formed by budding from the plasma membrane of an infected cell (reviewed in Refs. 35–38). This complex process is not well understood but must reflect the structural organization of the viral proteins in the infected cell and in the virion. The transmembrane viral glycoprotein is imbedded in the plasma membrane; it becomes, by far, the major protein exposed on the surface of infected cells and on the surface of the virion [2, 39, 40]. The M protein is localized to the inner surface of the virus membrane and may serve as a "bridge" between the G protein and the viral nucleocapsid; this consists of one molecule of viral RNA and the other three virus-encoded proteins, N, NS, and L [35, 37, 38, 41, 42].

The VSV budding process is not totally specific for the VSV G protein, since budding VSV cores can incorporate into virions glycoproteins of other, unrelated viruses such as retroviruses, and possibly also specific cell surface antigens [43–45]. Cellular glycolipids and various enzymes have also been identified in VSV particles (reviewed in Ref. 35).

We have recently accumulated evidence establishing that budding VSV will incorporate a discrete subset of host cell surface proteins into virions. This work has provided, we feel, considerable insight into the mechanism and specificity of formation of a budding virus particle.

In these studies, the proteins exposed on the cell surface were labeled by lacto-peroxidase-catalyzed iodination, and the cells were subsequently infected by wild-type VSV. About 2% of the cell surface [^{125}I]radioactivity was incorporated into particles which had the same sedimentation velocity as authentic VSV particles labeled with [^{35}S]methionine (Fig. 6).

Three lines of evidence demonstrated that the [^{125}I]-labeled material is indeed localized in VSV virions. First, no labeled material of similar sedimentation velocity was released by cells not infected with virus (Fig. 6, top panel). Second [^{125}]-labeled material cosedimenting with VSV also had the same density as authentic VSV particles in both equilibrium density gradients of sucrose (Fig. 7a, b) and potassium tartrate (data not shown). Third, these [^{125}I]-labeled particles were precipitated by antiserum specific for VSV G protein as efficiently as were authentic [^{35}S]methionine-labeled virions (Table I). Preimmune rabbit serum did not precipitate either class of labeled particles. These results show that cellular [^{125}I]-labeled surface proteins are incorporated into particles that are of the same size and density as VSV virions and that also contain VSV G protein.

Vero cells contain at least 10–15 polypeptides that are accessible to iodination with lactoperoxidase and [^{125}I] and that can be resolved by one-dimensional polyacrylamide gel electrophoresis (Fig. 8c). The profile of labeled host surface proteins incorporated into purified VSV virions (Fig. 8d) is clearly different from that of the total pupulation of labeled cell surface polypeptides. Some proteins are incorporated relatively poorly into virions, while others – especially a protein of about 20,000 molecular weight – are incorporated very efficiently. The profiles of [^{125}I]proteins in VSV grown on iodinated Vero cells and L cells (Fig. 8b, d) are broadly similar, but more detailed analysis would be necessary to determine if there are any relationships between members of the two classes of virion polypeptides.

HOST CELL SURFACE PROTEINS IN VSV PARTICLES LACKING VSV G PROTEIN

The following experiment shows that specific incorporation of host cell surface proteins into budding VSV particles does not depend on concomitant incorporation of viral G protein. In cells infected at 39.5°C by VSV mutants in complementation group V, such as tsL513 (V), maturation of G is defective; VSV G is synthesized but does not mature to the cell surface. Budding from these cells are particles lacking G, but containing all other VSV proteins and RNA in normal proportions [31, 32]. These noninfectious particles have a lower than normal density in sucrose gradients. Figure 7d shows that these light-density particles contain [^{125}I] radioactivity derived from the host cell surface. No [^{35}S]methionine-labeled G protein could be detected in this material, but these particles do contain [^{35}S]-labeled N, M, and L proteins (Ref. 48 and unpublished data). Further, these particles, whether labeled with [^{35}S]methionine or [^{125}I], are not precipitated by anti-G serum (Table I). The ratio of [^{125}I] to [^{35}S] radioactivity in these particles (0.040) is only slightly greater than that found in wild-type particles (0.024; Fig. 7a, b) produced in the same experiment. The spectrum of host [^{125}I]-labeled polypeptides in these particles is the same as in wild-type virions (cf Fig. 8).

SPECIFICITY IN RECOGNITION OF SURFACE PROTEINS DURING FORMATION OF A BUDDING VSV PARTICLE

Although it is difficult to calculate exact numbers of host polypeptides in a single VSV virion [46], it is nonetheless clear that certain surface proteins in the infected cells are selectively incorporated into budding VSV particles. VSV G protein is the major iodinatable cell surface protein at late times after infection [2]. As determined by surface labeling with pyridoxal phosphate, it is also by far the predominant surface protein in the virions, although minor species were detected which might be host surface glycoproteins [39]. How this specificity is achieved is not clear. Each surface VSV glycoprotein spike is apparently composed of a single molecule of G [35] but some type of lateral side-by-side aggregation of surface G protein may well occur. The ability of different host surface proteins to interact with G might determine the extent of their incorporation into virions. Alternatively, the extent to which different viral or cellular surface proteins are incorporated

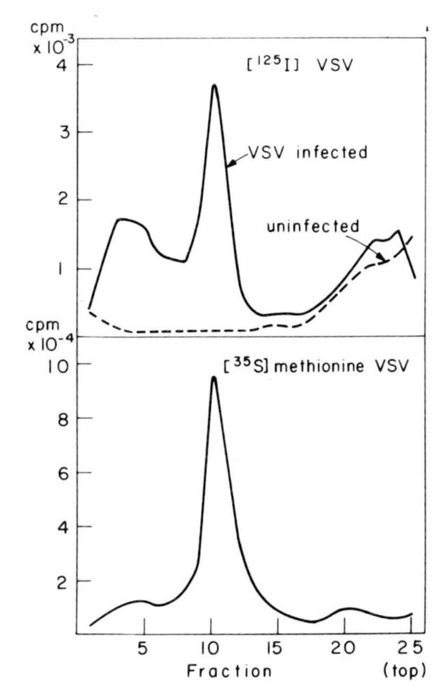

Fig. 6. Purification by velocity gradient centrifugation of VSV containing labeled host surface proteins. Monolayers (70 cm^2) of Vero cells were iodinated with [^{125}I], catalyzed by lactoperoxidase, glucose, and glucose oxidase [2]. They were then infected with 10 plaque-forming units of VSV and incubated for 8 h at 37°C. Control cultures (top panel) were not infected. In parallel, cultures were infected with VSV and labeled with [^{35}S]methionine 4–8 h. The cell medium was filtered through a 0.22 μ Millipore filter; then virus was pelleted through 20% sucrose by centrifugation for 60 min at 135,000g and 4°C using a SW50.1 rotor in a Beckman L5-65 ultracentrifuge. The viral pellets were then resuspended in PBS and layered on a linear 15–40% sucrose gradient, made up in PBS. Centrifugation, right to left, was for 45 min at 40,000 rpm and 4°C in the SW40 Beckman rotor. Fractions were collected from the bottom and aliquots counted directly in a scintillation counter.

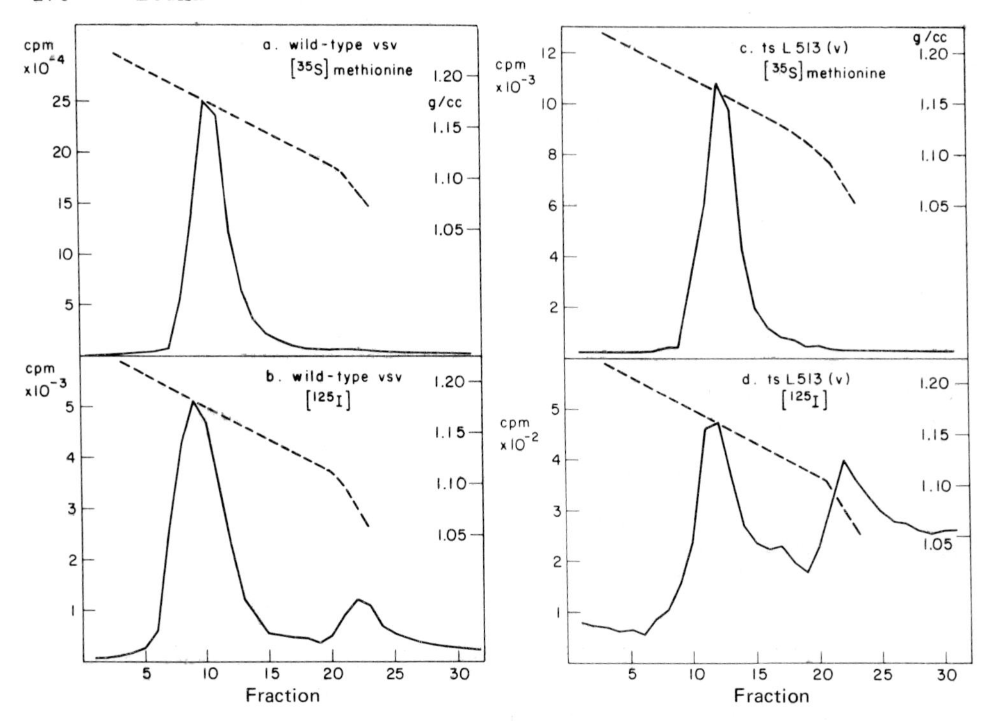

Fig. 7. Equilibrium centrifugation of [^{125}I]-labeled VSV particles and incorporation of cell surface proteins into VSV particles which lack viral G protein. Vero cells were labeled with [^{125}I] and infected at 39.5°C with either wild-type VSV (B) or mutant tsL513 (V) (D). In parallel, cells were infected with wild-type (A) or mutant (C) virus and labeled after infection with [^{35}S]methionine. The particulate matter released into the medium at 8 h was layered directly onto a 14-ml 20–50% sucrose gradient and centrifuged to equilibrium in the Beckman SW40 rotor. Aliquots of the fractions were precipitated with trichloroacetic acid and counted in the scintillation counter. The density of the peak fractions of wild-type particles is 1.177 (A) and 1.180 gm/cm^3 (B). That of mutant particles is 1.162 (C) and 1.160 gm/cm^3 (D). Solid line, radioactivity; dotted line, density.

TABLE I. Reactivity of [^{125}I]- and [^{35}S]methionine-Labeled Viral Particles With Anti-G Serum

		Percentage cpm recovered in immune precipitate	
Virus	Radioactive label	Control serum	Anti-G serum
Wild-type	[^{125}I]	0.13	0.87
	[^{35}S]	0.15	0.93
tsL513	[^{125}I]	0.08	0.13
	[^{35}S]	0.06	0.09

Aliquots of peak fractions from equilibrium gradient analysis of [^{125}I]-labeled VSV particles (grown on iodinated cells) or [^{35}S]methionine-labeled particles (labeled until 8 h of infection at 39.5°C) similar to that of Figure 7 were used for immunoprecipitation. An antiserum specific for the VSV G protein and preimmune rabbit serum were utilized, as detailed previously [46]; protein A was employed to precipitate antigen-antibody complexes. Indicated here is the fraction of added radioactivity recovered in the washed precipitates.

into virions might depend on the specificity of their interaction with the underlying matrix (M) protein. Although it is not possible to decide unequivocally between these alternatives, several types of evidence suggest that the second type of interaction may be the most important.

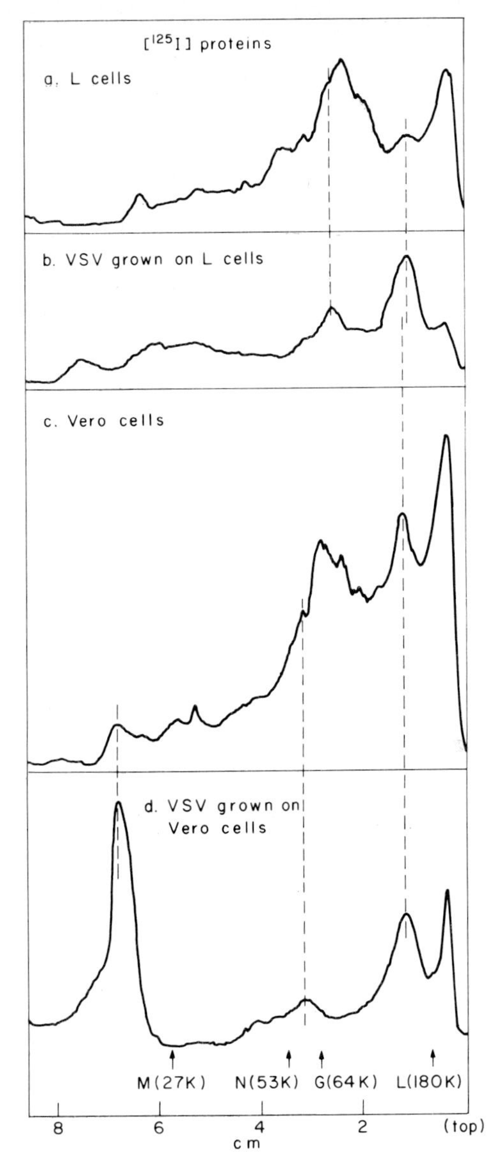

Fig. 8. Polyacrylamide gel analysis of [^{125}I]-labeled cellular and viral proteins. Monolayers of L or Vero cells were labeled with lactoperoxidase and [^{125}I]; proteins were directly extracted from one set of cells and analyzed by SDS gel electrophoresis. In parallel, VSV was grown on replicate cultures of [^{125}I]-labeled cells, and purified by both velocity (Fig. 6) and equilibrium gradient (Fig. 7) centrifugation. A portion of peak fractions representing the purified virions was analyzed in parallel by SDS gel electrophoresis. VSV virion proteins, labeled with [^{35}S]methionine, were also run in parallel slots as molecular weight standards. Shown here are scans (1.29 OD full-scale pen deflection) of radioautographs of the fixed, dried slab gel.

First, cells infected at the nonpermissive temperature by ts mutants in complementation group V release VSV particles that lack G but otherwise appear normal [31, 32]. These particles contain a normal proportion of host cell surface proteins (Fig. 7). Further, the size distribution of [^{125}I] surface proteins incorporated in these particles is the same as is found in infectious, wild-type particles that contain G (cf Fig. 8). Thus the presence of viral G protein on the cell surface is not essential for the selective incorporation of host cell surface proteins into budding VSV particles.

Second, the ratio of viral G protein to M protein in budding particles need not be constant among all released VSV particles (Ref. 46, and unpublished data), nor is the ratio of cell surface [^{125}I]proteins to internal viral proteins constant in all released particles (Fig. 6).

Finally, that cells infected at nonpermissive temperatures by certain ts VSV mutants in complementation group III (the group determining the viral matrix protein) produce VSV particles containing G and M but no nucleocapsid protein or RNA is a fact that suggests that the nucleocapsid is not essential for formation of a budding VSV particle [47]. Taken together with the result that G, too, is not essential for particle formation, this finding implicates M as the key polypeptide in budding of a virus; possibly formation of a submembrane M virus "skeleton" is all that is essential for budding from the plasma membrane.

The reduced fluidity of the lipid bilayer above a VSV M skeleton [48] could select for certain viral and cellular surface glycoproteins. Alternatively, the presence of a cellular or viral surface protein in a budding particle might depend on a specific interaction with the submembrane M skeleton.

ACKNOWLEDGMENTS

This study was supported by grants AI-08814 and AM-15322 from the US National Institutes of Health.

REFERENCES

1. Katz FN, Rothman JE, Knipe DM, Lodish HF: J Supramol Struct 7:353–370, 1977.
2. Knipe DM, Baltimore D, Lodish HF: J Virol 21:1128–1139, 1977.
3. Katz FN, Rothman JE, Lingappa V, Blobel G, Lodish HF: Proc Natl Acad Sci USA 74:3278–3282, 1977.
4. Lodish HF, Froshauer S: J Cell Biol 74:358–364, 1977.
5. Rothman JE, Lodish HF: Nature 269:775–779, 1977.
6. Wirth DF, Katz FN, Small B, Lodish HF: Cell 10:253–263, 1977.
7. Sefton BM: Cell 10:659–668, 1977.
8. Leavitt R, Schlesinger S, Kornfeld S: J Biol Chem 252:9018–9023, 1977.
9. Robbins PW, Hubbard SC, Turco SJ, Wirth DF: Cell 12:893–900, 1977.
10. Garoff H, Simons K, Dobberstein B: J Mol Biol 124:587–600, 1978.
11. Tomito M, Marchesi U: Proc Natl Acad Sci USA 72:2964–2968, 1975.
12. Springer TA, Strominger JL: Proc Natl Acad Sci USA 73:2481–2485, 1976.
13. Katz FN, Lodish HF: J Cell Biol 80:416–426, 1979.
14. Lingappa V, Katz FN, Lodish HF, Blobel G: J Biol Chem 253:8667–8670, 1978.
15. Rothman JE, Katz FN, Lodish HF: Cell 15:1447–1454, 1978.
16. Devillers-Theiry A, Kindt T, Scheele G, Blobel G: Proc Natl Acad Sci USA 72:5016–5020, 1975.
17. Garoff H, Soderland H: J Mol Biol 124:535–549, 1978.
18. Mowshowitz D: J Virol 11:535–543, 1973.
19. Rosemond H, Sreevalson T: J Virol 11:399–415, 1973.
20. Cancedda R, Villa-Komaroff L, Lodish HF, Schlesinger MJ: Cell 6:215–222, 1975.

21. Clegg JCS, Kennedy ST: J Mol Biol 97:401–411, 1975.
22. Lachmi BE, Kaariainen L: Proc Natl Acad Sci USA 73:1936–1940, 1976.
23. Wirth DF, Lodish HF, Robbins PW: (In press).
24. Simmons DT, Strauss JH: J Mol Biol 86:397–409, 1974.
25. Knipe D, Lodish HF, Baltimore D: J Virol 21:1121–1127, 1977.
26. Hunt LA, Etchison JR, Summers DF: Proc Natl Acad Sci USA 75:754–758, 1978.
27. Li E, Tabas I, Kornfeld S: J Biol Chem 253:7762–7770, 1978.
28. Kornfeld S, Li E, Tabas I: J Biol Chem 253:7771–7778, 1978.
29. Lafay F: J Virol 14:1220–1228, 1974.
30. Knipe DM, Baltimore D, Lodish HF: J Virol 21:1149–1158, 1977.
31. Lodish HF, Weiss RA: J Virol (In press).
32. Schnitzer T, Dickson C, Weiss R: J Virol 25:185–195, 1979.
33. Rose JK: Proc Natl Acad Sci USA 74:3672–3676, 1977.
34. Lodish HF, Porter MP, Zilberstein A: (In preparation).
35. Wagner RR: In Fraenkel-Conrat H, Wagner RR (eds): "Comprehensive Virology." New York: Plenum, 1975, vol 4, pp 1–94.
36. Pringle CR: In Fraenkel-Conrat H, Wagner RR (eds): "Comprehensive Virology." New York: Plenum, 1977, vol 9, pp 239–289.
37. Lenard J, Compans RW: Biochim Biophys Acta 344:51–94, 1974.
38. Lenard J: Ann Rev Bioeng 7:139–165, 1978.
39. Eger R, Compans RW, Rifkin DB: Virology 66:610–615, 1975.
40. Atkinson PH, Moyer SA, Summers DF: J Mol Biol 102:613–631, 1976.
41. Wagner RR, Prevec L, Brown F, Summers DF, Sokol F, Macleod R: J Virol 10:1228–1230, 1972.
42. Nakai T, Howatson AF: Virology 35:268–281, 1968.
43. Zavada J, Zavodska E: Intervirology 2:25–32, 1973/1974.
44. Hecht T, Summers DF: J Virol 19:833–845, 1976.
45. Weiss RA, Boettinger D, Murphy HM: Virology 76:808–825, 1977.
46. Lodish HF, Porter MP: Proc Natl Acad Sci (In press).
47. Schnitzer T, Lodish HF: J Virol 29:443–448, 1979.
48. Landsberger FR, Compans RW: Biochemistry 15:2356–2360, 1976.

Biological Recognition and Assembly 275–291 (1980)

Maturation of the Hemagglutinin-Neuraminidase and Fusion Glycoproteins of Two Biologically Distinct Strains of Newcastle Disease Virus

Glenn W. Smith, James C. Schwalbe, and Lawrence E. Hightower

Microbiology Section, Biological Sciences Group, University of Connecticut, Storrs, Connecticut 06268

We have compared the glycoproteins of two biologically distinct virulent strains of Newcastle disease virus. Cells infected by either strain AV or HP produce infectious virions and the cellular surfaces have hemadsorbing activity; however, only cells infected by strain AV undergo fusion from within. We fractionated chicken embryo cells and monitored the incorporation of radioactive proteins into virions to study the sites of synthesis, cellular locations, and kinetics of virion assembly for the hemagglutinin-neuraminidase (HN) and the fusion glycoproteins of each strain. We found that the HN glycoprotein of both strains was synthesized on rough endoplasmic reticulum (ER), accumulated in low-density membranes derived from smooth ER and the plasma membrane, and appeared in virions after a 30-min delay. The fusion glycoprotein of both strains was synthesized as a precursor in rough ER and subsequently processed to the active form. However, the sites of accumulation of the fusion glycoproteins were strain-dependent. The larger subunit F_1 of the fusion glycoprotein of strain AV was detected in subcellular fractions enriched for plasma membranes, while that of strain HP accumulated in denser fractions which contained internal membranes. This result suggests that differential compartmentation of viral glycoproteins may influence the expression of biologic activities such as fusion on cellular surfaces. Despite their different sites of accumulation, the F_1 glycoproteins or F-related polypeptides of similar size were assembled into virions of both strains. The kinetics of incorporation of this protein into virions appear to be too rapid for migration through internal membrane systems. Furthermore, tunicamycin did not block the incorporation of F-related polypeptides into virions. A hypothesis is presented to explain the unusual behavior of the F-related proteins.

Key words: paramyxovirus, Newcastle disease virus, glycoproteins, tunicamycin, virion assembly

Received April 27, 1979; accepted June 8, 1979.

INTRODUCTION

The avian paramyxovirus Newcastle disease virus (NDV) is assembled at the surface of infected cells [1]. The viral envelope which is derived from the host plasma membrane contains two antigenically [2] and physically [3, 4] distinct kinds of external projections. One type is composed of the hemagglutinin-neuraminidase glycopolypeptide HN. This glycoprotein is responsible for the hemagglutinating, receptor-binding (attachment), and neuraminic acid-cleaving properties of virions [2–5]. The other spike contains the two disulfide-linked fusion glycopolypeptides $F_{1,2}$ [6]. The activity of this projection is required for hemolysis, infectivity (penetration), and cellular fusion [2, 7].

The remaining structural proteins, all internal components of the virion, include: the most abundant protein of the nucleocapsid NP, a core-associated large protein L, a 47,000-dalton protein of uncertain location, and a nonglycosylated protein M, which may be both membrane- and core-associated [8–10].

Elegant ultrastructural studies of infected cells have revealed the general features of paramyxoviral maturation [1, 11, 12]; however, the molecular details of assembly of glycoprotein projections, folding of the nucleocapsid, transmembrane interactions between structural proteins, and envelopment of the ribonucleoprotein core are largely unknown. Recently several groups [10, 13] have used bifunctional reagents that cross-link proteins to study the molecular organization of virions, and evidence that HN may be a transmembrane protein was obtained [13]. Our laboratory is studying the dynamics of virion assembly. In this report we present kinetic analyses of the incorporation of polypeptides into virions of strains Australia-Victoria (AV) and Israel-HP, two biologically distinct virulent strains of NDV. As part of our analysis, we evaluated the role of glycosylation in assembly of the viral glycoproteins using tunicamycin (Tm), an antagonist of dolichol pyrophosphate-mediated glycosylation of polypeptides [14–16].

Cellular surfaces which have incorporated newly synthesized paramyxoviral proteins may acquire hemadsorbing and cell-fusing activities. The latter activity is called fusion from within (FFWI) [17] to distinguish it from the fusion activity directly associated with virions, fusion from without (FFWO) [18]. The expression of FFWI and the production of infectious virus are both strain-dependent events in cultured chicken embryo (CE) cells. The fusion glycoprotein is required for viral infectivity [7] and probably for FFWI [19], although direct evidence of the latter role has not been reported. This glycoprotein is initially synthesized as an inactive precursor glycopolypeptide F_0 [20–22]. In CE cells infected by most virulent strains, including strain AV used in the present study, F_0 is cleaved proteolytically to form biologically active $F_{1,2}$. The fusion glycoproteins are incorporated into infectious virions and also accumulate in plasma membranes, where they probably cause FFWI [19]. In contrast, the F_0 glycoprotein of avirulent strains is not cleaved in CE cells. FFWI does not develop in the infected cultures, and progeny virions which incorporate F_0 have only low levels of infectivity and hemolytic activity. Trypsin can cleave the virion-associated F_0 and restore full activity [5, 7].

Bratt and Gallaher [18, 23] have described a virulent strain of NDV, Israel-HP, which has an unusual phenotype. CE cells infected by strain HP produce infectious virions, but the cellular surface does not acquire fusion activity. Here we show that impairment of the proteolytic activation step cannot account for the absence of FFWI because the processed form of the fusion glycoprotein is present in infected cells. Evidence that the fusion glycoproteins of strains AV and HP accumulate at different sites in infected cells is presented. This observation may explain the inability of HP-infected cells to undergo FFWI.

METHODS

Cells and Medium

Primary cultures of chicken embryo cells were prepared from 10-day-old embryos as previously described [24]. To produce secondary cultures confluent primary cultures were trypsinized, resuspended in NCI medium (Gibco) supplemented with 5% calf serum, and incubated at 40°C under an atmosphere of 5% CO_2. Confluent secondary cultures of CE cells were used in all experiments except for the assay of cellular fusion, in which subconfluent cultures were used.

Virus Preparation, Assay, and Infection

Strains AV (Australia-Victoria, 1932) and HP (Israel-HP, 1935) of Newcastle disease virus were grown in embryonated chicken eggs as previously described [25]. The allantoic fluids were removed from infected eggs and immediately purified according to published methods [26, 27]. Purified stocks of virus were stored at −70°C.

Plaque assays were performed by routine procedures. Secondary cultures of CE cells were infected and incubated for two to three days at 40°C in NCI medium supplemented with 5% calf serum and 0.8% agarose.

In all experiments except the assay for cellular fusion, cell cultures were infected at an input multiplicity of 5 plaque-forming units (PFU) per cell and incubated at 40°C for 6 h in Eagle minimal essential medium (MEM) supplemented with 5% calf serum (standard medium).

Isolation and Characterization of Membranes From Infected Cells

The procedures used by Knipe et al [28] to fractionate suspension cultures of Chinese hamster ovary cells were applied with modifications to CE cells growin in 100-mm tissue culture plates. Infected cultures were scraped into a buffer containing 250 mM sucrose, 50 mM KCl, 2 mM $MgCl_2$, 1 mM $CaCl_2$, and 10 mM Tris-HCl, pH 7.4. This and all subsequent steps were carried out at 4°C. The cell suspension was placed in a cell disruption bomb (Parr Instrument Co.), allowed to equilibrate for 5 min under 250 psi of N_2 and disrupted by cavitation. Nuclei and any intact cells remaining in the homogenate were removed by low-speed centrifugation (1,000g for 5 min). To collect membranes, the supernatant was centrifuged at 100,000g for one h in a Spinco type 65 rotor. The membranes were washed once in low-salt buffer (1 mM Tris-HCl, pH 8.0; 1 mM EDTA), resuspended in 55% sucrose (w/v) in low-salt buffer, and included as the bottom layer in a discontinuous sucrose gradient. The gradient was centrifuged for 14 h at 80,000g in a Spinco SW27.1 rotor and fractions were pooled as shown in Figure 3. The pooled gradient fractions were diluted with low-salt buffer and subjected to ultracentrifugation at 100,000g for 30 min, and the resulting membrane pellets were either dissolved in polyacrylamide gel sample buffer (0.0625 M Tris-HCl, pH 6.8, 2.3% sodium dodecyl sulphate (SDS), 5% mercaptoethanol, 10% glycerol, 0.001% bromphenol blue) or resuspended in the appropriate assay buffer for further characterization.

Isolated membrane fractions were characterized by the following methods. Cellular surface membranes were labeled prior to disruption using ^{3}H-wheat germ agglutinin (3.7 Ci/mmole, New England Nuclear Corp.) according to the procedure of Hunt and Summers [29]. Membrane-associated radioactive ribosomes were used as markers for rough endoplasmic reticulum. The ribosomal RNA of marker ribosomes was allowed to incorporate 5,6-^{3}H-uridine (40 Ci/mmole, New England Nuclear Corp.) for 24 h prior to infection and cellular fractionation [28]. Galactosyltransferase activity, a marker for Golgi membranes, was assayed according to Fleischer et al [30]. NADH-cytochrome c reductase activity, assayed by the method of Wallach and Kamat [31], was used to locate endoplasmic reti-reticulum. Protein was measured by a colorimetric protein assay (Bio-Rad) with bovine serum albumin as the standard [32].

Preparation of Radioactive Proteins

For experiments involving the isolation of membranes, infected cultures were exposed to 20 μCi ^{35}S-methionine (500–1,300 Ci/mmole; New England Nuclear Corp., Amersham Corp.) per milliliter of standard medium containing 1% (0.15 mg/liter) of the normal methionine concentration and 2% dialyzed calf serum. After a 2-min incubation at 40°C, the radioactive medium was removed and replaced by either cold phosphate buffered saline without divalent cations (PBS) to stop incorporation of radioactive methionine or standard medium containing ten times (150 mg/liter) the normal methionine concentration of MEM. Following a 30-min chase period, the cultures were washed with cold PBS and prepared for N_2 cavitation as described above.

To monitor the incorporation of radioactive proteins into virions, infected cultures were rinsed twice with prewarmed MEM containing 1% (0.15 mg/liter) of the normal concentration of methionine and supplemented with 2% dialyzed calf serum. The washed cultures were then exposed to 100 μCi of ^{35}S-methionine per milliliter of the same medium. After a 10-min labeling period, the radioactive medium was removed, the cultures were washed twice with prewarmed standard medium containing a tenfold excess (150 mg/liter) of methionine, and then incubated in the same medium. Following various periods of chase, the medium was collected from replicate plates and centrifuged at 10,000g for 10 min to remove cellular debris. Radioactive virions were then collected by differential centrifugation at 100,000g for 1 h in a Spinco SW50.1 rotor. The resulting virus pellets were dissolved either in polyacrylamide gel sample buffer for subsequent analysis by polyacrylamide gel electrophoresis or in Tris buffer (0.0625 M Tris-HCl, pH 6.8, 2% SDS, 5% 2-mercaptoethanol) and then mixed with Aquasol (New England Nuclear Corp.) to prepare the sample for counting in a Packard liquid scintillation spectrometer.

In order to radioactively label glycoproteins, infected cultures were exposed to 50 μCi of ^{3}H-mannose (18.4 Ci/mmole, New England Nuclear Corp.) per milliliter of NCI medium (without glucose) supplemented with 5 mM sodium pyruvate and 2% dialyzed calf serum between three and nine hours after infection. Radioactive virions were collected from the culture medium as described above.

Polyacrylamide Gel Electrophoresis

Radioactive proteins extracted from virions, cells, or isolated membranes were analyzed by electrophoresis in polyacrylamide slab gels using the procedure of Laemmli [33]. Electrophoresis was performed at a constant current of 20 mA per gel. Optimum separation of the polypeptides of strain AV was achieved on either 9% (electrophoresed for 8 h) or 11.5% (electrophoresed for 16 h) polyacrylamide gels; however, better separations were obtained for strain HP with 7% polyacrylamide gels electrophoresed for 7 h. After electrophoresis the gels were fixed, stained with coomassie blue dye, and prepared for fluorography by the methods of Laskey and Mills [34]. The dried gels were exposed to Kodak XR-5 X-Omat R film for two to six days at −70°C.

Assay of Fusion From Within

Subconfluent cultures were infected with an input multiplicity of 0.7 PFU per cell and incubated for 12 h at 40°C in NCI medium supplemented with 5% calf serum under a 5% CO_2 atmosphere (pH 7.5). The cultures were fixed with 95% methanol and stained with Giemsa, and fusion events were quantified by Kohn's modification [17] of the procedure of Okada and Tadokoro [35]. Approximately 200 cells per plate were counted on duplicate plates. The fusion index (average number of nuclei per cell) was calculated by dividing the total number of nuclei by the total number of cells counted. The average number of fusion events per cell was determined by subtracting the fusion index of mock-infected cells from the index for infected cultures [36].

RESULTS

Strain Dependence of Fusion From Within

Bratt and Gallaher [18] observed that AV-infected CE cells fused much more frequently than HP-infected cells when incubated at 43°C in medium at pH 8.3, that is, under optimum conditions for FFWI. We have quantified fusion in cultures infected by these same strains incubated at 40°C in medium at pH 7.4, conditions routinely used in the fractionation protocol and kinetic analyses described herein. Subconfluent cultures of CE cells were either mock-infected (Fig. 1a) or infected at a low multiplicity with strain AV (Fig. 1b) or with strain HP (Fig. 1c). By 12 h after infection, AV-infected cultures contained many large multinucleate syncytia, while HP-infected cultures had fewer areas of fusion, most of which contained binucleate or trinucleate cells. The average number of fusion events per cell in AV-infected cultures was 2.19 compared with 0.08 for cultures infected by strain HP. Thus, the amount of FFWI which occurred in CE cells maintained under our conditions was also strongly strain-dependent.

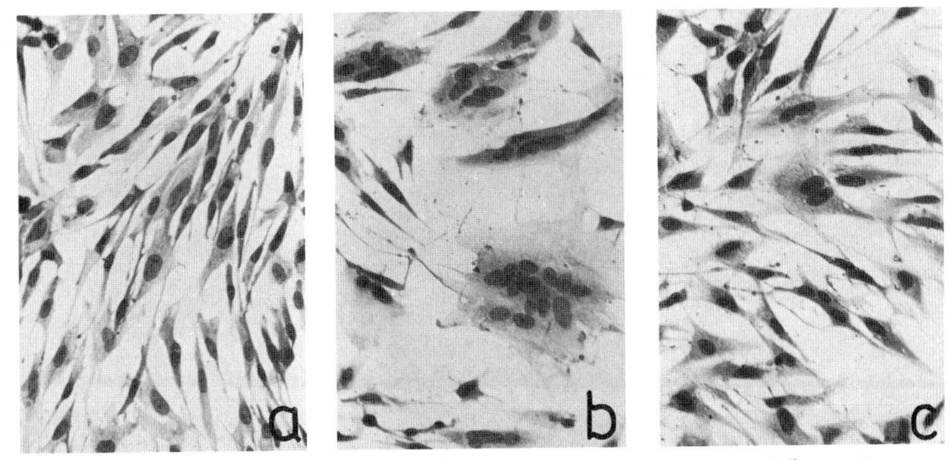

Fig. 1. Photomicrographs of infected and uninfected cultures. At 12 h after infection, CE cell cultures which had been mock-infected (a) or infected with strain AV (b) or strain HP (c) were fixed and stained. Representative microscopic fields are shown. × 250.

Accumulation of the Fusion Glycoprotein in AV- and HP-Infected Cells

If the precursor of the fusion glycoprotein F_0 was either not cleaved or aberrantly processed and rapidly degraded, HP-infected cultures would not be expected to undergo FFWI. To evaluate these possibilities we compared the radioactive proteins extracted from HP-infected cells to those from AV infection, which is permissive for FFWI. All of the major polypeptides of strain AV were separated by electrophoresis in 11.5% polyacrylamide gels in the presence of SDS (Fig. 2a). With the exception of the small F_2 glycopolypeptide which was not retained by these gels, the remaining major viral polypeptides were readily detected. The relatedness of F_0 and F_1 was established previously by tryptic peptide analysis [22].

A similar complement of viral proteins was obtained from HP-infected cells with the exception of F_1, which was not detected on 11.5% gels. However, the F_1 glycopolypeptide of strain HP could be separated from nucleocapsid protein (NP) in a lower percentage polyacrylamide gel (Fig. 2b). The F_1 of strain HP had a lower electrophoretic mobility than the comparable glycoprotein of strain AV; however, it appeared to be the final cleavage product, since the cell-associated form comigrated with F_1 from infectious virions of strain HP (data not shown).

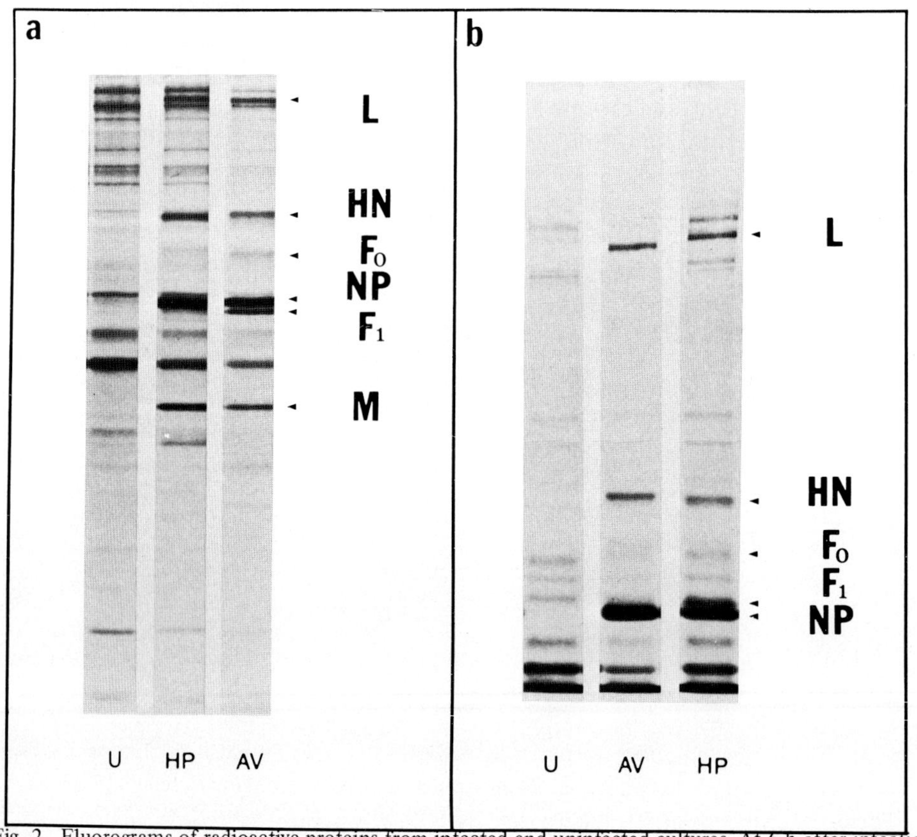

Fig. 2. Fluorograms of radioactive proteins from infected and uninfected cultures. At 6 h after infection, uninfected cells (U) and cultures that had been infected with strain AV (AV) or strain HP (HP) were exposed to ^{35}S-methionine for 30 min. The cultures were then dissolved in polyacrylamide gel sample buffer, and equivalent amounts of protein were applied to each slot of either 11.5% polyacrylamide gels (a) or 7% gels (b). The polyacrylamide slabs were processed for fluorography as described under Methods.

Characterization of Cellular Membranes

Having determined that F_1 accumulated in both AV- and HP-infected cells, we considered the possibility that the fusion glycoprotein of strain HP was blocked in its migration to the plasma membrane. We developed a cellular fractionation protocol to study the maturation pathways of viral proteins. For the characterization of membrane fractions, infected cultures were harvested and disrupted as described under Methods. Cellular membranes were collected by differential centrifugation and incorporated into the 55% sucrose layer of a discontinuous gradient. On the basis of the assays described under Methods, we determined that the material applied to the gradient contained about 80% of the plasma membranes originally present in the homogenate, 70% of the rough endoplasmic reticulum (ER) and Golgi activity, and 50% of the total endoplasmic reticulum. After centrifugation, fractions of the gradient were collected and pooled as shown in Figure 3. Five gradient fractions containing membranes (M1–M5) and the pellet (MB) were analyzed for protein content, radioactive ribosomes associated with membranes as a marker for rough ER, galactosyltransferase activity for Golgi Membranes, NADH-cytochrome c reductase activity as a general marker for endoplasmic reticulum, and binding of ^{3}H-wheat germ agglutinin to plasma membranes. Over 80% of the rough ER was concentrated in MB and in the dense membrane fractions M1 and M2. Golgi activity was located mainly in fractions M2 and M3. Fraction M3 was also the peak of cytochrome c reductase activity associated with internal membranes of intermediate density. About 50% of the plasma membrane marker was located in fraction M5. The low-density membranes in M5 contained only 2–3% of the total endoplasmic reticulum-associated cytochrome c reductase activity of the homogenate.

The distributions of both total protein and radioactive wheat germ agglutinin bound to plasma membranes in the discontinuous gradient were similar for fractionations of either AV- or HP-infected cells. In addition, polyacrylamide gel electropherograms of fractions of comparable density showed the same unique complement of cellular polypeptides. Therefore, similar fractions were obtained from cells infected by either strain.

Location of the Viral Glycoproteins in Infected Cells

Infected cultures were exposed to radioactive methionine for 2 min. The labeled cells were collected and disrupted, and cellular membranes were separated in discontinuous sucrose gradients. Radioactive proteins were extracted from each of the major gradient fractions and analyzed by polyacrylamide gel electrophoresis (data not shown). The HN and F_0 glycoproteins of both strains AV and HP were localized in fractions M1 and M2, indicating that each was synthesized in close association with rough ER. Detectable amounts of the cleavage product F_1 did not accumulate during the brief labeling period.

In order to identify the sites at which viral proteins accumulated, infected cultures were labeled with radioactive methionine for 2 min and then transferred to nonradioactive medium containing excess methionine for 30 min. Following the chase period, the cultures were fractionated, cellular membranes were isolated, and the radioactive proteins obtained from each fraction were separated by polyacrylamide gel electrophoresis (Fig. 4). The HN glycoprotein of both strains was detected in samples which contained high-, intermediate-, and low-density membranes, suggesting that the glycoprotein migrated from its site of synthesis in rough ER to both smooth ER and plasma membranes.

Unlike HN glycopolypeptide, the distribution of both F_0 and F_1 in subcellular fractions depended on the strain. The F_0 glycoprotein of strain AV was detected in all five fractions. Its cleavage product F_1 was concentrated in light-density membranes.

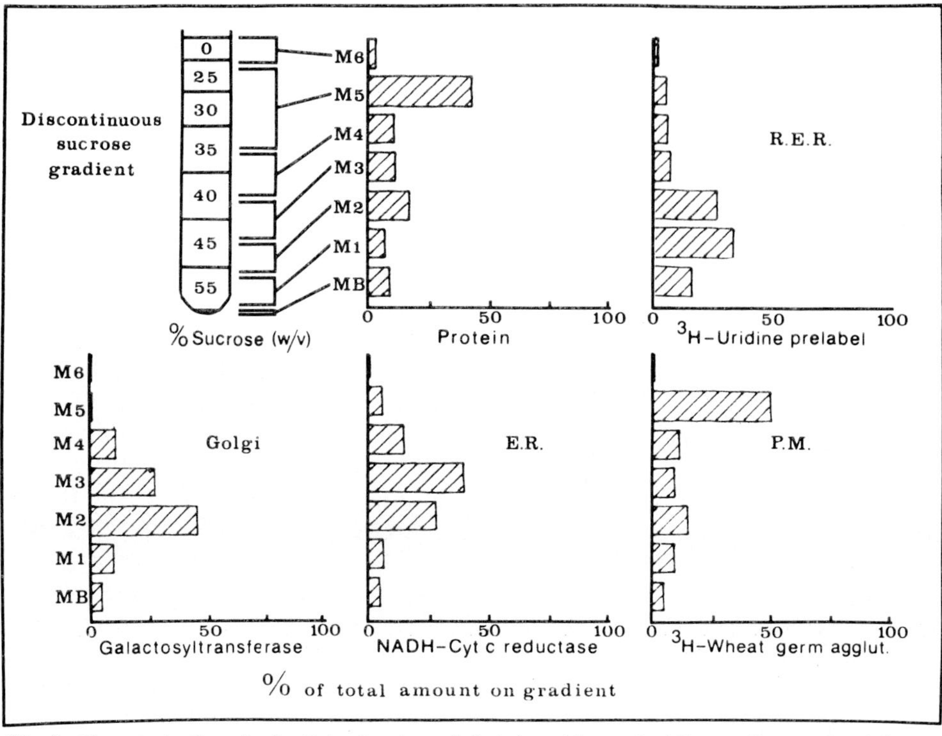

Fig. 3. Characterization of subcellular fractions. Cells infected by strain AV were disrupted and the nuclei were removed. Membranes and other particulate material were collected and incorporated into the 55% layer of a discontinuous sucrose gradient as described under Methods. After centrifugation, five membrane fractions (M1–M5), a lipoprotein layer (M6), and the bottom wash of the centrifuge tube (MB) were obtained. The histograms show the distribution of the various marker activities expressed as a percentage of the total activity recovered from the gradient. The distributions of protein, rough endoplasmic reticulum (RER), Golgi, endoplasmic reticulum (ER), and plasma membrane (PM) were determined as described under Methods.

In contrast, the F_0 and F_1 glycopolypeptides of strain HP accumulated mainly in M2 which contained rough ER and Golgi. Lower amounts were detected in M3, an intermediate density ER fraction which included Golgi. Neither form of the fusion glycoprotein was detected in either the major rough ER fraction M1 or in fraction M5, which contained plasma membranes. We conclude that the F_1 glycoprotein of strain AV accumulated in low- to intermediate-density membranes derived from the plasma membrane and smooth ER, while the F_1 glycoprotein of strain HP was located in fractions containing denser internal membranes. In addition to internal membranes fraction M2 probably contains nucleocapsids and any completed virions which remained cell-associated after the cultures were washed and harvested.

Assembly of Glycoproteins Into Virions

A radioisotopic pulse-chase experiment (Fig. 5) was performed to determine the effects of the strain-specific differences in glycoprotein distribution on the kinetics of virion assembly. Infected cultures were exposed to radioactive methionine for 10 min and then incubated in chase medium containing excess nonradioactive methionine. At various

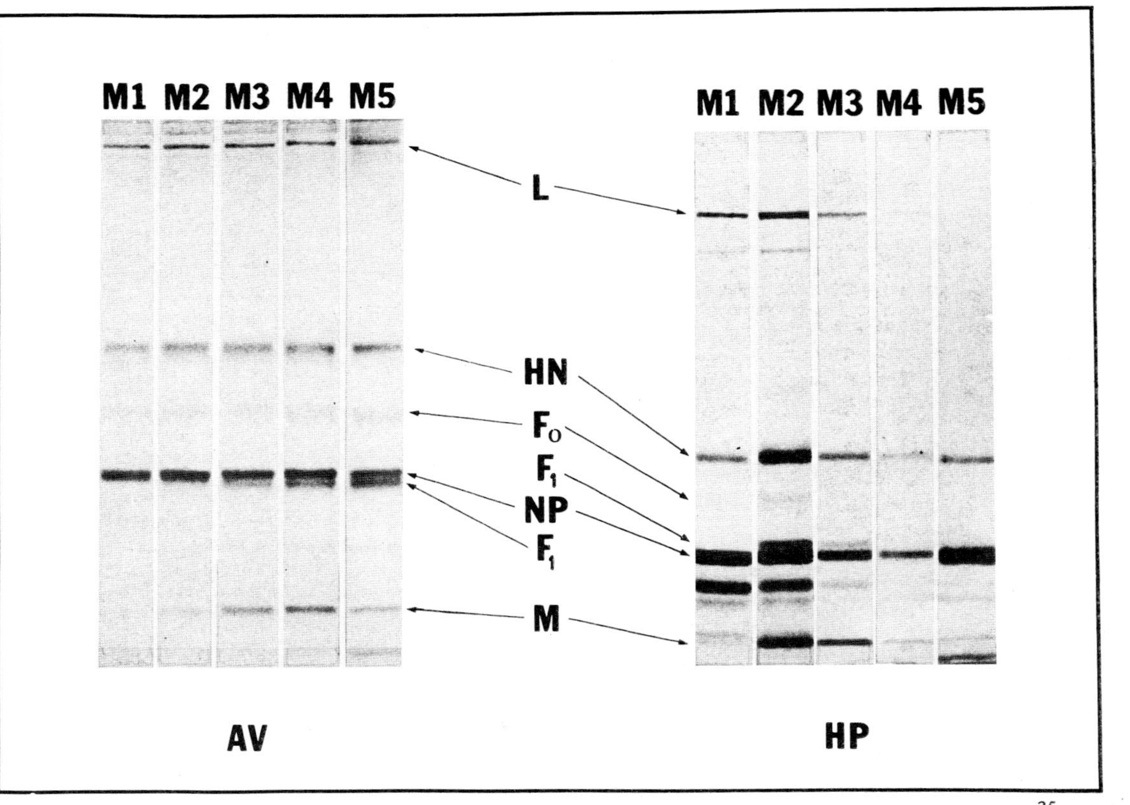

Fig. 4. Fluorograms of radioactive proteins from membrane fractions of infected cells. Infected cultures were labeled with ^{35}S-methionine for 2 min and then chased with excess methionine for 30 min. The cultures were subjected to cellular fractionation as described under Methods. The ^{35}S-methionine-labeled proteins extracted from isolated membrane fractions of AV- and HP-infected cells were separated on 11.5% and 7% polyacrylamide gels, respectively. Approximately equal amounts of radioactivity were applied to each slot; therefore, only the relative proportions of viral proteins within a channel can be compared. However, we have also quantified the absolute amounts of the radioactive viral proteins in each membrane fraction. The membrane fractions containing the highest relative amounts of the fusion glycoproteins also had the highest absolute levels.

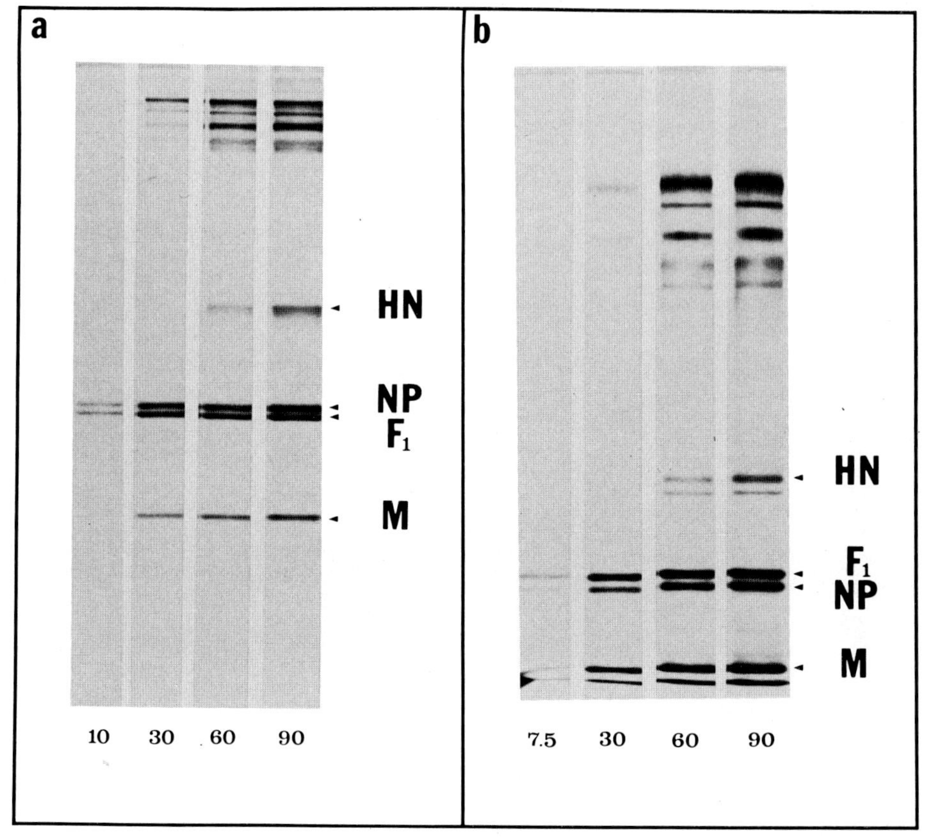

Fig. 5. Fluorograms of radioactive proteins from virions. Infected cultures were exposed to ^{35}S-methionine for 10 min. After various chase periods (listed in minutes below each gel slot), virions of strain AV (a) and strain HP (b) were collected from the culture medium. Radioactive proteins were extracted and analyzed by polyacrylamide gel electrophoresis and fluorography as described under Methods.

intervals, the virions released into the medium over replicate cultures were collected as described under Methods. Radioactive polypeptides were extracted from each sample and analyzed by polyacrylamide gel electrophoresis. Several large polypeptides incorporated ^{3}H-glucosamine and appeared to be mainly aggregates of cellular glycoproteins which presumably included fibronectin and other extrinsic glycoproteins [37]. They could be removed either by a mild protease treatment or by sedimenting the virions through a layer of 20% (w/v) sucrose (data not shown). Except for these large glycoproteins, the patterns of viral proteins shown in Figure 5 are similar to those obtained after additional purification of virions by equilibrium density ultracentrifugation. These additional purification steps were usually omitted to assure a more rapid and quantitative recovery of radioactive virions.

Following a short chase, both newly synthesized nucleocapsid protein (NP) and a polypeptide with the electrophoretic mobility of F_1 appeared in released virions of both strain AV (Fig. 5a) and strain HP (Fig. 5b). The M polypeptide was also detectable early, and it accumulated more rapidly in virions of strain HP than strain AV. The almost immediate availability of M for assembly of virions is characteristic of other negative strand viruses as well [28, 38, 39].

Radioactive HN was detected in virions of both strains only after 30–60 min of chase. The large glycoproteins which cosedimented with virions appeared after a similar lag. They provided convenient populations of cellular glycoproteins for comparison with the viral glycoproteins.

The drastic difference in the times of appearance of HN and F_1 in virions was unexpected. In order to analyze the kinetic behavior of the two glycoproteins in greater detail, fluorograms similar to those shown in Figure 5 were quantified for strain AV. The fluorograms were scanned with a densitometer, peak areas were calculated, and the percentage of the total virus-specific optical density contributed by each of the major viral polypeptides was determined. For each chase interval, these percentages were multiplied by the total radioactivity in released virions to generate the individual accumulation curves shown in Figure 6.

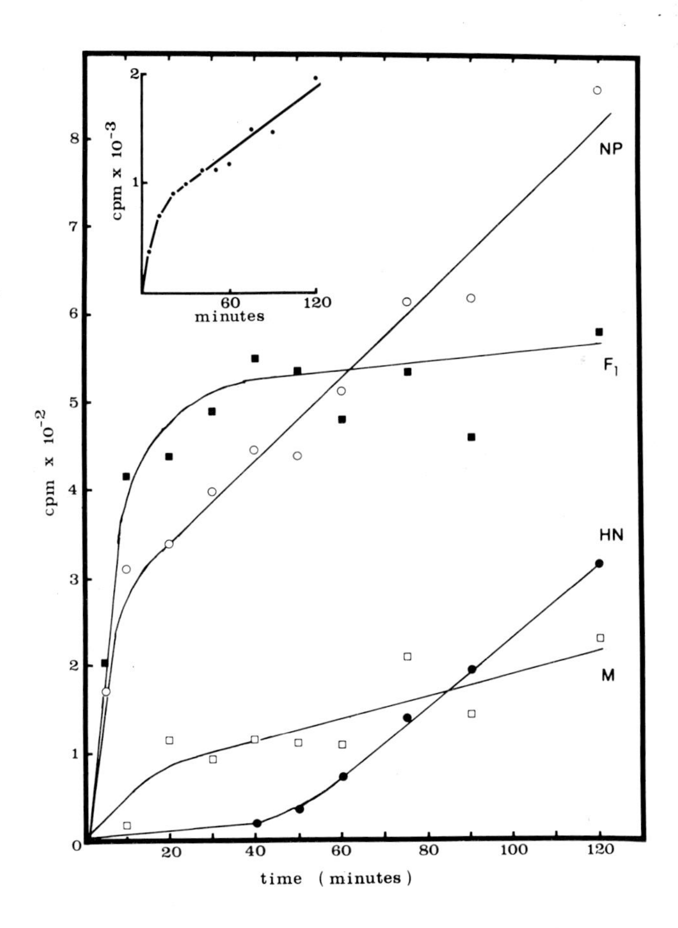

Fig. 6. Kinetics of incorporation of radioactive proteins into virions of strain AV. Fluorograms such as those in Figure 5 were quantified as described under Results. The inset shows the accumulation of radioactive virions released into the culture medium after various chase periods. The contribution of aggregates of large cellular proteins was subtracted for each chase period. The linear portions of all curves are the least-squares fit.

Within 5 min following a 10-min labeling period a radioactive species with the electrophoretic mobility of F_1 was detected in released virions. In other experiments, this polypeptide appeared in virions after a 2-min exposure of cells to radioactive methionine followed by a 5-min chase. We have established the identity of the F_1-size polypeptides which accumulated rapidly in virions by the mapping procedure of Cleveland et al [40]. The pattern of peptides generated by detergent-limited proteolysis of the rapidly incorporated species was the same as that of F_1-size polypeptides extracted from infected cells (data not shown).

After about 15 min of chase only a slight increase in the amount of F-related polypeptides in virions was detected, suggesting that the available pool of this protein was rapidly exhausted. In contrast HN glycoproteins did not appear in virions until about 30–40 min after synthesis. The HN which was synthesized during the radioisotopic pulse was available for virion assembly for at least 2 h.

The inset in Figure 6 shows the kinetics of accumulation of total radioactive protein in released virions. These experiments were performed during the steady-state phase of infection, a period during which we expected virions to be assembled at a constant rate [41]. However, instead of linear kinetics, the time course of accumulation of both infectious and radioactive virus was a roughly biphasic curve. We suspect that the rapid initial release of virions was due to a transient stimulation of cellular surface activity which may have been caused by the washing procedure used to remove unincorporated radioisotope.

Identification of Viral Glycoproteins

The accumulation of F-related polypeptides in virions was too rapid to accommodate the normal transit of glycoproteins through internal membranes. Therefore, we needed to verify that the F_1 glycoprotein of strain AV was indeed a carbohydrate-containing polypeptide. Radioactive glucosamine was used as a carbohydrate label in previous studies which showed that the virions of NDV contained three glycopolypeptides: F_1, F_2, and HN [6, 42]. In the experiment shown in Figure 7a, infected cells were exposed to radioactive mannose between 6 h and 12 h after infection. Released virions were collected from the culture medium and the radioactive proteins were extracted for analysis by polyacrylamide gel electrophoresis. The pattern of mannose-labeled polypeptides from virions (channel V) was compared to methionine-labeled marker proteins from infected cells (channel C). Two size classes of radioactive glycopolypeptides were detected, a major carbohydrate-containing species which comigrated with HN and a less radioactive species with the expected mobility of F_1. The third glycoprotein F_2 was not retained by these gels.

Effects of Tunicamycin on the Maturation of Viral Glycoproteins

The previous experiment showed that virions have a mannose-containing glycoprotein F_1 with the same mobility as the polypeptide which is rapidly assembled into virus particles. However, it was not possible to use radioactive mannose to determine if these properties belong to the same polypeptide because brief pulses and chases of radioactive carbohydrates are very inefficient in animal cells. Therefore, we approached this problem indirectly by determining the effect of tunicamycin, an inhibitor of glycosylation, on the assembly of the F-related polypeptides.

AV-infected cells were pretreated with Tm (1 μg/ml) for 2 h and then labeled with either radioactive methionine for 10 min or radioactive mannose for 6 h. Cells and virions from cultures which had incorporated ^{3}H-mannose were collected at the end of the labeling period and prepared for analysis on polyacrylamide gels. No radioactive mannose-containing

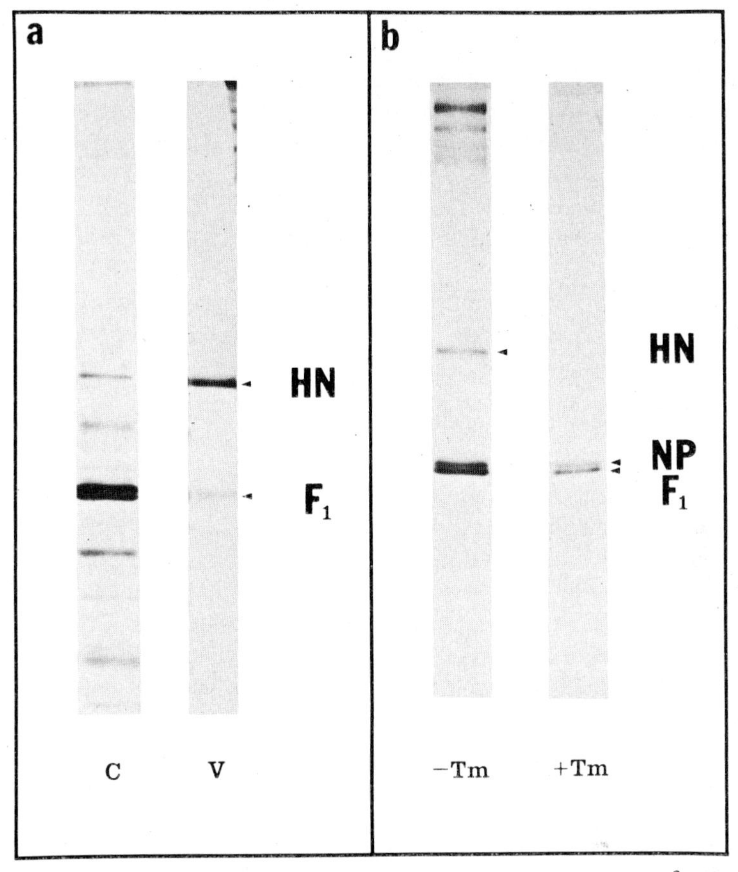

Fig. 7. Analysis of viral glycoproteins on polyacrylamide gels. a: Fluorogram of the ^{3}H-mannose-labeled proteins extracted from virions of strain AV (slot V) is compared to that of ^{35}S-methionine-labeled proteins extracted AV-infected cultures (slot C). b: Fluorograms of ^{35}S-methionine proteins extracted from virions of strain AV released from either control cells (−Tm) or tunicamycin-treated cells (+Tm) are compared.

polypeptides were detected in fluorograms of proteins extracted from either virions or cells (not shown). Therefore, Tm effectively blocked the dolichol pyrophosphate-mediated glycosylation of HN, F_0, and F_1 under our conditions of pretreatment.

Virions which had incorporated ^{35}S-methionine were collected after a 1-h chase period and were purified as described under Methods. Fluorograms of the ^{35}S-methionine-labeled proteins extracted from the virions of treated cultures (Fig. 7b, +Tm) and control cultures (Fig. 7b, −Tm) were compared. The NP and F_1-size polypeptides were incorporated into virions from both cultures as was the M protein, which could be detected after a longer exposure of the gels to X-ray film. In contrast, neither HN nor its unglycosylated polypeptide was detected in virions released from the treated cells.

The fluorograms shown in Figure 7b accurately reflect the fact that antibiotic treatment caused an approximatley threefold reduction in the accumulation of radioactive protein in virions. Despite this decrease, the ratio of radioactive F_1-size polypeptides to NP was similar in virions produced by both Tm-treated and control cultures. Therefore, Tm had very little selective effect on the accumulation of F_1-size polypeptides in virions.

DISCUSSION

The maturation pathways of paramyxoviral proteins have not been investigated as extensively as those of other negative strand viruses. In previous studies the intracellular distribution of proteins was determined for BHK21 cells infected by the virulent Italian strain of NDV [19] and for MDBK and CE cells infected by Sendai virus [43]. Detailed analyses of the incorporation of structural proteins into virions have been limited to Sendai virus [39, 44]. We have undertaken a comparative analysis of the glycoproteins of two virulent strains of NDV which have different biologic properties. The sites of synthesis, subcellular localization, and assembly of the glycoproteins into virions were compared. Because the hemagglutinin-neuraminidase and fusion glycoproteins possessed such distinctive characteristics, we have discussed each one separately.

The Hemagglutinin-Neuraminidase Glycoprotein

The HN glycoproteins of both strains AV and HP were synthesized on rough ER and accumulated in all of the membrane systems that we isolated. These data agree with other studies that showed that HN of paramyxoviruses migrated through internal membranes and into plasma membranes [19, 43]. The surfaces of cells infected with either strain are capable of hemadsorption, an observation which is consistent with the presence of HN in plasma membranes. Like the glycoproteins of vesicular stomatitis virus (VSV) [28] and influenza virus [38], the newly synthesized HN glycoprotein appeared in virions only after a lag of about 30 min. The common delay for all of these glycoproteins probably reflects the time required for migration along similar routes through internal membrane systems.

Takatsuki and Tamura showed that Tm blocked the incorporation of radioactive carbohydrates into polypeptides in NDV-infected cells [45]. It is known now that the antibiotic blocks the formation of N-acetylglucosamine-lipid intermediates necessary for the transfer of core sugars, including mannose, to certain asparagine residues in polypeptides [14–16]. We found that Tm inhibited the incorporation of radioactive mannose into HN and that the unglycosylated form could not be detected in virions. T. G. Morrison and co-workers recently observed that AV-infected Chinese hamster ovary (CHO) cells produced virus-like particles in the presence of Tm. The unglycosylated forms of both HN and an F-related polypeptide were detected in released particles (T.G. Morrison, personal communication). For fowl plague virus, Schwarz and co-workers [46] showed that the unglycosylated forms were more susceptible to proteolytic degradation than their glycosylated counterparts. Degradation of unglycosylated HN may explain the lack of this component in virions produced in our system in the presence of Tm.

The Fusion Glycoprotein

Like HN, the precursor of the fusion glycoprotein F_0 was synthesized on rough ER in both AV- and HP-infected cells. However, the cellular sites of accumulation of both F_0 and its cleavage product F_1 depended upon the strain. In fractionations of AV-infected cells, F_0 was distributed throughout all gradient fractions which contained membranes, while F_1 was concentrated in fractions enriched for smooth ER and plasma membranes. This pattern closely followed the distribution of the fusion glycoproteins previously reported for the Italian strain of NDV [19]. In contrast, both F_0 and F_1 accumulated in fractions containing intermediate-density membranes during HP infection. The distribution of the fusion glycoproteins among isolated cellular membranes closely paralleled the localization of galactosyltransferase activity and was different from the distribution of rough ER. Thus, it is possible that the fusion glycoproteins accumulated in the Golgi membranes of HP-infected cells. Another possibility that we are considering is that the fusion glycoprotein of HP accumulated in cell-associated virions which became detached from the cell during cavitation and copurified with dense internal membranes. Experiments are under way to distinguish between these two alternatives.

The different cellular distributions of the fusion glycoproteins of the two strains could explain their different capacities to mediate FFWI. We suggest that AV-infected cells can undergo FFWI because the fusion glycoprotein accumulates at the cellular surface. HP-infected cells are unable to participate in FFWI because most of the fusion glycoproteins either accumulate in internal membrane systems or become sequestered in cell-associated virions. Therefore, in addition to proteolytic activation of F_0, differential compartmentation of the fusion glycoproteins in infected cells may control the expression of biologic activity at the cell surface.

Despite the marked differences in the fractionation patterns of the fusion glycoproteins of the two strains, the kinetics of incorporation of F-related polypeptides into virions of strains AV and HP were remarkably similar. Newly synthesized F-related proteins were available for assembly into virions almost immediately after synthesis in CE cells. Similar kinetic patterns were obtained in MDBK cells; therefore, the rapid incorporation of the F-related polypeptides into virions was not limited to avian cells. In a study of the assembly of Sendai virus, Portner and Kingsbury [39] observed that newly synthesized F_0 was incorporated into virions with about one-half the lag time for HN. They suggested that HN and F_0 might follow different intracellular pathways. The rapid assembly of the F_1-size polypeptide of NDV could be a more exaggerated version of the Sendai pattern. However, the 15-min delay in the appearance of Sendai F_0 in virions could still accommodate a route through internal membranes. This does not seem possible for the NDV polypeptide.

In addition to its rapid assembly into virions, the response of the F_1-size polypeptides to Tm was unusual for a viral glycoprotein. The antibiotic had little effect on either the electrophoretic mobility or kinetics of assembly of F_1-size polypeptides into virions. We showed that virions of strain AV do have mannose-containing glycopolypeptides with the expected electrophoretic mobility of F_1 and that Tm blocked the incorporation of radioactive mannose into both F_0 and F_1 in infected cells. This suggests that the dolichol pyrophosphate carrier was responsible for the addition of the oligosaccharide core to these proteins. The rapidly incorporated F_1-size polypeptide has been identified as an F-gene product by peptide mapping. Our working hypothesis to explain these observations is that another F-related polypeptide having the same electrophoretic mobility as F_1 exists. Unlike the glycopolypeptide F_1, this second polypeptide would be nonglycosylated and rapidly incorporated into virions, and its assembly would be insensitive to Tm. F_1 might then compose only a portion of the 53,000-dalton size class of polypeptides extracted from virions. The presence of both an unglycosylated and a glycosylated form of the same gene product in virions would be unusual; however, the plausibility of our hypothesis is strengthened somewhat by the recent finding by Edwards and Fan [47] that cells infected with Moloney murine leukemia virus synthesize both a glycosylated and an unglycosylated form of the precursor of the major internal structural protein gag of the virus. Experiments designed to separate the putative forms of the F-related polypeptides of NDV are in progress.

ACKNOWLEDGMENTS

We thank Gayle Hightower and Liz Jean for aid in preparing the manuscript; Dana Astheimer and Pam Chatis for technical assistance; Dr. P.W. Robbins for tunicamycin; Drs. M.A. Bratt, T.G. Morrison, and P.W. Robbins for helpful discussions; Public Health Service (HL23588-01) and the National Science Foundation (PCM 78-08088) for research support. This work benefited from use of a Cell Culture Facility supported by the National Cancer Institute (CA 14733).

REFERENCES

1. Bang FB: Bull Johns Hopkins Hospital 92:291, 1953.
2. Seto JT, Becht H, Rott R: Virology 61:354, 1974.
3. Scheid A, Choppin PW: J Virol 11:263, 1973.
4. Seto JT, Becht H, Rott R: Med Microbiol Immunol 159:1, 1973.
5. Nagai Y, Klenk HD: Virology 77:125, 1977.
6. Scheid A, Choppin PW: Virology 80:54, 1977.
7. Nagai Y, Klenk HD, Rott R: Virology 72:494, 1976.
8. Colonno RJ, Stone HO: J Virol 19:1080, 1976.
9. Miller TJ, Stone HO: Abstr Am Soc Microbiol 1978, p 235.
10. Nagai Y, Yoshida T, Hamaguchi M, Iinuma M, Maeno K, Matsumoto T: Arch Virol 58:15, 1978.
11. Choppin PW, Compans RW: In Fraenkel-Conrat H, Wagner R (eds): "Comprehensive Virology." New York: Plenum, vol 4, 1975, p 95.
12. Rentier B, Hooghe-Peters EL, Dubois-Dalcq M: J Virol 28:567, 1978.
13. Miyakawa T, Takemoto LJ, Fox CF: In Baltimore D, Huang AS, Fox CF (eds): "Animal Virology." New York: Academic, 1976, p 485.
14. Takatsuki A, Kohno K, Tamura G: Agr Biol Chem 39:2089, 1975.
15. Tkacz JS, Lampen JO: Biochem Biophys Res Commun 65:248, 1975.
16. Struck DK, Lennarz WJ: J Biol Chem 252:1007, 1977.
17. Kohn A: Virology 26:228, 1965.
18. Bratt MA, Gallaher WR: Proc Natl Acad Sci USA 64:536, 1969.
19. Nagai Y, Ogura H, Klenk HD: Virology 69:523, 1976.
20. Kaplan J, Bratt MA: Abstr Am Soc Microbiol 1973, p 243.
21. Samson ACR, Fox CF: J Virol 12:579, 1973.
22. Hightower LE, Morrison TG, Bratt MA: J Virol 16:1599, 1975.
23. Gallaher WR, Bratt MA: In Fox CF (ed): "Membrane Research." New York: Academic, 1972, p 383.
24. Carver DH, Marcus PI: Virology 32:247, 1967.
25. Bratt MA, Rubin H: Virology 33:598, 1967.
26. Hightower LE, Bratt MA: J Virol 13:788, 1974.
27. Collins PL, Hightower LE, Ball LA: J Virol 28:324, 1978.
28. Knipe DM, Baltimore D, Lodish HF: J Virol 21:1128, 1977.
29. Hunt LA, Summers DF: J Virol 20:637, 1976.
30. Fleischer B, Fleischer S, Ozawa H: J Cell Biol 43:59, 1969.
31. Wallach DFH, Kamat VB: Methods in Enzymol 8:164, 1966.
32. Bradford MM: Anal Biochem 72:248, 1976.
33. Laemmli UK: Nature 227:680, 1970.
34. Laskey RA, Mills AD: Eur J Biochem 56:335, 1975.
35. Okada Y, Tadokoro J: Exp Cell Res 26:108, 1962.
36. Gallaher WR, Levitan DB, Blough HA: Virology 55:193, 1973.
37. Olden K, Pratt RM, Yamada KM: Cell 13:461, 1978.
38. Hay AJ: Virology 60:398, 1974.
39. Portner A, Kingsbury DW: Virology 73:79, 1976.
40. Cleveland DW, Fischer SG, Kirschner MW, Laemmli UK: J Biol Chem 252:1102, 1977.

41. Hightower LE, Bratt MA: J Virol 15:696, 1975.
42. Mountcastle WE, Compans RW, Choppin PW: J Virol 7:47, 1971.
43. Lamb RA, Choppin PW: Virology 81:371, 1977.
44. Famulari NG, Fleissner E: J Virol 17:605, 1976.
45. Takatsuki A, Tamura G: J Antibiot 24:785, 1971.
46. Schwarz RT, Rohrschneider JM, Schmidt MF: J Virol 19:782, 1976.
47. Edwards SA, Fan H: J Virol 30:551, 1979.

NOTE ADDED IN PROOF

We now have conclusive evidence that the unglycosylated protein which comigrated with F_1 is a unique viral protein with properties of the P protein of other paramyxoviruses. Our hypothesis concerning the behavior of F_1 in cells should be interpreted with caution until the subcellular location of P is defined.

Biological Recognition and Assembly 293–300 (1980)

The Assembly of Papaya Mosaic Virus Coat Protein With DNA

John W. Erickson and J.B. Bancroft

Department of Plant Sciences, University of Western Ontario, London Ontario, Canada

Products of specific (pH 8.0–8.5) and nonspecific (pH 6.0) assembly reactions of papaya mosaic virus (PMV) coat protein with DNA are described. The strandedness, topology, and sugar moiety of the nucleic acid are important parameters for assembly in nonspecific conditions. The linear, single-stranded form of λ DNA, but not the double-stranded form, reacted with PMV protein to form multiply initiated particles whose helical segments apparently annealed to produce continuous, tubular particles. With the circular, single-stranded DNA of ϕX174, partially tubular, partially extended particles were made. Poly(dA), unlike poly(A) [Erickson JW, AbouHaidar M, Bancroft JB: Virology 90:60, 1978], was not encapsidated by PMV protein under specific assembly conditions. With all DNAs tested, extended particles were the only products formed in specific conditions at pH 8.5.

Key words: papaya mosaic virus, virus assembly, protein-nucleic acid recognition

The assembly of PMV has been studied in detail, as have reactions of the coat protein with RNA from various sources [1–4]. PMV coat protein forms helices, which have the structure of the native virus, only with its own RNA or that from related viruses at pH 8.0 to 8.5. This reaction is termed specific and occurs subsequent to attachment of the protein to the RNA to form narrow, extended particles. The formation of these unusual particles is nonspecific at pH 8.0 or 8.5, occurring with RNAs from a variety of sources. At pH levels below 8.0, nonspecific reactions also occur, but instead of extended particles being formed, particles with normal, albeit segmented or kinked, helices are made.

In this paper, we describe the products of PMV protein–DNA assembly reactions in conditions of specificity and nonspecificity and compare them with those made with RNA.

Received April 4, 1979; accepted August 16, 1979.

MATERIALS AND METHODS

Preparation of Virus, Protein, and RNA

PMV, PMV coat protein, and PMV RNA were prepared as described elsewhere [1]. Extinctions of $E^{0.1\%}_{260\ nm}$ = 25 for the RNA [1] and $E^{0.1\%}_{280nm}$ = 0.75 for the protein [5] were used.

DNAs

Purified λ DNA, at 250 μg/ml in 0.01 M Tris buffer, pH 8.0, was the generous gift of Dr. G. Mackie. Purified φX174 DNA and poly(dA) were obtained from Miles Laboratories. Extinctions of $E^{0.1\%}_{260\ nm}$ of 23.0 and 30.0 for φX174 DNA and poly(dA), respectively, were used.

Assembly

Assembly experiments were performed at 25°C at a protein: nucleic acid ratio of 20:1 (w/w), as before [1], at pH 6.0 (0.01 M MES) or pH 8.0–8.5 (0.01 M Tris).

Electron Microscopy

Electron micrographs of perparations negatively stained with 1% uranyl acetate were obtained with a Philips EM201.

RESULTS

Assembly With Linear Double- and Single-Stranded DNA

PMV protein mixed with native, double-stranded λ DNA formed thin nucleoprotein structures at both pH 6.0 (Fig. 1) and pH 8.5 (not shown).

When native λ DNA was denatured by heating to 100°C for 5 min, followed by rapid cooling, the type of product depended on the pH. In nonspecific conditions helical virus-like particles were formed (Fig. 2A). In specific conditions, only thin, extended particles were observed (Fig. 2B), which resemble the extended particles made under certain conditions with RNA (inset, Fig. 2B) [1]. This means that single-stranded DNA behaves like heterologous RNA in regard to specificity [4].

Limited Multi-Initiation With Single-Stranded DNA

The apparently continuous helical structure formed by PMV coat protein with linear single-stranded DNA at pH 6.0 suggested that initiation of the growth of the helix occurred at a single site on the DNA, because the protein helix did not show the gaps indicative of multi-initiation found with RNA (inset, Fig. 2A). However, multi-initiation could still have occurred if the helices were originally discontinuous but subsequently annealed. This possibility was assessed by examining reaction products made soon after mixing the DNA and protein. Within seconds after mixing, thin, extended-type particles were observed (Fig. 3A), followed by segmented particles after 2 min (Fig. 3B). The latter annealed within 30 min to form the mainly continuous particles shown in Figure 2A.

Assembly With Circular DNA

There is no requirement in the assembly process with DNA for a free end because single-stranded, covalently closed DNA circles from φX174 also form normal helical

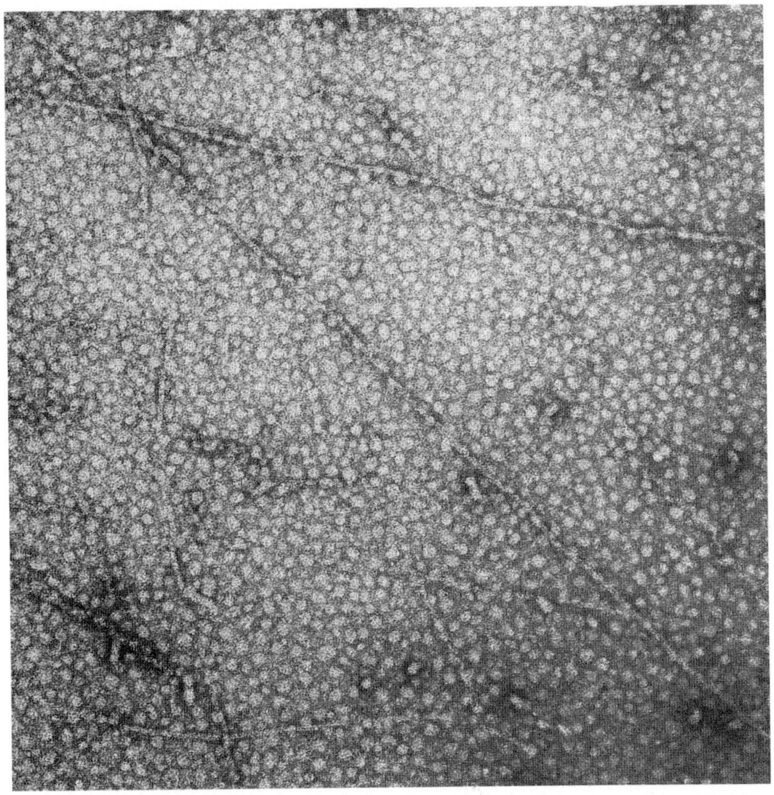

Fig. 1. Electron micrograph of extended particles formed with native λ DNA (double-stranded) and PMV protein at pH 6.0 at 25°C after 30 min (magnification, × 90,000).

particles with PMV protein (Fig. 4). Since the circular DNA was not fully encapsidated by a virus-like helix, there is presumably a physical constraint on the degree of bending that the helix may endure.

Assembly With Poly(dA)

The helix initiation site at the 5′ end of PMV RNA is enriched with adenylic acid [3]. Poly(A) is recognized and encapsidated at both pH 6.0 and 8.5 [4]. We wished to determine whether the sugar moiety of the nucleic acid had an effect on assembly and accordingly reacted PMV protein with poly(dA) at pH 6.0 and 8.5. Virus-like helices were made at pH 6.0 (Fig. 5A) but not at pH 8.5 (Fig. 5B).

DISCUSSION

The various assembly reactions observed with DNA and RNA are shown in Table I, and the general pattern is clear. With unrelated nucleic acids (except poly(A) and poly(C) [4]), narrow, extended particles are the final assembly products made under specific reaction conditions. Preliminary optical diffraction studies of these particles show that they possess helical symmetry and are of more than one kind. In nonspecific conditions, how-

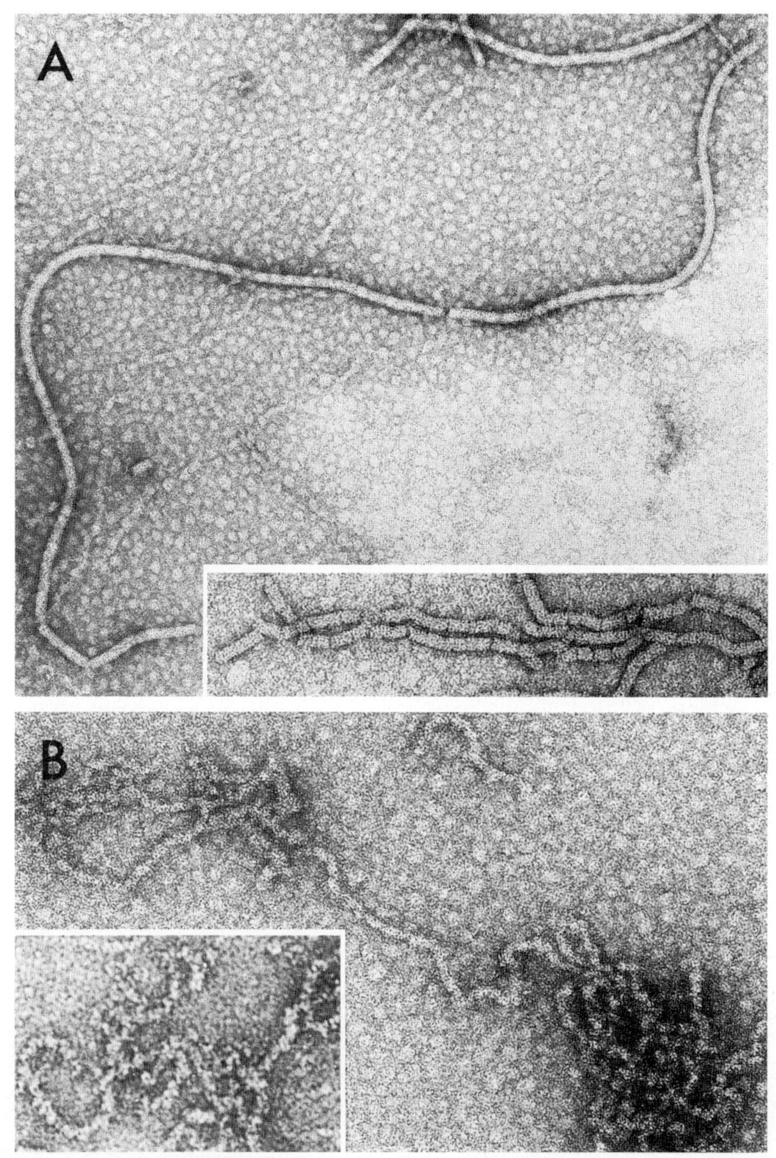

Fig. 2. Electron micrographs of products formed with denatured λ DNA (single-stranded) and PMV protein at pH 6.0 (A) and pH 8.5 (B) at 25°C after 30 min (magnifications, × 90,000 for A; × 135,000 for B). Insets display kinked, segmented (upper), and extended (lower) particles formed with PMV protein and RNA (magnifications, × 135,000 upper; × 110,000, lower.

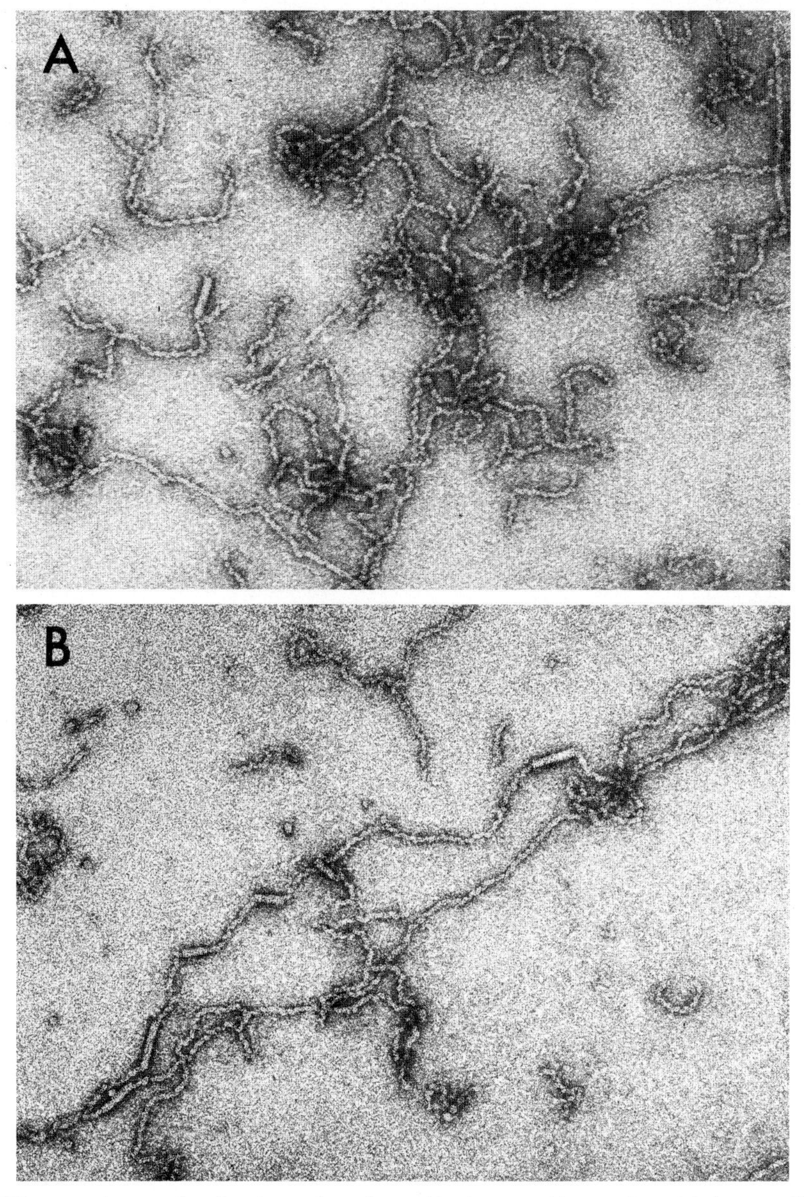

Fig. 3. Electron micrographs of early products formed in the assembly reaction with denatured λ DNA (single-stranded) at pH 6.0 at 25°C after A) several seconds and B) 2 min (magnifications, × 90,000).

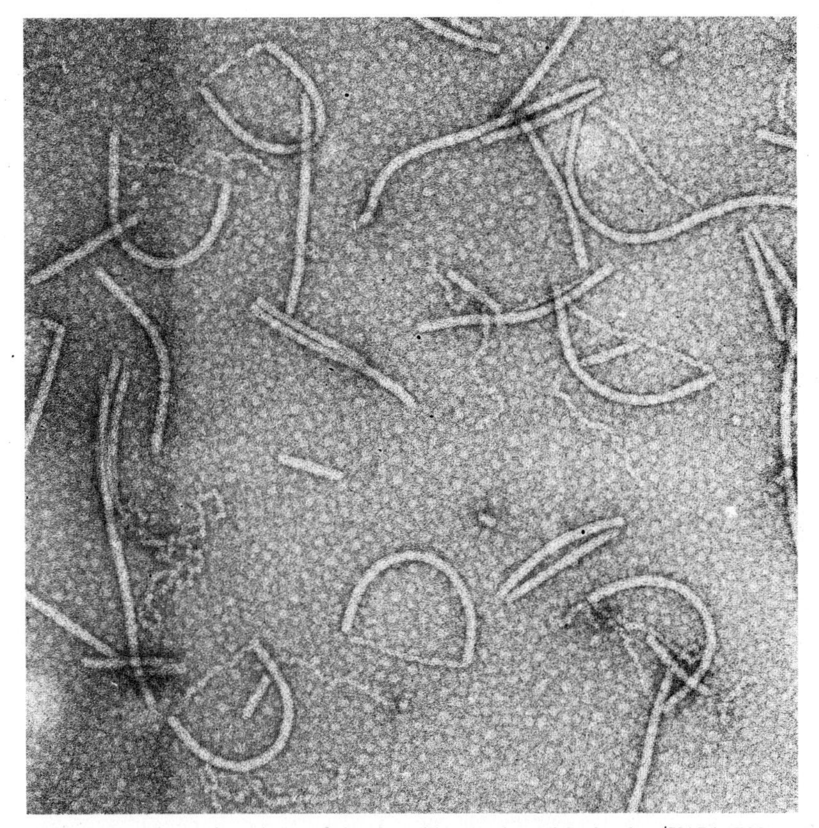

Fig. 4. Electron micrograph of products of the assembly reaction with circular ϕX174 (DNA (single-stranded) at pH 6.0 at 25°C after 30 min (magnification, × 90,000).

ever, tubes are formed with single-stranded nucleic acids. These are segmented with RNA or continuous with DNA.

The strandedness of DNA (and probably RNA) is critical for the helix-forming reaction with PMV protein, because mainly extended particles were formed with double-stranded λ DNA in nonspecific conditions. These results are consistent with our finding that in vitro assembly is inhibited by agents that increase the secondary structure of PMV RNA [5].

The helical tubes made with single-stranded λ DNA at pH 6.0 possess the same pitch and true repeat (unpublished optical diffraction) as the virus [6] and are the same as tubes made at pH 6.0 with RNA, except that the latter are segmented. Helices made at low ionic strength at pH 6.0, with both DNA and RNA, can be initiated at a variety of sites, but apparently only those made with DNA can anneal to form continuous helices resembling those made with homologous RNA in specific conditions. The reason for this difference is not known.

A role for the sugar moiety in the initiation of helical tube formation in specific conditions is evident. PMV protein will form virus-like helices with poly(A) [4] but not with poly(dA) at pH 8.5. Possibly, the 2′–OH group of the ribose is involved in recognition specificity.

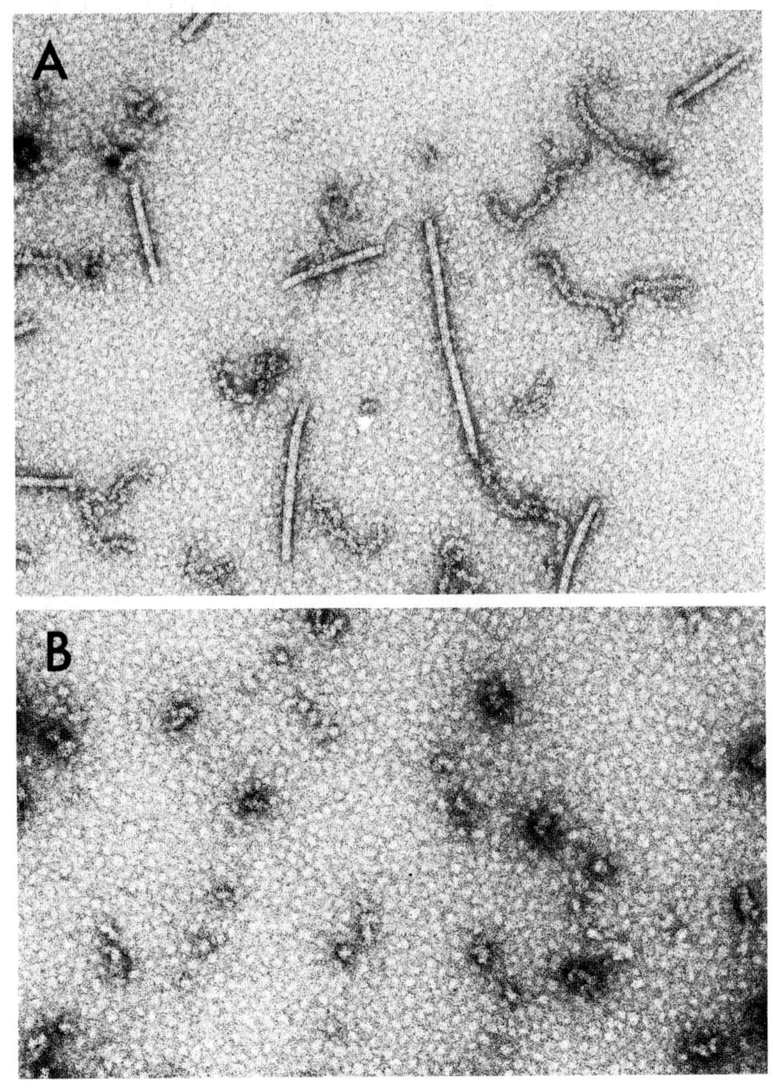

Fig. 5. Electron micrographs of assembly products made with poly(dA) and PMV protein at 25°C after 30 min at A) pH 6.0 and B) pH 8.5 (magnifications, × 90,000).

TABLE I. Final Assembly Products Made With PMV Coat Protein in Specific and Nonspecific Reaction Conditions

	Final product[a]		
Nucleic acid	Specific[b]	Nonspecific[b]	Reference
Homologous RNA[c]	Tubes	Kinked tubes	[1]
Heterlogous RNA	Extended particles	Kinked tubes	[4]
Poly(A)	Tubes	Kinked tubes	[4]
Poly(dA)	Extended particles	Tubes	
λ DNA, single-stranded	Extended particles	Tubes	
λDNA, double-stranded	Extended particles	Extended particles	
φX174 DNA, circular, single-stranded	Extended particles	Partially tubular, partially extended	

[a]"Final product" refers to the *major* assembly product observed in the electron microscope after 30 min.
[b]"Specific" and "nonspecific" refer to assembly at pH 8.0–8.5 and pH 6.0, respectively.
[c]"Homologous RNA" means RNA from PMV and related viruses.

ACKNOWLEDGMENTS

We wish to thank Dr. R.R. Shivers for generously making available to us a Philips 201 electron microscope for this study.

This work was supported by grants from the NSERC and from the University of Western Ontario Academic Development Fund.

REFERENCES

1. Erickson JW, Bancroft JB: Virology 90:36, 1978.
2. Erickson JW, Bancroft JB: Virology 90:47, 1978.
3. AbouHaidar M, Bancroft JB: Virology 90:54, 1978.
4. Erickson JW, AbouHaidar M. Bancroft JB: Virology 90:60, 1978.
5. AbouHaidar M, Erickson JW, Bancroft JB: Virology 98:116, 1979.
6. Tollin P, Bancroft JB, Richardson J, Payne N, Beveridge T: Virology 98:108, 1979.

Biological Recognition and Assembly 301–310 (1980)

Molecular Packing of Nucleic Acids in Spherical Viruses

Stephen C. Harrison

Gibbs Laboratory, Harvard University, Cambridge, Massachusetts 02138

Electron microscopy and X-ray scattering show that double-stranded DNA is wound in a regular way in the heads of phage such as P22 and λ. DNA is "pumped" into such heads, which may simply act as passive containers. In contrast, the protein-coat subunits of tomato bushy stunt virus have a flexibly tethered N-terminal domain that appears to be a major locus of interactions with the compactly folded RNA. Both modes of packaging permit considerable latitude in the detailed arrangement of the nucleic-acid chain, while still ensuring specific and efficient incorporation.

Key words: **tomato bushy stunt virus, packing in TBSV, packing in phage heads, bacteriophage T4, bacteriophage P22, bacteriophage lambda**

The packaging of nucleic acids by viral-coat proteins presents both geometric and chemical problems. The most obvious is the symmetry mismatch between a protein surface lattice and a nonrepeating linear polymer. In simple helical cases, this can be overcome by a binding site that is approximately sequence independent. In tobacco mosaic virus, three nucleotides fit into precise sites on each subunit, and in the assembled virus, the RNA backbone conforms to the symmetry of the protein helix [1]. Specificity for viral nucleic acid is provided by a special nucleating region on the RNA; aspects of sequence and secondary structure in this region appear together to provide exceptional affinity for the 34-subunit protein disk aggregate [2]. In other cases, less exact but equally well-determined arrangements are found. Two examples – double-stranded DNA in phage heads [3] and RNA in a spherical virus [4] – are described here. They are presented as illustrations of a more general property of larger biological structures, ie, precise assembly without the requirement for strict order in all components. In bacteriophage λ, it is evolutionarily convenient not to make excessive demands on DNA size. In tomato bushy stunt (TBSV), it is possible to package RNA without disrupting helical regions, and the protein subunits appear to have sufficient flexibility to accomodate various patterns of secondary structure in the nucleic acid.

Received August 29, 1979, accepted August 30, 1979.

DNA IN PHAGE HEADS

The double-stranded DNA phages package DNA by insertion into a preformed head [5]. The precursors are generally somewhat smaller in diameter than the mature head, and they contain core structures that are variously processed (and lost) prior to or concomitant with DNA incorporation.

X-ray scattering and electron microscopy together show that DNA is wound in a relatively uniform way inside the head [3, 6–8]. Micrographs of negatively stained particles disrupted on a grid show "donut" or toroidal windings of DNA relatively spherical in overall shape when derived from isometric phage like λ, and elongated in an axial direction when derived from T4 [6]. The structrues from disrupted T4 giants are particularly striking, and suggest that the toroid could have some overall supertwist in situ (Fig. 1) [7].

X-ray scattering from phage in solution gives a more precise view of the packing and its variation. The way in which information is derived from such experiments is shown in Figure 2. Diffraction in the 25-Å region gives information about the side-to-side packing of DNA in the head, and the rippled "sampling" of this diffraction shows that there is some degree of long-range order. The detailed arguments are presented elsewhere [3], and we may summarize the conclusions as follows. 1) DNA is wound into the head, with adjacent stretches of double-helix running essentially parallel to each other. The local side-to-side spacing is relatively constant throughout (Fig. 3). In P22 and wild-type λ, this spacing is similar to the tightest spacings found in DNA condensed by PEG or controlled dehydration [9]. 2) The orientation with respect to the phage-tail of the windings diagrammed in Figure 3 is not specified by X-ray studies of isometric phage, but in elongated heads such as those of T4, the coil axis is normal to the tail axes (Fig. 1), imposing minimum curvature on the packed DNA [7]. These conclusions are based on elec-

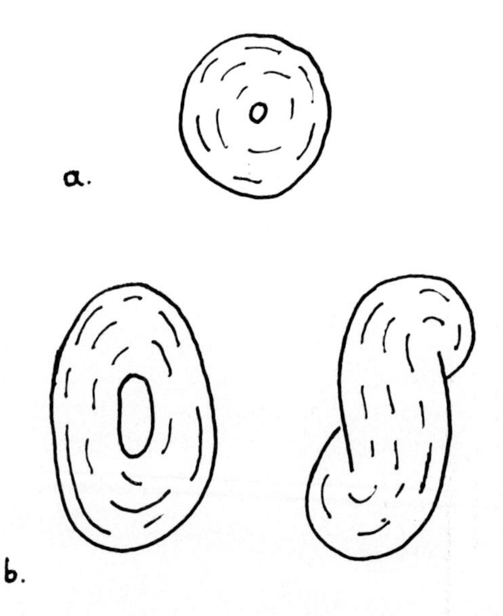

Fig. 1. Drawings of DNA coils from disrupted phage heads, visualized by electron microscopy. These are redrawn from references 6 and 7. a) Isometric; b) T4 giants, tail axis vertical.

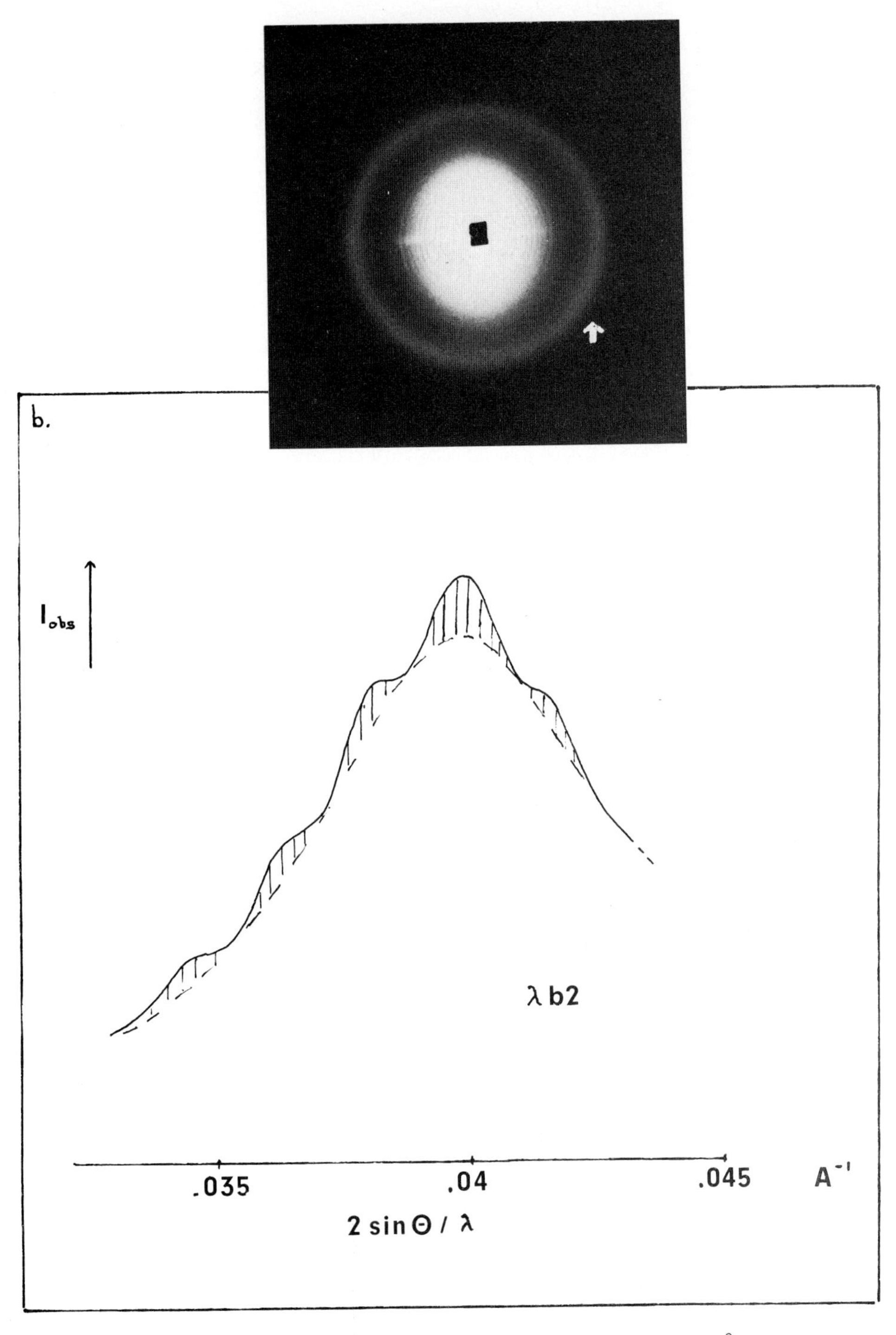

Fig. 2. a) X-ray diffraction from λb2 in solution. The ring at a spacing of about 25 Å (arrow) arises from DNA in the phage heads. b) A densitometer trace of this ring, showing both a diffuse and a rippled component.

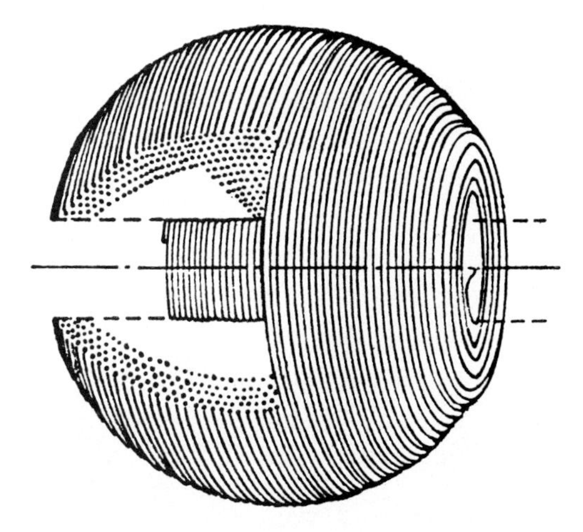

Fig. 3. Schematic drawing of DNA in an isometric phage head.

tron microscopy and X-ray diffraction of T4 giants, and they are consistent with much earlier flow-birefringence measurements on normal T4 [10]. Electric dichroism studies of isometric P22 and T7 give values of opposite signs than expected from a uniquely oriented coil with axis normal to the tail, but of considerably smaller absolute value than expected from a simple coil coaxial with the tail [11]. The interpretation depends upon the assumption that electric anisotropy is imparted by the tail, rather than by the DNA coil itself. There may, in any case, be no single, well-determined orientation of DNA with respect to the tail, but rather a distribution both of axial direction and of overall supertwist (if any). 3) If the DNA content of the head is varied (by deletion or insertion in the case of λ, which packages by a unique sequence cleavage rather than a headful mechanism), the uniformity of winding and regularity of spacing is preserved, and the local side-to-side distance increases or decreases so that a roughly constant volume, the internal volume of the head, is occupied by the length of DNA in question [3]. This observation shows that under the ionic conditions used (essentially physiological, and intracellular phage have also been examined), electrostatic repulsion of adjacent DNA helices, rather than resistance to bending, dominates the energetics of packing. In effect, the shorter molecules in large deletion mutants prefer to preserve tighter winding near the center of the head, and to "dilute" the coil uniformly, rather than to remain tightly packed in larger-diameter windings. The results also show that energy is required to package DNA, since work must be done against the repulsions when going, eg from a 78% full head (a b221 equivalent) to a full wild-type particle. Mechanisms involving passive "condensation" of DNA – eg by polyamines [12] – do not appear to be sufficient to account for the observed packing in λ deletions, and an active DNA "pump" seems implicated. ATP is indeed split during in vitro packaging of λ DNA [13].

Is there any specific interaction between head protein and DNA? The conclusions just summarized actually suggest relatively little interaction, as might be expected if injection is to proceed easily. This suggestion follows from the simplest models of "pumping" DNA into the head, which involve some shifting of DNA along the inside surface. Consider packaging in λ for example, which appears to proceed by polarized, processive incorporation from *cos* site to *cos* site along a concatemer [14]. If, after a 78% wild-type complement has entered, a *cos* site is encountered (deletion mutant b221), cleavage occurs and packaging is complete. Adjacent turns of the DNA molecule are spaced about 30 Å apart. If the DNA is really wild-type, additional material enters, and the final spacing is about 27.5 Å. Assuming that DNA inside the head is "ignorant" of when the next *cos* site will appear, and that it assumes an equilibrium structure at any stage during packaging, adjacent turns must move closer together as filling proceeds (eg from 78% to 100% full – the b221 to wild-type arrangement), and some turns must shift aside to allow a turn from an inner layer to squeeze outwards. Since such turns must increase in length, longitudinal sliding is also required. Such motions would not be completely incompatible with protein-DNA interactions, especially if they were mediated by flexibly tethered sites as in TBSV. Moreover, more elaborate models could involve an essentially invariant arrangement in the outermost layer of DNA windings, with shifting and sliding occurring only in inner layers. There is however, no strong suggestion of such nonuniformity in the x-ray data (see below), and in such a case one might expect that the spacing of inner windings in deletions would be slightly larger than expected for filling a constant volume (since the outermost layer would be more tightly packed), whereas the opposite is true [3]. These implications are, of course, not decisive.

The following model calculation shows that the simplest picture of DNA packing is consistent with the x-ray observations. We imagine that DNA pumped into an isometric phage head packs in an essentially "random" manner, constrained simply by the shape of the head, by electrostatic repulsions, and by its own stiffness. (The leading end may or may not be attached to a site on the head). A cross section of the final coiled structure can be modelled by allowing spherical beads to pack in a dish of appropriate diameter (Fig. 4a). The diameter of the beads represents the side-to-side DNA spacing (electrostatic repulsion), scaled appropriately to the size of the dish. There is a strong tendency, evident in Figure 4, for beads to lie in layers concentric with the rim of the dish. Successive layers, moving inwards, must contain increasing numbers of packing faults, in order to preserve a constant side-to-side distance, and the packing is essentially hexagonal well away from the rim. If we approximate the DNA coil by a set of concentric rings, with a cross section represented by the packing of beads as shown, then the spherically averaged diffraction from such a structure can be calculated in a straight-forward way, with the result shown in Figure 4b. Note that the 25 Å diffraction from phage heads (Fig. 2) includes both "sampled" ("rippled") and diffuse components, with their ratio being essentially a measure of long-range order in DNA packing. The agreement of calculated and observed intensities (Fig. 2b, 4b) is evident, although the overall linewidth is sharper in the calculated curve and the ratio of sampled to diffuse intensity somewhat higher: DNA in a phage head packs less regularly against its neighbors than glass beads in a dish.

Details of the packaging mechanism are at present uncertain, although several specific proposals have been presented [15, 16]. It is known that in λ, the left-hand end of the molecule enters first and the right-hand end is attached to the tail [14, 17]. It would be useful to know which part of the molecule (or whether, randomly, all of it) lies immediately adjacent to the head, as well as to know whether and where the leading end is attached [18].

RNA IN TBSV

The incorporation of single-stranded RNA into a spherical virus appears to present problems quite distinct from the packaging of DNA in phage. Flexibility of the RNA molecule releases constraints on packing, but self-assembly requires strong interaction with

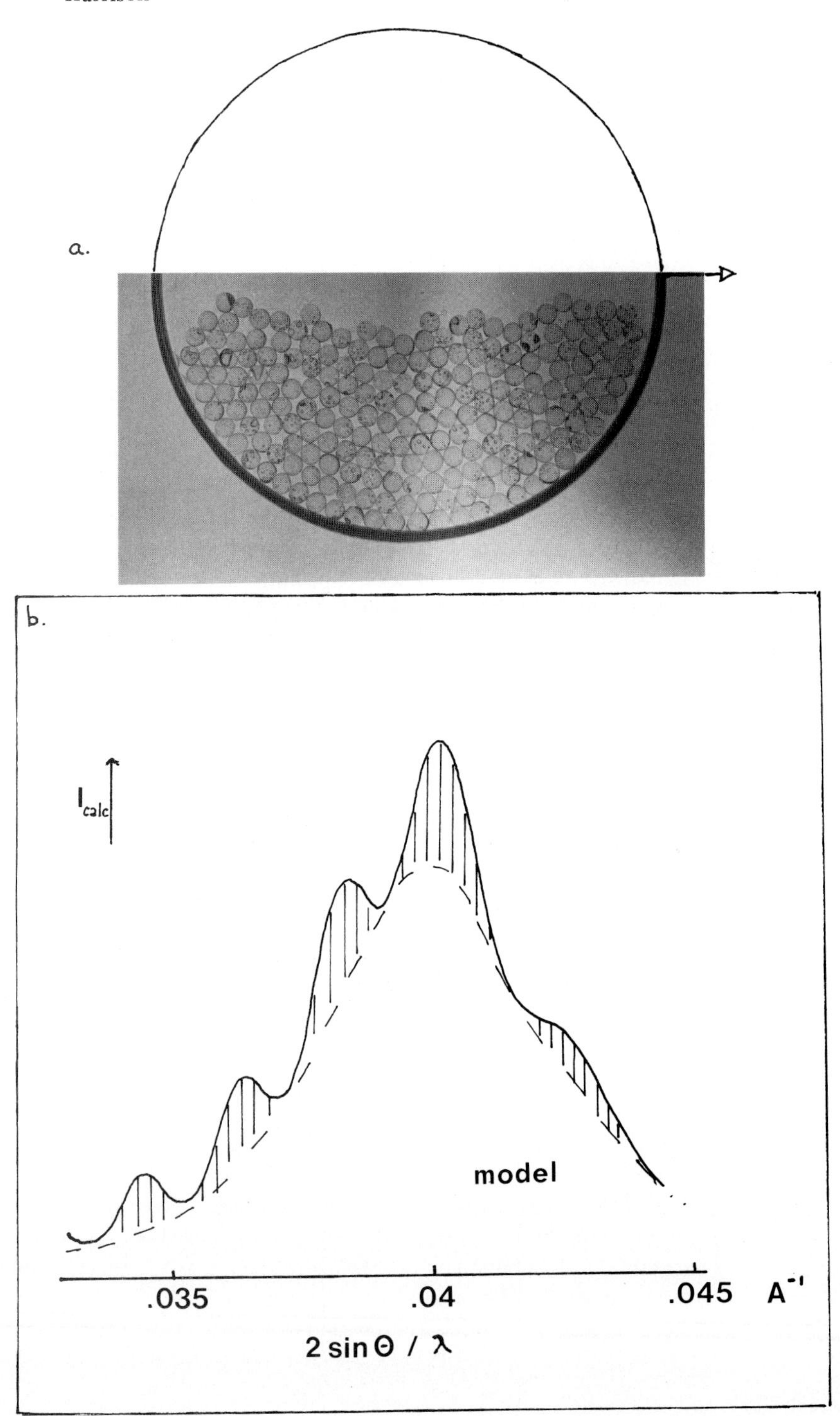

Fig. 4. Illustration of model calculation, designed to show that both rippled and diffuse diffraction can arise from DNA wound tightly, but without specific interaction, into a spherical head. a) Beads allowed to pack randomly but closely into a disk. Note that near the rim, concentric layers tend to form. Imagine rings generated by rotating this packing about the axis shown. A set of such rings can be taken as an approximation to the coil in Figure 3, and the diffraction calculated exactly. Note that the bead packing is taken as an analog of random assortment within a precise container, given a fairly sharp local-packing radius of individual segments. b) Results of the calculation.

coat protein. A fixed binding site on each protein subunit would force the nucleic acid molecule to adopt a piecewise icosahedrally symmetric arrangement, with connecting pieces looping in a more disordered fashion between the bound segments. In the TBSV particle, the flexibility appears instead to be taken up on the protein subunit itself: The RNA bonding region is not positionally fixed with respect to the part of the subunit rigidly held in the surface lattice [4].

The general architecture of the TBSV particle is presented diagrammatically in Figure 5. The picture is based primarily on a high-resolution X-ray crystallographic-structure determination. [4] The particle is composed of 180 coat-protein subunits (MW ~ 40,000), probably one chain of an 80,000 MW protein, and a molecule of single-stranded RNA (4,800 nucleotides). The coat subunit, containing about 280 amino acid residues, folds into three regions: a projecting domain (P); a domain that forms a more tightly connected shell (S); and an N-terminal arm of about 100 residues (a). Subunits accommodate to the three

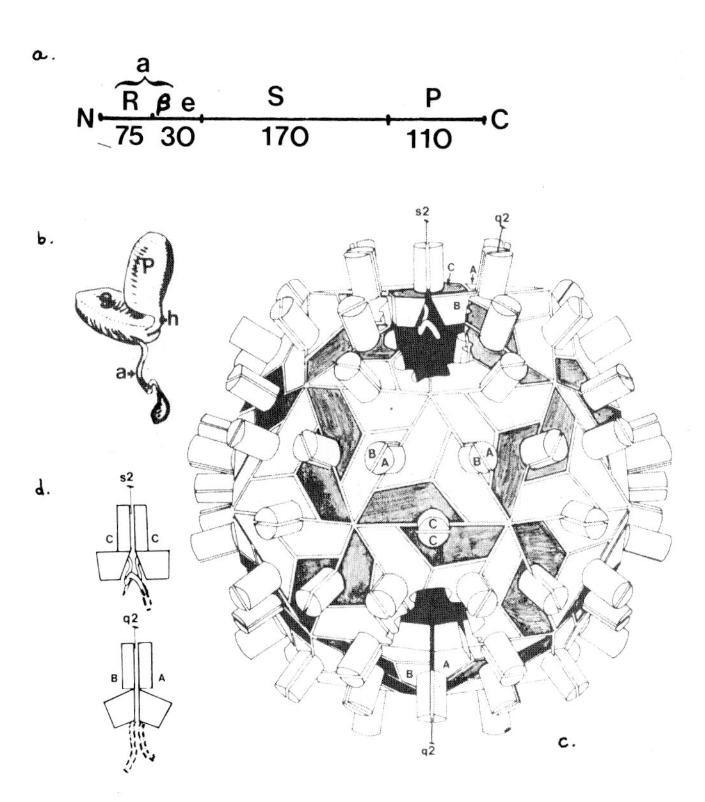

Fig. 5. Packing of protein subunits in TBSV. a) Arrangement of domains in sequence. b) Schematic subunit, showing P and S domains, the interdomain hinge (h) and N-terminal arm (a). c) Arrangement of subunits in the virus particle. A, B, and C denote the three packing environments of the subunit; outer surfaces of C subunit S domains are shaded. S domains of A subunits pack around fivefold axes; S domains of B and C alternate around threefolds. The quasi-threefold axis relating S domains of an ABC trimer is nearly parallel to the adjacent strict twofold (s2). Trimers therefore present a rather flat surface across the strict dyad and a distinctly sharper dihedral angle (about 40°) across the quasi dyad (q2). This difference in local curvature can be seen at the two places where the shell has been cut away to reveal S domain packing near strict (top) and quasi (bottom) dyads. d) The two states of the TBSV subunit found in this structure, viewed as dimers about strict (s2) and local (q2) twofold axes. Subunits in C positions have the interdomain hinge 'up' and a cleft between twofold-related S domains into which fold parts of the N-terminal arms. Subunits in the quasi-twofold-related A and B positions have hinge 'down', S domains abutting, and a disordered arm.

symmetrically distinct environments (A, B, C) in the T=3 icosahedral surface lattice by flexion at the hinge between S and P and by an ordering or disordering of part of the arm. Subunits at positions A and B (60 of each) have one hinge configuration, and the entire arm region appears to be spatially disordered. Subunits at positions C (60 in all) have another hinge position, and that part of the arm designated "β,e" in Figure 5 is folded in an ordered way along the bottom of the S domain. The portion marked R is spatially disordered in C subunits as well as in A and B. (By "spatially disordered" I mean that these regions adopt enough distinct positions with respect to the fixed S and P domains and hence to the crystal lattice, that they appear "smeared out" in the X-ray electron-density map. The packing is tight enough that these parts of the molecule are probably *not* actually moving about.) All the RNA is spatially disordered.

How is RNA packed in the TBSV particle? Since actual contacts are not discernible in the electron density map, we must take an indirect approach. Small-angle X-ray and neutron scattering show that RNA lies immediately within the S-domain shell, and also that some protein penetrates to much deeper radii (Fig. 6) [19, 20]. The volume within the S-domain shell (where the cavity has an outer radius of about 112 Å) is just sufficient to accommodate RNA with a mean hydration of 1.5 g H_2O/g protein [4]. These figures correspond to extremely efficient packing, comparable to DNA in phage heads or to tRNA in crystals. Hypochromicity of RNA in situ suggests that about 65% of the bases are stacked in helical stems (Harrison, 1975, unpublished), as in most sorts of RNA in solution. Inspection of "flower" diagrams of sequenced phage

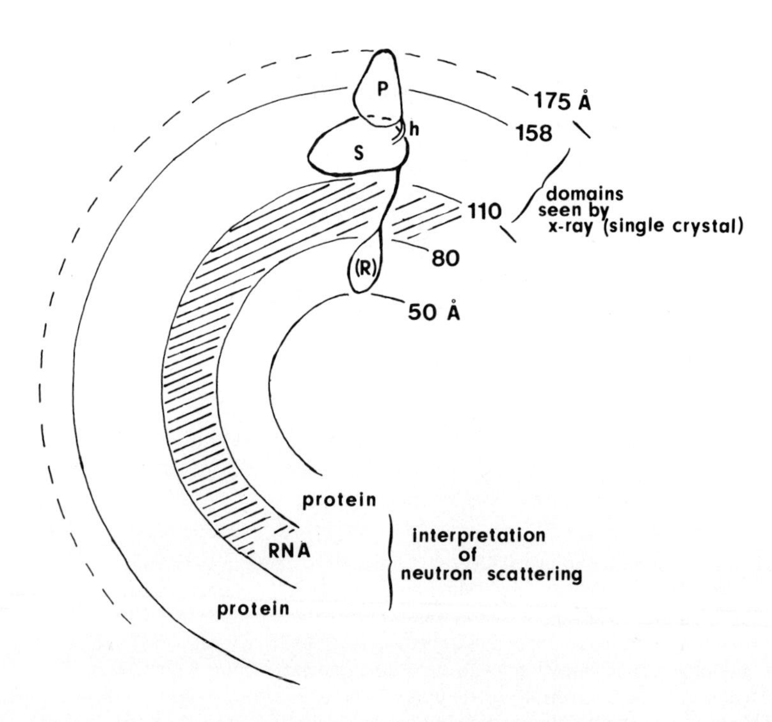

Fig. 6. Radial location of components in TBSV. The boundaries are established (approximately) from small-angle neutron scattering [reference 19].

RNA molecules [21] suggests that such stems are likely to be 10 to 20 base pairs long – ie 30 to 60 Å – and considering this distribution, the radii on Figure 6, and the likely irregularity of the connecting loops, it seems plausible that stem-like portions are tangentially oriented – packed against the inward-facing S-domain surface in a layer one or two stems thick.

How does RNA interact with the protein subunits? The inward-facing part of the S-domain, clearly seen in the x-ray map, does not appear to be a good candidate for RNA interactions [4]. This part of the domain is an extended, four-strand β-sheet. None of the inward-facing side chains are clearly identifiable as arginine or lysine, but there are a few candidates for disordered basic residues, one of which has been confirmed to be lysine by initial sequence data. Thus, even if this residue binds an RNA phosphate (and indeed, given the proximity, it may well do so), the precise location of the phosphate must vary from subunit to subunit, requiring some disorder in the side chain itself. A better candidate for tight RNA binding is the extreme N-terminal region of the molecule ("R" in Fig. 5). Preliminary sequence data (Sauer, 1979, unpublished), show about nine basic residues in the first 75 amino acids, and only one acidic. A plausible picture thus involves an approximately 70-residue, "R-domain", tethered flexibly to the β/e region, which is itself capable of considerable conformational rearrangement. On A and B subunits, an inwardly extended β/e arm could easily position R domains in a 50–70 Å shell (Fig. 6). The extent to which this picture is accurate needs to be tested by crosslinking experiments and by direct visualization of the putative R domain (cleavage, crystallization) if possible.

The significance of a flexibly linked RNA-binding domain becomes clear if we consider the probable structures encountered in assembly. From physico-chemical results on viral RNA of comparable size [22], the TBSV RNA molecule in solution can be pictured as a flexibly jointed and highly-branched collection of rodlets of varying length, with an overall radius of gyration of about 250 Å at physiological ionic strength. This collection of rodlets must be condensed into about one-tenth the volume with a quite efficient packing, as described above. It is not clear whether there is a single, uniquely stable pattern of stems and loops in the RNA (ie, a unique "flower" pattern), although the longer and more GC-rich helices are probably well determined. It may therefore be important to package a distribution of stem/loop arrangements with equal efficiency. Fixed binding sites on the S-domains would constrain particular portions of the RNA molecule to lie in a rigid, more or less triangular lattice (due to the symmetry of the icosahedral shell). Imcompatibilities with particular distributions of stem lengths could arise. Moreover, if stems were bound, the overall packing efficiency would be reduced, since they would be held in nonparallel arrays.

We do not yet have a detailed view of the assembly of any RNA containing spherical virus. It is reasonable to assume that in TBSV initial interaction between protein and RNA occurs at a nucleating sequence having particular affinity for a protein binding site. This binding site is likely to be the N terminal regions of the coat subunit, but the as yet uncharacterized 80,000 MW component could have a role in the process as well. What state of association the protein has attained when this occurs is also unknown. We can infer that most or all of the subunits actually do bind to RNA from the tenacious interaction between these components in TBSV and TCV. If so, the flexible linkage between the RNA binding site and the part of the protein locked into the surface lattice makes it much easier to accomodate the unsymmetrical RNA molecule in an icosahedral shell.

ACKNOWLEDGMENTS

My colleagues, W.C. Earnshaw, G. Bricogne, F.K. Winkler, C.E. Schutt, and A. Olson are responsible for most of the work from which these thoughts were drawn; they should be allotted no blame for the speculations. The work was supported by NIH grant CA-13202 and by an Alfred Sloan Fellowship.

REFERENCES

1. Caspar DLD: Adv Protein Chem 18:37, 1963.
2. Zimmern D: Cell 11:463, 1977.
3. Earnshaw WC, Harrison SC: Nature 268:598, 1977.
4. Harrison SC, Olson A, Schutt CE, Winkler FK, Bricogne G: Nature 276:368, 1978.
5. Casjens S, King J: Ann Rev Biochem 44:555, 1975.
6. Richards K, Williams RC, Calendar R: J Mol Biol 78:255, 1973.
7. Earnshaw WC, King J, Harrison SC, Eiserling FA: Cell 14:559, 1978.
8. North ACT, Rich A: Nature 191:1242, 1961.
9. Maniatis T, Venable JH Jr, Lerman LS: J Mol Biol 84:37, 1974.
10. Gellert M, Davies D: J Mol Biol 8:341, 1964.
11. Kosturko LD, Hogan M, Dattagupta N: Cell 16:515, 1979.
12. Gosule LC, Schellman JA: Nature 259:333, 1976.
13. Kaiser D, Syvanen M, Masuda T: J Mol Biol 91:175, 1975.
14. Emmons SW: J Mol Biol 83:511, 1974.
15. Hendrix RW: Proc Natl Acad Sci USA 75:4779, 1978.
16. Black LW, Silverman DJ: J Virol 28:643, 1978.
17. Chattoraj DK, Inman RB: J Mol Biol 87:11, 1974.
18. Syvanen M, Yin J: J Mol Biol 126:333, 1978.
19. Chauvin C, Witz J, Jacrot B: J Mol Biol 124:641, 1978.
20. Harrison SC: J Mol Biol 42:457, 1969.
21. Fiers W et al: Nature 260:500, 1976.
22. Witz J and Strazielle C: In Timasheff S, Fasman G (eds): "Subunits in Biological Systems, part B." New York: Dekker, 1976.

Biological Recognition and Assembly 311–317 (1980)

Nucleic Acids-Protein Interactions: Structural Studies by X-Ray Diffraction and Model Building

Sung-Hou Kim

Department of Biochemistry, Duke University Medical Center, Durham, North Carolina 27710

X-ray diffraction and model-building studies suggest that the α-helix and β-structure of protein may interact with double-helix nucleic acids. In the former case basic side chains of amino acid residues in the α-helix can interact with phosphate groups of the double helix, and in the latter case the peptide backbone in a β-ribbon can hydrogen-bond to the backbone of the double helix of nucleic acids. These structurally compatible interactions between well-known secondary structures of protein and the double helix provide models for understanding specific interaction between nucleic acid and protein during general recognition, or preliminary stages before base sequence-specific recognition.

Key words: **DNA-protein interaction, RNA-protein interaction, β-ribbon–double-helix interaction, α-helix–double-helix interaction, nucleoprotamine structure, X-ray diffraction studies of nucleic acid–protein interaction, model building of nucleic acid–protein interaction, structural basis of regulation in sperm maturation**

INTRODUCTION

It has been 26 years since the double-helix model of Watson-Crick for DNA was proposed. However, it is still not known, at the three-dimensional (3-D) structural level, how most of nucleic acid-associated proteins interact specifically with double-helix DNA. There are basically two classes of proteins: One class of protein recognizes particular DNA base sequences, and the other recognizes the repeating secondary structure of nucleic acid independently of base sequence in the double-stranded DNA. As of today, no 3-D structural studies of the first category are known. This article reviews the past 3-D structural studies of the second category of interactions and presents some recent results.

Dr. Kim is now at the Department of Chemistry, and Laboratory of Chemical Biodynamics, University of California, Berkeley, California 94720

Received March 28, 1979; accepted April 26, 1979.

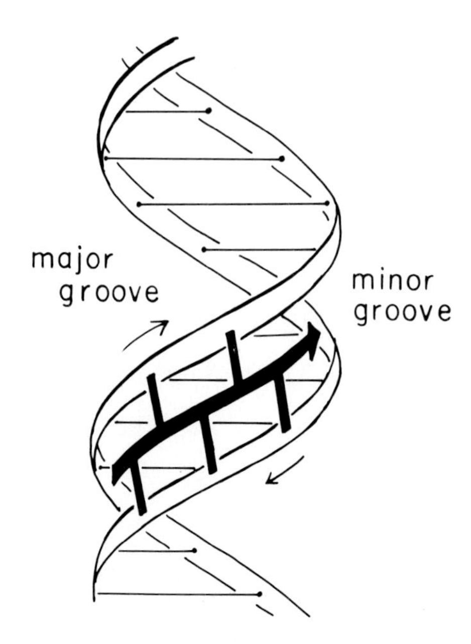

Fig. 1. Schematic drawing of β-chain–double-helix interaction model of Feughelman et al [1], showing how protamine might bind to DNA double helix. The extended polypeptide chain in the β-conformation, as indicated by the dark strips, winds around the minor groove of the DNA helix. The phosphate groups on the DNA double helix are in contact with the basic ends of arginine side chains, as indicated by the thin black strips. Nonbasic residues are presumed to bulge out from the main polypeptide chain.

β-Chain—Double-Helix Interaction Model

In 1955 Fueghelman and his colleagues [1, 2] proposed a model for nucleoprotamine, a complex between double-helix DNA and protamine, based on their X-ray diffraction studies of the material in fibrous state. In this model a protamine molecule, a short protein rich in basic residues, is fully extended as a β-chain and laid along the minor groove of DNA B form. Interactions in this model are between the positively charged basic residue side chains and the negatively charged phosphates of DNA, with no interactions involving the backbone of the peptide. The schematic drawing of this model is shown in Figure 1.

β-Ribbon—Double-Helix Interaction Model

In 1974 Carter and Kraut [3] extended the observation of Chothia [4] (that β-pleated sheets in proteins often adopt a right-handed twist) to model building of a β-ribbon (a pair of antiparallel β-chains) into double-helix RNA. Subsequently Church et al [5] showed that a similar model can be built for β-ribbon and double-helix DNA. These two models relating β-ribbon and double-helix RNA or DNA are built on the basis of certain structural elements that characterize both β-ribbon and double-helix nucleic acid: a) Double-helix DNA and RNA have two types of approximate twofold axes, one on the base pair plane and the other between adjacent base pairs (see Fig 2b); there are also two types of twofold axes in antiparallel β-ribbon as shown in Figure 2a; b) the distance between two adjacent phosphate groups on the double-

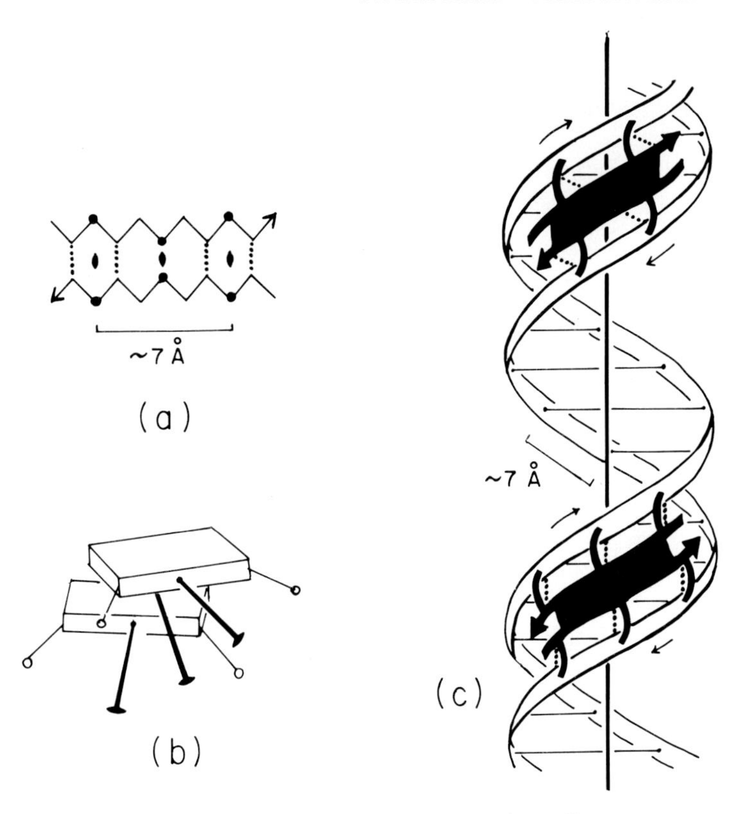

Fig. 2. a. Diagrammatic representation of a pair of antiparallel chains in a β-ribbon. The backbone of polypeptides is indicated by zig-zag lines and hydrogen bonds are shown as dotted lines. The locations of two types of pseudo-twofold axes are indicated by oval-shaped signs. The distance between alternate α-carbons along the polypeptide is about 7 Å. b. Two adjacent base pairs in double-helix DNA or RNA are illustrated as two thick slabs with glycosyl bonds attached to them (by lines with circles at the end.) This schematic drawing shows two different types of pseudo-twofold axes (indicated by dark rods with ellipses) in double-helix nucleic acids: One on the base pair plane and the other between two base pair planes. c. Two types of β-ribbon–double-helix interaction models are shown. The antiparallel β-ribbons are indicated by black ribbons and hydrogen bonds between peptide and nucleotide backbones are shown by dotted lines. Curved black strips indicate the basic residues from the β-ribbon reaching and neutralizing the phosphate groups on polynucleotide backbones. Two types of models can be built: One (top) where polarity of the DNA is parallel to the polarity of the peptide; the other (bottom), where the two polarities are opposite each other. The distance between two adjacent phosphate groups on the outside along each strand of the double helix is about 7 Å.

helix backbone of nucleic acids on the outside is approximately 7Å; the distance between alternate α-carbon atoms on polypeptides are also approximately 7 Å. These symmetry and dimensional similarities between the double-helix nucleic acid and the β-ribbon, in addition to the right-handed twist, of both secondary structures are the basis of the specific models described above. The schematic drawings of these interactions are shown in Figure 2c, and the details of the proposed hydrogen-bonding schemes are described in Carter and Kraut [3] and Church et al [5]. One interesting consequence of these models is that, if alternating residues of a polypeptide are basic residues, then all the basic side chains will be on the same side of the β-ribbon and they can neutralize the negative charges of phosphate at the same time that the peptide backbone is forming hydrogen bonds, as schematically shown in Figure 2c. One such example is histone H2b of calf thymus, in which, near the N-terminus of this protein, the basic side chain is found in approximately every other residue.

α-Helix–Double-Helix Interaction Model

More recently a single-crystal X-ray diffraction technique has been used to study interaction between protamine and double-helix RNA [6]. Such studies were possible because there are several amino acid sequences of protamines that are available, and three-dimensional structures of transfer RNA (for a review see Kim [7]) shows that the molecule is composed of approximately two double-helix regions at 90° to each other in the shape of an L.

Protamines are extremely basic, small proteins of molecular weights of about 4,200 daltons, and they bind tightly to DNA or RNA (for a review see Ando et al [8]). They are found tightly associated with DNA in fish spermatozoa. Nearly two-thirds of amino acid residues in protamines are basic, and these basic residues are usually found clustered four or five in a sequence. Thus, protamine-nucleic acid interaction provides one of the simplest systems for the study of interaction of basic protein–nucleic acid complexes.

The details of experimental conditions and the analyses have been presented by Warrant and Kim [6]. The summary of experimental procedures is as follows. First, single crystals

Fig. 3. Right: Difference electron density map between a crystal of protamine-yeast phenylalanine tRNA and a tRNA crystal without protamine. The contouring level was chosen to show only those regions where the electron density is greater than five times the average electron densities of the entire unit cell. An α-helix of ten amino acid residues is fitted into this electron-dense segment. Left: Schematic drawing showing the location of an α-helix segment bound to the shallow (minor) groove of the T stem of yeast phenylalanine tRNA crystal structure. The α-helix segment is shown as a cylinder. The shallow grooves in the tRNA structure are shaded.

of a yeast phenylalanine transfer RNA were formed. Into the solutions which contained tRNA crystals were added protamine at a ratio of four protamine molecules to one tRNA, which corresponds on the average to one arginine residue per phosphate. X-ray diffraction data at 5.4-Å resolution have been collected for the tRNA crystals and for tRNA crystals soaked in protamine solutions. Using the knowledge of the three-dimensional structure of this tRNA in this particular crystal form and the differences of diffraction intensities between native tRNA crystals and protamine-soaked crystals, one can calculate maps of electron density due to the protamine alone. The electron density map thus calculated shows a small, cylindrical region of high electron-density lying on the minor groove of a double-helix region of the tRNA (Fig. 3). Furthermore, this electron-

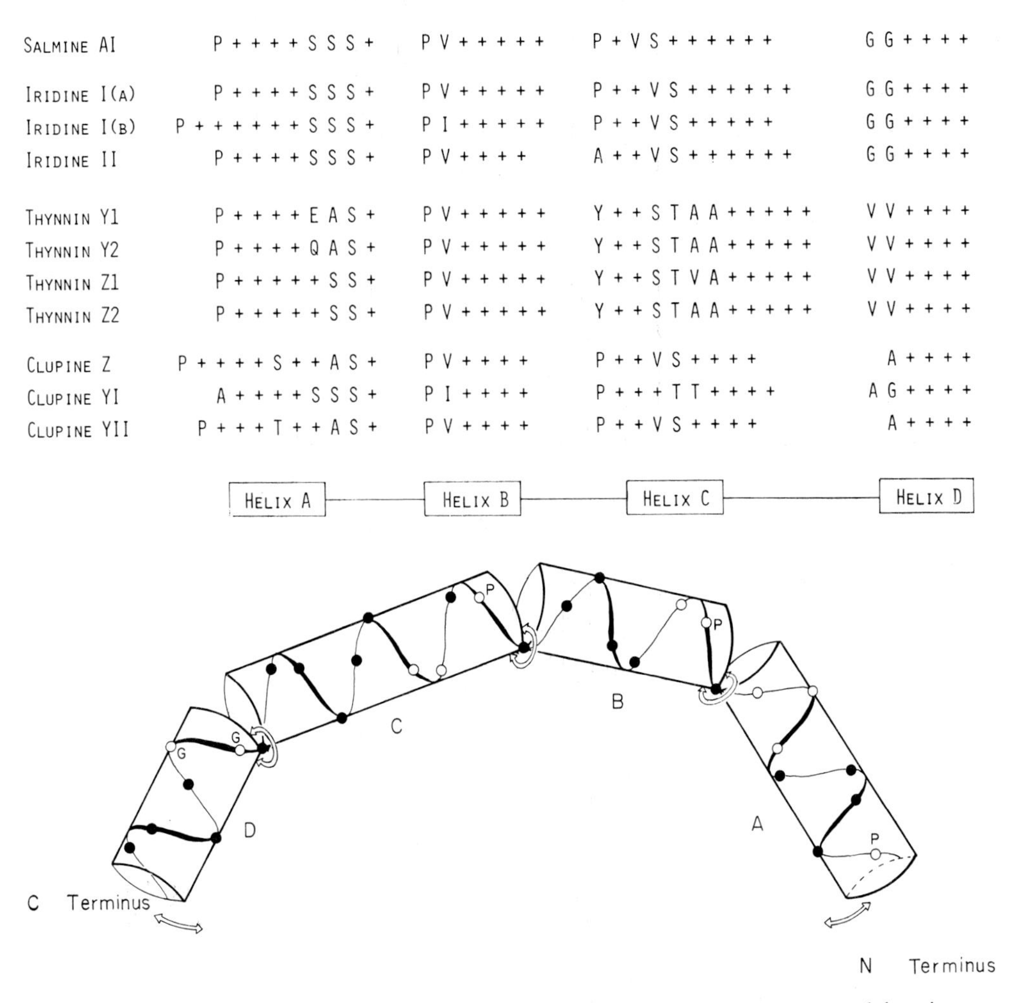

	Helix A	Helix B	Helix C	Helix D
Salmine AI	P + + + + S S S +	P V + + + + +	P + V S + + + + + +	G G + + + +
Iridine I(A)	P + + + + S S S +	P V + + + + +	P + + V S + + + + + +	G G + + + +
Iridine I(B)	P + + + + + + S S S +	P I + + + + +	P + + V S + + + + +	G G + + + +
Iridine II	P + + + + S S S +	P V + + + +	A + + V S + + + + + +	G G + + + +
Thynnin Y1	P + + + + E A S +	P V + + + + +	Y + + S T A A + + + + +	V V + + + +
Thynnin Y2	P + + + + Q A S +	P V + + + + +	Y + + S T A A + + + + +	V V + + + +
Thynnin Z1	P + + + + + S S +	P V + + + + +	Y + + S T V A + + + + +	V V + + + +
Thynnin Z2	P + + + + + S S +	P V + + + + +	Y + + S T A A + + + + +	V V + + + +
Clupine Z	P + + + + S + + A S +	P V + + + +	P + + V S + + + +	A + + + +
Clupine YI	A + + + + S S S +	P I + + + +	P + + + T T + + + +	A G + + + +
Clupine YII	P + + + T + + A S +	P V + + + +	P + + V S + + + +	A + + + +

Fig. 4. Top: All the known sequences of protamines [9] are arranged to show four proposed domains. Abbreviations: A, alanine; E, glutamic acids, G, glycine; I, isoleucine; P, proline; Q, glutamine; S, serine; T, threonine; V, valine; Y, tyrosine; +, arginine. Bottom: Diagrammatic representation of a protamine structure. The protamine is assumed to be composed of four α-helix domains joined by three partially flexible joints at the residues just prior to two internal prolines and one joint before a pair of glycine residues. Arginines are shown as dark circles; P, proline; G, glycine.

dense region can be interpreted as due to an α-helix structure formed by about eight amino acid residues. The distance between the axis of the α-helix and the phosphates of the RNA double-helix is compatible with the interpretation that two consecutive, fully stretched arginine side chains are hydrogen-bonding to the two phosphates across the minor groove of the double-helix RNA and neutralizing the charges at the same time. Therefore, each domain with four or five consecutive arginines acts as a "divalent cross-linking" agent, with a pair of consecutive arginines binding to one double helix across a groove, and the other pair binding to the other double helix.

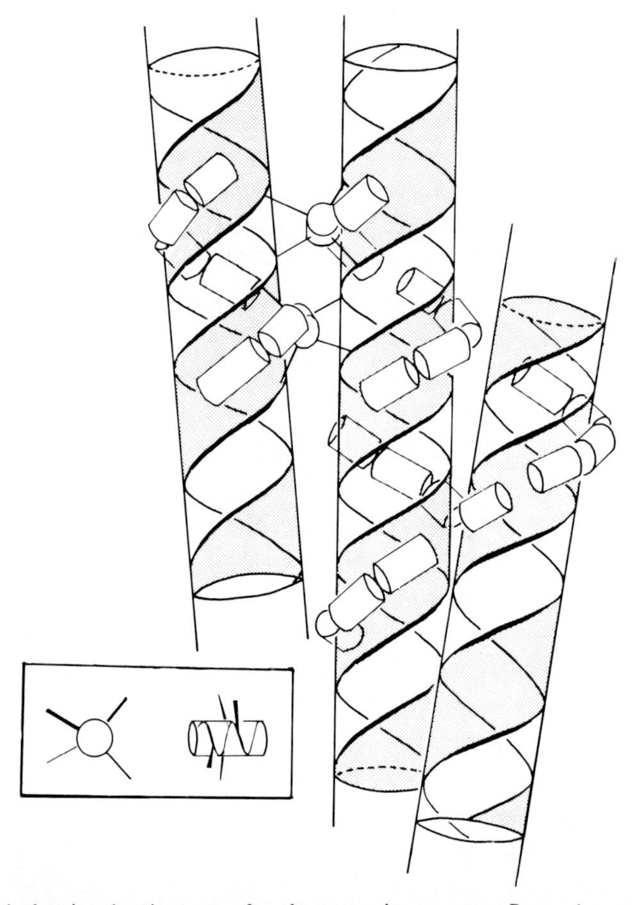

Fig. 5. Schematic drawing showing parts of nucleoprotamine structure. Protamine molecules are represented by several connected cylinders: Each corresponds to an α-helix domain. They wrap around the major groove (shaded) of a DNA double helix shown as a long, large cylinder. Inset: An α-helix domain with four consecutive arginine residues (represented as four rods radiating out from the cylinder) shown in two orthogonal orientations. Two types of possible cross-linking are shown. Between the left and middle DNA double helices there are two α-helix segments with arginine side chains shown as straight lines. Each segment forms two contacts across a major groove of one DNA double helix and two across a minor groove of the other DNA double helix (or forms contacts across two major grooves from two DNA double helices). Between the middle and the right DNA double helices, three α-helix segments from one protamine molecule are in contact on the major groove of the middle DNA double helix and the fourth segment on the major groove of the right DNA double helix. One protamine can also wrap around more than one DNA helix.

An examination of known amino acid sequences of the protamines [9] strongly suggests that the protamine molecule is probably composed of four α-helix domains, each domain containing an average of eight amino-acid residues, as schematically shown in Figure 4.

When this protamine model is fitted to the DNA B form, it fits very nicely on the major groove. This is because the major groove of the DNA B form is about the same size as the minor groove of the double-helix region of the tRNA. The schematic drawing of DNA-protamine complex is shown in Figure 5. This nucleoprotamine model is consistent with the X-ray fiber diffraction results of Feughelman et al. [1], the chemical protection experiment of Mirzabekov et al [10], and others. (For further discussion, see Kim [6].)

One of the interesting consequences of the α-helix—double-helix model for nucleoprotamine is that it provides a structural basis for understanding the regulation of sperm maturation. The following structural interpretation of the known sequence of events is suggested: The protamine molecules are synthesized in the cytoplasm and then serine residues are phosphorylated. The negatively charged phosphate group can sequester a pair of arginines nearby, thus changing the "divalent cross-linking α-helix domains" of protamine into "monovalent" domains. The fully phosphorylated molecules then enter the nucleus and displace the histones. At this point, the histone displacement can occur effectively without condensing DNA which may render chromatin less accessible to protamines. When histones are displaced, the protamines get dephosphorylated (thus, each α-helix domain becomes a "divalent cross-linking" agent again); concomitant condensation of DNA occurs and stops all transcription, a final step in fish sperm maturation.

DISCUSSION

Although there are no 3-D models for base sequence-specific recognition as yet, X-ray diffraction and careful model-building studies show that there are structural features common to secondary structures of proteins, such as α-helix and β-ribbon, that are complementary to features in double-helix DNA and RNA. Such complementarity suggests several modes of interaction between proteins and double-helix nucleic acids for general recognition as well as for setting a stage for the base sequence-specific recognition.

ACKNOWLEDGMENTS

Results from the author's laboratory cited in this article have been supported by grants from the National Institutes of Health (CA-15802, K4-CA-0032) and the National Science Foundation (PCM76-4248).

REFERENCES

1. Feughelman M, et al: Nature 175:834–838, 1955.
2. Wilkins MHF: Cold Spring Harbor Symp. Quant Biol 21:75–90, 1956.
3. Carter C, Kraut J: Proc Natl Acad Sci USA 71:283–287, 1974.
4. Chothia C: J Mol Biol 75:295–302, 1973.
5. Church GM, Sussman JL, Kim SH: Proc Natl Acad Sci USA 74:1458–1462, 1977.
6. Warrant RW, Kim SH: Nature 271:130–135, 1978.
7. Kim SH: In Meister A (ed): "Advances in Enzymology." New York: Wiley, 1978, vol 46, pp 279–315.
8. Ando T, et al: "Protamines: Isolation, Characterization, Structure and Function." New York: Springer, 1973.
9. Croft LR: "Handbook of Protein Sequences." Oxford: Joynson-Bruvvers, 1973 (Supplement A, 1974).
10. Mirzabekov A, San'ko D, Kolchinsky A, Melnikova A: Eur J Biochem 75:379–389, 1977.

Biological Recognition and Assembly 319–326 (1980)

The Role of Antibody Multivalency in Immune Effector Processes

David M. Segal, Steven K. Dower, Joye F. Jones, and Julie A. Titus

Immunology Branch, National Cancer Institute, National Institutes of Health, Bethesda, Maryland

The preparation of covalently crosslinked oligomers of antibody molecules is described. It is shown that these oligomers can be used to study quantitatively the effects of multivalency on binding to cell surface Fc receptors. It is also shown that three effector systems differ in the ways in which they interact with multivalent immune complexes.

Key words: **antibody-dependent cell-mediated cytolysis (ADCC), antibodies, bivalent affinity labels, complement, Fc receptors, immunoglobulins, passive cutaneous anaphylaxis (PCA)**

Immunoglobulins possess a vast array of complementarity determining regions with which they can bind virtually any foreign antigen that the immune system is likely to encounter. Subsequent to antigen binding, immunoglobulins can then interact with various effectors, for example, complement, phagocytes, or "killer cells." Interactions between antibodies and effectors occur between specific sites on the constant portions of the immunoglobulin molecules and soluble or cell-bound receptor proteins of the effector systems. Antibody molecules thus serve as adapters between antigens, which bind at one end of the immunoglobulin molecule, and effectors, which bind at the other end.

The mechanisms by which effector systems recognize antibodies complexed with antigens versus those that are free have been the subject of many studies over the past several years [1–3]. One school holds that upon combination with antigen the antibody would undergo an allosteric transition in the constant region of the immunoglobulin. This antigen-induced conformational change would give rise to a structure that would bind tightly to the receptor proteins of the immune effector (Fig. 1, top).

A second model (Fig. 1, bottom) proposes that the important feature in antigenic recognition is the formation of multivalent immune complexes. The linking of several immunoglobulin molecules might either enhance the affinity with which a multivalent

Abbreviations: ADCC, antibody-dependent cell-mediated cytolysis; $(BADL\text{-}Pro)_2$-EDA, bis (α-bromoacetyl-ϵ-DNP-Lys-Pro) ethylenediamine; BDPE, bis-DNP-pimelic ester; DNP, 2, 4-dinitrophenyl; IgG, immunoglobulin G; PCA, passive cutaneous anaphylaxis; TNP, 2,4,6-trinitrophenyl.

Received May 25, 1979; accepted May 31, 1979.

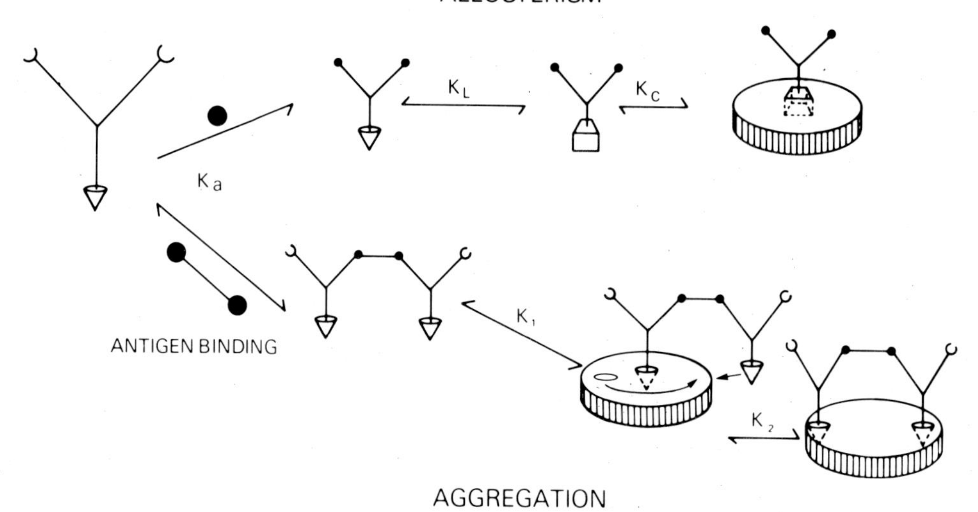

Fig. 1. Diagram of allosteric and aggregational mechanisms of immune activation. Antibody molecules are represented by Y-shaped figures, with antibody-combining sites at ends of arms of Y and cell-binding portion at bottom of stem. Antigens are drawn as filled circles. In the allosteric model (top) antigen binds to combining sites on Fab portions with association constant K_a and induces a conformational change in the cell-binding region of the Fc. The equilibrium constant for the conformational transition is K_L. The new conformation in the cell-binding region then binds tightly (with binding constant = K_C) to the (complementary) receptor on the cell surface. In the aggregational model (bottom), bivalent antigen links two antibody molecules together. The dimer first binds univalently with equilibrium constant K_1. A second cell surface receptor moves into position, and the second subunit of the dimer binds with constant K_2.

effector would bind to the antibodies or, alternatively, might serve as a signal by cross-linking receptor molecules.

Differences between allosteric and multivalency models often show up in ligand-protein dose response curves. In Figure 2B the saturation of hemoglobin with oxygen is shown to follow a transition that is much sharper than expected for a simple non-cooperative process [4]. This abrupt transition arises because the binding of an oxygen molecule by one hemoglobin subunit provokes a conformational change in a second subunit, with the result that the second oxygen molecule binds more tightly than the first. In contrast, when antibodies interact with multideterminant antigens (Fig. 2a), the dose-response curves are biphasic, reaching a maximum when the largest number of crosslinks is formed [5]. In Figure 2a the parameter measured is the fraction of antibody that precipitates when a multideterminant antigen is added. However, both complement fixation and binding to effector cells show similar behavior. A number of studies [1, 2] suggest that antibodies of the IgG and IgE classes are recognized by immune effectors because of the formation of multivalent complexes. In this paper we explore some of the mechanisms by which multivalent immune complexes might interact with different effector systems.

THE PRODUCTION OF MODEL IMMUNE COMPLEXES

Natural antigen-antibody complexes are heterogeneous with respect to size, and

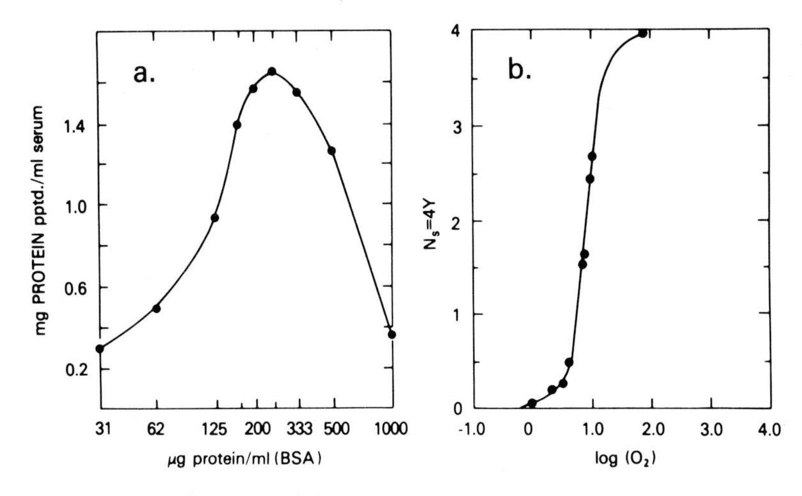

Fig. 2. Dose-response curves for aggregating and allosteric systems. A. The precipitation of rabbit anti-bovine serum albumin antiserum with bovine serum albumin (data taken from reference 5). B. The binding of oxygen by hemoglobin. N_S is the number of oxygen molecules bound per hemoglobin molecule. log (0_2) is the log of the oxygen pressure in mm (data taken from reference 4).

are therefore not well suited for studying the effects of antibody multivalency upon immune effector processes. In the past few years this problem has been largely overcome by the production of bivalent affinity labeling reagents. These compounds (Fig. 3) bind to antibodies and form covalent bonds with residues near the combining sites. Because the reagents are bivalent, they can form multivalent complexes that resemble natural immune complexes in that crosslinking occurs at the antibody combining sites (Fig. 4). However, unlike natural immune complexes, the affinity labeled antibodies are covalently crosslinked. Stable oligomers of defined size can thus be prepared with the bivalent affinity labels.

Two such reagents have been reported [6, 7], both of which specifically crosslink anti-DNP antibodies. One, $(BADL\text{-}Pro)_2$-EDA (Fig. 3), produces oligomers with occupied combining sites [6]. This reagent can extend about 19 Å in length. The oligomers formed with $(BADL\text{-}Pro)_2$-EDA are stable only in solutions containing free hapten, and when hapten is removed, they will polymerize. The other reagent, bis-DNP-pimelic ester (BDPE) (Fig. 3), forms oligomers [7] with free sites (Fig. 4). Oligomers formed with BDPE can be further polymerized by adding antigens that are multivalent in DNP, and they will bind to cells coated with DNP or TNP groups.

Monomer, dimer, and trimer fractions have been isolated by gel filtration from anti-DNP antibodies crosslinked with both BDPE and $(BADL\text{-}Pro)_2$-EDA. Larger oligomers are more difficult to purify, and the isolation of homogeneous tetramer has not yet been reported. The structures of the oligomers are, of course, not fixed by the crosslinking process, and because of the segmental flexibility of the IgG molecule [8], the oligomers are probably highly motile both in solution and when bound to cells. In this respect it is not clear whether the specifically crosslinked oligomers resemble natural immune complexes. For this reason it is best to study oligomers crosslinked with both $(BADL\text{-}Pro)_2$-EDA and BDPE, as well as oligomers crosslinked nonspecifically with dimethylsuberimidate (Figs. 3 and 4). Strong configurational effects would most likely be apparent from differences in the ways in which equal-sized oligomers crosslinked with different reagents

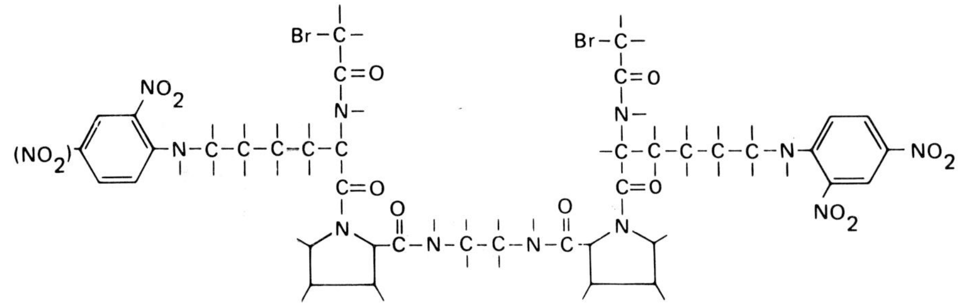

Br–C–
C=O
N–
NO$_2$
(NO$_2$)–N–C–C–C–C–C
C=O
N
C–N–C–C–N–C
C=O
N
–C–C–C–C–C–N–NO$_2$
C=O
N–
Br–C–
NO$_2$

(BADL-pro)$_2$- EDA

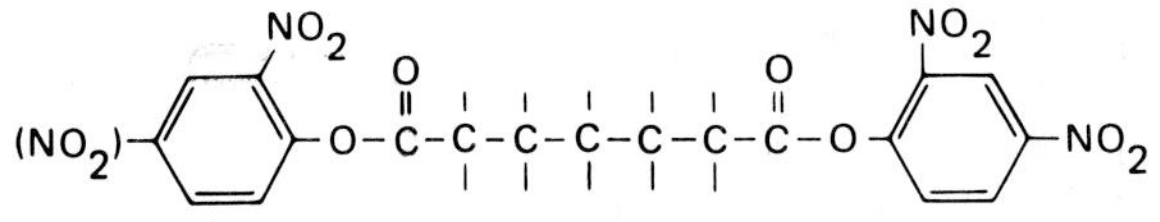

NO$_2$ NO$_2$
(NO$_2$)–O–C–C–C–C–C–C–C–O–NO$_2$

Bis-Dnp-Pimelic Ester

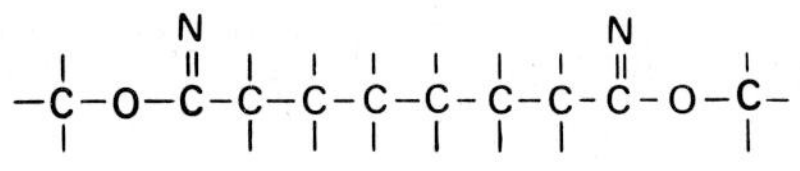

N N
–C–O–C–C–C–C–C–C–C–C–O–C–

Dimethyl Suberimidate

Fig. 3. The chemical structure of three bivalent crosslinking reagents. (BADL-Pro)$_2$-EDA and bis-DNP-pimelic ester (BDPE) crosslink anti-DNP antibodies at their combining sites. Dimethylsuberimidate crosslinks two adjacent lysyl residues on the same or different protein molecules.

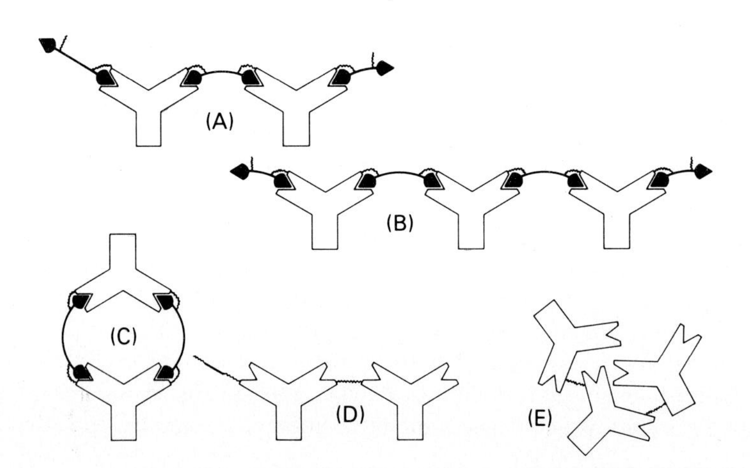

Fig. 4. Schematic representations of types of oligomers formed by the reagents shown in Figure 3. Anti-DNP antibodies are indicated by the Y-shaped structures, and the combining sites are represented by the V-shaped notches in the arms of the Y. DNP groups are designated as black sectors on the bivalent reagents. A. A dimer formed with (BADL-Pro)$_2$-EDA. Note that the DNP combining sites remain occupied with this reagent. B. A (BADL-Pro)$_2$-EDA trimer. C. A cyclic dimer formed with (BADL-Pro)$_2$-EDA. D. A dimer formed with BDPE. Note the combining sites are free. E. A trimer formed with dimethylsuberimidate. Crosslinking can occur between any portions of the molecules, and the combining sites remain unoccupied.

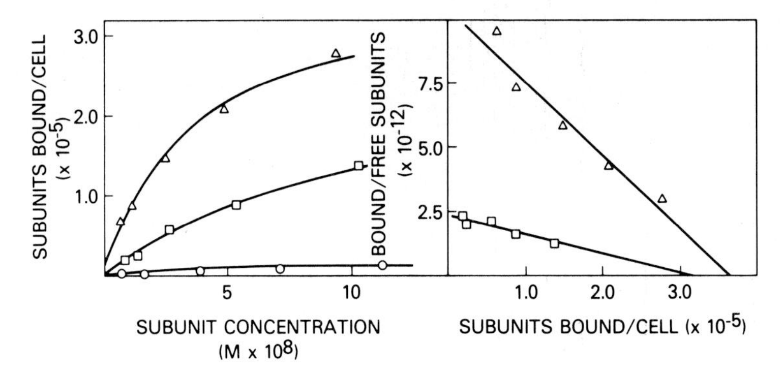

Fig. 5. The binding of oligomers of rabbit IgG to $P388D_1$ cells at 30°. A subunit is a 7S constituent of an oligomer. The oligomers used were $(BADL\text{-}Pro)_2$-EDA; ○, monomer; □, dimer; and Δ, trimer. Left. Binding curves for the three oligomers. The dimer and trimer curves are calculated from the parameters obtained from the Scatchard analyses illustrated in the right panel. The monomer binding is too weak to be analyzed quantitatively, and thus the curve was drawn by hand. Right. Scatchard plots for dimer and trimer binding. The best fit straight lines shown give: For the trimer, 3.7×10^5 receptors per cell, $K_a = 2.9 \times 10^7\ M^{-1}$, for the dimer, 3.2×10^5 receptors per cell, $K_a = 7.5 \times 10^6\ M^{-1}$.

interact with immune effectors, for example, in fixing complement or in binding to effector cells.

The Binding of Immunoglobulin Oligomers to Cells Bearing Fc Receptors

The interaction of immune complexes with effectors of the immune system involves the binding of the Fc region of immunoglobulins to sites on the effectors. Many of these effectors are cells that interact with immune complexes by binding the Fc regions of immunoglobulins via specific receptors (known as "Fc receptors") on the cell surface [9].

Figure 5 shows the binding of IgG oligomers to Fc receptors on cells from the macrophage line, $P388D_1$ [10]. It can be seen that the affinity for the cells increases with the size of the oligomer. In other studies [11, and Titus and Segal, unpublished results] it has been observed that this relationship between oligomer size and binding affinity is independent of both the mode of crosslinking and the presence of antigen. Since the only factor affecting the affinity appears to be the size of the oligomer, it is highly probable that oligomeric complexes bind more tightly than monomeric immunoglobulin molecules because they form multiple points of attachment with Fc receptors on the cell surface.

The multivalent interactions are reflected in the kinetic behavior of the system. Figure 6 shows the rates of binding and dissociation of trimeric IgG from $P388D_1$ cells. When the cells are incubated with trimer, the uptake of oligomer reaches a plateau as the system approaches equilibrium. After removal of unbound oligomer from the medium by washing, the trimer experiences an initial rapid rate of dissociation, followed by an extremely slow phase. The slow release of the bound trimer can be understood in terms of the equilibria occurring on the cell surface. In order for the trimer to be released from the cell all of the Fc regions must sequentially dissociate from the Fc receptors to which these are bound. At any stage in this process free receptors may "recapture" the trimer, Figure 6 shows that this "recapture" can be prevented by the addition of a large excess

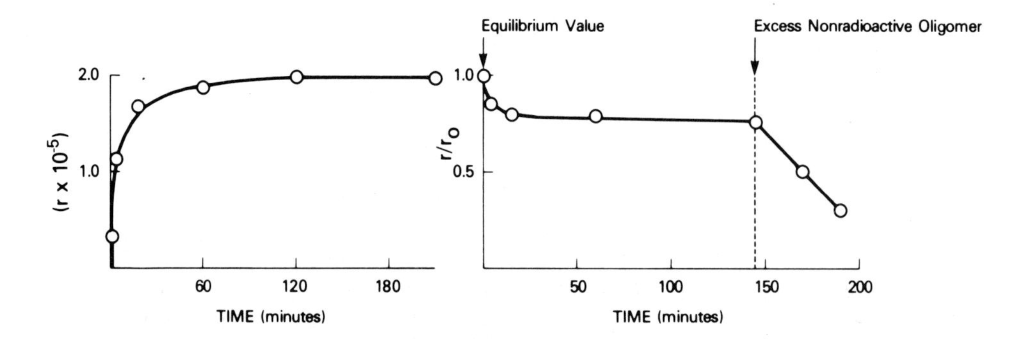

Fig. 6. The kinetics of the interaction of IgG trimer with P388D$_1$ cells at 0°. Left. P388D$_1$ cells were incubated with 8.0×10^{-8} M of subunits as a function of time. Right. P388D$_1$ cells were allowed to equilibrate with 3.0×10^{-8} M of ^{125}I subunits. At zero time the cells bound 1.1×10^5 molecules of subunit per cell. The release reaction was initiated by washing the cells with medium at 0°C. At the indicated time, unlabeled oligomer was added to a final concentration of 3 mg/ml; r refers to the number of IgG subunits bound per cell.

of oligomer, which saturates the free receptors. Under these conditions the multivalent effect is abolished and the trimer dissociates rapidly from the cells.

Thus, both the equilibrium binding and the kinetic behavior of this model system are markedly affected by the multivalency of the interacting species. Since many of the molecular interactions of the immune system are multivalent, it should be expected that these types of interactions will affect the ways in which immune effectors process antigen-antibody complexes.

Effects of Antibody Multivalency on Immune Functions

In general, the interactions of immune effectors with antigen-antibody complexes require that the complexes be multivalent with respect to antibody. The one known exception to this rule, the fixation of complement by antibodies of the IgM class [12], may require some sort of conformational change to elicit this effector function [13, 14]. Studies on IgM antibodies have been carried out in several laboratories and will not be discussed further in this paper.

Oligomers formed from anti-DNP antibodies with BDPE or dimethylsuberimidate (Fig. 3) have free combining sites and can bind to TNP-coated erythrocytes [15]. In this way target cells can be produced which have antibody monomers, dimers, trimers, etc distributed randomly over the cell surface. These target cells have been useful in determining the distributional requirements of IgG molecules for eliciting effector responses.

In one group of experiments whole guinea pig complement was used to lyse trinitrophenylated (TNP) sheep erythrocytes that had been sensitized with oligomers of anti-TNP IgG [15]. The results of such an experiment (Fig. 7A) demonstrate that lytic efficiency increases continuously with oligomer size and with the number of 7S IgG subunits bound.

A second system, antibody-dependent cell-mediated cytotoxicity (ADCC), was investigated using TNP-chicken erythrocytes coated with BDPE oligomers as targets and normal murine splenocytes as effectors [16]. The data from one experiment are shown in

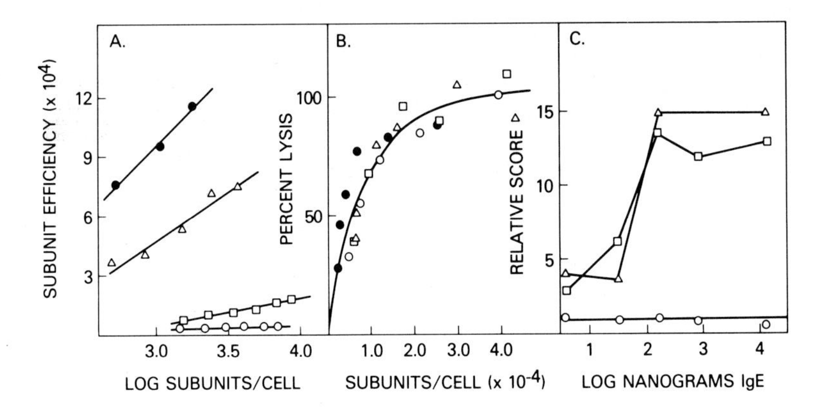

Fig. 7. Effect of oligomers on three immune functions. Oligomers used in the studies were monomeric (○), dimeric (□), trimeric (Δ), or larger (●) fractions of crosslinked immunoglobulins. A. Complement-mediated lysis of TNP-sheep erythrocytes coated with BDPE oligomers of rabbit anti-TNP IgG. Subunit efficiency is defined by $E_S = Z/N_S$, where Z is the number of lytic sites per cell, and N_S is the number of 7S IgG subunits bound per cell [15]. B. ADCC of TNP-chicken erythrocytes sensitized with BDPE oligomers of rabbit anti-TNP IgG. Normal murine splenocytes were used as effectors. Data are presented as the percent of maximum lysis [16]. C. PCA reactions in the skin of a rat injected with oligomers of rat IgE. The injection sites were scored as described previously [17]. The relative scores, plotted here, were derived by dividing the scores of the sites injected with IgE by the scores of the sites injected with solvent alone.

Figure 7B. In this system, lysis increases with the number of antibody molecules bound to the target cells, but is unaffected by the size of the oligomer; monomers induce as much lysis as the larger fractions.

Histamine release by mast cells, a third system, was studied via passive cutaneous anaphylaxis (PCA) in the skins of rats injected with various sized oligomers of IgE [17]. The data in Figure 7C show that monomer induces no PCA reaction, whereas dimers and trimers induce reactions equally well, indicating that dimeric IgE can act as a unit signal for mast cell degranulation.

In all three systems illustrated in Figure 7 antibody multivalency is an important factor. With both cell- and complement-mediated lysis, target cells must have numerous IgG molecules on their surfaces to be lysed. In order to elicit histamine release, mast cells must bind an unknown number of IgE dimers. The systems differ most strikingly in the effect that crosslinking the antibodies has on immune functions.

With complement-mediated lysis, protein molecules of the complement system interact multivalently with antibody molecules [18]. Since protein molecules are small relative to cells, multivalent interactions are much more likely to occur if the antibody molecules are located in close proximity on the cell surface. In ADCC, the antibody molecules on the target cell interact with Fc receptors on the surface of the effector cell. Since Fc receptors can move laterally in the cell membrane, they can readily bind antibody anywhere on the target cell surface and do not need to have antibody molecules near one another for multivalent interactions. Mast cell degranulation appears to require the formation of IgE-Fc receptor doublets on the cell surface; dimers of IgE are thus required as a minimal unit signal for eliciting this response. With ADCC, no such unit signal for the initiation of lysis is readily apparent.

These results demonstrate that model immune complexes formed by covalently crosslinking immunoglobulins can serve as useful probes for studying various antibody-dependent immune systems. Clearly, they show that different systems are not affected equally by oligomerization.

REFERENCES

1. Metzger H: Adv Immunol 18:169, 1974.
2. Metzger H: Contemporary Topics in Molecular Immunology 7:119, 1978.
3. Winkelhake JL: Immunochemistry 15:695, 1978.
4. Rossi-Fanelli A, Antonini E, Caputo A: J Biol Chem 236:397, 1961.
5. Campbell DH: "Methods in Immunology: A Laboratory Text for Instruction and Research." New York: WA Benjamin and Co, 1970.
6. Segal DM, Hurwitz E: Biochemistry 15:5253, 1976.
7. Plotz PH, Kimberly RP, Guyer RL, Segal DM: Molec Immunol (in press).
8. Cathou RE, Dorrington K: In Fasman GD, Timasheff SN (eds): "Biological Macromolecules–Subunits in Biological Systems," Vol 7. New York: Marcel Dekker, 1975, p 91
9. Dickler HB: Adv Immunol 24:167, 1976.
10. Segal DM, Hurwitz E: J Immunol 118:1338, 1977.
11. Segal DM, Titus JA: J Immunol 120:1395, 1978.
12. Borsos T, Rapp HJ: Science 150:505, 1965.
13. Feinstein A, Rowe AJ: Nature 205:147, 1965.
14. Brown JC, Koshland ME: Proc Natl Acad Sci USA 72:5111, 1975.
15. Segal DM, Guyer RL, Plotz PH: Biochemistry 18:1830, 1979.
16. Jones JF, Plotz PH, Segal DM: Molec Immunol (in press).
17. Segal DM, Taurog JD, Metzger H: Proc Natl Acad Sci USA 74:2993, 1977.
18. Reid KBM, Porter RR: In Inman FP, Mandy WJ (eds): "Contemporary Topics in Molecular Immunology." New York: Plenum Press, 1975, p 1.

Biological Recognition and Assembly 327–330 (1980)

Thermodynamic and Kinetic Aspects of Assembly

H. P. Erickson, Workshop Organizer

Anatomy Department, Duke University, Durham, North Carolina 27710

NUCLEATION OF ASSEMBLY OF 2-D AND TUBULAR POLYMERS: ROLE OF SUBUNIT ENTROPY IN COOPERATIVE POLYMERIZATION

Harold P. Erickson, Duke University

The reason that nucleation is thermodynamically unfavorable is that small polymers have a larger fraction of subunits at an edge and therefore a smaller number of bonds per subunit. An essential requirement for nucleation is that each small polymer in the pathway of assembly must be sufficiently stable to exist, in equilibrium with the given pool of free subunits, at a "reasonable" concentration, estimated to be on the order of 10^{-15} M or greater. The equilibrium concentrations have been calculated for the case of assembly of subunits onto the microtubule lattice, using simple principles of thermodynamics outlined below.

The basis for this calculation is simply illustrated by the schematic diagram of the microtubule lattice in Figure 1, which shows three possibilities for addition of a subunit to a growing 2-D polymer. It can attach a) to the side of the sheet, forming a lateral bond with energy e_a; b) to the end of one of the longitudinal filaments, forming a bond of energy e_b; or c) to a "cozy corner" or niche, when one exists, forming both a lateral and a longitudinal bond, of energy $e_a + e_b$. Each of these three associations may be represented by an association constant, and since there are only two bond energies involved K_c must be determined if K_a and K_b are specified.

The relation of the three association constants to each other and to the two bond energies is given in the equations below.

Received May 1, 1979; accepted May 1, 1979.

$$K_a = \exp \frac{1}{RT} (e_a - TSi) \tag{1a}$$

$$K_b = \exp \frac{1}{RT} (e_b - TSi) \tag{1b}$$

$$K_c = \exp \frac{1}{RT} (e_a + e_b - TSi) \tag{1c}$$

$$K_c = K_a K_b \exp \frac{TSi}{RT} \tag{1d}$$

The crucial point in this formalism is the separation of the free energy of association into two terms. The bond energy, e, includes all forces favoring association (including the solvent entropy effects that drive the hydrophobic bonding). The term TSi is the free energy corresponding to the rotational and translational entropy of the free subunit that is lost when the subunit is polymerized. TSi therefore corresponds to the free energy required to immobilize a subunit in the polymer. An important assumption implicit in the equations is that this is the same, regardless of whether the subunit forms a lateral or longitudinal bond or both at once.

In a new derivation the magnitude of TSi is estimated to be as low as 2.1 kcal/mole or as high as 6.9 kcal/mole, depending on whether the subunit is flexibly or rigidly attached to the polymer ($\pm 30^\circ$ or $\pm 1^\circ$ bending about the intersubunit bond for these two extreme cases). It should be noted that even the maximum value of 6.9 kcal/mole is much smaller than the value of 20–30 kcal/mole estimated by Chothia and Janin (Nature 256:705–708, 1975). The details of the new derivation and the problem with the earlier estimate were presented in the poster session and will be published elsewhere.

Values of the association constants appropriate for the case of microtubule assembly are $K_a = 0.37\ M^{-1}$, $K_b = 8{,}100\ M^{-1}$, and $K_c = 10^5\ M^{-1}$. These estimates are based on the assumptions that the "critical concentration" $= K_c^{-1} = 10^{-5}$ M, $e_b/e_a = 5$, and TSi = 2.1 kcal/mole. The bond energies e_a and e_b are 1.5 and 7.5 kcal/mole, respectively.

An important principle governing any self-assembly reaction is that assembly must proceed in steps of bimolecular reaction. These can be either stepwise addition of single subunits to the growing polymer, or pairwise interaction of larger, preformed intermediates.

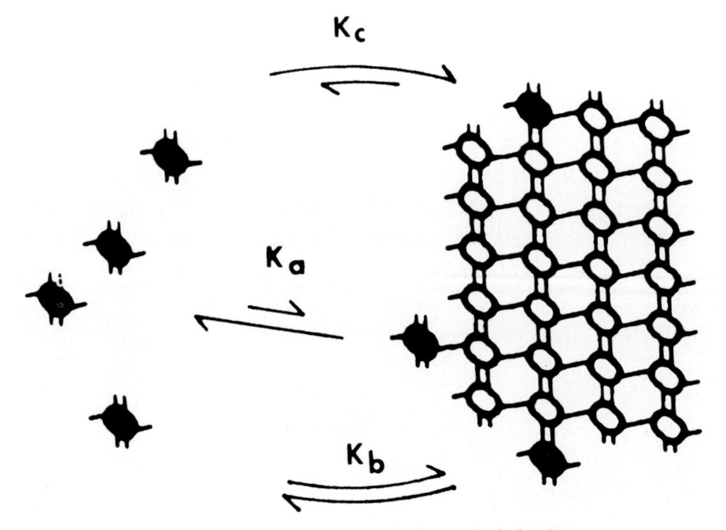

Fig. 1. Schematic diagram of microtubule lattice.

Using the values of bond energies given above, the equilibrium concentrations of different small filaments and 2-D polymers were calculated (the equations are similar to those given for the three association constants). A pathway of bimolecular reactions can be specified such that each intermediate exists at a "reasonably high" concentration, greater than 10^{-12} M, provided that the concentration of free subunits is supersaturated, ie, several times higher than the critical concentration. The calculations show further that assembly of longitudinal filaments and longitudinal growth of a 2-D polymer can be obtained by stepwise addition of single subunits, but that the less favorable lateral growth probably involves addition of preformed longitudinal filaments, five to ten subunits long.

A MODEL OF MICROTUBULE ASSEMBLY

Richard Weisenberg, Temple University

This model is based on the following specific assumptions: 1) The probable structure of a microtubule is inconsistent with simple helical assembly mechanisms; 2) microtubule-associated proteins (MAPs) promote assembly by inducing the formation of tubulin oligomers that can add cooperatively to microtubules as segments of protofilaments; 3) guanosine triphosphate (GTP) stabilizes intertubulin bonds in the microtubule, and the effect of polymerization-coupled GTP hydrolysis is to destabilize microtubules; 4) the hydrolysis of GTP during polymerization depends on the number and geometry of intersubunit contacts. There is experimental evidence for all but the last of these assumptions.

This model appears to be able to explain a number of observations on microtubule assembly. These include polarity of microtubule growth, stimulation of assembly by MAPs, stimulation of assembly by substoichiometric concentrations of GTP, subunit flow through microtubules, and assembly and stability of microtubules in GDP. With additional assumptions it can explain the behavior of microtubules in colchicine. The most important consequence of the model is the prediction of highly cooperative behavior in microtubule assembly. This may explain the rapid disassembly of microtubules by calcium ions and the ability of agents such as guanosine diphosphate and colchicine, under some conditions, to inhibit polymerization but not to disassemble already formed microtubules. The latter phenomenon may be important in the cellular regulation of microtubules, since it allows microtubule lengths to be regulated by agents that block elongation but do not induce depolymerization.

ASSEMBLY OF SICKLE CELL HEMOGLOBIN

William A. Eaton, National Institutes of Health

Sickle hemoglobin has many features in common with other protein self-assembly systems, and much of the structural and physical chemical information on sickle hemoglobin is directly relevant to understanding more complex protein polymerizing systems. Sickle hemoglobin polymerizes into multistranded helical fibers, which spontaneously align to form an ordered phase. The overall process of polymerization and alignment is commonly referred to as gelation. The thermodynamics of sickle hemoglobin gelation have been studied both by sedimentation and calorimetric methods [1–7]. The sedimentation studies show that at equilibrium the sickle hemoglobin gel is well described as a two-component, two-phase system in which the ordered polymer phase may be treated as a protein crystal, and the hemoglobin in the solution phase, which exists entirely as monomers (ie, $\alpha_2\beta_2$ molecules), may be regarded as a solubility. The positive enthalpy change and relatively large heat capacity change accompanying gelation indicate that hydrophobic interactions are a major factor in maintaining the thermodynamic stability of the polymer. A principal difference between sickle hemoglobin and other protein self-assembly systems is that the intermolecular interactions in the polymer are much weaker, as evidenced by the high solubility (>15 gm/100 ml) in physiologic solvents. As a result of the high protein concentrations nonideality from excluded volume effects plays a major role in the thermodynamics. The simple crystallization model is readily extended to multiple components and accounts very well for the equilibrium results on sickle hemoglobin partially saturated with ligand and on mixtures of sickle hemoglobin with other hemoglobins.

The kinetics of sickle hemoglobin gelation have also been well studied. A wide variety of physical techniques have been used to monitor the process, including linear birefringence, light scattering and turbidity, nuclear magnetic resonance relaxation times and linewidths, and calorimetry. All of the techniques show the same basic result for the thermally induced process. Prior to the appearance of polymer there is a pronounced delay, which depends inversely on a very high power (30–50) of the initial concentration. The delay time also depends on approximately the same high power of the equilibrium solubility, indicating that the rate of polymerization is under nearly equilibrium control. These findings are readily explained by a kinetic model in which polymerization is limited by the rate of formation of critical nuclei. The basic idea of the model is that aggregates grow initially by thermodynamically unfavorable, bimolecular addition steps to form the critical nucleus. As the aggregates grow the number of bonds per molecule continuously increases, and this stabilizing effect just balances the destabilizing, net loss of translational and rotational entropy for the formation of the critical nucleus. To go from monomer to polymer, then, the system must pass over a free-energy barrier. The critical nucleus, which is the least stable (and therefore least abundant) species, need not have any "special" structure, but may simply be regarded as a small piece of the infinite polymer. By assuming constant equilibrium between critical nucleus and monomer, a nucleation rate constant is derived, and the formalism of Oosawa [8] can then be applied to obtain an analytic expression for the characteristic time of the reaction. This kinetic model, which includes the large effects of nonideality, is also readily extended to provide a self-consistent explanation for the concentration, temperature, and composition dependence observed in solutions containing multiple components. A next step in the development of this kinetic model for nucleated polymerization is to quantitatively describe the free-energy barrier. This requires accurate estimates of the various contributions to the free-energy change that accompany aggregation, which is a general problem in protein association reactions.

REFERENCES

1. Hofrichter J, Ross PD, Eaton WA: Kinetics and mechanism of deoxyhemoglobin S gelation: A new approach to understanding sickle cell disease. Proc Natl Acad Sci USA 71:4864, 1974.
2. Ross PD, Hofrichter J, Eaton WA: Calorimetric and optical characterization of sickle cell hemoglobin gelation. J Mol Biol 96:239, 1975.
3. Hofrichter J, Ross PD, Eaton WA: Supersaturation in sickle cell hemoglobin solutions. Proc Natl Acad Sci USA 73:3035, 1976.
4. Ross PD, Hofrichter J, Eaton WA: Thermodynamics of gelation of sickle cell deoxyhemoglobin. J Mol Biol 115:111, 1977.
5. Eaton WA, Hofrichter WA: Successes and failures of a simple nucleation theory for sickle cell hemoglobin gelation. In Caughey WS (ed): "Biochemical and Clinical Aspects of Hemoglobin Abnormalities." New York: Academic, 1978, p 443.
6. Hofrichter J: Ligand binding and the gelation of sickle cell hemoglobin. J Mol Biol 128:335, 1979.
7. Sunshine HR, Hofrichter J, Eaton WA: Gelation of sickle cell hemoglobin in mixtures with normal adult and fetal hemoglobins. J Mol Biol (In press).
8. Oosawa F, Asakura S: Thermodynamics of Polymerization of Proteins, Academic Press, New York, 1975.

Biological Recognition and Assembly 331–332 (1980)

Immune Electron Microscopy of Enzyme Complexes and Cellular Organelles

L. Kahan and D. Glitz

Department of Physiological Chemistry, University of Wisconsin, Madison, Wisconsin 53706 (L.K.) and Department of Biological Chemistry, University of California, Los Angeles, California 90024 (D.G.)

WORKSHOP REPORT FROM 1979 ICN-UCLA SYMPOSIA ON BIOLOGICAL RECOGNITION AND ASSEMBLY

This workshop was based upon the open discussion of the problems and limitations (as well as the accomplishments and potential) of immune electron microscopy. Experimental results considered were focused on immune electron microscopy of E coli ribosomal subunits and glutamine synthetase. The participants included about sixty individuals, most of whom were interested in the use of the technique but had not yet attempted such work.

RESEARCH AREAS DISCUSSED

Consideration of immune electron microscopy using anti-protein antibodies led to discussion of: 1) Immunochemical specificity of the technique when rabbit antisera are the source of the antibodies utilized. 2) Lack of exposure of antigenic determinants of some proteins in the E coli ribosome. 3) Interpretation of two-dimensional transmission images as three-dimensional structures. Discussion of the use of anti-hapten antibodies centered on: 4) Immunogen selection and preparation for the induction of useful antibodies. 5) The specificity and affinity of anti-hapten antibodies for elements of macromolecular complexes such as the ribosome.

STATE OF RESEARCH WITHIN THE SUBFIELD FOR RESEARCH AREAS DISCUSSED ABOVE

1) The antibody preparations which have been utilized in the mapping of E coli ribosomes have given fairly reliable information, though some specific locations such as protein S4 are incorrect. These antibodies do not have the monoclonal specificity desirable for detailed mapping.

3) Interpretation of electron micrographs at present does not provide high resolution map locations for ribosomal proteins. In large part the interpretation of images in terms of a model depends upon the expertise of the observer in interpreting the position of pro-

Received and accepted August 30, 1979.

tuberances, etc, as images of orientations of a particular model. There is no objective, quantitative method of determining a map location from a set of images. Although the development of three-dimensional models of the ribosome which explain the two-dimensional transmission images observed in the electron microscope represents a significant step in solving the structure of the ribosome, the models are subject to dispute and have not been stringently tested. Methods of testing these models by analysis of a large number of images would appear to be lacking, although there is active work in this area. The most obvious solution to the interpretation problem would come from electron diffraction analysis of two-dimensional ribosome arrays; however, such arrays have been reported only for eukaryotic ribosomes.

4) Use of anti-hapten antibodies is sometimes limited by their lower affinity (relative to anti-protein antibodies) for the organelle, or by uncertainties regarding their potential crossreactivity with nontarget ribosomal components. As a result crosslinked subunit dimers are more difficult to demonstrate, and greater reliance may be placed on observations of subunit monomer-antibody complexes. Chemical cross linking of antibodies to ribosomal subunits might help reduce the problem.

STATE OF THE ART

This subfield has come into existence only within the last few years. Although several limitations are now becoming apparent, the potential is sufficiently clear to indicate that a much larger number of investigators should be entering the field, and that a greater variety of systems are amenable to study. Some likely areas of progress during the next five years include: 1) Production of monoclonal antibodies to ribosomal proteins through the hybridoma technique, in order to improve the specificity of the mapping and decrease sensitivity to contaminating antibodies. 2) Further testing of present models of the E coli small and large ribosomal subunits. 3) Completion of the mapping of the proteins of the E coli ribosome. 4) Mapping of modified nucleosides of the ribosomal RNA within the ribosome structure. 5) Localization of sites of chemical modification of ribosomal proteins and RNA. 6) Application of immune electron microscopy to studies of eukaryotic ribosome structure.

Biological Recognition and Assembly 333–334 (1980)

Lipoproteins

L.J. Banaszak

Department of Biological Chemistry, Washington University Medical School, St. Louis, Missouri 63110

WORKSHOP REPORT FROM 1979 ICN-UCLA SYMPOSIA ON BIOLOGICAL RECOGNITION AND ASSEMBLY

Because of the theme of the conference, it was noted at the start of this workshop that the concept of recognition and assembly implies a degree of specificity which may or may not be present throughout a macromolecular system containing both lipid and protein. Realizing that practically nothing is known about the nonbonding interactions which might occur between some form of lipid and its related apoprotein, the two formal presentations of the Lipoprotein Workshop concentrated on the general structural organization of complex soluble lipoproteins.

The current state of knowledge on the structural organization of the serum lipoproteins VLDL, LDL, and HDL was described by David Atkinson from Boston University. At the start, Atkinson noted that only a limited number of phases are possible in mixed-lipid systems and summarized the properties of each phase as has been characterized by x-ray methods. A comment from a participant stressed the lack of information of the effect of protein on such phase diagrams. Atkinson then summarized the amino-acid sequence information available on the various serum apoproteins noting that methods for predicting secondary structure suggest that some of the proteins might fold into repeating segments of an alpha helical conformation. These so-called amphipathic helixes contain side chains of negative, positive, and hydrophobic character such that they could easily interact with phospholipid. A participant noted that such side chain sequences in segments of alpha helix are similar to those noted in other globular protein structures and hence may be shaky evidence for their presence in lipoproteins. A series of illustrations showing spherical lipid core models of serum-HDL and-LDL lipoproteins, combined with measured and calculated small-angle x-ray scattering curves, was used by Atkinson to argue for this form of lipid-protein organization. Differential scanning calorimetry indicated thermal transitions in these molecules which were similar to those found in simple lipid micellar systems, the data also supporting the spherical core model.

An alternate model for any lipoprotein could have the overall organization consisting of subunits most likely arranged with some form of point group symmetry. Such a model has been found for the yolk lipoprotein system and the results of the group at Washing-

Received and accepted August 30, 1979.

ton University including D. Ohlendorf, R. Wrenn, and J. Ross was presented by Len Banaszak. Although the yolk lipoprotein complex has considerably less lipid than any form of serum lipoprotein, a more detailed low resolution model has been obtained by using a microcrystalline form present in vivo. Using x-ray powder diffraction and image reconstruction methods, a model of the yolk lipoprotein complex to a resolution of about 20 Å showed that the molecule has an elongated shape consisting of two subunits arranged around a local two-fold symmetry axis. Ridges on the molecular surfaces suggest domains that could be attributed to the known polypeptide components. Because of the rather typical globular-like structure of this lipoprotein, the suggestion was made that it is doubful that any unique form of secondary structure, such as amphipathic helixes, comprise the three-dimensional folding. Banaszak also noted that, based on size and shape arguments, it was extremely unlikely that the lipid was present in any form of micellar arrangement.

Discussion which followed these two presentations was relatively brief. Questions were raised about factors determining specific assembly in lipoprotein models which have a spherical lipid core. Definitive evidence about the overall organization of the serum lipoproteins is lacking, although the present data are clearly interpretable in terms of a model with a simple spherical core. In the yolk-lipoprotein system only a subunit model can be used to interpret the present three-dimensional maps.

Because of the late hour, the closing period, devoted to a discussion of the afternoon poster sessions, was very brief. During the poster session, Woods and Roth (Abstract #302)* from the University of Maryland, Baltimore County, presented evidence for the solubilization of the phosvitin receptor from chicken oocytes using Triton X-100. Grobovsky and Brewer (Abstract #303), from the University of Southern California Medical School, showed data describing the in vitro synthesis of a viral protein from PM2 and its insertion into suitably prepared cytoplasmic membranes of its host, Alteromonas especjisna. Martinez-Carrion et al (Abstract #304), working at the Medical College of Virginia Commonwealth University, had prepared a photo-activated chemical label believed to distribute in the hydrophobic regions of membrane lipids and had studied the effects of this label on the polypeptide chains of the acetylcholine receptor. One of the polypeptide chains of the acetylcholine receptor was not labelled by the nitrene formed upon photoactivation.

Norris and co-workers, at the University of Wisconsin (Abstract #305), have purified an olfactory receptor for napthoquinone repellants from the insect Periplanta Americana using Triton X-100. The purified form of the receptor could interact electrochemically with the odorant. Atkinson, Shipley and Small (Abstract #306) presented small-angle x-ray and neutron data on the complexes formed between apo-HDL and phospholipid. Such complexes may represent a lipoprotein similar to the nascent "discoidal" particles synthesized by the liver prior to incorporation of cholesterol esters.

In summary, the Lipoprotein Workshop included both a formal and poster sessions on soluble lipoproteins and reconstituted membrane protein-lipid systems. For the serum and yolk lipoproteins, protein-lipid recognition was not yet definable at the molecular level but two organizational models have been proposed. Reports on the development of new biochemical and chemical techniques for studying membrane-protein systems indicated steady progress in this area. The purification and study of biologically active forms of membrane proteins appears to be an area that will be particularly productive in the near future. Direct determination of the molecular structure of membrane or soluble lipoproteins using diffraction methods still seems to be hampered by the lack of highly ordered specimens. The lack of crystalline lipid-protein systems is likely to remain a serious limiting factor in high-resolution diffraction analysis in the immediate future.

*Abstracts of the 8th Annual ICN–UCLA Symposia on Molecular and Cellular Biology, J Supramol Struct, Supplement 3, New York: Alan R. Liss, Inc., 1979.

Biological Recognition and Assembly 335–336 (1980)

Electron Diffraction and Membrane Proteins

N. Unwin

MRC Laboratory of Molecular Biology, Hills Road, Cambridge, England

There are several membrane proteins which form ordered assemblies and hence are amenable to analysis by electron microscopy/diffraction. They fall into three classes: 1) those which normally are ordered in vivo, eg, bacteriorhodopsin; 2) those which sometimes become ordered in vivo, eg, gap-junction protein; and 3) those which can be induced to order artificially, eg, cytochrome oxidase.

Examples from each class were discussed at the Workshop (gap junctions, matrix protein from E coli, cytochrome oxidase, and bacteriorhodopsin). The interest was in using electron microscopy/diffraction to 1) determine high resolution structure directly, 2) locate specific sites on the membrane proteins, and 3) detect conformational changes, and hence obtain an idea about how a particular membrane protein functions.

With gap junctions the changes involved between the coupled and uncoupled states may be quite large, because they seem to allow transport of molecules up to molecular weight of 1,000 or to block transport altogether. Hence, with this system there is a good chance of detecting conformational changes even at quite low resolution. Zampighi and Unwin find two slightly different forms which may be related to the coupled and uncoupled states. They are being analysed in three dimensions, using negative stain. Matrix protein, described by A. Steven, also forms a large transmembrane channel and could therefore be analysed in a similar fashion. So far, however, only the low-resolution projected structure has been determined.

Cytochrome oxidase is being studied by several groups by both stained and unstained methods. Using single-layer sheets, S. Fuller and R. Capaldi (in collaboration with R. Henderson and J. Deatherage) showed that it is a Y-shaped transmembrane molecule, with the top part of the Y being exposed to the matrix side of the inner mitochondrial membrane. This analysis makes it possible to understand the details of the collapsed vesicle form [W. Goldfarb and J. Frank] in terms of single molecules. Antibody-labeling experiments [T. Frey and S. Chan] are beginning to tell us where the individual polypeptide chains are located. The resolution of the structure analysis ($\sim$25Å) and of the antibody labeling ($\sim$40Å) is not yet sufficient to throw light on the character of the secondary structure or on how the molecule functions. It is limited by the negative stain and by the size and "floppiness" of the antibody fragments.

Received and accepted August 30, 1979.

An important development of technique was illustrated by S. Hayward in discussing purple membrane. He finds that its resistance to radiation damage is markedly improved at low temperatures (factor of about 5). This should enable better data to be obtained, for instance, in heavy-atom labeling experiments [M. Dumont and W. Wiggins]. Almost certainly, several different heavy-atom sites will be located on bacteriorhodopsin over the next few years.

The general impression gained from the Workshop was that in some cases a combination of approaches (eg, sequence information, heavy-atom labeling, and low-temperature observation) may lead to about as detailed a picture as one has come to expect from x-ray crystallography of protein crystals. In other cases, greater emphasis might have to be placed on other techniques, (eg, x-ray diffraction of oriented pellets), because the degree and extent of order in the microscope preparations is insufficient to permit high resolutions to be attained. The low-temperature approach, the development of labeling methods, and improvements in specimen preparation are likely to provide the key to future advances.

Biological Recognition and Assembly 337–343 (1980)

Regulation of Glutamine Synthetase

B. Magasanik and E. Stadtman

Department of Biology, Massachusetts Institute of Technology, Cambridge, Massachusetts 02139 (B.M.) and Laboratory of Biochemistry, National Heart, Lung, and Blood Institute, National Institutes of Health, Bethesda, Maryland 20205 (E.S.)

WORKSHOP REPORT FROM 1979 ICN-UCLA SYMPOSIA ON BIOLOGICAL RECOGNITION AND ASSEMBLY

From detailed hydrodynamic measurements, electron microscopic examination, and x-ray defraction studies, it is evident that the glutamine synthetase (GS) of Escherichia coli and other gram-negative bacteria is composed of 12 apparently identical 50,000 dalton subunits, which are arranged in two hexagonal arrays, one superimposed upon the other in a face-to-face fashion. Extensive enzymic studies have demonstrated that the catalytic activity of GS in E coli is regulated by at least five different mechanisms, including 1) cumulative feedback inhibition by eight or more end products of glutamine metabolism; 2) divalent cation-dependent interconversion of the enzyme between catalytically active (taut) and inactive (relaxed) configuration; 3) covalent modification mediated by an elaborate bicyclic cascade system, in which an adenylyltransferase (ATase) catalyzes the cyclic adenylylation and deadenylylation of a tyrosyl residue in each subunit of GS, and in which modulation of the adenylylation-deadenylylation cycle is facilitated by the cyclic interconversion of a small regulatory protein (P_{II}) between uridylylated (P_{IID}) and unmodified (P_{IIA}) forms, as catalyzed by a uridylyltransferase/uridylyl-removing enzyme complex (UT/UR); 4) positive and negative metabolite control of the ATase- and UT/UR-catalyzed modification reactions; and 5) repression and derepression of GS synthesis by metabolite-sensitive genetic factors.

These structural and regulatory features of GS have been extensively reviewed [Stadtman and Ginsburg, The Enzymes, 10:755 (1974); Stadtman and Chock, Current Topics in Cellular Regulation, 13:53 (1978); Magasanik et al, Current Topics in Cellular Recognition, 8:119 (1974)] .Although genetic studies on the regulation of GS synthesis and biochemical studies on the regulation of GS activity have been of common interest to all investigators in these fields, studies in these two areas have involved widely different groups of investigators with different scientific backgrounds and technical expertise. Accordingly, the workshop was divided into two parts. The first part was concerned with the relationship between structural features of GS and its regulation by feedback inhibition, and by the interconvertible enzyme cascade system. The second part was concerned with genetic studies on the regulation of GS synthesis.

Received and accepted August 30, 1979.

PART I

Major discussants in the first part were A. Ginsburg, D. Eisenberg, E. Stadtman, J. Villefranca, and F. Wedler. They addressed the following questions: 1) How many separate ligand-binding sites exist on a single subunit of E coli glutamine synthetase? 2) To what extent do subunit interactions affect the catalytic parameters of the enzyme? 3) What is known about the interaction of GS with the ATase and the P_{II} regulatory protein?

The major conclusions are summarized as follows:

1) Site Specificity for Various Allosteric Effectors

The suggestion that there are separate binding sites on GS for each of eight different feedback inhibitors was based on the original observation of Woolfolk and Stadtman showing that a high concentration of any one of the inhibitors caused only partial inhibition of the enzyme, and that the residual activity obtained in the presence of one inhibitor was subject to further inhibition by addition of any one of the other inhibitors. Furthermore, the fraction of residual activity that was inhibited by the second inhibitor was the same as the fraction of GS activity inhibited by the second inhibitor when it was tested by itself. In other words, the effects of multiple inhibitors were cumulative. Such cumulative behavior could most easily be explained by the existence of a separate, noninteracting binding site on each subunit for each one of the effectors. This possibility was further supported by kinetic experiments showing that L-alanine, adenosine monophosphate (AMP), and carbamyl-P are noncompetitive inhibitors with respect to any one of the substrates, and that glycine, cytidine triphosphate (CTP), and L-tryptophan are partially competitive with respect to glutamate, whereas L-histidine and glucosamine-6-P are partially competitive with respect to NH_4^+. It was therefore inferred that there is a minimum of three different kinds of inhibitory sites. Furthermore, studies by Shapiro and Stadtman showed that the inhibitors within each of these three kinetic classes could be differentiated by their ability to effect the rate of inactivation of relaxed enzyme by organic mercurials. Thus, one inhibitor from each group (namely AMP, tryptophan, and L-histidine) protects the enzyme from inactivation, whereas carbamyl-P and CTP stimulated inactivation, and L-alanine, glycine, and glucosamine-6-P were without effect. In addition, it was observed that the various inhibitors could be differentiated by their effects on the rate of reassociation of urea-dissociated subunits (Ciardi and Stadtman).

In the meantime, a more critical theoretical analysis of multiple inhibitor effects on enzymatic activity by Chock and Stadtman (unpublished) showed that, under certain conditions, the observed cumulative inhibition effects could be explained by nonexclusive binding of multiple effectors at a common allosteric site. Moreover, results of more sophisticated kinetic experiments by Stadtman, Park, and Smyrniotis failed to yield convincing evidence for the existence of multiple allosteric sites on the enzyme. Furthermore, results of neutron magnetic resonances (NMR) studies by Dalquist and Purich led these workers to conclude that alanine inhibition is due to its direct competition with glutamine for binding at the substrate site. On the basis of these data, they proposed that the effects of all eight feedback inhibitors might be explained by direct competition for binding at substrate sites, rather than to separate allosteric sites.

In view of these considerations, the question of how many, if any, separate allosteric sites exist on a subunit of GS has recently received a lot of attention, and has been examined by the use of several different experimental approaches. The results of these various investigations were summarized by the participants of the workshop. Villefranca noted that from studies of the effects of feedback inhibitors on the amplitude of the intrinsic fluorescence of GS as measured with a stopped-flow fluorometer, it has been established by Chock and co-workers that the following pairs of amino acids can bind simultaneously to the unadenylylated form of GS: L-Ala + Gly; L-Ala + Glu; L-Ala + D-Ala; L-Ala + D-Val; Gly + Glu; Gly + D-Val; Glu + D-Val. However, the binding of D-Val and D-Ala is mutually exclusive. In addition, the binding of either L-Ala, D-Val, or Gly is enhanced by the presence of ATP. These results indicate that there are separate binding sites for L-Ala, Glu, ATP, and for D-amino acids to which either D-Ala or D-Val can bind; Gly can bind simultaneously to both the L-Ala site and the D-amino acid site. A kinetic analysis of intrinsic fluorescence changes associates with the substrate interactions (Rhee and Chock) established that the catalytic cycle involves the formation of two different transition state complexes, I_1 and I_2. From the effects of feedback inhibitors, it was deduced that the binding of L-Ala to its allosteric site inhibits interaction of I_2 with NH_3 in the catalytic cycle; D-Val inhibits the formation of I_1; Gly can inhibit both I_1 formation and I_2 conversion to product; AMP inhibits the conversion of I_1 to I_2. It follows from the latter effect that AMP and ATP must bind at separate sites.

On the basis of NMR studies carried out in Villefranca's laboratory, it has been determined that the binding of L-Ala to the $GS_{1.0}$-$(Mn)_2$ ADP complex does not affect significantly the $1/T_{1m}$ values for protons of bound Glu; likewise, the $1/T_{1m}$ values for protons of bound L-Ala are not affected by the binding of Glu. These results support the conclusion that L-Ala and Glu can coexist on the enzyme. Similar studies show that bound Gly cannot be displaced from $GS_{1.0}$-$(Mn)_2$ ADP by Glu, but can be displaced by a mixture of L-Ala and D-Val. These and other NMR data confirm the conclusion based on the fluorescence studies showing that there are separate binding sites for Glu, L-Ala, and D-Val, and that Gly can bind to both the D-Val and L-Ala sites.

By means of ^{13}C-NMR studies, Villefranca, Rhee, and Chock have shown that AMP binds to a site which is $\geqslant 20$Å away from the catalytic site, ie, the n_2-Mn binding site, suggesting that AMP binds at a site different from the ATP or ADP substrate binding site.

From proton NMR experiments, it could be calculated (Villefranca et al) that the distances between the tight divalent metal ion binding site (n_1 site) and the α, β, and γ protons of Glu are 8.0, 7.4, and 7.0 Å respectively, and the distances of these same protons to the n_2 metal ion site are 7.8, 7.6, and 6.9 Å respectively. In addition, it was shown that the distances between the n_1 site and the methyl group and the α-carbon atom of L-Ala and of the methylene group of Gly are 9.6, 9.2, and 8.7 Å respectively. Ann Ginsburg pointed out that from calorimetric measurements by Shrake and Ginsburg, it was found that the heat released by the simultaneous binding of Glu and L-Ala to $GS_{1.0}$-$(Mg)_2$ ADP is greater than the heat released by the binding of each ligand independently; moreover, Glu exhibits an antagonistic effect with respect to the binding of Pi to $GS_{1.0}$-$(Mg)_2$ ADP, whereas L-Ala exhibits a synergistic effect on the binding of Pi. These data also indicate the existence of separate binding sites for Glu and L-Ala. In similar studies, it was established that the heat released by the simultaneous binding of L-Trp and AMP to $GS_{2.3}$-Mn is equal to the sum of the heats of binding of each ligand alone. This shows that there are separate binding sites on the enzyme for Trp and AMP. This conclusion is also supported by the equilibrium dialysis studies of Ginsburg and co-workers, showing that there are 12 tight binding sites on GS for both Trp and AMP, and that the binding of AMP is not significantly affected by the presence of high concentrations of other feedback inhibitors including Trp, Gly, His, or the substrates Glu and NH_4^+, or of closely related mononucleotides such as UMP, CMP, and GMP, or by mixtures of all of these compounds. The fact that 12 equivalents each of AMP and Trp are bound per mole of GS further indicates that there are separate sites for each of these ligands on each subunit of the enzyme.

Finally, it should be noted that since histidine is able to inhibit only the Mn^{2+} form of GS, there was some question as to whether the effect of this ligand might be partly, if not entirely, attributed to its capacity to chelate Mn^{2+}. However, the studies of Eisenberg et al showing that His protects GS from proteolytic attack by various proteases indicate that there is a binding site on GS for His.

In summary, the evidence obtained by diverse physical-chemical techniques provides substantial proof that there are separate binding sites on GS for L-Ala, Trp, Glu, AMP, ATP, and D-amino acids, and that Gly is able to bind to both the L-Ala site and the D-amino acid site.

2) Subunit Interactions and Catalytic Activity

Although GS is composed of 12 apparently identical subunits, except for the fact that dissociated subunits have little or no activity, efforts to demonstrate a contribution of homologous subunit interactions on catalytic parameters have been generally unsuccessful. Positive cooperativity with respect to ligand interactions has been observed only with tryptophan. Increasing concentrations of this ligand elicit a sigmoidal response with respect to both equilibrium binding and the inhibitory response. Otherwise, only slight negative cooperativity has been seen with some ligands, especially in the activity response of transferase activity to increasing concentrations of glutamine. At this symposium Ann Ginsburg et al presented a poster showing that the binding of L-Methionine (SR)-sulfoximine (MSOX, considered to be a transition state analog of glutamine synthetase) to GS can be monitored by UR spectroscopy. Whereas spectrophotometric titration of Mn-enzyme with MSOX yielded Hill numbers (n_H) of 0.9–1.0, titration of the Mg-enzyme yielded an n_H value of 0.5; this indicates that rather strong negative interaction occurs with the Mg form but not the Mn form of the enzyme.

In other experiments by Rhee, Chock, Wedler, and Sugiyama, it was observed that the rate of irreversible inactivation of the γ-glutamyl transferase activity and the concomitant rate of formation of MSOX-phosphate and ADP, which occurs when GS is incubated with ATP and MSOX, decrease progressively with time from the expected first-order rate. This indicates that inactivation of one subunit retards the reactivity of other subunits with ATP and MSOX, ie, there appears to be a strong negative subunit interaction in response to MSOX-phosphate-enzyme complex formation. Older studies of Ginsburg et al and more recent studies of Stadtman et al have shown that heterologous interactions between adenylylated and unadenylylated subunits of partially adenylylated GS molecules affect the sensitivity of the enzyme to denaturing agents (viz, urea) and the apparent Km of the enzyme for glutamate.

3) Interactions of GS With the Regulatory Proteins

Adenylylation of GS involves the transfer of an adenylyl group from ATP to a single tyrosyl residue on each subunit and the concomitant formation of PPi, whereas the deadenylylation of GS involves phosphorlytic cleavage of the AMP-tyrosyl bond to form ADP and unmodified GS. Both reactions are catalyzed by the same adenylyltransferase (ATase). However, the ability of ATase to catalyze the adenylylation reaction is dependent upon the presence of the unmodified form of Shapiro's regulatory protein (P_{IIA}), whereas the ability to catalyze the deadenylylation reaction is dependent upon the presence of the uridylylated form of the regulatory protein (P_{IID}). This raises two questions: 1) Do the adenylylation and deadenylylation reactions compete directly with one another, ie, are they both catalyzed at the same site on the ATase? 2) Do P_{IIA} and P_{IID} exert their effects by interacting with ATase or with GS? With respect to the first question, the available evidence indicates that the adenylylation and deadenylylation reactions are catalyzed at separate catalytic sites on the single 115,000 dalton polypeptide chain of ATase. This is indicated by the kinetic experiments of Brown and Segal, and of Rhee et al, showing that the rate of adenylylation of unadenylylated enzyme ($GS_{\overline{0}}$) is not affected by the addition of even high concentrations of fully adenylylated enzyme ($GS_{\overline{12}}$), and is further supported by the fact that a 64,000-dalton fragment (proteolytic cleavage product?) of ATase can catalyze the adenylylation reaction, but not the deadenylylation reaction (Anderson et al). It is noteworthy that likewise the uridylylation and deuridylylation of the P_{II} protein are catalyzed at separate sites on the UT/UR enzyme or enzyme complex. Thus, the uridylylation and deuridylylation activities are enriched simultaneously by various purification steps, but each can be selectively inactivated by appropriate denaturing conditions (Magni and Mangum; Huang). Nevertheless, both activities are lost simultaneously by a single point mutation in the Gln-D gene (Garcia, Kustu, and Rhee).

With regard to the second question above, direct evidence that the P_{II} protein reacts with ATase has been obtained by showing that the intrinsic fluorescence of ATase is enhanced by the presence of P_{II} (Caban and Ginsburg). Whether or not P_{II} can also react directly with GS has not been investigated.

PART II

The second portion of the workshop was devoted to a discussion of the regulation of glutamine synthetase formation and of the putative role of GS as a regulator of the synthesis of enzymes of nitrogen metabolism in different bacteria.

These problems have been most extensively explored in the enteric organisms. Of the participants in the workshop, Greg Pahel and Bonnie Tyler of M.I.T. reported on Escherichia coli, Forrest Foor and Boris Magasanik of M.I.T. on Klebsiella aerogenes, and Jean Brenchley of Purdue on Salmonella typhimurium. There was general agreement that in all three organisms the level of GS can vary more than a hundredfold in response to changes in the composition of the growth medium. The highest level is found in cells growing on a good source of carbon, such as glucose and a growth-rate-limiting source of nitrogen (essentially any utilizable nitrogen compound other than ammonia). The lowest level of GS is found in cells growing in medium containing a poor source of carbon, such as histidine, as well as ammonia and glutamine. The structural gene for GS is *glnA*, located on the chromosome of each of the three organisms in a position corresponding to position 85 on the E coli map.

The expression of the *glnA* gene requires in all three organisms the product of another unlinked gene, *glnF*, in a position corresponding to 68 on the E coli map. Mutants lacking this product because of a mutation in the *glnF* gene, or, in the case of E coli, because of the insertion of the transposon Tn10, produce GS at a level corresponding to that found in the wild-type strain under the most repressing condition. It was previously shown that, in K aerogenes, a mutation closely linked to *glnA* suppressed the requirement for glutamine resulting from the *glnF* mutation, but did not establish normal regulation of GS synthesis; rather it resulted in the production of GS at a low constitutive level, irrespective of the presence or absence of a functional *glnF* gene. This GlnR phenotype was completely recessive. Similar *glnA*-linked suppressors in S typhimurium were then described by Brenchley. It was reported by Pahel and Tyler that E coli mutants with the characteristic recessive GlnR phenotype can be obtained by the insertion of phage Mu into a site closely linked to *glnA*. Apparently, the loss of the product of a *glnA*-linked gene, *glnG*, relieves the GS from the repression that results from the loss of the *glnF* product or from cultivation in energy-poor, glutamine-rich medium. The fact that the *glnG* mutants fail to produce GS at the high level characteristic of the wild-type strain cultivated in an energy-rich, nitrogen-poor medium suggests that the *glnG* product is also required for the activation of GS synthesis.

The regulation of GS synthesis by repression and activation requires the existence of target sites linked to *glnA* for the repressor (operator) and the activator (initiator). Magasanik reported the isolation by David Rothstein of a putative operator mutation at the *glnA* site in K aerogenes. This mutation was isolated as a dominant suppressor of the glutamine requirement resulting from the *glnF* mutation. It affects only the expression of the linked *glnA* gene (*cis*-dominant). It never allows GS to be produced at a level lower than that characteristic of the GlnR phenotype. However, it does allow production of GS at a high level in the presence of a functional *glnF* gene when the strain is grown in the energy-rich, nitrogen-poor medium.

The expression of the *glnA* gene is also subject to regulation by the product of the *glnB* gene, found in K aerogenes at a position corresponding to position 55 of the E coli map. A mutation in this gene, *glnB3*, affects the protein P_{II} of the GS-adenylylation system in such a way that it cannot be converted by UTase to an effector of GS-deadenylylation by ATase. The *glnB3* mutant produces highly adenylylated GS at a low level and requires glutamine for growth. Foor and Reuveny have shown that this glutamine requirement can be suppressed by insertion of the transposon Tn5 either into a site closely linked to *glnB3* or into an as yet unmapped site or sites, resulting in the loss of ATase. Mutants lacking ATase produce under all conditions nonadenylylated GS, but exhibit normal regulation of GS synthesis only in the presence of the wild-type *glnB* allele. Extracts of mutants carrying the *glnB3*-linked Tn5 insertion are entirely devoid of P_{II} activity. These mutants produce GS in all media at the high level characteristic for the wild-type strain grown in an energy-rich, nitrogen-poor medium (GlnC phenotype). The *glnB3*-linked insertion also results in a lower rate of adenylylation and deadenylylation of GS in the intact cell, but does not prevent these processes. The formation of GS in this mutant requires the presence of a functional *glnF* gene. These results suggest that activation of GS synthesis by the products of the *glnF* and *glnG* genes is antagonized by the product of the *glnB* gene, presumably P_{II}. This antagonism might be eliminated by the uridylylation of P_{II} by UTase, a reaction which depends on the high level of 2-ketoglutarate and low level of glutamine found in cells growing on an energy-rich, nitrogen-poor medium. Such a mechanism may account for the regulation of the level of GS in the cell by the availability of ammonia and glutamine.

The glutamine requirement resulting from the *glnB3* mutation can also be suppressed by *glnA*-linked mutations. Strains carrying the *glnA*-linked suppressor have the GlnC phenotype, irrespective of the presence or absence of the *glnB3* mutation, and require a functional *glnF* product for GS synthesis. Similar mutants of S typhimurium were described by Brenchley. In all organisms the *glnA*-linked mutations resulting in the GlnC phenotype have pleiotropic effects. In K aerogenes the mutation eliminates the repressive effect of ammonia on a variety of enzymes of nitrogen metabolism, such as histidase, proline oxidase, and urease. In S typhimurium, the mutation increases the ability of the cell to transport glutamine and arginine.

The properties of glutamine-requiring mutants of B subtilis were described by Dennis

Dean of Purdue. All of the mutants were closely linked, none were completely devoid of GS activity, and all contained considerable GS antigen. Many of them had 10–20 fold more than the wild-type strains. The fact that some of these leaky GS-antigen overproducing mutants were isolated by virtue of their ability of utilize D-histidine as source of nitrogen, suggests the possibility that GS antigen activates the formation of a transport system for D-histidine. However, none of the mutants was altered in the regulation of the synthesis of histidase, arginase, of glutamine transport, or sporulation. The inability to isolate mutants totally lacking GS activity suggests the possibility that GS plays an essential role in which it cannot be replaced by the provision of glutamine, the product of its enzymatic activity.

An apparent separation of the enzymatic and regulatory activities of GS was discovered by Robert Ludwig of M.I.T. in strain 32H1 of Rhizobium. This organism contains two proteins with GS activity. GSI closely resembles the GS of the enteric bacteria in being subject to adenylylation and in its stability at 50°. GSII is thermolabile and, though it can be inactivated by ammonia, seems not to be subject to adenylylation, and its synthesis is much more sensitive to repression by ammonia than that of GSI.

The organism can utilize ammonia as sole of nitrogen, though poorly. The utilization of ammonia for glutamate synthesis depends on GSII and glutamate synthase. The loss by mutation of either activity results in a glutamate requirement. However, the only glutamine-requiring mutants isolated so far are affected in both GSI and GSII: the former is always present in adenylylated form, and the latter is inactive. These glutamine requirers are unable to form nitrogenase, whether grown as free living cells or in a root nodule. Two classes of revertants to glutamine independence could be isolated: those that had regained GSII activity but still contained only adenylylated GSI, and those that contained only nonadenylylated GSI but still lacked GSII. Only the latter class of revertants had regained the ability to produce nitrogenase. These results suggest that the primary role of GSII is to produce glutamine as a substrate for glutamate synthase in the pathway of ammonia utilization for glutamate synthesis, and that the primary role of GSI is to regulate the formation of nitrogenase in response to the level of ammonia. Apparently either enzyme can supply the cell with sufficient glutamine for protein synthesis.

Biological Recognition and Assembly 345–346 (1980)

Ciliary and Microtubular Organization

P. Satir

Department of Anatomy, Albert Einstein College of Medicine, Bronx, New York 10461

This workshop focused mainly on the organization and function of components of the ciliary axoneme. The axoneme, a bundle of 9+2 microtubules, has proven to be a model which provides suggestions regarding microtubular organization and function in general, similar to the manner in which striated muscle has been used as a model for actin-myosin-based cell motility – and with similar reservations. The complexity of axonemal organization points out that basic principles of interaction in microtubule assembly remain unsolved, including especially: 1) specificity of interaction between tubulin and other proteins, and 2) polarity considerations. The nine ciliary doublet microtubules of the axoneme are apparently based on a standard assembly of tubulin subunits into the A subfiber, and present information suggests that the lattice is formed by equivalent or quasiequivalent interactions between tubulin dimers. Ciliary tubulins may, of course, differ slightly from other cytoplasmic tubulins. However, this picture of assembly is grossly oversimplified since onto the A subfiber several different proteins or multiprotein complexes assemble in a spatially nonrandom way.

The dynein arms. Dynein is an adenosine triphosphatase (ATPase) that transiently links adjacent microtubules, causing them to slide relative to one another. The precise interactions of dynein with the microtubule lattice, the exact force generation cycle, and the synchrony vs asynchrony of arm activity on different microtubules remain important unsolved problems. This is an area of great current interest.

The radial spoke complex. Spokeless mutants have been used to define this multiprotein structure. The spokes lie perpendicular to the arms. They are presumed to interact via the spoke head with projections from the central microtubules and to act to control sliding, perhaps by forming transient attachments with the central complex.

Interdoublet links. These lie parallel to the inner row of arms and connect adjacent microtubules, possibly forming an elastic girdle around the axoneme which is part of the mechanical control of beat coordination.

The spokes and links form a system that is digested by trypsin, about which relatively little is known, but which clearly is part of the mechanism that transduces sliding into bending. This system restricts the overall magnitude of sliding displacements and coordinates microtubule activity during beat. In the presence of ATP and Mg^{2+}, after trypsin treatment, all axonemal doublets slide actively with a single polarity and displacements

Received and accepted August 30, 1979.

of tens of micrometers are seen. In bending cilia, however, the maximum displacement between doublets is roughly 0.3 μm and all sliding does not have a single direction, which suggests that active sliding is asynchronous during beat and that some doublets are passively moved.

Current work seeks to identify switch points within the axoneme which turn active sliding off and on. These may be sensitive to agents that arrest beat, such as Ca^{2+} and vanadate. The mode of arrest by these agents is being sought.

Membrane links. Structural links have been seen between microtubules and membrane. Attempts are being made to characterize these biochemically.

Biological Recognition and Assembly 347–348 (1980)

Helical Viruses

D. Marvin and L. Makowski

European Molecular Biology Laboratory, 6900-Heidelberg, Federal Republic of Germany (D.M.) and Rosenstiel Basic Medical Sciences Research Center, Brandeis University, Waltham, Massachusetts 02154 (L.M.)

The two most important helical virus systems that were discussed at this workshop were tobacco mosaic virus (TMV) and filamentous bacterial viruses (FV).

TMV

The detailed x-ray diffraction work on the structure of the virus and on the structure of crystals of virus coat protein is coming to fruition. X-ray data on crystals of protein give high-resolution information on the large-radius regions of the protein, but information is lacking on the small-radius parts of the protein and on the RNA. X-ray data on oriented gels of virus give lower resolution information on the whole virus. By combining data of the two types, Holmes and Stubbs have learned something about the RNA-protein interaction. The three RNA nucleotides associated with each protein form a "lobster claw" structure which grasps a stretch of α-helical protein between the planes of the nucleotide bases.

Detailed physical-chemical studies of the assembly of TMV protein are continuing, especially in Shuster's lab. This is a well-defined system in which to make such studies.

FV

The general structure of the Pfl strain of FV is clear, but there is some disagreement about two different aspects of the structure: symmetry and subunit shape. Marvin's group feels that the distinction between two symmetry options (4.4 or 5.4 units in one 15-Å turn of the helix) has not yet been rigorously made; Makowski and Caspar are certain that there are 5.4 units/turn. Marvin points out that in any case the detailed interactions between adjacent protein subunits will be virtually the same for the two symmetry options. Marvin's group feels that the detailed shape of the protein subunit is not yet known, so the best approximation is the simplest one, ie, a single continuous rod of α-helix; Makowski and Caspar are certain that there is a break in the middle of the α-helix.

Marvin reported that the Pfl virion is not just a static structure, but can undergo a temperature-induced structural transition involving a slight rearrangement of the local packing of subunits (motions of a few tenths of an angstrom are involved) which is am-

Received and accepted August 30, 1979.

plified over the length of the virion to create a rotation of one end with respect to the other of many tens of turns, and a translation of one end with respect to the other of over 1,000 Å. Even more dramatic rearrangements of proteins can be induced by hydrophobic solvents.

Day proposed a model for FV DNA that has common features for Pf1, Xf, and fd strains of FV, in spite of the apparently different nucleotide:protein stoichiometry between the strains (1:1 for Pf1, 2:1 for Xf, and 2.3:1 for fd).

Gray presented high-resolution electron micrographs indicating a knob-and-stalk structure for the adsorption protein at the end of the fd virion and presented a model of how this protein could fit onto the end of the coat protein assembly.

McPherson presented x-ray data on the crystal structure of the gene 5 DNA prepackaging protein of fd and also on the structure of the complex between the protein and oligonucleotides. This work is at an exciting stage and promises to be richly rewarding, both for understanding DNA-protein interactions and for understanding detailed processes of molecular assembly.

Author Index

Subject Index